Clean Energy and Environment

Clean Energy and Environment

Edited by **Marrianne Fox**

SYRAWOOD
PUBLISHING HOUSE
New York

Published by Syrawood Publishing House,
750 Third Avenue, 9th Floor,
New York, NY 10017, USA
www.syrawoodpublishinghouse.com

Clean Energy and Environment
Edited by Marrianne Fox

International Standard Book Number: 978-1-68286-367-1 (Hardback)

Printed in China.

Contents

Preface

This book has been an outcome of determined endeavour from a group of educationists in the field. The primary objective was to involve a broad spectrum of professionals from diverse cultural background involved in the field for developing new researches. The book not only targets students but also scholars pursuing higher research for further enhancement of the theoretical and practical applications of the subject.

The rising levels of pollution and depleting resources have given rise to the concept of clean energy. The modern times demand for a massive amount of energy to power the industries. This book delves into the varied sources of clean energy along with discussing their availability and effects on the environment. It discusses the tools and techniques required to harness clean energy and how it can be employed as a power source. Some interesting concepts such as smart grids, electric vehicles, power system planning and protection are discussed in detail. This book aims to provide the readers with an advanced understanding of this rapidly expanding field. It encompasses researches from all parts of the world to provide the readers with a global outlook on the importance of clean energy and the existing production methods and modules. Researchers, scholars and students will find this book beneficial.

It was an honour to edit such a profound book and also a challenging task to compile and examine all the relevant data for accuracy and originality. I wish to acknowledge the efforts of the contributors for submitting such brilliant and diverse chapters in the field and for endlessly working for the completion of the book. Last, but not the least; I thank my family for being a constant source of support in all my research endeavours.

Editor

1

Small signal stability analysis of power systems with high penetration of wind power

Ping HE, Fushuan WEN (✉), Gerard LEDWICH, Yusheng XUE

Abstract The integration of large amount of wind power into a power system imposes a new challenge for the secure and economic operation of the system. It is necessary to investigate the impacts of wind power generation on the dynamic behavior of the power system concerned. This paper investigates the impacts of large amount of wind power on small signal stability and the corresponding control strategies to mitigate the negative effects. The concepts of different types of wind turbine generators (WTGs) and the principles of the grid-connected structures of wind power generation systems are first briefly introduced. Then, the state-of-the-art of the studies on the impacts of WTGs on small signal stability as well as potential problems to be studied are clarified. Finally, the control strategies on WTGs to enhance power system damping characteristics are presented.

Keywords Wind turbine generators, Wind farm integration, Small signal stability, Damping characteristic

P. HE, South China University of Technology,
Guangzhou 510640, China
e-mail: hplkz@126.com
P. HE, Zhengzhou University of Light Industry,
Zhengzhou 450002, China
F. WEN, G. LEDWICH, Queensland University of Technology,
Gardens Point, Brisbane, Australia
(✉) e-mail: fushuan.wen@gmail.com
G. LEDWICH
e-mail: g.ledwich@qut.edu.au
Y. XUE, State Grid Electric Power Research Institute,
Nanjing 210003, China
e-mail: xueyusheng@sgepri.sgcc.com.cn

1 Introduction

The energy policies in many industrialized countries place the emphasis on the large-scale development of renewable energy-based power generation. The ever-increasing power generation from renewable energy sources, especially wind energy, would be helpful for reducing the emissions of pollutions, and hence mitigate global warming and improve environment. The wind power reserve is huge around the globe. It is reported in [1] that the technical potential of onshore wind energy is very large, around 2.74×10^{12}–5.0×10^{12} KWh (kilowatts-hour). It is estimated by the Global Wind Energy Council that about 2×10^7 MW of electricity could be utilized, which is ten times larger than the available water energy. In the past decades, more and more wind power is integrated to power systems around the globe, and the trend is expected to continue in the foreseeable future [2].

The wind energy reserve in China is abundant, the onshore wind potential reaches 3.226×10^7 KWh per year, in which about 2.53×10^6 KWh can be utilized; while the offshore wind energy potential reaches 7.5×10^6 KWh [2]. According to global wind report 2011 [3] from European Wind Energy Association (EWEA), around 2×10^4 MW [3] of new wind energy capacity were installed in China in 2011, and the total wind installed capacity reached about 6.5×10^4 MW by the end of 2011.

In recent years, the installation of wind power energy has moved from small wind farms with a few wind turbines to large wind farms with more than hundreds of MW of capacity, and the connection point of wind farm have developed from low-voltage-level electricity distribution system to high-voltage-level electricity transmission system. With the increasing penetration level, the impact of wind power on power system dynamics and stability become increasingly complicated. The increasing utilization of wind power gives rise to a new set of problems [4] in frequency fluctuations, harmonics pollution as well as concerning the safe and stable operation, the peaking and frequency modulation and the economical dispatch of power systems. Consequently, the impact of wind power penetration on power system stability [5–20] has drawn great attention to the power system engineers and scientists. Meanwhile, the large-scale interconnected power system is forming in China, and the problem of the weak damp or negative damp status between inter-network or local network with wind power integrated to power systems is critical. Besides, the increasing capacity and different connection point of wind power maybe worsen the damping performances of power systems, meanwhile, wind power resources are generally located far from large centers of consumption, which may involve transmitting energy for long distances through congested lines and perhaps a significant change of generation profile and typical power flows, and may have significant impact on small signal stability. Therefore, it's necessary to systematically analyse the influence of wind power integration on power system oscillation mode and damping characteristic.

The purpose of this paper is to analyze the impacts of wind power on small signal stability and corresponding control strategies to improve power system damping characteristic and get a summarized study on the hot topics. The paper is organized as follows: Section 2 introduces the characteristics of WTGs including the different types and technologies of wind power grid connected. The principles and methods of power system small signal stability analysis are described in Section 3. Section 4 summarizes the effects of large scale wind power on small signal stability, including the analysis of the modeling and simulation, the research on power system oscillation mode and damping characteristic, and the auxiliary control strategies on WTGs to reinforce small signal stability. The conclusion is given in Section 5.

2 Characteristics of WTGs

WTGs are used to extract the kinetic energy from the wind and then convert it into electricity. Wind turbines usually rotate at a speed of 3–30 miles/h. WTGs system normally consists of wind turbine, generator, grid interface converters, wind blades, yaw control, brake, gear box, control device, and an anemometer, etc.

For the moment, there are three prevailing kinds of WTGs in the market, including the squirrel cage induction wind power generator (SCIG), the doubly-fed induction wind power generators (DFIG), and the direct-driven permanent magnet wind power generators (DDPMG). In this section, the characteristic of three popular types [1, 2] of WTGs and the technologies [8–12] of wind power grid connected are introduced briefly.

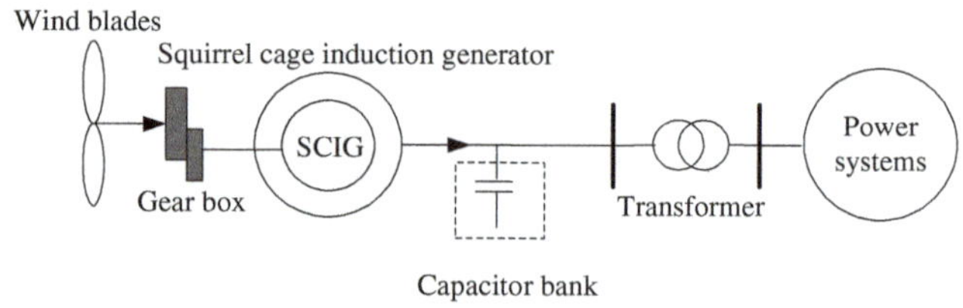

Fig. 1 Schematic diagram of SCIG system

2.1 SCIG

The wind turbine rotor is connected to the generator through a gear box and the stator is directly coupled to the grid as shown in Fig. 1. SCIG always has high reactive power consumption and hence uses capacitor banks to maintain the required voltage level at the grid. The slip, and hence the rotor speed of a SCIG varies with the amount of power generated. These rotor speed variations are very small, approximately 1 to 2 percent. Therefore, this wind turbine type is normally referred to as a constant speed or fixed speed turbine. Nonetheless, they can run at two different speeds by changing the number of pole pairs of the stator windings.

2.2 DFIG

The schematic diagram of a grid-connected DFIG system is depicted in Fig. 2. The DFIG system includes the wind turbine, the drive train, the gear box, the induction generator and the back-to-back pulse width modulation (PWM) converters. The stator is directly coupled to the grid and the rotor is connected to the gird through the PWM converters. The active power and reactive power of DFIG can be decoupled control by adjusting the rotor side converter modulation coefficient, which can improve the energy conversion efficiency, enhance the power factor of wind farms and the voltage stability. This makes DFIG popular in the recent years [1, 9, 10]. However, because the internal resistance of DFIG is very small, the capacity of DFIG damped oscillation is limited.

2.3 DDPMG

The DDPMG wind power system involves the wind turbine generator, the drive train and the full scale back-to-back PWM converters, as presented in Fig. 3. The rotor of DDPMG is directly connected to the generator and the stator is coupled to the grid through a full-scale PWM converter. By comparing to DFIG, DDPMG saves the gear box, the

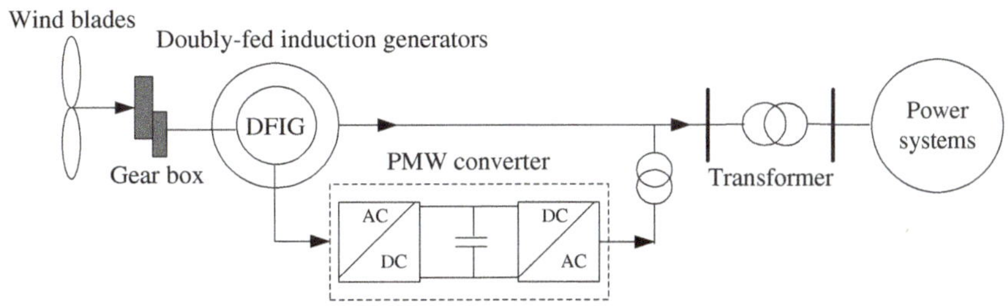

Fig. 2 Schematic diagram of DFIG system

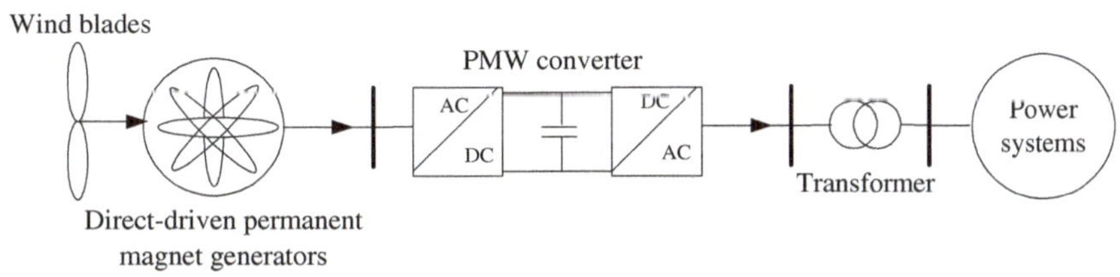

Fig. 3 Schematic diagram of DDPMG system

brush and the slip ring, which improves the operation reliability, reduces the maintenance cost, however, it needs full-scale converter. The output electromagnetic power of DDPMG is controlled by the pitch angle control and the full-scale PWM converter, which makes DDPMG and the electrical grid completely decoupled.

2.4 Performance comparisons

Each of these WTGs has its own advantages and disadvantages as elaborated in [1, 5, 8, 12]. They can be summarized [8] using Table 1. The main strength of DFIG is that more energy can be generated for a specific wind range, and the more efficient aerodynamic, besides, there is a reduced mechanical stress as the rotor acts as a flywheel thereby reducing the drive train torque variations. The major drawbacks of DDPMG are that they are large, heavy and complex ring generators and need large power electronic converter.

Table 1 Comparison of different WTGs

	Advantages	Disadvantages
SCIG	Less expensive, electrically efficient simple and robust construction	Noisy, high mechanical stress, aerodynamically less efficient, requires gear box
DFIG	Less noisy, aerodynamically efficient small rating converter	It is expensive and electrically less efficient, requires gear box
DDPMG	Less noisy, aerodynamically efficient, no gear box, less mechanical stress	Needs large rating converter, expensive, heavy and large generator

3 Power system small signal stability

Small signal stability (or small-disturbance stability) in a power system is the ability of the system to ascertain a stable operating condition following a small perturbation around its operating equilibrium [7, 21]. Of particular importance in the analysis of small-signal stability is the determination of the electromechanical modes of oscillation (EMO). The electromechanical modes involve the rotors of individual generators or of groups of generators oscillating or swinging against each other, and they can be subdivided into inter-area, local-area and intra-station modes [7].

Small signal stability analysis is carried out based on the system nonlinear equations, describing the dynamic behavior of the system, linearized around a chosen operating point. Based on the motion of the rotors and the torque characteristics of generator, the mechanism of small signal stability is analyzed, such as the negative damping mechanism, the resonance mechanism, nonlinear singular and chaos mechanism [22–24].

It has been known that there are two factors in influencing the small signal stability of power systems.

1) Lack of sufficient synchronizing torque, which is proportional to the power angle increment, resulting in an increasing of the rotor angle of the generator.
2) Lack of sufficient damping torque, which is proportional to the generator rotational speed increment, resulting in an increasing oscillation of the rotor.

At present, there are two main methods for analyzing power system small signal stability.

1) Eigenvalue analysis method based on the state space model;
2) Frequency domain analysis method based on the transfer function matrix.

Eigenvalue analysis method, which based on the linear system theory and Lyapunov stability theory, has been recognized as one of the broadest methods to analyze small signal stability. In this method, the complex power system can be linearized around a stable operating point which can give a close approximation to the system to be studied [7]. And then, stability analysis is done by computing eigenvalues as well as the right and left eigenvectors of its state matrix, hence the judgment information of system stability, such as the oscillation frequency, the attenuation factor, the impact factor, and the strongly correlated state variables, can be obtained. At present, eigenvalue analysis method is a powerful tool to research the electromechanical mode and the damping performances of power systems, and it has been successfully used in small signal stability evaluation, the damping controller siting determination [25] and parameter optimization [26, 27] and so on.

The frequency domain model of power systems described by the transfer function matrix, i.e. $\boldsymbol{y}(s) = \boldsymbol{G}(s)\boldsymbol{u}(s)$, (where, $\boldsymbol{y}(s)$ is the output vector; $\boldsymbol{G}(s)$ is the transfer function matrix; $\boldsymbol{u}(s)$ is the input vector), can be obtained by appropriately choosing the input and output variables. The system is stable under small disturbance, which is equivalent in the negative real part of all poles of $\boldsymbol{G}(s)$, and it can be examined by the multivariable Nyquist stability criteria. The scale of power system had little impact on the frequency domain analysis method, which makes it reliable to analyze the large scale power systems. Whereas compared eigenvalue analysis method, it provides insufficient information. However, with the development of modern multivariable control system frequency domain theory, frequency domain analysis method gets more and more attentions by the electricity workers in recent years, and then achieves further development. The frequency domain analysis model of power system small signal stability, including the high-voltage direct current (HVDC), is built in [28]. And in [29], it is applied to the robust stability analysis of power systems.

The well-known power system stabilizer (PSS) [21, 25, 26] is used to generate supplementary control signals for the excitation system of a generator so as to damp low frequency oscillations in the power system concerned. Intelligent algorithm is introduced to the PSS design [27] in recent years, and some simple structure, adaptable, and better robustness PSS are used to control the large scale power system stability.

4 Impacts of wind power integration on small signal stability

4.1 Modeling and simulation

The increasing wind power penetration on power systems requires the development of adequate wind farm models for representing the dynamic behavior of wind farms on power systems. A lot of researching and analyzing works [10, 13–15, 30–33] by the scholars have been done on modeling and simulation for power system small signal stability with wind power integration, which include modeling of three popular WTGs and the power system small signal analysis with penetration of wind power, and the focus of research is DFIG.

A reduced order dynamic modeling of large (MW) capacity fixed speed and doubly-fed asynchronous generator wind turbine is discussed in [10], in the model, speed control characteristics and converter protection of the DFIG are implemented. Meanwhile, a dynamic model is also derived in [13], which can be used to simulate the DFIG wind turbine using a single-cage and double-cage representation of the generator rotor, as well as a representation of its control and protection circuits. This model can also be used to investigate transient stability of large power systems.

Reference [14] gives the modeling of the DFIG considering operating conditions below and above synchronous speed, which are actually achieved by means of a double-sided PWM converter joining the machine rotor to the grid. And the stator-flux-oriented vector control is applied to decouple the active and reactive powers generated by the generator. Likewise, the steady-state analysis of DFIG in the subsynchronous and supersynchronous regions is also discussed in [15]. An optimum control strategy, which maximizes the total electrical power output of the double output induction generator is determined, and a equivalent circuit is given, but in which core losses and harmonics are ignored.

Based on the relationship of flux linkage, potential and electric current, dynamic mathematics model of DFIG is deduced in [30], and this model neglects the stator transients and the harmonics of electrical magnitudes. The development of equivalent wind farm models can reduce the model order and the computation time under the impact of wind farms on power systems. Two equivalent wind turbines are developed in [31] by using MATLAB/Simulink: one for aggregated wind turbines with similar winds, and another for aggregated wind turbines under any incoming wind. A wind park dynamic model together with a base methodology for its application to power system studies is presented in [32], which is applied to the operating conditions of the selected sets of wind turbine experimental benchmark data from actual wind parks, and the results show its effectiveness. And a unified, modular, small-signal dynamic model of DFIG is presented in the PSCAD/EMTDC environment [33]. Similarly, a SCIG simulation model with thyristor-based soft-start module is presented in [20].

From the present literatures, it can be seen that the modeling techniques of WTGs based on aerodynamic have been relatively mature. Many simulation softwares, such as DIgSILEN/Power Factory, PSASP, MATLAB, PSS/E, PSCAD/EMTDC, include the WTG module. In the actual research, the detailed or simplified model can be built according to the aim of the research to study the small signal stability of power systems with wind power integration.

4.2 Influence of wind power integration on power system oscillation mode and damping characteristics

With the increase in penetration of these wind turbines, the power system dominated by synchronous machines will experience a change in dynamics and operational characteristics. Given this assertion, Slootweg and Klind initiated the issue that the impact of constant speed WTGs and variable speed WTGs on small signal stability of power systems in 2003 [11]. Later on more researchers started to concentrate on this subject by the means of modal analysis or time-domain simulations, and then the impact of large scale wind power integration on power system oscillation modes and damping performances has become a hot research topic.

The impact of DFIG based wind farms on the system small signal stability is investigated in [12], which showed that the DFIG based wind generating system has negative damping under changed operating condition (wind speed), thus this paper suggest the additional damping controller should be adopted.

The impact of increasing wind power on the damping performance of the New Zealand power system is assessed in [16]. The damping performance of different scenarios is compared, based on the eigenvalues of a linearized model; the studies indicate that the damping performance of the system is not materially affected by the types of wind energy converter technology (SCIG, DFIG, or DDPMG). Moreover, [17] investigates the impact of wind power integration in Norway on the damping of interarea mode oscillations in the Nordic grid. The impact on the interarea mode oscillations is tested for various types of generators. The results show that SCIG improves the damping of the interarea mode oscillation between Norway and Sweden, while DFIG and DDPMG decrease the damping.

Table 2 Different impacts of WTGs on dampings

	Positive	Negative
SCIG	✔	
DFIG	✔	✔
DDPMG	✔	✔

According to the linearization-based eigenvalue analysis, time domain simulations, and probabilistic eigenvalue analysis, [18] shows that large wind power integration can have positive or negative impacts on system damping depending on the location of the wind power plant, the amount of the conventional generation replaced by wind power and the stress level of the power system.

The effect of increased penetration of DFIG on the stability of inter-area oscillations in a small typical power system is examined by the PSS/E program in [19]. It is shown that the effect is generally favorable, but sometimes the inter-area mode is less damped or can even be destabilized.

Power system oscillation damping with DFIG is investigated in [34, 35], and the transient and small signal stability of DFIG are analyzed in [36] by using the sensitivity of the eigenvalues with respect to inertia. Both beneficial and detrimental impacts with the increased penetration of DFIG are found. Meanwhile, [37] advocates that DDPMG without the controllers will have no adverse effect on the stability.

It is widely agreed that SCIGs, whose stators are directly coupled to the grid, have positive impacts on system damping, as shown in Table 2; DFIGs and DDPMG have both positive and negative impacts on the damping performance of the electromechanical oscillation modes. In practice, the effect of WTGs on power system small signal stability should be discussed according to the types of oscillations, the capacity of integrated wind power and different connection points.

4.3 Control strategies on WTGs to improve power system damping characteristic

According to the analysis above, WTGs require additional controllers to damp power system oscillations. In this section, control strategies on WTGs to improve power system damping characteristic are reviewed.

During the last few decades, many different maximum power point tracking (MPPT) control strategies have been developed in [38] and [39]. This enabled the selection of the optimal MPPT for each wind energy conversion system (WECS) project. And a prototype version of the control strategy of a 20 KW DDPMG for maximum power tracking is proposed in [40] for the WECS.

The auxiliary control loop for oscillation damping that adjusts the active power command to damp the interarea oscillation is proposed in [35]. Moreover, a power system stabilizer (PSS) using a speed deviation is proposed in [41]. It is reported that the presence of the PSS in the DFIG system improves the damping of the oscillations in the network. Nevertheless, it is very important to optimize the controller parameters of the PSS to achieve the best performance.

Reference [42] focuses on the impact of a damping controller on the super/subsynchronous operation of DFIG, which would help in proper tuning of the DFIG controllers to enhance the system small-signal stability. Based on the concept of dynamic frequency character, the phase and amplitude conditions for wind farms to produce the positive damping effects on their connected power systems are proved in [43], and a damping control strategy of DFIG is studied on 2-area and multi-area power systems by means of Power System Analysis Software Package (PSASP). Meanwhile, a control strategy for the multilevel frequency converter of DDPMG is proposed to augment the low voltage ride through (LVRT) requirement as well as to minimize voltage fluctuation in [44].

Recently, a flux magnitude and angle controller (FMAC) scheme is put forward for DFIGs, by which the terminal voltage and stator power can be controlled via the magnitude and angle of the rotor flux vector decouply [45]. For DFIG, eigenvalue sensitivity [46] is used to design damping controllers aimed at damping electromechanical oscillations. Additionally, multi-objective optimal controller design using differential evolution (DE) of DFIG is presented in [47]. Eigenvalue analysis and time-domain simulations are performed on a single machine infinite bus system as well as a nine-bus multi-machine system to illustrate the control performance of DFIG with the optimised controller parameters. Particle swarm optimisation (PSO) [48] is also applied for optimising parameters of controllers of DFIG.

5 Conclusions

The influence of large scale wind power integration on power system small signal stability and the corresponding control strategies are analyzed in this paper as follows.

1) Nowadays, there are three major types of WTGs on the market: SCIG, DFIG, and DDPMG. It is universally accepted that SCIGs have a positive effect on the damping characteristic while the pros and cons impacts of DFIG and DDPMG are presented. In fact, the factors, such as the types of WTGs, the way of grid-connected, the capacity of integrated wind power and the different connection point, will bring different effect on the system damping, which maybe beneficial or detrimental.
2) The topic research of power system small signal stability with wind power penetration focuses on the modal analysis and time-domain simulations, through which the impact of wind power on power system oscillation modes and damping performances are discussed.
3) It is found that the current hot focuses of the research of improving power system damping characteristic with wind power integration are mostly in the control strategies of the additional damping controller, including the controller design, optimization of the parameter of the damping controller, the coordinated tuning of different controllers, eigenvalue sensitivities, the input signal selection, PSS, and so on.
4) The simulation softwares used in researching wind power grid-connected systems mostly include DIgSILEN/Power Factory, PSASP, MATLAB, PSS/E, and PSCAD/EMTDC, which provide the user with a convenient WTG module for further study.

Acknowledgments This work is supported by a discovery project of Australia Research Council (No. DP120101345), and National Key Technology Research and Development Program of China (No. 2011BAA07B02).

References

[1] Joselin Herbert GM, Iniyan S, Sreevalsan E et al (2007) A review of wind energy technologies. Renew Syst Energy Rev 11(6):1117–1145
[2] Ackermann T (2005) Wind power in power systems. Wiley, Chichester
[3] Annual market update (2011) Global wind Report. Global Wind Energy Council (GWEC), Brussels, Belgium
[4] Sorensen P, Cutululis NA, Vigueras-Rodriguez A (2007) Power fluctuations from large wind farms. IEEE Trans Power Syst 22(3):958–965
[5] Sun YZ, Wang LX, Li GJ et al (2010) A review on analysis and control of small signal stability of power systems with large scale integration of wind power. In: Proceedings of the 2010 international conference on power system technology (POWERCON'10), Hangzhou, 24–28 Oct 2010, p 6
[6] Saifur R, Pipattanasomporn M (2011) Operating impacts and mitigation strategies with large-scale wind power penetration in

the United States. Automat Electr Power Syst 35(22):3–11 (in Chinese)
[7] Kundur P (1994) Power system stability and control. McGraw-Hill, New York
[8] Chedid R, LaWhite N, Ilic M (1993) A comparative analysis of dynamic models for performance calculation of grid-connected wind turbine generators. Wind Energy 17(4):168–182
[9] Yang LH, Xu Z, Ostergaard J et al (2012) Advanced control strategy of DFIG wind turbines for power system fault ride through. IEEE Trans Power Syst 27(2):713–722
[10] Holdsworth L, Wu XG, Ekanayake JB et al (2003) Comparison of fixed speed and doubly-fed induction wind turbines during power system disturbances. IEE P-Gener Transm Distrib 150(3):343–352
[11] Slootweg JG, Klind WL (2003) The impact of large scale wind power generation on power system oscillations. Electr Power Syst Res 67(1):9–20
[12] Yateendra M (2008) Advances in power system small signal stability analysis considering load modeling and emerging generation resources. Dissertation, The University of Queensland, Brisbane
[13] Ekanayake JB, Holdsworth L, Wu XG et al (2003) Dynamic modeling of doubly fed induction generator wind turbines. IEEE Trans Power Syst 18(2):803–809
[14] Tapia A, Tapia G, Ostolaza JX et al (2003) Modeling and control of a wind turbine design doubly fed induction generator. IEEE Trans Energy Convers 18(2):194–204
[15] Cadirci I, Ermis M (1992) Double-output induction generator operating at subsynchronous and supersynchronous speeds: steady-state performance optimisation and wind-energy recovery. IEE P-Electr Power Appl 139(5):429–442
[16] Vowles DJ, Samarasinghe C, Gibbard MJ et al (2008) Effect of wind generation on small-signal stability—A New Zealand example. In: Proceedings of the 2008 IEEE power and energy society general meeting—conversion and delivery of electrical energy in the 21st century, Pittsburgh, 20–24 Jul 2008, p 8
[17] Hagstrøm E, Norheim I, Uhle K (2005) Large scale wind power integration in Norway and effect on damping in the Nordic grid. Wind Energy 8(3):375–384
[18] Rueda JL, Erlich I (2011) Impacts of large scale integration of wind power on power system small-signal stability. In: Proceedings of the 4th international conference on electric utility deregulation and restructuring and power technologies (DRPT'11), Weihai, 6–9 Jul 2011, pp 673–681
[19] Tsourakisa G, Nomikosb BM, Vournas CD (2009) Effect of wind parks with doubly fed asynchronous generators on small-signal stability. Electr Power Syst Res 79(1):190–200
[20] Quinonez-Varela G, Cruden A (2008) Modelling and validation of a squirrel cage induction generator wind turbine during connection to the local grid. IET Gener Transm Distrib 2(2):301–309
[21] Kundur P, Paserba J, Ajjarapu V et al (2004) Definition and classification of power system stability IEEE/CIGRE joint task force on stability terms and definitions. IEEE Trans Power Syst 19(3):1387–1401
[22] Demello FP, Concordia C (1969) Concepts of synchronous machine stability as affected by excitation control. IEEE Trans Power Appl Syst 88(4):316–329
[23] Gomes S, Martins N, Portela C (2003) Computing small signal stability boundaries for large-scale power systems. IEEE Trans Power Syst 18(2):747–752
[24] Srivastava KN, Srivastava SC (1998) Elimination of dynamic bifurcation and chaos in power systems using FACTS devices. IEEE Trans Circ Syst I 45(1):72–78
[25] Tse CT, Tso SK (1988) Design optimization of power system stabilizers based on modal and eigen sensitivity analyses. IEE P-Gener Transm Distrib 135(5):406–415
[26] Wang KW, Chung CY, Tse CT et al (2000) Multimachine eigenvalue sensitivities of power system parameters. IEEE Trans Power Syst 15(2):741–747
[27] He P, Wang KW, Tse CT et al (2007) Studies of the improvement of probabilistic PSSs by using the single neuron model. Int J Electr Power Energy Syst 29(3):217–221
[28] Osauskas CM, Hume DJ, Wood AR (2001) Small signal frequency domain model of an HVDC converter. IEE P-Gener Transm Distrib 148(6):573–578
[29] Anderson PM, Fouad AA (2002) Power system control and stability. Wiley, Chichester
[30] Feijóo A, Cidrás J, Carrillo C (2000) A third model for the doubly-fed induction machine. Electr Power Syst Res 56(2):121–127
[31] Fernández LM, Saenz JR, Jurado F (2006) Dynamic models of wind farms with fixed speed wind turbines. Renew Energy 31(8):1203–1230
[32] Estanqueiro AI (2007) A dynamic wind generation model for power systems studies. IEEE Trans Power Syst 22(3):920–928
[33] Tabesh A, Iravani R (2008) Small-signal model and dynamic analysis of variable speed induction machine wind farms. IET Renew Power Gener 2(4):215–227
[34] Tsourakis G, Nomikos BM, Vournas CD (2009) Contribution of doubly fed wind generators to oscillation damping. IEEE Trans Energy Convers 24(3):783–791
[35] Fan LL, Miao ZX, Osborn D (2008) Impact of doubly fed wind turbine generation on inter-area oscillation damping. In: Proceedings of the 2008 IEEE power and energy society general meeting—Conversion and delivery of electrical energy in the 21st century, Pittsburgh, 20–24 Jul 2008
[36] Gautam D, Vittal V, Harbour T (2009) Impact of increased penetration of DFIG-based wind turbine generators on transient and small signal stability of power systems. IEEE Trans Power Syst 24(3):1426–1434
[37] Wu F, Zhang XP (2009) Small signal stability analysis and control of the wind turbine with the direct-drive permanent magnet generator integrated to the grid. Electr Power Syst Res 79(12):1661–1667
[38] Chedid R, Mrad F, Basma M (1999) Intelligent control of class of wind energy conversion systems. IEEE Trans Energy Convers 14(4):1597–1604
[39] Mohamed AZ, Eskander MN, Ghali FA (2001) Fuzzy logic control based maximum power tracking of wind energy system. Renew Energy 23(2):235–245
[40] Tan K, Islam S (2004) Optimum control strategies in energy conversion of PMSG wind turbine system without mechanical sensor. IEEE Trans Energy Convers 19(2):392–399
[41] Hughes FM, Anaya-Lara O, Jenkins N et al (2006) A power system stabilizer for DFIG-based wind generation. IEEE Trans Power Syst 21(2):763–772
[42] Mishra Y, Mishra S, Li FX et al (2009) Small-signal stability analysis of a DFIG-based wind power system under different modes of operation. IEEE Trans Energy Convers 24(4):972–982
[43] Hao Z, Yu Y, Zeng Y (2011) Control strategy for increasing power system damping with wind turbine-driven doubly-fed induction generator. Autom Electr Power Syst 35(15):25–29 (in Chinese)
[44] Muyeen SM, Takahashi R, Murata T et al (2010) A variable speed wind turbine control strategy to meet wind farm grid code requirements. IEEE Trans Power Syst 25(1):331–340
[45] Anaya-Lara O, Hughes FM, Jenkins N et al (2006) Rotor flux magnitude and angle control strategy for doubly fed induction generators. Wind Energy 9(5):479–495
[46] Sigrist L, Rouco L (2009) Design of damping controllers for doubly fed induction generators using eigenvalue sensitivities. In: Proceedings of the power systems conference and exposition (PSCE'09), Seattle, 15–18 Mar 2009, p 7

[47] Yang L, Yang GY, Xu Z et al (2010) Optimal controller design of a doubly-fed induction generator wind turbine system for small signal stability enhancement. IEE P-Gener Transm Distrib 4(5):579–597

[48] Wu F, Zhang XP, Godfrey K et al (2007) Small signal stability analysis and optimal control of a wind turbine with doubly fed induction generator. IEE P-Gener Transm Distrib 1(5):751–760

Author Biographies

Ping HE received her B.E. and M.E. degrees from Zhengzhou University, China, in 2003 and 2006, respectively. She has been a lecturer at Zhengzhou University of Light Industry since 2006. She is concurrently pursuing the Ph.D. degree in South China University of Technology. Her research interests include power system stability analysis and control, wind power generation.

Fushuan WEN received his B.E. and M.E. degrees from Tianjin University, China, in 1985 and 1988, respectively, and Ph.D. from Zhejiang University, China, in 1991, all in electrical engineering. He joined the faculty of Zhejiang University in 1991, and has been a full professor and the director of the Institute of Power Economics and Information since 1997, and the director of Zhejiang University-Insigma Joint Research Center for Smart Grids since 2010. He had been a university distinguished professor, the deputy dean of the School of Electrical Engineering and the director of the Institute of Power Economics and Electricity Markets in South China University of Technology (SCUT), China, from 2005 to 2009. His current research interests include power industry restructuring, power system alarm processing, fault diagnosis and restoration strategies, and smart grids.

Gerard LEDWICH received his Ph.D. degree in electrical engineering from the University of Newcastle, Newcastle, Australia, in 1976. He has been a Chair Professor in electrical asset management at Queensland University of Technology (QUT), Brisbane, Australia, since 1998. His interests are in the areas of power systems, power electronics, and controls. Prof. Ledwich is a Fellow of I.E. Australia and a senior member of IEEE.

Yusheng XUE received his Ph.D. degree in Electrical Engineering from the University of Liege (Belgium) in 1987. He became a Member of Chinese Academy of Engineering in 1995. He is now the Honorary President of State Grid Electric Power Research Institute (SGEPRI), State Grid Corporation of China. His research interests include nonlinear stability, control and power system automation.

2

Research on large-scale dispatchable grid-connected PV systems

Xiaonan HAN (✉), Xin AI, Yingyun SUN

Abstract This paper describes the research on a large-scale dispatchable grid-connected photovoltaic (PV) system for supplying power to the grid for dispatch instead of supplying the electricity to a local load. In order to maximise the value of the solar energy, a hybrid electricity storage consisting of batteries and supercapacitors is used with the PV system. This paper proposes a control strategy focusing on the DC power at the DC link rather than at the grid-connected inverter. Two typical sets of real data, collected from existing sites, are used to demonstrate the practicality of the system. Finally, the simulation results are used to demonstrate the good performance and feasibility of the proposed system together with the proposed control strategy.

Keywords Large-scale PV system, Dispatchable generation, Centralized grid-connected, Battery, Supercapacitor

1 Introduction

Since the energy crisis and the serious environmental pollution are regarded as the greatest challenges for humanity in recent years. Renewable energy is increasingly being developed in different fields worldwide and in particular for electricity generation.

The electricity power system is accepted as one of the main sources of CO_2 emissions, and therefore the development of renewable energy, which is considered to be non-polluting for generating electricity, has been a hot topic in recent years. Solar energy is recognized as the most promising renewable energy and is quickly being developed due to its advantages: inexhaustible supply, inexpensive and easy to maintain, etc.

Large-scale grid-connected photovoltaic (PV) systems have become a major development area for research although there are still many barriers, it is expected to have a major influence on the future grid [1]. More than 100 large-scale, above 10 MW, grid-connected PV plants have been built in the world and many countries are working on the 100 MW PV plants [2]. There are grid-connected PV systems and distributed network connected PV systems [3]. The grid-connected systems can also be divided into systems which can be dispatched and those which cannot. This is decided by whether or not the system is equipped with storage system [3].

In this paper, a large-scale dispatchable grid-connected PV system is proposed and simulated using MATLAB/Simulink. The storage system consists of both batteries and supercapacitors to achieve better performance. The simulation results verify the control strategy and main target of providing a constant power output, 10 MW, which is required by the grid for centralized dispatch.

2 System design and modeling

2.1 Structure of the system

When the large-scale centralized PV system is connected with the grid, the grid should be able to regard it as a

X. HAN, X. AI, Y. SUN, State Key Laboratory for Alternate Electrical Power System with Renewable Energy Sources, North China Electric Power University, Changping District, Beijing 102206, China
(✉) e-mail: hanxn0421@gmail.com
X. AI
e-mail: aixin@ncepu.edu.cn
Y. SUN
e-mail: sunyy@ncepu.edu.cn

normal generation plant able to deliver a rated capacity and therefore be centrally dispatched [4]. Consequently, the PV plant should be equipped with an energy storage system in order that it can supply constant power to the grid all the time, including days and nights. This enables the PV plant to deliver power as required by the dispatcher instead of only supplying energy to the load when the sun shines.

A hybrid electricity storage of battery and supercapacitor has been selected in this paper with the target of providing a higher efficiency and better electricity quality. There are various devices which could be used for storage including batteries, flywheels, supercapacitors, and fuel cells (FC), all of which have different specific values of energy density and power density making them applicable for short-term or long-term energy storage [5].

The electrochemical battery, with its high energy density, is one of the most cost-effective energy storage devices, but its power density is relatively poor [6]. The supercapacitor, also called the electrical double layer capacitor, has a correspondingly high power density but low energy density. Subsequently, the complementary characteristics of batteries and supercapacitors make them an attractive choice for a hybrid electricity storage system in which the battery is chosen as the long-term storage device to meet the demand and the supercapacitor provides the short-term storage device to absorb the fast, high-frequency power fluctuations of the PV. Also, the supercapacitor can operate at any state of charge, since it stores energy by physically separating positive and negative charges, which means that it can be operated fully charged or fully discharged [7]. However, for the batteries, the state of charge *SOC* must be controlled to maintain an appropriate level to ensure a long cycle-life. The supercapacitor can therefore be used to protect the battery from short duration transients and help prolong the life of the battery. The hybrid electricity storage can therefore provide a better performance in terms of economy and generation efficiency [5].

2.2 System modeling

2.2.1 PV array

A PV cell is the smallest component within a PV plant which is used to build PV modules. The PV array consists of several PV modules in series or parallel. If N_s is the number of the PV cells and N_{ss} and N_{pp} refer to the numbers of the PV modules in series and parallel respectively, the voltage of the PV array is N_{ss} times the voltage of a PV module and the current capability is N_{pp} times that of an individual cell. The equivalent circuit PV cell is shown in Fig. 1 and the mathematical model of the PV array is given in (1) [8, 9].

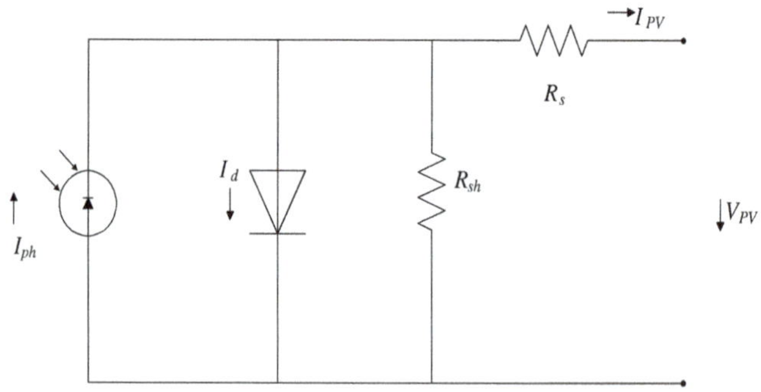

Fig. 1 Equivalent circuit of a PV cell

$$I_{PV} = N_{pp}\left\{I_{ph} - I_d\left[exp\frac{q\left(\frac{V_{PV}}{N_sN_{ss}} + \frac{I_{PV}R_s}{N_{pp}}\right)}{AKT} - 1\right] - \frac{\frac{V_{PV}}{N_sN_{ss}} + \frac{I_{PV}R_s}{N_{pp}}}{R_{sh}}\right\}, \quad (1)$$

where I_{ph} is photovoltaic current; I_d is diode reversed saturation current; R_s is equivalent series resistance of the PV array; R_{sh} is equivalent parallel resistance of the PV array; I_{PV} is output current of the PV array; V_{PV} is terminal voltage of the PV array; A is diode ideal factor of the P-N junction (dimensionless constant); K is Boltzmann constant (1.38×10^{-23} J/K); T is ambient temperature (°C); q is electronic charge (1.6×10^{-19} C).

The photovoltaic current, I_{ph}, is dictated by several factors of which the solar irradiance and the ambient temperature are the most significant. The diode reversed saturation current, I_d, is dependent on the diode characteristics. The values can be calculated by the (2) and (3).

$$I_{ph} = \frac{S}{S_{ref}}\left[I_{ph,ref} + k_o\left(T - T_{ref}\right)\right], \quad (2)$$

$$I_d = I_0\left\{\left[\frac{qU_d}{AKT}\right] - 1\right\}, \quad (3)$$

where S is solar irradiance (W/m^2); S_{ref} is reference value of the solar irradiance (1000 W/m^2); $I_{ph,ref}$ is reference value of photovoltaic current; k_o is short-circuit current temperature coefficient related to the cell material; T_{ref} is reference value of the ambient temperature (25°C or 273K); I_0 is reversed saturation current through the P–N junction of the equivalent diode; and U_d is terminal voltage of the equivalent diode.

It can be demonstrated that, using the mathematic equations above, the output of the PV system is affected by adverse weather conditions and behaves randomly. More importantly, the output has a nonlinear relationship with its principal influencing factors, irradiation and temperature.

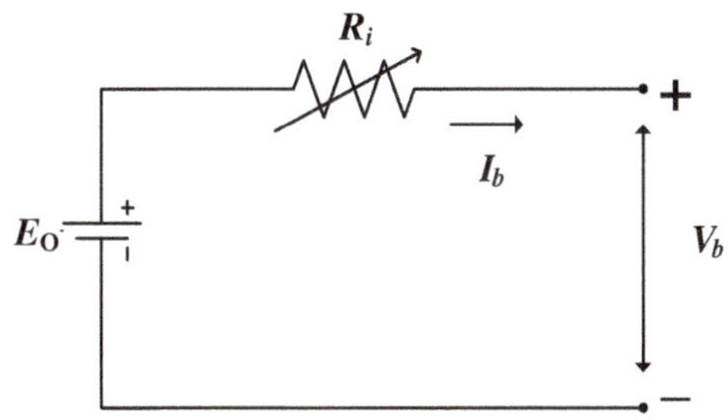

Fig. 2 Equivalent circuit of the improved simple battery model

2.2.2 *Hybrid electricity storage*

Although it is difficult to simulate the true behavior of the lead-acid battery, several models have been developed by researchers including the simple battery model and the Thevenin model [10]. Within the MATLAB/Simulink software, an existing battery model was chosen in which the type of battery and its significant parameters can be set easily. The structure of this existing model is very complex but the basic theory is based on the improved simple battery model whose equivalent circuit is shown in Fig. 2.

$$R_i = \frac{R_0}{S^k}, \tag{4}$$

$$S = 1 - \frac{\text{Ah}}{C_{10}}, \tag{5}$$

$$V_b = E_0 - I_b R_i, \tag{6}$$

$$SOC = SOC_0 - \frac{1}{C_b}\int_0^t I_b d_\tau, \tag{7}$$

where R_i is non-constant internal resistance; R_0 is internal resistant when fully charged; Ah is unit of amp-hour; C_{10} is 10-h capacitor under the reference temperature (Ah); k is coefficient related to the capacitor and the discharge rate of battery; S is constant (equal to 0 when fully discharged and 1 when fully charged); V_b is terminal voltage of the battery; E_0 is voltage of the ideal voltage source; I_b is current through the battery; SOC_0 is initial state of charge (equal to 1 when fully charged); and C_b is battery capacitor (Ah).

When the battery is fully charged or fully discharged, the *SOC* is 100% or 0% respectively. In practical terms, the battery should never be fully discharged to ensure battery longevity. The minimum allowable *SOC* is typically about 20%.

Figure 3 shows the equivalent circuit of a basic supercapacitor in which the variable capacitance C plays the most important role and the resistance R_2 describes the self-discharge of the supercapacitor. The capacitance C_p together with the resistance R_p represents the fast dynamic changes of the supercapacitor. C_p is lower than C. The resistance R_1 which is connected in series represents the energy loss during the charging or discharging period, while the resistance R_3 is added in case of over-voltage. The variable capacitances C and C_p should be chosen and regulated by the size the supercapacitor [11].

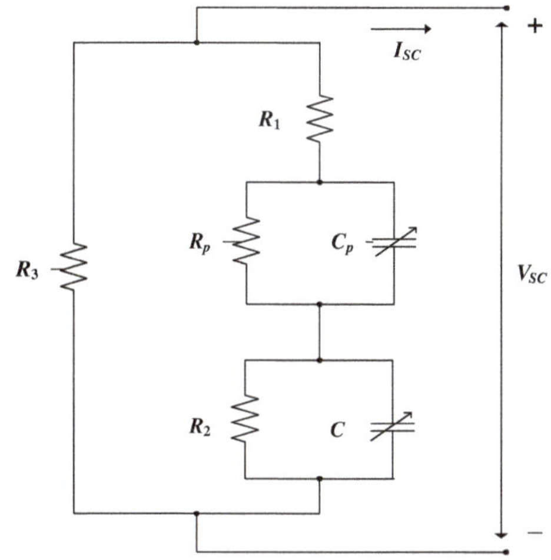

Fig. 3 Equivalent circuit of the supercapacitor model

$$i_{SC}(t) = C_{SC}\frac{d}{d_t}u_{SC}(t), \tag{8}$$

where i_{SC} is current through supercapacitor; C_{SC} is overall capacitance of supercapacitor; and u_{SC} is terminal voltage of supercapacitor.

2.2.3 *Converters*

In the SimPower Library of MATLAB/Simulink, there are models of inverters whose parameters and power electronic devices can be defined. The inverter is used to convert DC voltage into AC voltage. Consequently, the single-phase full-bridge inverter using insulated gate bipolar transistor (IGBT) and Diodes has been selected for this system.

The equivalent circuits of the Boost DC/DC converter and the Buck-Boost DC/DC converter are shown in Fig. 4. The Boost DC/DC converter is used to increase the input DC voltage. As a bi-directional converter, the Buck-Boost DC/DC converter has two operating modes as its name suggests. The two modes can be switched to regulate the duty of the MOSFET, as part of the control strategy. The capacitors are for filtering in other to obtain consequent output voltage V_{out}. Attention has to be paid that the output voltage V_{out} since it has an opposite polarity to the input voltage V_{in}.

3 Control strategy

3.1 MPPT of the PV array

The maximum power point tracking (MPPT) of a PV system is an indispensable control to ensure that the PV array is working at the maximum power point no matter how the solar irradiation and ambient temperature change. The efficiency of the electricity generation of PV system can therefore be kept at a maximum.

Several MPPT algorithms have been proposed by previous researchers including the Perturb and Observe (P&O), the Constant Voltage tracking (CV) and the Incremental Conductance (INCond) [12]. Compared with other algorithms, the Incremental Conductance (INCond) has become the most widely used. It has many advantages; one being that the higher tracking accuracy can be maintained even when the weather conditions change rapidly [12]. Since there is no disturbance around the maximum power point, it is able to provide better stability than the P&O method with its additional perturbation. Accordingly, the INCond version of MPPT was chosen for the control strategy in this paper.

Since there is only one maximum power point on the P–V curve under fixed solar irradiation and ambient temperature, the gradient at that point is zero. Therefore, when the incremental conductance is equal to a negative value of the instantaneous conductance, the PV system is operating at the maximum power point. The basic mathematic theory is described by the following equations [12].

$$\begin{cases} P_{PV} = I_{PV} V_{PV} \\ \frac{dP_{PV}}{dV_{PV}} = I_{PV} + \frac{dI_{PV}}{dV_{PV}} V_{PV} = 0 \\ \frac{dI_{PV}}{dV_{PV}} = -\frac{I_{PV}}{V_{PV}}, \end{cases} \tag{9}$$

where P_{PV} is output power of PV array; I_{PV} is output current of PV array; and V_{PV} is output voltage of PV array.

The flow chart of the INCond is shown in Fig. 5. The output voltage and current of the PV array were constantly monitored and the values are compared with the previous ones. The incremental conductance is then compared with the negative value of the instantaneous conductance in order to find the point at which it satisfies (9). The reference value of PV array terminal voltage is then adjusted to the output of the MPPT control which is the reference voltage of PV array V_{ref}. This is the voltage corresponding to the maximum power point. This is used as the input of a double-loop control to create the control signal for the DC/DC converter.

3.2 Control strategy of the battery

The battery is the main storage device and should supply enough power to ensure that the constant output power is provided to the grid. In particular, this is required during the evenings and cloudy days when there is no solar irradiation for the PV array.

When the output power of the PV system is higher than the required value, the surplus power is stored in the battery for later use.

The flow chart in Fig. 6 illustrates the control strategy for the battery. Based on the target power capacity, the first

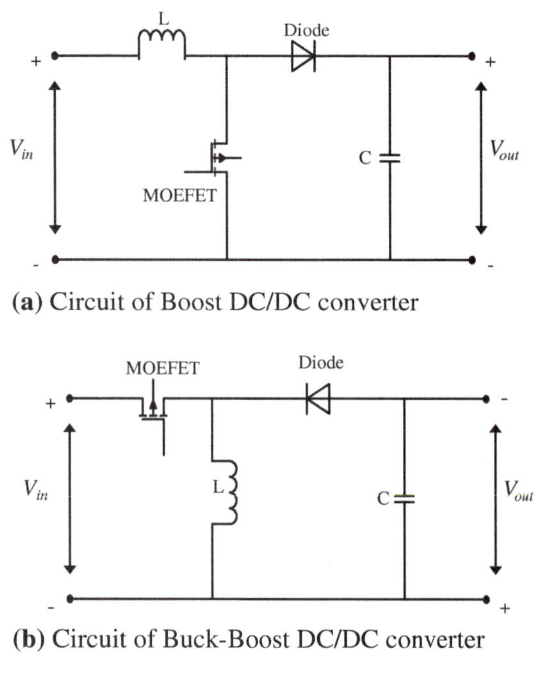

Fig. 4 Circuits of Boost DC/DC converter and Buck-Boost DC/DC converter

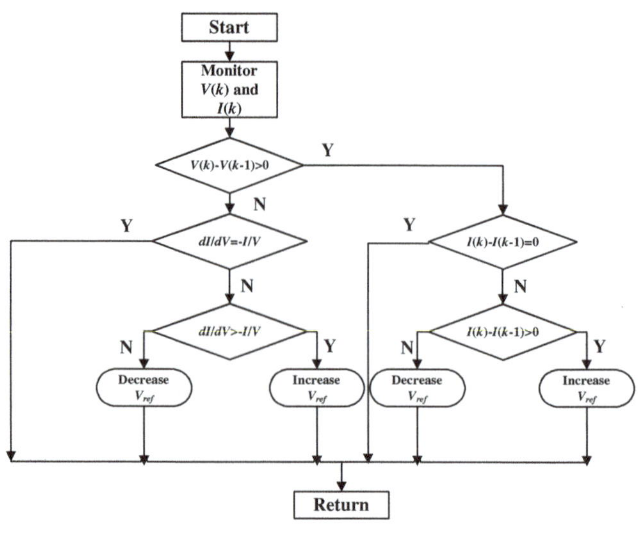

Fig. 5 Flow chart of the INCond MPPT

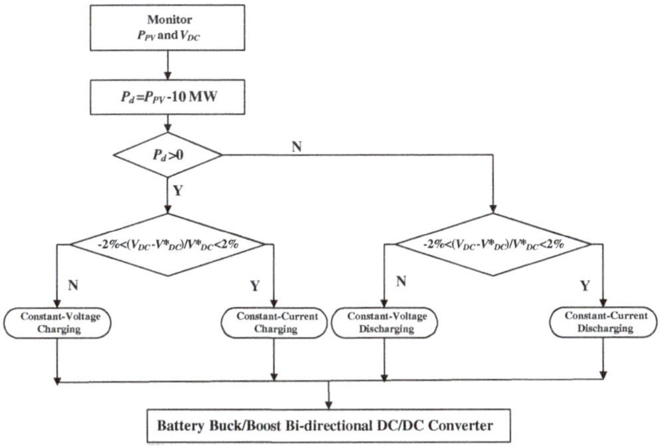

Fig. 6 Control strategy of the battery

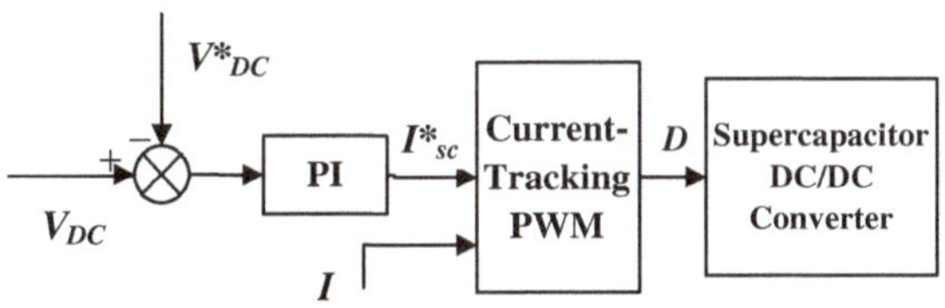

Fig. 7 Control strategy of the supercapacitor

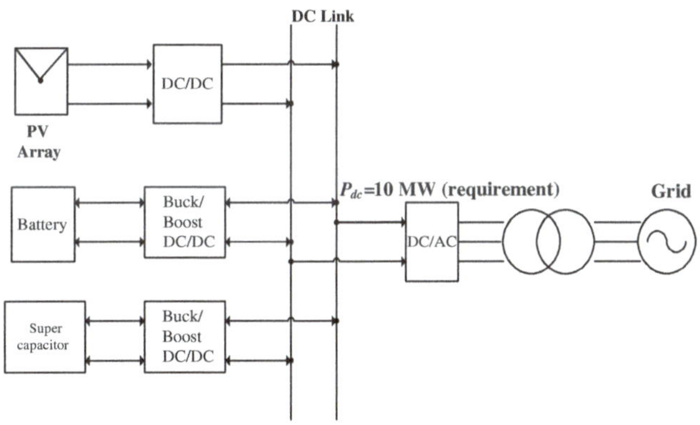

Fig. 8 Structure of the large-scale dispatchable grid-connected PV system

step is to monitor the difference, P_d, between the output power of the PV array, P_{PV}, and the set output power, in this case which is 10 MW. When this is positive, there is surplus power which can be stored in the battery. When it is negative, power is required to be supplied from the battery to the grid.

There are several operating modes of the battery during charging and discharging which are defined as constant current or constant voltage. In order to reduce the voltage fluctuation at the junction point, the battery should be switched to the required operating mode depending on the DC voltage, as shown in Fig. 6. The acceptable voltage fluctuation is 2% and the V^*_{DC} in the flow chart is the reference value for the DC link.

3.3 Control strategy for the supercapacitor

Due to the high power density and the low energy density, the discharging time of the supercapacitor is very short and its response is very fast. The supercapacitor plays the role of absorbing the high-frequency power fluctuations from the PV and maintaining the voltage of the DC link in a proper range. The random nature of these fluctuations is a key factor of the PV system and the instantaneous change in the output power is unpredictable. The supercapacitor always charges or discharges in a shorter time than the battery therefore the battery is protected from the high frequency fluctuations.

Figure 7 shows the control strategy for the supercapacitor. The voltage of the DC link is monitored and compared with the reference value, V^*_{DC}, resulting in the error signal which is the input of the PI controller. The reference current of the supercapacitor, I^*_{SC}, is used with the monitored supercapacitor current, I_{sc}, to create the control signals for the Buck/Boost DC/DC converter through the current-tracking PWM.

3.4 Special operation modes

The system's set power output to be delivered to the grid is typically 50% of the rated power of the grid-connected PV plant in order to ensure a good performance. Other settings can be used dependant on the weather conditions. Generally, these operation modes can be classified into three modes according to the following equations.

$$\begin{cases} SOC = 20\% \\ 0 < P_{PV} < 10, \end{cases} \tag{10}$$

$$\begin{cases} SOC < 20\% \\ P_{PV} = 0, \end{cases} \tag{11}$$

$$\begin{cases} SOC = 100\% \\ P_{PV} > 10. \end{cases} \tag{12}$$

In the above equations, 20% is considered to be the acceptable lowest value of *SOC* of the battery. In the first mode based on (10), the power from the PV arrays is not enough to satisfy this requirement and the battery cannot supply the desired energy level due to its low *SOC*. Under this circumstance, the PV plant should be disconnected with the grid since the PV system cannot meet the desired power output for the grid. During this time, the PV plant is used to charge the battery until the *SOC* reaches a correspondingly high value when it can be switched to the normal operation mode.

As shown in (11), the second mode of operation, there is no output from the PV arrays and the battery cannot

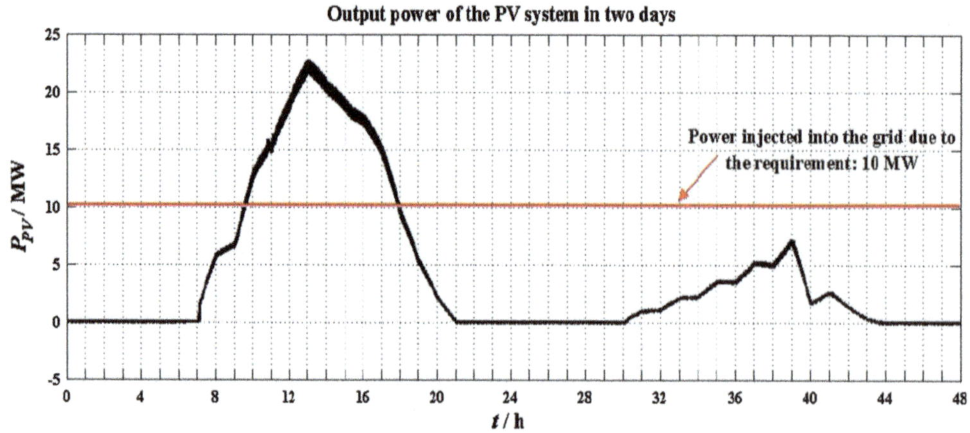

Fig. 9 Output power of the PV system

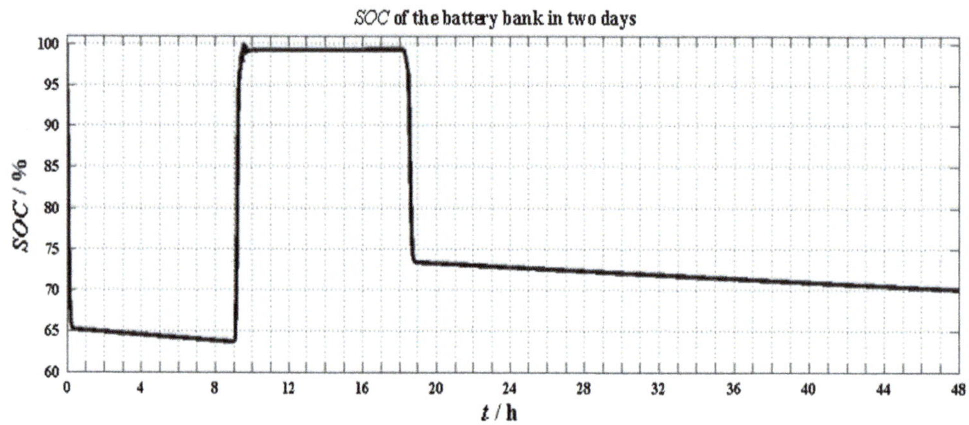

Fig. 10 *SOC* of the battery

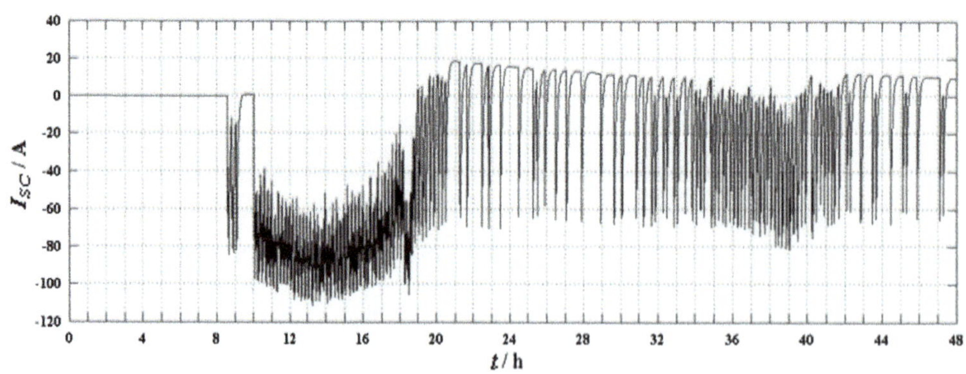

Fig. 11 Current through the supercapacitor

provide any output. The PV plant should be disconnected from the grid and left idle until good weather returns.

In the last mode, represented by (12), there is redundant solar energy which the battery cannot store since it is fully charged. Considering the proper design of the plant, this situation is considered rare. When this happens, the PV system should dump the redundant power into the grid, if possible. To ensure that the correct mode for the PV plant is used, the output power of the PV array and the *SOC* of the battery are continuously monitored.

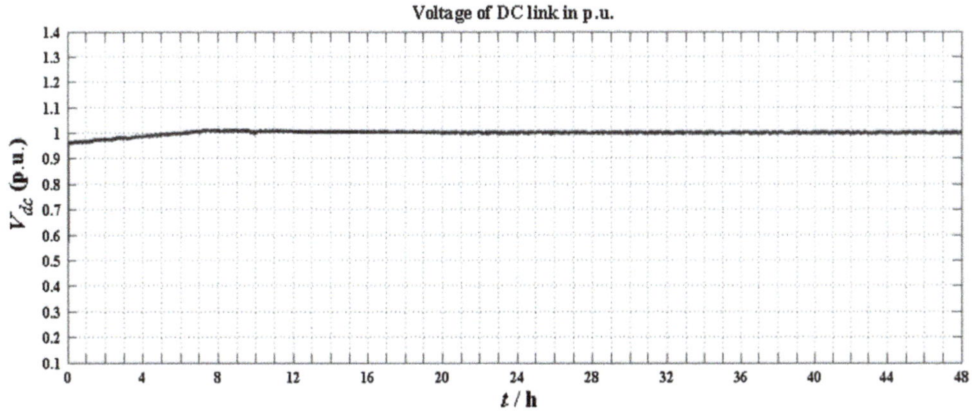

Fig. 12 Voltage of the DC link

4 Simulation results and discussion

The system has been modeled in MATLAB/Simulink. Weather data for one-year collected from a 100 MW PV plant in Qinghai province was used to provide irradiation patterns for two consequent days of operation to demonstrate the good performance of the system. The first day was a sunny day with adequate solar irradiation during the daytime and the second day was cloudy with much less sunshine. The control strategy of the grid-connected inverter was an important feature of the study and has been described in another research from [2]. This study focused on the performance of the DC power flows at the junction point providing the input of the grid-connected inverter, as shown in Fig. 8.

The output power of the PV system for the two days of the test is shown in Fig. 9. The output power in the first sunny day was much higher than that for the second cloudy day. The output power follows that of the solar irradiation derived from the weather data. This demonstrates the good performance of the MPPT control function.

During the first day, the peak output power from the PV array was higher than the set power output level. During the second day, however, it was lower than the set level. The surplus during the first day was redundant power and was stored in the storage system. During the second day, power was required to be supplied from the storage.

The *SOC* of the battery during the two days is shown in Fig. 10. The initial *SOC* was 100%, showing that the battery was fully charged at the beginning of the test. During the first day, the battery was first discharged until sunrise and then almost fully charged again at around noon due to the high output power of the PV system. During the second day, the battery is continually discharging because of the shortage of the solar irradiation.

The differences in the *SOC* are due to the different discharging or charging current according to the system's operation mode. Figure 11 illustrates the current through the supercapacitor during the two days. The positive current represents the discharging of the supercapacitor and vice versa.

In order to test the quality of supply at the DC link, the DC voltage V_{DC} was monitored as shown in Fig. 12. It is presented in per unit values, p.u. The simulation results show that the DC voltage was maintained at around the reference value with only small fluctuations. There was an obvious error at the start of the simulation and this was attributed to settling down of the simulation algorithms. It shows the stability in the DC power prior to input to the grid-connected inverter. A higher level of functions was expected as a result of the influence on grid on the PV system [13].

5 Conclusions

With the development of the renewable technologies, large-scale grid-connected PV plant has become a hot topic.

The conclusions are:

1) A centralized, large-scale grid-connected PV system with a battery-supercapacitor hybrid electricity storage has been described. Taking advantage of their complementary characteristics, the batteries and supercapacitors have been used as the main and secondary storage devices to provide a high quality of supply and efficiency. The PV system was modeled in the MATLAB/Simulink and the control strategies demonstrated.
2) Two sets of weather data were collected from an existing PV site and used as the input for the system to demonstrate its feasibility. The first day was sunny, while the second was cloudy.
3) The simulation studies demonstrated that the proposed system could operate smoothly over the 48 hours test

period. Using the storage system, the PV system was able to supply a continuous 10 MW of power to the grid as set by the centralized dispatch schedule. This was achieved for both the sunny and the cloudy day. The voltage of the DC link was maintained within an acceptable range.

Acknowledgment This work was supported by National High Technology Research and Development Program, "863 key technologies and development on large-scale grid-connected PV plants" (No. 2011AA05A301).

References

[1] Gao MJ, Hui D, Gao ZH et al (2013) Presentation of national wind/photovoltaic energy storage and transmission demonstration project and analysis of typical operation modes. Automat Electr Power Syst 37(1):59–64 (in Chinese)
[2] Ai X, Han XN, Sun YY (2013) Grid-connection characteristics of large-scale photovoltaic power station and its low-carbon operation and control technology. Power Syst Technol 37(1):15–23 (in Chinese)
[3] Chen W, Ai X, Wu T et al (2013) Influence of grid-connected photovoltaic system on power network. Electr Power Automat Equip 33(6):26–32 (in Chinese)
[4] Bao XN, Zhang JC, Xu M et al (2013) Active power hierarchical control strategy of interconnected photovoltaic station based on hybrid energy storage. Automat Electr Power Syst 37(1):115–121 (in Chinese)
[5] Zhang Y, Jiang ZH, Yu XW (2008) Control strategies for battery/supercapacitor hybrid energy storage systems. In: Proceedings of the 2008 IEEE energy 2030 conference, Atlanta, GA, USA, 17–18 Nov 2008
[6] Hu XL, Tseng KJ, Srinivasan M (2011) Optimization of battery energy storage system with super-capacitor for renewable energy applications. In: Proceedings of the 2011 IEEE power electronics and ECCE Asia (ICPE & ECCE), Jeju, Republic of Korea, 30 May–3 Jun 2011, pp 1552–1557
[7] Zhang JC, Ji LT (2010) An effective hybrid energy storage system based on battery-EDLC for distributed generation systems. In: Proceedings of the 5th industrial electronics and applications (ICIEA'10), Taichung, China, 15–17 Jun 2010, pp 819–824
[8] Kasera J, Chaplot A, Maherchandani JK (2012) Modeling and simulation of wind-PV hybrid power system using Matlab/Simulink. In: Proceedings of the 2012 IEEE students conference on electrical, electronics and computer science (SCEECS'12), Bhopal, India, 1–2 Mar 2012
[9] Zhang Y, Lu GQ, Zhang Y et al (2011) Influence of grid connection of photovoltaic power station on power grid reliability. East Chin Electr Power 38(5):700–706 (in Chinese)
[10] Sun K, Shu QF (2011) Overview of the types of battery models. In: Proceedings of the 30th Chinese control conference (CCC'11), Yantai China, 22–24 Jul 2011, pp 3644–3648
[11] Johansson J (2008) Comparison of simulation programs for supercapacitor modelling. MA Dissertation, Chalmers University of Technology, Gothenburg, Sweden
[12] Cui Y, Cai BH, Li DY et al (2006) Comparative studies on the MPPT control algorithms of solar energy photovoltaic system. Acta Energy Solar Sin 27(6):535–539 (in Chinese)
[13] Ai X, Han XN, Sun YY (2013) The development status and prospect of grid-connected photovoltaic generation and its related technologies. Mod Electr Power 30(1):1–7 (in Chinese)

Xiaonan HAN, graduate student of North China Electric Power University. Her research interests include renewable energy generation, smart grid, and analysis and control of the power system.

Xin AI, doctoral tutor and professor in North China Electric Power University. His research interests include smart grid and micro grid, analysis and control of the power system.

Yingyun SUN, doctor and associate professor in North China Electric Power University. His research interests include power system optimization and grid-connected renewable energy system.

3

Frequency control ancillary service provided by a wind farm: dual-BESS scheme

Dailin YAO (✉), San Shing CHOI, King Jet TSENG, Tek Tjing LIE

Abstract Network frequency control function is incorporated into a grid-connected wind farm-dual battery energy storage system (BESS) scheme. The design of the scheme takes advantage of the rapid response characteristics of the BESS and the in-built short-term overloading capability of the associated power conversion devices. A control strategy to regulate the BESS output power is then proposed. It is shown that the frequency control action offered by the BESS complements amicably with that of conventional generators in enhancing the frequency regulation attribute of the grid system.

Keywords Battery energy storage systems, Wind power generation, Frequency control ancillary service

1 Introduction

One major impediment to the large-scale penetration of wind generation into grid systems is due to the concern on its impact on the network security and quality. Depending on the extent of the fluctuations in the generated wind power, variations in network frequency and voltages could reach unacceptable levels if appropriate corrective actions are not taken. Indeed, satisfactory power quality can be achieved as part of an ancillary service offered by the generators through the trading of real and reactive powers, as explained in [1, 2].

With the above considerations in mind, a dual battery energy storage system (BESS) scheme to alleviate the impacts of wind generation on grid systems has been proposed in [3]. The scheme involves the application of two identical BESS modules, denoted as BESS Module 1 and BESS Module 2 in Fig. 1. Each BESS module is interconnected to a DC/DC converter of the same capacity. In the figure, BESS Module 2 is shown being charged by the wind turbine generators (WTG) and the BESS module is said to be in the stand-by mode where the WTG could represent a cluster of wind turbine generators in a wind farm. The charging power is denoted as P_c. Concurrently, the so-called in-service BESS (BESS Module 1 in this instance) is discharging power to the AC grid. The two BESS will interchange their roles when BESS Module 1 is unable to sustain the delivery of the scheduled power due to, for example, the battery low state of charge (SOC). BESS Module 2 will then take over the in-service role by supplying energy to the grid while BESS Module 1 is in turn to be charged by the WTG. In this manner, the dual-BESS scheme has allowed the unsteady input wind power dynamics to be decoupled from the grid.

Comparison of the dual-BESS scheme with other schemes shown in the literature has been included in [3]. The proposed scheme is to participate in power market operation and is based on the assumption that deep-cycle batteries have been used. As reported in [3–5], the advantages of the proposed dual-BESS scheme can be summarized as follows:

- Allows the wind generation to participate in the power market and makes a firm commitment on its dispatch;
- Profit optimization is also achievable;

D. YAO, S. S. CHOI, K. J. TSENG, Center for Smart Energy Systems, School of Electrical and Electronic Engineering, Nanyang Technological University, Singapore 639798, Singapore
(✉) e-mail: yaod0002@e.ntu.edu.sg
T. T. LIE, Auckland University of Technology, Auckland 1102, New Zealand

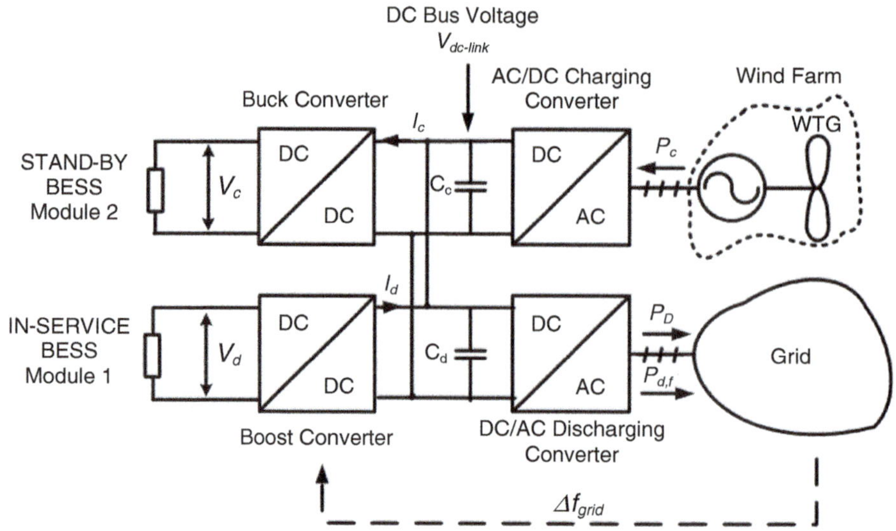

Fig. 1 Schematic of the proposed wind farm—dual BESS scheme inter-connected to grid

- Battery lifetime can be better estimated as the BESS are designed to operate in full charge-discharge cycles;
- The charging/discharging process of the BESS can be flexibly controlled by this scheme.

The study in [3] focuses on proposing a method to schedule the dispatched power of the wind farm-dual BESS scheme, so as to meet the prevailing power market requirements. However, various studies have shown that BESS is an excellent power system asset for providing system frequency regulation function [6–11]. Indeed frequency regulation is even recognized as the most profitable application of energy storage in power system [12]. It is in the context of providing frequency control that the proposed scheme has excellent potential. The variable nature of wind has often resulted in the dispatched output power of the dual-BESS scheme at levels much less than the rated power of the wind farm/converters. Therefore, the capacity margin between the dispatched power and the rated power of the converters can be used to support the frequency regulation function: an ancillary service which is also profitable to the wind farm operator, in addition to the sale of the dispatched power [1].

This paper investigates the possibility of applying the dual-BESS scheme for providing frequency control service, while the basic function of power dispatching is still being taken into consideration. The frequency regulation service provided by the dual-BESS scheme is described and compared with other schemes in Section 2. By extending the well-established conventional frequency control system model to include the wind power dual-BESS scheme, the design of the dual-BESS scheme for frequency control is then shown in Section 3. Section 4 demonstrates that the capacity of the discharging converter can be reduced by capitalizing on the short-term overloading capability of the converter. Several illustrative examples are then presented in Section 5. The main findings are summarized in Section 6.

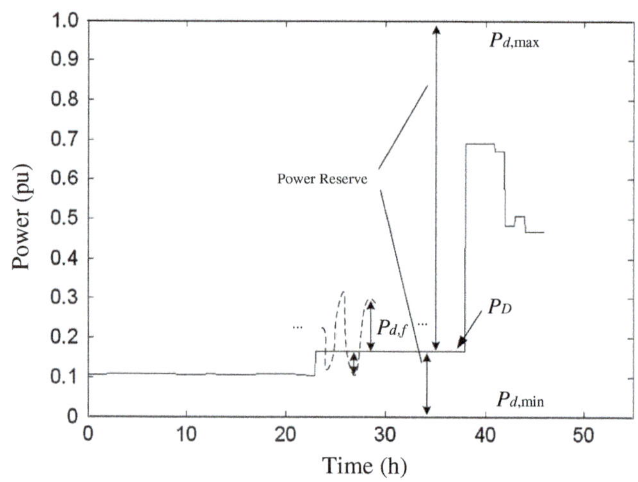

Fig. 2 Dispatch and reserve power: BESS power rating is assumed to be 1 pu

2 Role of the dual-BESS scheme in network frequency regulation

As explained in Section 1, the main motive of the proposed dual-BESS scheme considered in [3] is for the wind farm to achieve comparable level of power dispatchability as that of a conventional power station. The frequency control function discussed in the present paper is to be added as an ancillary service offered by the wind farm-BESS power station.

2.1 Proposed frequency regulation scheme

With reference to Fig. 2, power reserve of the wind farm-dual BESS scheme is defined as the ability of the scheme to supply power higher or lower than the scheduled level P_D. Interested readers may wish to refer to [3, 4] to understand how P_D is determined. In the present investigation, the dual-BESS scheme is to operate as a power reserve source as well: in response to the grid frequency deviation Δf_{grid} as shown in Fig. 1, a command shall be generated to adjust the output power of the in-service BESS. In this manner, the dual-BESS scheme will not only meet the power schedule but also provide the frequency regulation service. This would be possible provided that the sum of the net level of P_D and the frequency regulating power component, denoted as $P_{d,f}$ in Fig. 2, does not exceed the discharging converter rating. In this way, the wind farm dual-BESS scheme functions in a manner similar to a conventional power station in providing frequency regulation ancillary service.

There is, however, a distinct difference between the dual-BESS scheme and a conventional wind farm. In [3], the authors have attempted to match the behaviors of the in-service and standby BESS when designing the dispatch strategy for the dual-BESS scheme. The final outcome is that when the state of charge (SOC) of the in-service BESS is approaching its lower limit, the SOC of the stand-by BESS shall reach its maximum level. In this way, the total amount of stored energy in the two BESS is approximately equal to the capacity of a full BESS. This unique feature of having one full-BESS reserve capacity is observed even though the wind farm output is totally weather-dependent. As the capacity of the BESS is designed in [3] to cater for power dispatch purpose, typically the amount of stored energy is sufficient to sustain the BESS full power discharge for several hours. Thus the dual-BESS scheme shall have more than adequate reserve in providing primary frequency control function for grid systems. For instance, under the jurisdiction of European Union for the Co-ordination of Transmission of Electricity (UCTE), or its recent replacement ENTSO-e, generating units are required to sustain for 15 minutes on primary frequency control [9]. While such technical criteria could be subject to regular review, the intent of this paper is to adopt this frequency support requirement to illustrate the proposed approach.

An important observation is concerning the frequency regulation ability of the dual-BESS scheme: the regulating task is most strenuous when P_D is high while wind generation is at the maximum. In this instance and in accordance to the dispatch strategy of [3], the scheduled dispatch power of the wind farm is likely to approach the boost converter power rating. Therefore, there is only a limited amount of spare converter capacity for the BESS to use in regulating system frequency. Furthermore, reserve power of other generating units in the grid also tends to be low too under this high load demand period. This will place a limit on the frequency regulation ability of the dual-BESS scheme.

2.2 Comparison with other schemes

Much recent research attention has been directed towards frequency regulation by using the stored kinetic energy in the wind turbines [13–17]. The main drawback of this approach is that the wind turbines have to operate at speeds which will not allow the maximum harnessing of energy from the wind.

Alternatively, the use of battery energy storage systems for primary frequency control has been studied by many researchers [6–11] and different BESS schemes have been adopted. For instance, the authors in [11] consider the installation of two BESS in different locations of an isolated power system. During a sudden demand change, the two BESS are to share portions of the load change based on their rated capacity. The configuration and operation of their system are fundamentally different from the dual-BESS scheme examined in the present investigation.

The BESS schemes in [8] and [9] use single BESS to achieve frequency regulation. The BESS capacity may not be enough to absorb or discharge power because of the practical limit placed on the SOC of the BESS. The proposed dual-BESS scheme does not encounter such a problem. Furthermore, the operational strategies of the BESS in [8, 9] are such that the schemes commit a fixed amount of stand-by power reserve. Also unlike [8, 9], the power reserve in the dual-BESS is not committed for the long term. As proposed in [3], the dispatch power of the dual-BESS is traded in a daily power market. Therefore, its capability to provide frequency regulation varies with the scheduled P_D.

Generally, when the frequency regulation is considered as an additional function of the energy storage scheme, it would pose additional constraints on the design of the control strategy. The authors of [10] propose the use of batteries in parked electric vehicles (EV) to contribute to the grid frequency regulation, an ancillary function similar to that offered by the dual-BESS scheme. The grid connection is to charge up the EV batteries within a specified time. As such, the EV should be readily charged within a duration set by the customer, whereas in [3], the committed dispatch power of the dual-BESS shall be guaranteed. Apart from the similarity on the constraints when treating frequency regulation as an ancillary function, the EV scheme takes the form of distributed energy storage control: thus, the optimization of the vehicle owner benefits distinguishes the EV operational strategy from that of the dual-BESS scheme.

3 Control system design for frequency regulation

Fig. 3 shows the load-frequency regulation scheme of the power system, in which the dual-BESS scheme has been incorporated. The portion of the figure with thicker lines corresponds to the power system model whereas the remaining portion corresponds to the control system. By adopting the conventional modeling approach in load-frequency control such as that described in [18], the generators in the grid are aggregated to form a single equivalent unit. The aggregated generator is represented simply as a first-order governor in cascade with a first-order turbine model, in which the time constants τ_g and τ_T have been introduced. The conventional droop setting is denoted by r. In addition, a ramp rate limiter is included to reflect the mechanical constraint on the rate of change of the governor. Also, electro-mechanical device is often insensitive to small signals to effect a change in its output. Therefore a dead-band is also included in the control scheme to account for this phenomenon. An additional control loop to account for the action of the BESS has been added. Time lag in the BESS is represented by the time constant τ_{BESS}. The Power and Energy Limiters associated with the BESS shall be described later.

Although not considered in this study, α_1 and α_2 are the respective participation factors of the BESS and conventional generators in secondary frequency control. This control mode is commanded by power system control center and in accordance to factors such as the amount of reserves, economic dispatch and ramp rate.

This paper only focuses on the primary control loop, and the associated control blocks are thus described as follows.

3.1 Power limiters of the BESS

The dual-BESS is designed for both power dispatching and frequency regulation applications. Therefore, the discharge power level of the BESS is:

$$P_b = P_D + P_{d,f} \tag{1}$$

where P_b is the total output power of the BESS, P_D is the scheduled power dispatch from the in-service BESS and $P_{d,f}$ is the discharged/charged power required to regulate system frequency.

Recall from Fig. 1 that the BESS is composed of a battery module and the associated discharging/charging converter. The maximum discharge/charge power of the BESS is constrained by the battery cell design and characteristics, as well as by the power capacity of the power converter designed for specific application. For normal power dispatch, typically for NaS battery considered in this study, the BESS is expected to operate at relatively steady power level for several hours [3]. The lifetime of the BESS can be estimated based on the number of full charge/discharge cycles the battery has undergone. For short duration pulse-type output power application, however, NaS battery is reported to be capable of supplying up to 5 times of its rated power for 30 s [19]. Therefore, it can be assumed that the NaS BESS maximum discharging/charging power level is much greater than the converter rating, if the discharge/charge duration of interest is no more than 30 s. The maximum charge/discharge power level must therefore be considered in conjunction with the ratings of the power converters connected to the battery modules.

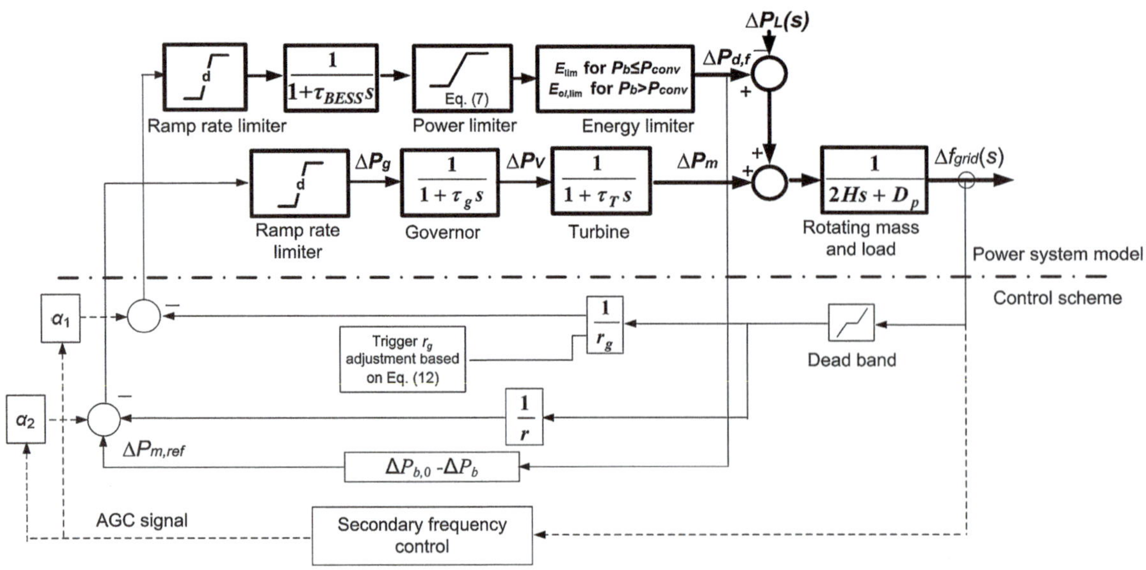

Fig. 3 Load-frequency regulation scheme with the inclusion of the dual-BESS

The current and voltage ratings of the semiconductor switches in power converters are normally designed with some overloading capability [20]. Such capability, typically valid for up to 20 s, is often given in the converter manufacturer's data sheet. The converter over-loading capability allows frequency regulation to be achieved through increasing the BESS charging/discharging power limit over a short duration. Thus the feasible operating range of P_b should be determined by taking into consideration the duration of P_b.

In view of the above, for long-term dispatch operation, one can express P_b to have limits

$$P_b \in [-P_{c,\max} \quad P_{d,\max}] \tag{2}$$

In (2), $P_{c,\max}$ and $P_{d,\max}$ are the respective maximum charging and discharging powers of the BESS module. Since the maximum charge/discharge power is limited to match the converter power rating, denoted as P_{conv}, therefore,

$$P_{c,\max} = P_{d,\max} = P_{conv} \tag{3}$$

On the other hand, for short-duration (pulse-type) application,

$$P_b \in [-P_{conv} - P_{ol} \quad P_{conv} + P_{ol}] \tag{4}$$

In (4), P_{ol} denotes the overloading capacity of the converter. As explained earlier, P_{ol} is a function of the duration of the pulse. For example, the authors have conducted a survey to assess what would be the typical overloading capability of the selected inverters. Fig. 4 shows the overload capabilities of the selected inverters observed from [21–25] differ due to the differences in their design and applications, although there are some overlaps among curves. It can be observed from the survey that there are two common features in the specifications of the inverters. Firstly, the inverters are designed to have a varying level of overload capability so that the reliability of the product can be increased, as has been considered in [20]. Secondly, the maximum withstand time for the overloading decreases with respect to the increase of overloading power.

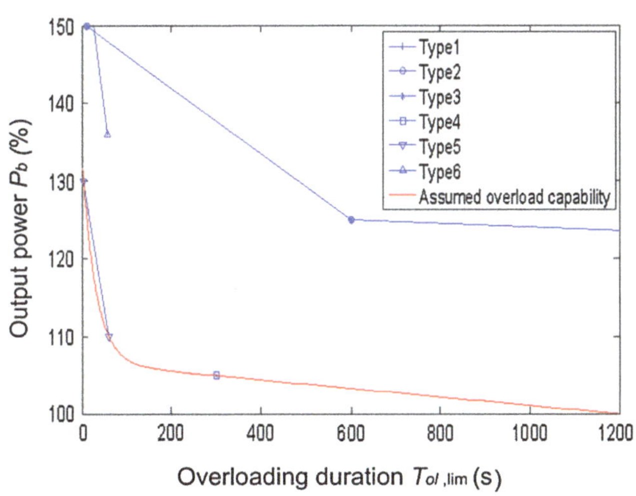

Fig. 4 Overloading capability limit curves of selected inverters

To incorporate these observations on P_{ol} into this study, a generic overload capability of inverter is assumed and this capability is shown by the curve labeled as "Assumed overload capability" in Fig. 4. The assumed curve is based on the lowest overload capability obtained in the survey and can be expressed in the following mathematical form:

$$P = 25.69e^{-3.3263\times10^{-2}T_{ol,\lim}} + 106.7e^{-5.42\times10^{-5}T_{ol,\lim}} \tag{5a}$$

In (5a), P is expressed in percentage of the converter rating and the overloading duration $T_{ol,\lim}$ is in second. As P_{conv} denotes the converter rating, therefore

$$P = P_{conv} + P_{ol} = \alpha_{ol} P_{conv} \tag{5b}$$

In (5b), α_{ol} is defined as the overloading factor.

Equations (5a) and (5b) allow the authors to study the general overloading capability of converter incorporated in the dual-BESS scheme. Based on (2) and (3), the viable range of $P_{d,f}$ for dual-BESS long-duration operation can be stated as

$$P_{d,f} \in [-P_{conv} - P_D \quad P_{conv} - P_D] \tag{6}$$

From (4) and for short-duration (pulse-type) operation, $P_{d,f}$ can be stated as

$$P_{d,f} \in [-P_{conv} - P_{ol} - P_D \quad P_{conv} + P_{ol} - P_D] \tag{7}$$

The last equation is as a consequence of the converter overloading capability expressed by (5b). The power limiter (7) therefore applies to BESS power flow control loop of Fig. 3 when frequency control action is initiated.

3.2 Energy limiter

The energy limiter in Fig. 3 is considered for two situations: inverter overloading and non-overloading periods. As shown in Fig. 4, the overloading capability of the inverter is only valid for a limited duration. In order to take both overloading power level and duration into consideration, an energy limiter must be included in the designed scheme. Also, even if the inverter is not overloaded, the BESS should not be continuously charged or discharged without bounds.

3.2.1 Energy limit ($E_{ol,\lim}$) during inverter overloading

Incorporating an energy limiter into the control scheme under the inverter overloading condition is to reflect the overloading duration $T_{ol,\lim}$ as a constraint. From Fig. 4, the maximum amount of energy ($E_{ol,\lim}$) associated with the overloading capability of the inverter can be expressed in terms of the overloading power and overloading duration, as follows.

$$E_{ol,\lim} = P_{ol} \times T_{ol,\lim} \quad (8)$$

A typical $E_{ol,\lim}$ plotted against the overloading factor and duration is shown in Fig. 5. It is based on (5a). However, as shown in Fig. 5a, the energy limit depends on the overloading factor and in turn the power level. In this study, the strategy is to ramp down the BESS output during the overloading period before it reaches $T_{ol,\lim}$, and after the system frequency has stabilized. An example is shown in Section 5 to illustrate this strategy.

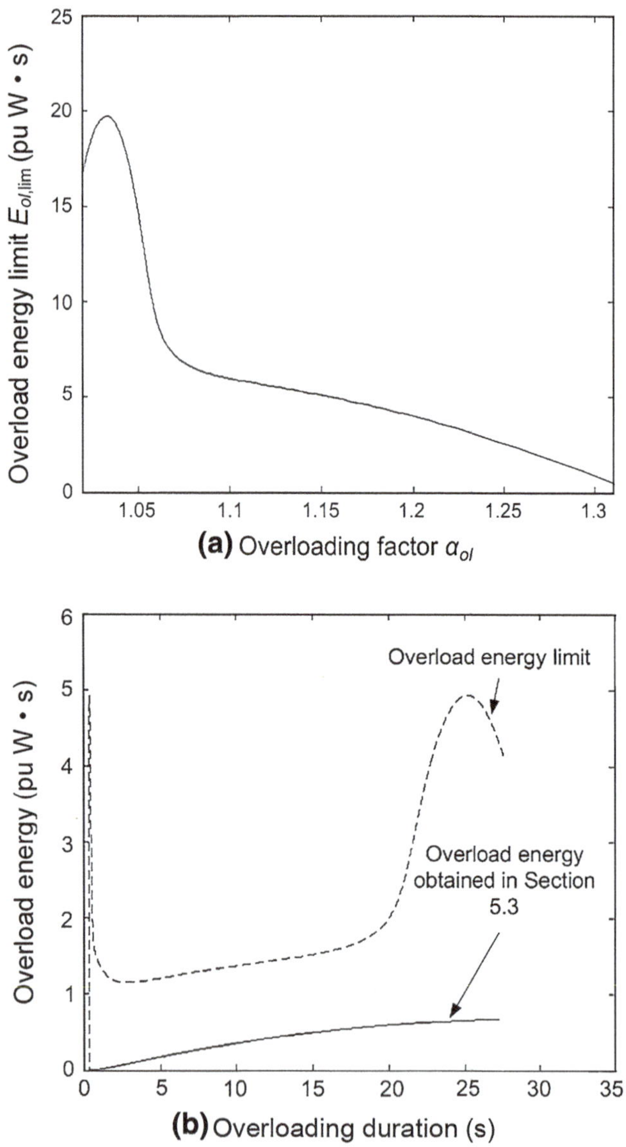

Fig. 5 (a) Overload energy limits plotted against the overloading factor α_{ol} (b) overloading duration

3.2.2 Energy limit ($E_{\lim}$) during inverter non-overloading period

During the normal operation, the amount of stored energy in the battery used for frequency regulation should also be constrained. This is because a proportion of the stored energy in the dual-BESS scheme is to be used to achieve power dispatchability. Hence the energy level in the BESS should be maintained to satisfy the dispatch commitment which is determined by the method described in [3]. Also, a minimum amount of primary reserve ($E_{\lim}$) is normally required under the grid code, for instance, 15 minutes under UCTE/ENTSO-e rule. For frequency control, charging/discharging energy of the BESS during non-overloading period should therefore be controlled to the minimum level, subject to meeting the grid code requirement.

3.2.3 Energy limiter settings

In view of the above, the energy limiter shall be set according to the following conditions:

When $P_b > P_{conv}$, the energy limiter should be set to $E_{ol,\lim}$, which is calculated using (8). When the energy limit is violated, the output power of the battery shall be adjusted.

When $P_b \leq P_{conv}$, the energy limiter should be set to $E_{\lim}$, which is described in Section 3.2.2. When the energy limit is violated, the output of battery power shall also be adjusted.

Section 3.3.2 shall explain how the battery output power can be adjusted.

3.3 BESS droop setting r_g

The BESS load-frequency control loop is added for the purpose of improving the network frequency regulation performance. As shown on Fig. 3, this loop involves the droop setting r_g. This section is pertaining to the selection of the value for r_g.

Consider the case of a generator trip. Since the BESS can be expected to respond much faster than the remaining on-line generating units to the resulting system frequency drop, the initial response of the overall power system shall be dominated by the BESS control loop. Suppose it is desirable for the steady-state frequency reduction following the generator trip is to be above a prescribed threshold value. However, due to the constrained over-loading capability of the BESS converter described earlier, the contribution from the BESS can only be sustained at high level for a limited period. Hence the manner by which the BESS subsequently reduces its output power should be based on the remaining on-line generators' ability to increase their output powers. The following section is to examine the extent of the control participation undertaken by the BESS through the judicious adjustment in r_g.

3.3.1 Initial setting of r_g

From Fig. 3 and by ignoring the initial contributions made by the remaining on-line generators, the steady-state frequency deviation at the setting of r_g can be readily computed as

$$\Delta w_{ss} = (-\Delta P_L)\frac{1}{D_p + 1/r_g} \tag{9}$$

Therefore, the larger the value of r_g, the greater shall be the steady-state frequency deviation Δw_{ss}. Suppose Δw_{ss} is to be controlled to be less than a pre-set limit $\Delta w_{ss,\lim}$ through this primary frequency control loop. Also assume the worst-case scenario of a power imbalance ΔP_L of (say) P_x pu. A practice often used by utility in designing load-frequency control is for ΔP_L to correspond to the tripping of the largest on-line unit in the power system. Then from (9), it can be readily seen that r_g must not exceed the value of $r_{g,0}$ where

$$r_{g,0} = 1\bigg/\left(\frac{-P_x}{\Delta w_{ss,\lim}} - D_p\right) \tag{10}$$

As a guide, r_g may therefore be set at the value of $r_{g,0}$ as (10) guarantees that Δw_{ss} shall not exceed $\Delta w_{ss,\lim}$ under the considered worst-case scenario.

Alternatively, the initial setting of r_g could be determined to ensure that the frequency deviations shall not trigger any under-frequency load shedding (UFLS) incident. The generation-load imbalance can be modeled as a step change ΔP_L and the system response can be analyzed in the time-domain. The frequency excursions shall be checked against the trigger setting of UFLS. This approach shall be illustrated in a numerical example in Section 5.

3.3.2 Adjustment of r_g

As explained earlier, the battery output power must be adjusted if the converter power and/or energy capacity limits are reached. Even if the limits have not been reached and the frequency drop has been arrested, it is desirable to reduce the BESS output power if the BESS is overloaded. The adjustment in the BESS output power can be achieved by increasing r_g to reduce the BESS contribution to the frequency regulation. Conversely, if the battery energy and power levels are within the allowable ranges but $r_g > r_{g,0}$, r_g should be reduced. This can be readily seen from Fig. 3: decreasing r_g causes the proportion of the power contribution from the BESS to increase.

As r_g changes and the BESS adjusts its output power after the initial period, the generators shall then respond by adjusting their output powers to compensate for the BESS output power change. To meet this requirement, the rate of change of $1/r_g$ must not exceed the maximum output power ramp up/down rate of the generators. Let α denotes the rate of change of $1/r_g$, thus α must satisfy the condition

$$\alpha \leq (1/r_{g,0})/T_{\min} \tag{11a}$$

In (11a),

$$T_{\min} = \frac{P_x}{\lambda_{\max}} \tag{11b}$$

where $\lambda_{\max}$ is the maximum ramp rate of the generators.

r_g may be adjusted in steps: $r_g(k)$ at the kth step is

$$\frac{1}{r_g(k)} = \begin{cases} \frac{1}{r_g(k-1)} + \alpha\Delta t, & \textit{if} \quad \text{(a)} \\ \frac{1}{r_g(k-1)} - \alpha\Delta t, & \textit{if} \quad \text{(b)} \end{cases} \tag{12}$$

where, (a) battery contribution is to increase; and (b) battery contribution is to decrease; $r_g(0) = r_{g,0}$ and Δt is the interval between two consecutive adjustments.

The above concept can be further illustrated using Fig. 6: Upon the onset of the generator trip, the BESS output power pulses up rapidly to a value $\Delta P_{b,0}$. Note that $\Delta P_{b,0}$ corresponds only to the change in the $P_{d,f}$ component of P_b. Subsequently, the output of BESS is to decrease at the ramp-down rate α' as shown in Fig. 6 through adjusting $1/r_g$. The decrease in ΔP_b shall be met by the corresponding increase ΔP_m in the generators' output. Fig. 6 shows the reference signal $\Delta P_{m,ref}$ for ΔP_m is generated such that $\Delta P_{m,ref} = \Delta P_{b,0} - \Delta P_b$. To avoid any possible adverse control outcome due to the generators' inability to make up for the decrease in ΔP_b, ramping-down time from $\Delta P_{b,0}$ to $\Delta P_b \approx 0$ should not be shorter than $T_{\min}$. Therefore, α can be determined based on (11a). As shown in (11b), $T_{\min}$ is calculated based on the worst case scenario when at the end of the primary frequency control stage, P_x shall be picked up by the remaining on-line generators and not the BESS.

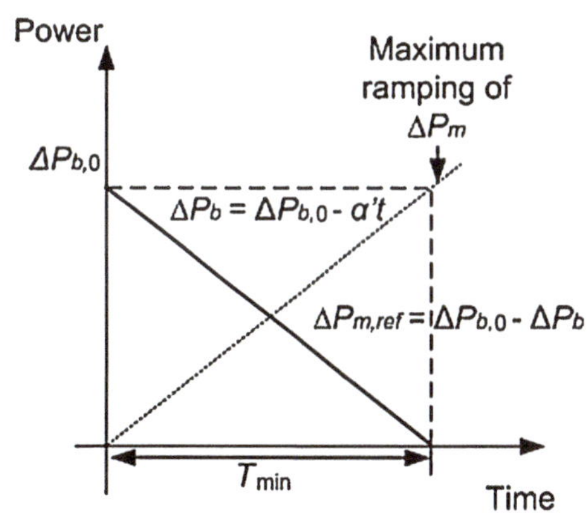

Fig. 6 Ramp down of BESS output power

Next, consider the closed-loop transfer function when ignoring the boundaries set by the limits, and when only the BESS is undertaking the frequency regulation function:

$$\frac{\Delta f_{grid}}{\Delta P_L} = \frac{-\frac{1}{(2Hs+D_p)}}{1+\frac{1}{r_g(2Hs+D_p)(\tau_{BESS}s+1)}} = -\frac{r_g\tau_{BESS}s + r_g}{2r_gH\tau_{BESS}s^2 + r_g\tau_{BESS}D_ps + 2r_gHs + r_gD_p + 1} \quad (13)$$

Based on Routh-Hurwitz analysis, the closed-loop system is seen to be stable if r_g is positive.

3.3.3 Control of ΔP_m

As the BESS adjusts its output power, the corresponding regulation in ΔP_m is through the manipulation in the power reference level $\Delta P_{m,ref}$ for the generators. As shown in Fig. 3, $\Delta P_{m,ref}$ is an additional control signal on the generators, and it is fed to the same summing junction where the conventional frequency droop control is effected.

Under the normal condition described under Section 3.2.2 in which the BESS converter overloading has not occurred, of course, $\Delta P_{m,ref} = 0$.

4 Analysis on reducing the required capacity of converter based on its overloading capability

This section is to illustrate that the required capacity of the converters can be effectively reduced by taking its overloading capability into the problem formulation.

As described in Section 3.1, if there is no additional grid code requirement on power dispatching, the dispatch power limits of the dual-BESS are set by the rating of the DC/AC converters i.e. the converter is allowed to operate within the range $[0, P_{conv}]$, where P_{conv} is the power rating of the converter.

4.1 Without considering converter overloading

If the overloading capability is not considered, the dual-BESS is not able to support frequency regulation when dispatching power at the rated value, and converters are not oversized. If the average system demand is 1 pu and the aggregated rating of the dual-BESS scheme is α_{bess} pu. Since load demand is time-varying and random, it would be appropriate to quantify load in statistical term. Suppose load variations are assumed to follow a normal distribution with variance of σ. The range [-3σ pu, 3σ pu] covers 99.9% of the possible load variations. Hence, to achieve 0.999 probability of meeting the load demand by the wind farm-dual BESS scheme under the maximum power dispatch scenarios, the converter should be up-sized at the very least by a factor β such that

$$\alpha_{bess} \times \beta = 3\sigma \Rightarrow \beta = \frac{3\sigma}{\alpha_{bess}} \quad (14)$$

Using the numerical example shown in Section 5, $\alpha_{bess} = 0.25$, and assume $\sigma = 0.01$, $\beta = 0.12$. That is, the converter should be up-sized by 12% to meet the requirements.

4.2 With overloading capability considered

If the overloading capability of power converter $P_{ol} = 20\%$, it is not necessary to up-size the converter since $\beta \leq 0.2$. If $\beta > 0.2$, the converter should only need to be up-sized by the factor $\beta - 0.2$.

5 Illustrated examples

As explained in Section 3, often the power system is designed to contain frequency excursion following the tripping of the largest on-line generating unit. This example serves to illustrate the use of the dual-BESS scheme to meet this requirement. Suppose $P_x = 0.05$ pu. System parameters obtained from [18, 26] are used: $\tau_T = 0.5$ s, $\tau_g = 0.2$ s, $D_p = 0.8$ pu MW/Hz, $r = 0.05$. Thermal generators are assumed subject to a ramp rate limit of $\pm$ 0.17% pu MW/s [19] (expressed in terms of the units' power ratings).

5.1 Determination of $r_{g,0}$

Also from [9], the authors have specified a non-critical frequency window in the form of a frequency deviation dead-band within which no primary frequency regulation is to function. Specifically, the dead-band is set at $\pm$ 0.001 pu, a level which is also adopted in this example. Steady-state frequency deviation limit $\Delta w_{ss,\lim}$ is assumed to be -0.004 pu Hz. The base frequency is 50 Hz. It is used to determine the initial setting $r_{g,0}$. By applying (10), it is seen that

$$r_{g,0} \leq 0.0812 \quad \text{pu Hz/pu MW} \quad (15)$$

Therefore, $r_{g,0}$ could be set at 0.08 pu Hz/pu MW. However, as explained in Section 3.3.1, the suitability of this setting should be evaluated further in case of violating UFLS frequency setting threshold, as follows.

Let ΔP_L be a step change of $P_x = 0.05$ pu, where the total installed capacity is taken as 1 pu. Substituting the known values of τ_{BESS}, H and D_p into (13) and setting $r_{g,0} = 0.08$, inverse Laplace yields

Fig. 7 (**a**) Frequency deviations with the integration of BESS. ▶ (**b**) Output of BESS. (**c**) Output of thermal generators. (**d**) Profile of $1/r_g$ following the generator trip at $t = 0$

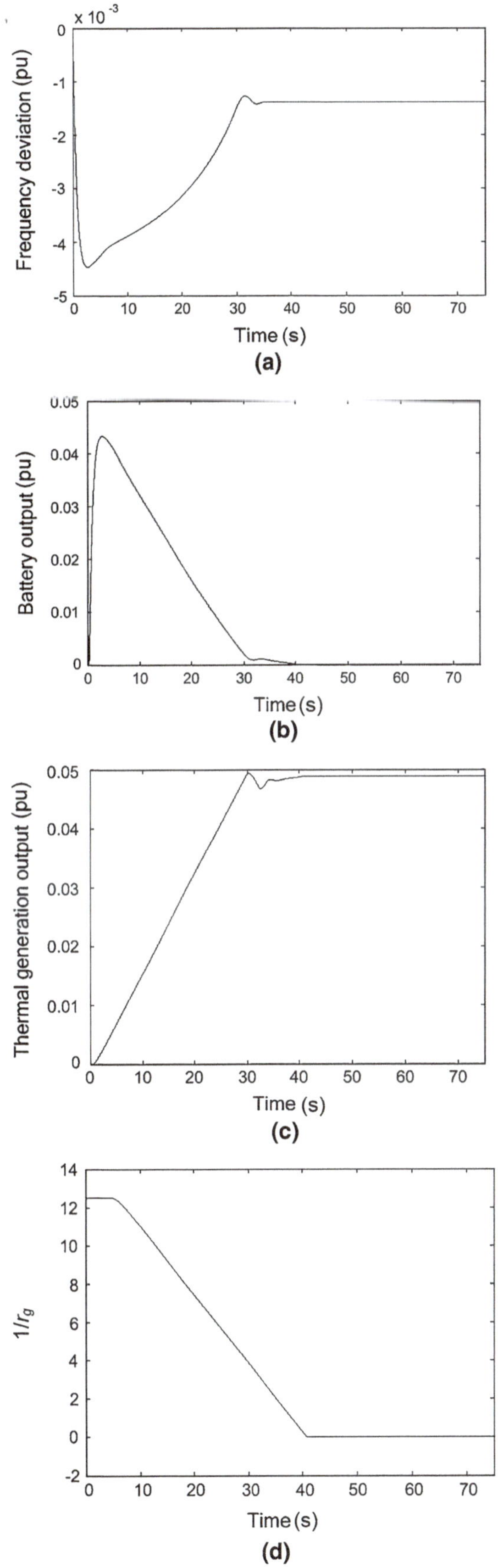

$$\Delta f_{grid}(t) = -0.0037594 - 0.00012496\ e^{-8.51874\,t} + 0.00388436\ e^{-1.56126t} \quad (16)$$

The minimum value of $\Delta f_{grid}(t)$ is -0.0037 pu. If the first stage UFLS frequency setting is -0.02 pu, a value which is quite typical in power systems [18], the selected $r_{g,0} = 0.08$ pu Hz/pu MW in this example will be quite suitable as it will prevent any load shedding event from occurring. If this condition is not met, then $r_{g,0}$ should be reduced until the UFLS event can be avoided.

5.2 Determination of α, $E_{\lim}$ and limits on $P_{d,f}$

Next, suppose the average load demand is 1 pu and the wind generation supplies 10% of the demand on average. Assume the wind output power corresponds to 40% of the installed capacity of the wind generation. Hence, $P_{conv} = 1 \times 10\%/40\% = 0.25$ pu.

Since wind generation supplies 10% of the demand on average, the remaining 90% of the load demand is supplied by other thermal generators. Assume these generators operate at 90% of their rating $P_{g,rated}$. Hence, $P_{g,rated} = 1 \times 90\%/90\% = 1$ pu. The ramp rate limit of these thermal units is therefore ± 0.0017 pu MW/s.

The appropriate rate of change of $1/r_g$ is calculated by the method describe in Section 3.3. The minimum elapsed time for the thermal units to pick up the sudden 0.05 pu loss of generation is 0.05/0.0017 or 29.4 s. Include a safety margin into the design, (say) the thermal unit pick-up time of 35 s, and since $r_{g,0} = 0.08$, from (11) and (12), $1/r_g$ is set to change at the rate $\alpha = 1/(0.08 \times 35) = 0.3571$ pu MW/s. In this way, it will take 35 s for the battery output to reduce from its initial loading level value to zero.

As explained in Section 2.1, in some countries, the primary frequency control should be able to sustain for 15 min. Therefore, in this example, the energy limit $E_{\lim}$ is set as $0.05 \times 60 \times 15$ pu MW·s = 45 pu MW·s.

The worst case is when the dispatched power has already reached the converter rating, i.e. $P_D = P_{conv}$. In this instance, the converter should only be overloaded for a short while. However, from (7), the discharge power to regulate frequency in the initial moment shall be within the range:

$$P_{d,f} \in [-2P_{conv} - P_{ol} \quad P_{ol}] \quad (17)$$

Substituting $P_{conv} = 0.25$ pu and $P_{ol} = 20\% \times 0.25$ pu $= 0.05$ pu into (15), the limits on the frequency regulation power component $P_{d,f}$ are $[-0.55, 0.05]$.

5.3 Dynamic simulation

Following the sudden 0.05 pu generation trip, the BESS can help to support network frequency by forcing its converter to operate at its overloading limits. Simulation study has been carried out to illustrate this strategy and the obtained results are as shown in Fig. 7.

Within five seconds following the generator trip, the fall in frequency (Fig. 7a) has been arrested but the battery output has to be reduced because the converter power limit is reached. The reduction in the battery output power is then picked up by the thermal generators. These operations are shown in Fig. 7b and 7c respectively. $1/r_g$ changes in a linear manner, as illustrated in Fig. 7d.

During the stage when the overloading capability of the BESS module is utilized, the corresponding energy profile can be obtained by interpolating the curve in Fig. 5a. Comparing the overload energy limits (shown as dotted line in Fig. 5b with the overload energy (solid line in the same figure), it shows that the proposed strategy to adjust r_g is prudent for this application. This is because the strategy has not resulted in the overload energy limit being breached.

Comparing the frequency deviation with that of the case when the BESS is not in use, as shown in Fig. 8a, it is not surprising to note that the frequency deviation has indeed been significantly reduced. The large variations in system frequency observed in Fig. 8a are due to the ramp-up and ramp-down limits of the thermal generators. The ramp rate limits of the thermal generators also prolong the damping time of the frequency deviations. Based on the proposed control scheme, when there is a frequency deviation in the system, a signal will be sent to the thermal units as their reference/target output power. As illustrated in Fig. 8b, the actual output of generator cannot track the reference signal. Therefore, it demonstrates that the dual-BESS scheme is indeed effective in improving the performance of network frequency regulation.

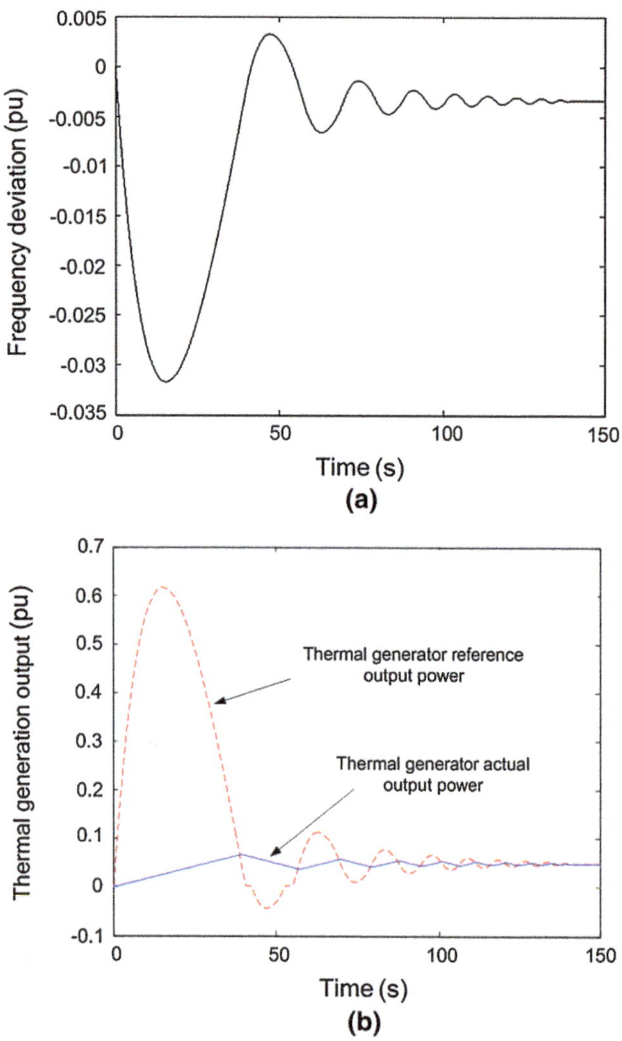

Fig. 8 (**a**) System frequency deviation without the BESS. (**b**) Thermal generation output without the BESS

6 Conclusions

The frequency regulation ancillary function of the dual-BESS scheme [3] has been proposed in this paper. It makes use of the fast-response characteristic of BESS for primary frequency control, to match with the control of conventional generators in which their ramping up/down rates are physically constrained. The overloading capability of the converter can be included into the design of control scheme which enables the capacity of converter to be minimized. Numerical results have shown that the dual-BESS scheme is beneficial in providing frequency support, as part of the network ancillary service.

References

[1] Rebours YG, Kirschen DS, Trotignon M et al (2007) A survey of frequency and voltage control ancillary services–Part II: economic features. IEEE Trans Power Syst 22(1):358–366

[2] Rebours YG, Kirschen DS, Trotignon M et al (2007) A survey of frequency and voltage control ancillary services–Part I: Technical features. IEEE Trans Power Syst 22(1):350–357

[3] Yao DL, Choi SS, Tseng KJ et al (2012) Determination of short-term power dispatch schedule for a wind farm incorporated with dual-battery energy storage scheme. IEEE Trans Sustain Energy 3(1):74–84

[4] Yao DL, Choi SS, Tseng KJ et al (2010) Determination of a dispatch strategy to maximize income for a wind turbine-BESS

power station. In: Proceedings of the international power and energy conference (IPEC'10), Singapore, 27–29 Oct 2010, 6p

[5] Yao DL, Choi SS, Tseng KJ et al (2012) Dynamic study of a battery change-over scheme of a wind farm containing dual BESS. In: Proceedings of the international power and energy conference (IPEC'12), Ho Chi Minh City, Vietnam, 12–14 Dec 2012, 7p

[6] Aditya SK, Das D (2001) Battery energy storage for load frequency control of an interconnected power system. Electr Power Syst Res 58(3):179–185

[7] Lu CF, Liu CC, Wu CJ (1995) Effect of battery energy storage system on load frequency control considering governor deadband and generation rate constraint. IEEE Trans Energy Conver 10(3):555–561

[8] Mercier P, Cherkaoui R, Oudalov A (2009) Optimizing a battery energy storage system for frequency control application in an isolated power system. IEEE Trans Power Syst 24(3):1469–1477

[9] Oudalov A, Chartouni D, Ohler C (2007) Optimizing a battery energy storage system for primary frequency control. IEEE Trans Power Syst 22(3):1259–1266

[10] Sekyung H, Soohee H, Sezaki K (2010) Development of an optimal vehicle-to-grid aggregator for frequency regulation. IEEE Trans Smart Grid 1(1):65–72

[11] Goya T, Omine E, Kinjyo Y et al (2011) Frequency control in isolated island by using parallel operated battery systems applying H∞ control theory based on droop characteristics. IET Renew Power Gener 5(2):160–166

[12] Oudalov A, Chartouni D, Ohler C et al (2006) Value analysis of battery energy storage applications in power systems. In: Proceedings of the 2006 IEEE PES power systems conference and exposition (PSCE'06), Atlanta, GA, USA, 2206–2211 29 Oct–1 Nov 2006

[13] Žertek A, Verbič G, Pantoš M (2012) Optimised control approach for frequency-control contribution of variable speed wind turbines. IET Renew Power Gener 6(1):17–23

[14] Teninge A, Jecu C, Roye D et al (2009) Contribution to frequency control through wind turbine inertial energy storage. IET Renew Power Gener 3(3):358–370

[15] Zhang ZS, Sun YZ, Lin J et al (2012) Coordinated frequency regulation by doubly fed induction generator-based wind power plants. IET Renew Power Gener 6(1):38–47

[16] Keung PK, Li P, Banakar H et al (2009) Kinetic energy of wind-turbine generators for system frequency support. IEEE Trans Power Syst 24(1):279–287

[17] Ma HT, Chowdhury BT (2010) Working towards frequency regulation with wind plants: combined control approaches. IET Renew Power Gener 4(4):308–316

[18] Bevrani H, Hiyama T (2011) Intelligent automatic generation control. CRC Press, Boca Raton, FL, USA

[19] Nichols DK, Eckroad S (2003) Utility-scale application of sodium sulfur battery. In: Proceedings of the 7th annual international stationary battery conference (Battcon'03), Marco Island, FL, USA, 28–30 Apr 2003, 9p

[20] Sugawara Y, Ogata S, Izumi T et al (2009) Development of a 100 kVA SiC inverter with high overload capability of 300 kVA. In: Proceedings of the international symposium on power semiconductor devices and IC's (ISPSD'09), Barcelona, Spain, 14–18 Jun 2009, 4p

[21] Industrial AC and DC drives.

[22] Statron IDS and IDT range of inverters with single phase or three phase output using advanced digital control technology for userfriendly operation flexible interfaces. Accessed 2000

[23] ASTRID Saturm 3/3-40 60 80 kVA UPS 100 125 technical data Accessed 2012

[24] MicroGram computers.

[25] Powerware series: Eaton 5110 UPS

[26] Saadat H (2004) Power system analysis. McGraw-Hill, Singapore

Dailin YAO received her B.E. degree in Electrical Engineering from Wuhan University, Hubei, China in 2007. She then received her M.Sc. and PhD degrees in Power Engineering from Nanyang Technological University, Singapore in 2008 and 2013 respectively. Her research interests are in energy storage systems, power system control and renewable energy systems.

San Shing CHOI received his B.E. and Ph.D. degrees from the University of Canterbury, New Zealand in 1973 and 1976, respectively. He had worked in the New Zealand Electricity Department, National University of Singapore and the State Energy Commission of Western Australia. Dr. Choi is now a Professor in the School of Electrical and Electronic Engineering, Nanyang Technological University, Singapore. His research interests are in power quality, power system control and renewable energy systems.

King JET TSENG was born in Singapore. He received the B.Eng. (First Class) and M.Eng. degrees from the National University of Singapore, Singapore, and the Ph.D. degree from Cambridge University, Cambridge, U.K. He is currently the Head of Power Engineering Division at the School of Electrical and Electronic Engineering, Nanyang Technological University, Singapore. He teaches electrical power engineering and power electronics at both undergraduate and postgraduate levels, and manages a number of research programs in the area of electrical energy and power.

Tek TJING LIE received his B.S. degree from Oklahoma State University, USA in 1986. He then obtained his M.S. and Ph.D. degrees from Michigan State University, USA in 1988 and 1992, respectively. He had worked in Nanyang Technological University, Singapore. Dr. Lie is now a Professor and Head of Dept. of Electrical and Electronic Engineering, Auckland University of Technology, New Zealand. His research interests include power system control, deregulated power systems and renewable energy systems.

4

Recent advancement on technical requirements for grid integration of wind power

Xiaoge LIU, Zhao XU (✉), Kit Po WONG

Abstract The installation of wind energy has experienced rapid development during recent years. As a result, the operation of power system can be greatly affected. Therefore, the operators of different countries have formulated the grid codes which reinforce technical requirements for wind power plants. In this paper, recent grid codes published in different countries have been carefully reviewed. The basic requirements of active power control and reactive power compensation, both of which have particular influence on wind power plants operation, are focused on. Based on this review, it is understood that with the increasing integration of wind energy, there is a worldwide trend to regulate the performance of wind power plants so that they can contribute to the stability and reliability of power system.

Keywords Wind power, Grid integration requirements, Low voltage ride through, Active power control

1 Introduction

Considering the increasing public awareness of energy crisis, renewable energy such as wind and solar energy have gained rapid development globally. During the past 15 years, the commercial installation of wind energy has increased continuously [1, 2].

Wind energy proved to be clean without CO_2 emission. However, due to the stochastic nature of wind, the characteristics of the power generated by wind turbines (WTs) are quite different from that provided by conventional generators. Consequently, the stability and reliability of the whole power system can be influenced by the increasing penetration level of wind power [3, 4]. In order to minimize the adverse impact of wind energy, the performance of wind power plants (WPPs) should be regulated.

For the conventional power plants (CPPs), which mainly consist of thermal synchronous generators, the operation principles have been well-established. The implementation of these principles regulates the behaviors of CPPs and takes the advantages of synchronous generators to ensure the stability of the power system.

Based on the existing technical regulations for CPPs, the grid codes, which provide basic requirements for WPPs, have been published by the transmission system operators (TSOs) in recent years. It is expected that WPPs can be similar in performance to CPPs [5].

According to these grid codes, the wind power plants should participate in frequency control and voltage control under normal conditions [5–8]. In the event of voltage drop, the requirements of Low Voltage Ride Through (LVRT) and supply of reactive power have been reinforced [5, 6]. As the contribution from wind power generation becomes increasingly larger, these grid codes may be revised and enhanced in the future [5].

For manufacturers and researchers, the grid codes can be regarded as important guide-lines for the development of WT techniques [6]. In recent years, many control schemes aiming at satisfying the grid codes requirements have been proposed. Several promising methods for frequency control have been provided in [7, 9, 10]. In addition, some LVRT strategies for individual WT have been developed in [11–13].

X. LIU, Z. XU, The Hong Kong Polytechnic University, Hung Hom, Hong Kong
(✉) e-mail: eezhaoxu@polyu.edu.hk
K. P. WONG, The University of Western Australia, Crawley, WA 6009, Australia

Table 1 Grid codes published in different countries

Country	TSO	Issue year	Refs
Denmark	Energinet.dk	2009	[15]
Germany	TenneT	2012	[16]
Canada	Aeso	2013	[17]
Ireland	EirGrid	2012	[18]
China	State Grid Corporation of China (SGCC)	2009	[19]

In this paper, the grid codes listed in Table 1 are reviewed [4, 5, 14]. Due to the rapid development of wind energy in the listed countries, these grid codes are considered to be representative. Detailed requirements of these grid codes have been discussed in Section 2 and Section 3. Specifically, the principles on active and reactive power control under normal operation are compared in Section 2. Also, the basic requirements in abnormal condition are analyzed in Section 3.

2 Requirements of performance under normal condition

2.1 Active power control

The requirements of active power control concern about the performance of active power output provided by WPPs. Traditionally, WPPs are simply required to generate active power as much as possible [20]. However, since the penetration level of wind energy has grown rapidly, TSOs formulate various control functions of active power such as frequency control and power gradient limit, which must be implemented [20].

Frequency control refers to the capability to participate in frequency stabilization. For WPPs, TSOs have determined their own frequency variation that the WPPs should tolerate. Figures 1–3 shows the detailed requirements of frequency control in Denmark and Germany. In Denmark, the normal range of frequency is between 49.5 and 50.2 Hz [15]. While in Germany, the normal variation of frequency is 49.00–50.5 Hz [16]. As for active power control, all the grid codes demands that the active power production should be adjustable in the event of the small deviation of frequency. In Germany, as Fig. 2 shows, active power provided by WPPs can be decreased only when frequency becomes higher than 50.2 Hz [16]. Figure 3 is the droop requirements of frequency response in Denmark, in which f2 and f3 are the boundaries of dead band. When frequency varies between f1 and f4, primary control for frequency should be implemented [14, 15]. Besides, the duration and

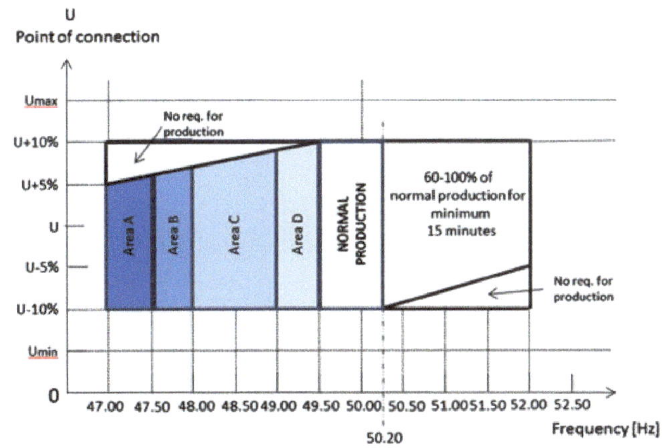

Fig. 1 Active power production requirements in Denmark [15]

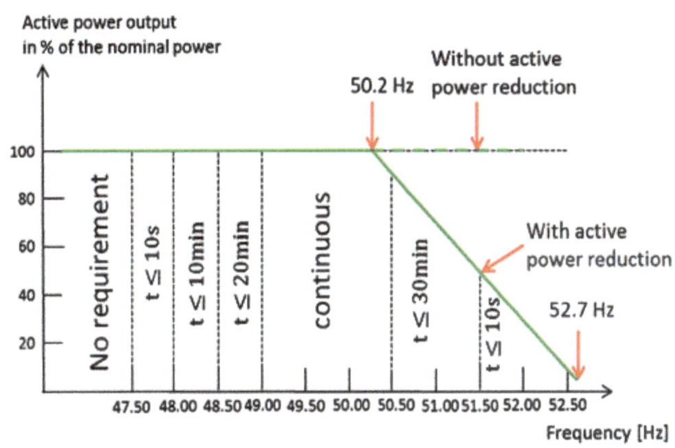

Fig. 2 Basic requirements for the active power output of a generation unit in German grid code [16]

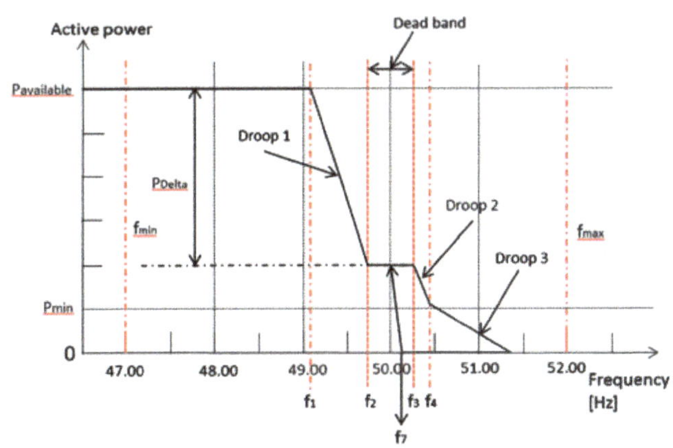

Fig. 3 Frequency control for wind power plants with a power output higher than 25 MW in Danish grid code [15]

accuracy of frequency control function have been regulated in Danish grid codes. Moreover, the Danish grid code, which includes the most specific requirements on active power control herein, demands WPPs should regulate the active power output considering the voltage of point of connection (POC) as well as frequency.

Table 2 Requirements of active power control for frequency stabilization in Canada [17]

Frequency (Hz)	Minimum time delay
>61.7	0 s
61.6 to 61.7	30 s
60.6 to <61.6	3 min
>59.4 to <60.6	Continuous Operation
>58.4 to 59.4	3 min
>57.8 to 58.4	30 s
>57.3 to 57.8	7.5 s
>57.0 to 57.3	45 cycles
≤57.0	0 s

Table 3 Constraints of active power output in Chinese grid code [19]

The installed capacity of wind farm (MW)	10 min maximum ramp rate (MW)	1 min maximum ramp rate (MW)
<30	10	3
30–150	Installed capacity/3	Installed capacity/10
>150	50	15

Similarly, the grid codes published in China, Ireland and Canada have established their own frequency boundaries for continuous operation [17–19]. In China, the WTs should operate continuously when frequency varies from 49.5 to 50.2 Hz [19]. The normal operation range in Ireland is 49.5–51.0 Hz [18]. It is necessary to point that the nominal frequency in the selected Canadian grid code is 60 Hz [17]. The details of Canadian requirements have been illustrated in Table 2 [17].

Furthermore, considering the deviation of frequency indicates the imbalance between the supply and demand of active power, WPPs are required to operate with a certain reserve capacity [4, 20]. The reserve power can be utilized for frequency stabilization [4, 20]. Besides, some papers proposed various strategies which recommend that the inertial energy stored in the rotor of WT can be exploited temporarily to participate in primary frequency control [7, 9, 10].

Power gradient limit determines the constraints of ramp rate to avoid frequency deviation, especially when active power provided by WPPs varies rapidly [21]. The specific requirements have been formulated by TSOs. In China, the recommended constraints of active power production have been listed in Table 3 [19]. The gradient limit in Ireland can be varied between 1 and 30 MW per minute [18]. Such requirements should consider the characteristics of conventional power plants so that the active power supply of the whole system can keep steady [20].

For individual wind turbine, the active power can be fluctuating when wind changes rapidly. To prevent the fluctuation of power and smooth the production, energy storage devices such as batteries and capacitors can be installed [22]. In addition, some coordinated control strategies such as adjusting pitch angle or controlling the DC-link voltage have been proposed [22, 23].

The frequency control and power gradient limit may be in conflict with the capture of wind energy, such as maximum power point tracking (MPPT) control. Consequently, the operation of WTs can be influenced. For one certain wind farm with centralized controllers, both the wind condition and the situation of each WT should be taken into account so that the reference signal generated by the centralized controllers can be appropriately dispatched to each WT [20]. Moreover, some autonomous control schemes, which are alternative for WPPs, are required to be developed further.

2.2 Reactive power control

Reactive power control concerns the stability of voltage [24]. In the selected grid codes, the requirements of reactive power control are often illustrated by power factor curves [24]. Particularly, a bonus or penalty regulation based on the variation of power factor has been established in Spain [25].

To formulate the requirements of reactive power control for WPPs, the technical characteristics of individual WT should be considered. In the 1990s, fixed-speed wind turbines (FSWTs) were widely installed [26, 27]. However, the induction generator equipped in FSWT can absorb reactive power from the power grid [27]. As a result, the reliability of power system can be decreased, especially when grid faults occur. Moreover, FSWT can hardly control the output of reactive power [27]. Therefore, such WTs cannot contribute to the stabilization of grid voltage automatically. Therefore, TSOs simply formulated general operating constraints of power factor for WPPs [24]. The Danish grid code published in 2000 even required that WPPs should operate with unity power factor [24, 28]. To fulfill these requirements, capacitor banks are required for reactive power compensation [26, 27].

Since 2001, DFIG has been a dominant type in industry [2, 26, 27]. For individual DFIG, rotor side converter, grid side converter and DC-link capacitor should be equipped. Hence, it is possible that generator in DFIG can be excited by controlling rotor side converter without absorbing reactive power from power grid [27]. Furthermore, the installed capacitor and converters can be used to provide reactive power for voltage stabilization.

Due to the development of WT techniques, the grid codes published recently formulate more control functions

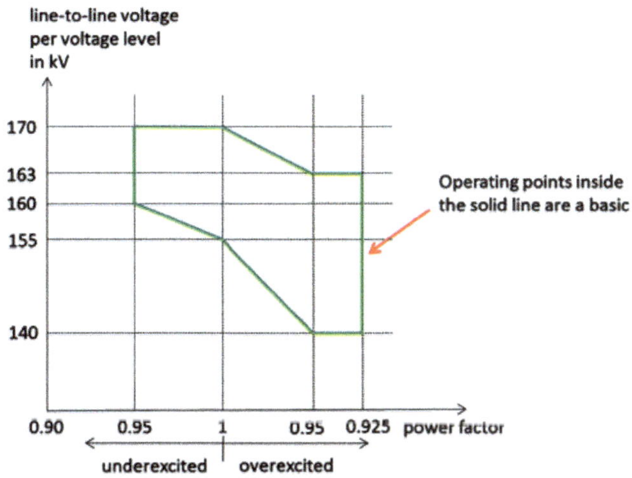

Fig. 4 Operation range of WPPs in Germany [16]

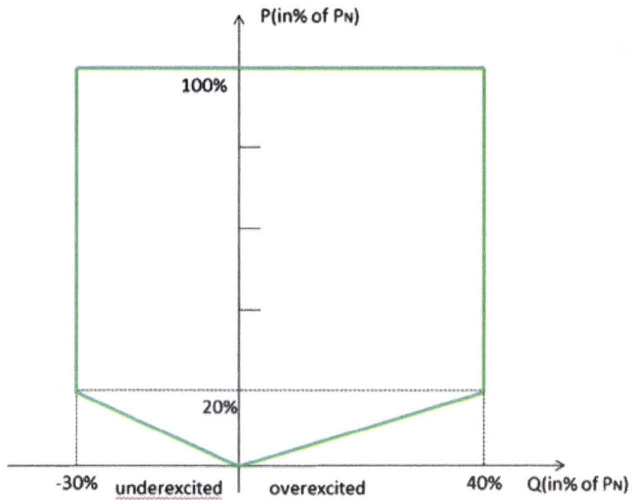

Fig. 5 Operation range of individual wind turbine in Germany [16]

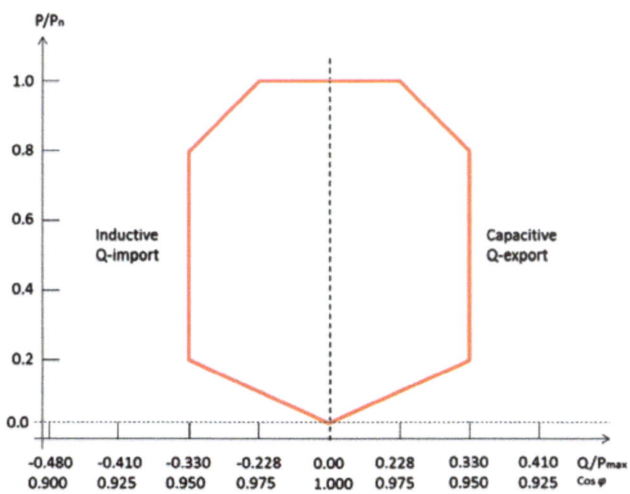

Fig. 6 Operation range of WPPs in Denmark [15]

Table 4 Range of power factor in Canada, Ireland and China [17–19]

TSO	Country	Power factor range
AESO	Canada	−0.95 to 0.9 [17]
EirGrid	Ireland	0.85 lagging to 0.85 leading [18]
SGCC	China	0.95 lagging to 0.95 leading [19]

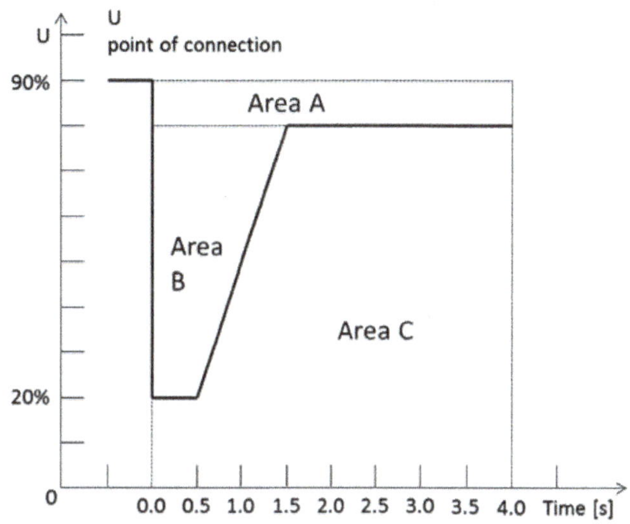

Fig. 7 LVRT requirements in Denmark [15]

of reactive power. For the wind farms established in Germany and Denmark, the basic requirements of power factor have been shown in Figs. 4–6. In Germany, when the voltage decreases from 170 to 140 kV, the power factor should vary from 0.95 unexcited to 0.925 overexcited [8, 16]. Moreover, the German grid code provides detailed requirements for individual wind turbine (as shown in Fig. 5). Besides, the operation regulations of power factor formulated by the TSOs in Canada, Ireland and China have been illustrated in Table 4.

The grid codes published in Canada, Ireland and China demand that WPPs provide the controllable reactive power depending on the supply of active power. It should be noted that the TSOs in some countries such as Germany and Denmark have provided specific requirements that the variation of reactive power production should depend on the change of voltage [8, 15, 16]. As presented in Fig. 4, these requirements are in terms of the relationship curves between voltage and power factor.

To implement the regulations above, the capability of WTs can be considered to be employed [8]. Besides, it is alternative to install extra reactive power source for WPPs [8]. In view of increasing integration level of wind energy, the specific control schemes, which aim at stabilizing grid voltage, can be investigated further.

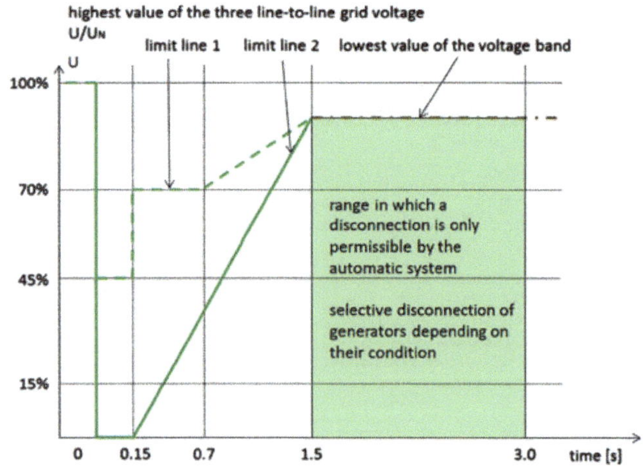

Fig. 8 LVRT requirements in Germany [16]

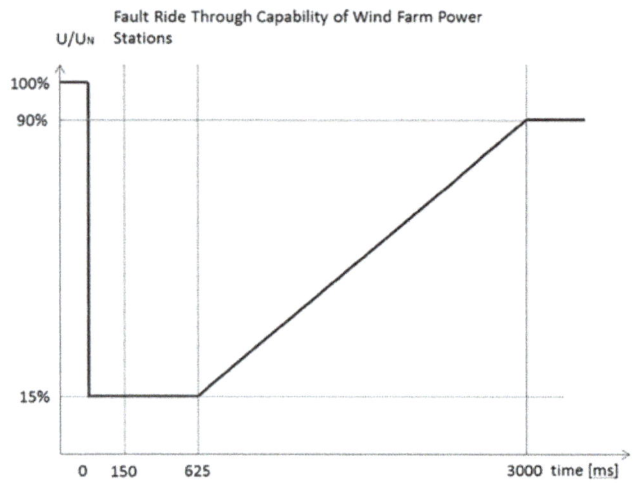

Fig. 9 LVRT requirements in Ireland [18]

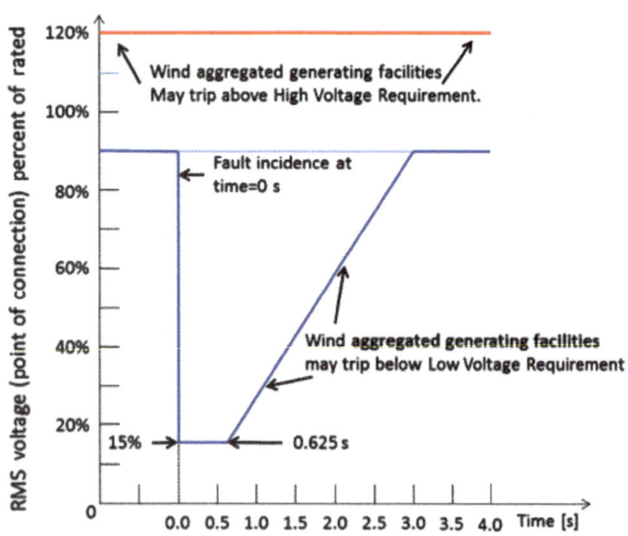

Fig. 10 LVRT requirements in Alberta, Canada [17]

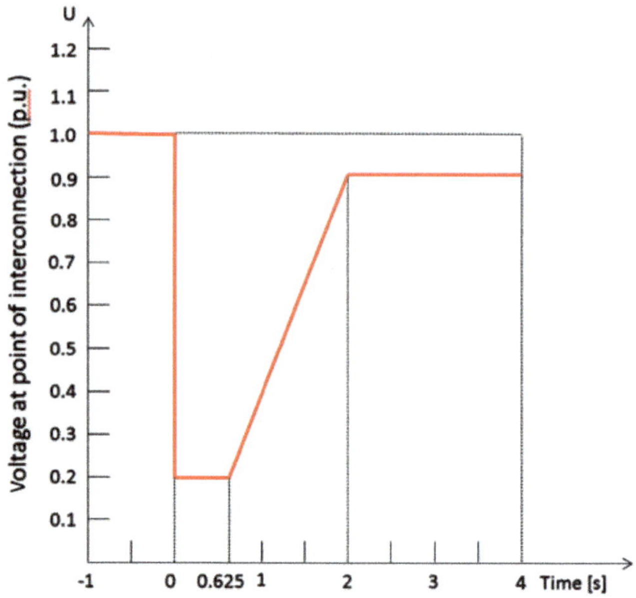

Fig. 11 LVRT requirements in China [19]

3 Requirements of performance under fault condition

3.1 Requirements of LVRT

When the penetration level of wind energy was quite low, the impact of wind generators was not considered by TSOs [14, 29]. At that time, wind turbines were allowed to trip when grid faults occurred [29]. As the integration level of wind energy increases, the premature tripping of WTs may result in voltage collapse [30]. Therefore, the requirements of low voltage ride through, which demand that wind turbines should stay online rather than trip immediately when voltage sags, has been reinforced by TSOs [29]. The characteristics of LVRT in each country have been shown in Figs. 7–11.

The LVRT requirements of Denmark must be implemented in the case of faults of one, two or three phases. As shown in Fig. 7, the wind power plant must stay online in Area A. In Area B, the wind power plants should connect to the grid and provide reactive power for voltage support. Area C is the range that the wind power plants can be tripped [15]. As described in Figs. 9–11, the requirements of LVRT ability regulated in China, Ireland and Canada are similar [15–19].

The LVRT requirements of Germany, which is the most complicated, have been presented in Fig. 8. As the figure shows, the power system cannot become unstable above Limit Line 1. As for the area above the Limit Line 2, the wind power plants should keep on connecting to the grid. If a wind turbine becomes unstable, it can be self-disconnected from the grid [16].

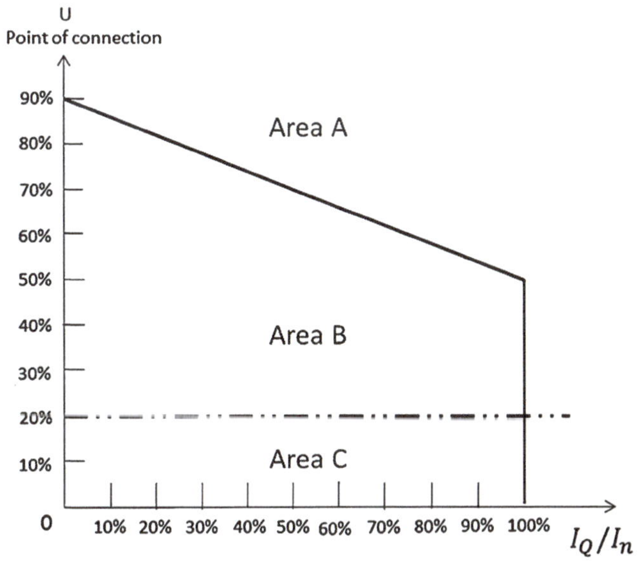

Fig. 12 Requirements of reactive current supply in Denmark [15]

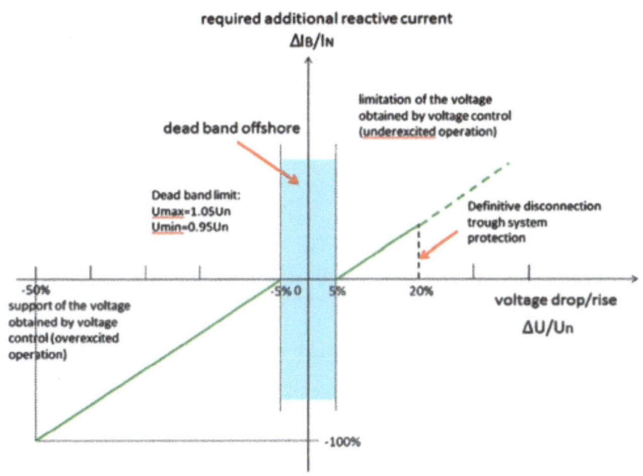

Fig. 13 Requirements of reactive current supply in Germany [16]

3.2 Requirements of reactive current supply

As analyzed in Section 2, FSWTs can hardly participate in voltage support. Since DFIGs have been widely installed recently, it is possible that the WPPs are able to provide available reactive current to enhance the stability of power system when grid faults occur [11].

Figures 12, 13 illustrate the requirements of reactive current supply in Denmark and Germany during the grid faults. Danish grid code demands that reactive current has a priority to be provided in Area B [15]. In Germany, wind turbines should provide reactive power no later than 20 ms after the voltage dip is detected [16].

Moreover, some countries such as Ireland encourage WPPs to generate active power during voltage dip [13, 18]. However, it should be observed that the demand of active power will decrease when the voltage sags. In order to avoid the occurrence of overcurrent in DFIGs, the active power production should be correspondingly reduced [12, 13]. Irish grid code claims that the value of active power provided by wind farm should be in proportion to the residual voltage under the faults [18].

When the voltage returns to normal range, the recovery rate of active power has been regulated in some grid codes so that the power surge can be avoided [24]. In Germany, the recovery rate should be between 10 % and 20 % of nominal power per second [16].

For individual DFIG, it is expected that the potential of reactive current supply can be exploited [12, 13]. Several control schemes proposed that the capacitors and converters installed inside should be utilized to fulfill these requirements [13]. The schemes are promising to be applied in industry. By adopting the schemes, it is possible that DFIGs can be similar in performance to STATCOM during the low voltage. Moreover, it is necessary to note that both the situation of each WT and the wind condition should be considered when the schemes are implemented.

4 Conclusions

Modern power system has been influenced by the significant penetration level of wind energy. The impact of wind energy should be carefully analyzed and reconsidered [29].

In order to ensure the secure and economic operation of power system, TSOs in different countries have published various grid codes. In this paper, the basic requirements of grid codes have been discussed. Generally speaking, the grid code published by TenneT, which is a German TSO, formulates the most detailed technical requirements.

Considering the continuous growth of penetration level of wind energy, the requirements of grid codes should be reviewed and revised continuously [29]. Based on the paper, it is understood that grid integration requirements are recently strengthened in several respects such as reactive power control and LVRT to enhance the security of overall power system.

For manufactures and researchers, these grid codes can serve as important guidelines to develop new control strategies so that the performances of wind power plants can meet fulfill the ever changing requirements [6].

References

[1] International Energy Agency (2012) World energy outlook 2012. OECD/IEA, Paris
[2] Musunuri S, Ginn HL (2011) A fast maximum power extraction algorithm for wind energy systems. In: Proceedings of the 2011 IEEE Power and Energy Society general meeting(PES'11), Detroit, 24–29 July 2011, San Diego, CA, USA, p 7
[3] Gautam D, Vittal V, Harbour T (2009) Impact of increased penetration of DFIG-based wind turbine generators on transient and small signal stability of power systems. IEEE Trans Power Syst 24(3):1426–1434
[4] Singh B, Singh SN (2009) Wind power interconnection into the power system: a review of grid code requirements. Electr J 22(5):54–63
[5] Altın M, Göksu O, Teodorescu R et al (2010) Overview of recent grid codes for wind power integration. In: Proceedings of the 12th international conference on optimization of electrical and electronic equipment (OPTIM'10), Basov, Russia, 20–22 May 2010, pp 1152–1160
[6] Tsili M, Papathanassiou S (2009) A review of grid code technical requirements for wind farms. IET Renew Power Gener 3(3):308–332
[7] de Almeida RG, Peas Lopes JA (2007) Participation of doubly fed induction wind generators in system frequency regulation. IEEE Trans Power Syst 22(3):944–950
[8] Wu Q, Xu Z, Østergaard J (2010) Grid integration issues for large scale wind power plants (WPPs). In: Proceedings of the 2010 Power and Energy Society general meeting (PES'10),Minneapolis, 25–29 July 2010, p 6
[9] Tarnowski GC, Kjar PC, Sorensen PE et al (2009) Variable speed wind turbines capability for temporary over-production. In: Proceedings of the 2009 IEEE Power and Energy Society general meeting (PES'09), Calgary, Canada, 26–30 July 2009, p 7
[10] Chang-Chien LR, Lin WT, Yin YC (2011) Enhancing frequency response control by DFIGs in the high wind penetrated power systems. IEEE Trans Power Syst 26(2):710–718
[11] Ullah NR, Thiringer T, Karlsson D (2007) Voltage and transient stability support by wind farms complying with the E.ON netz grid code. IEEE Trans Power Syst 22(4):1647–1656
[12] Yang L, Xu Z, Østergaard J et al (2012) Advanced control strategy of DFIG wind turbines for power system fault ride through. IEEE Trans Power Syst 27(2):713–722
[13] Xie D, Xu Z, Yang L et al (2013) A comprehensive LVRT control strategy for DFIG wind turbines with enhanced reactive power support. IEEE Trans Power Syst 28(3):3302–3310
[14] Mohseni M, Islam SM (2012) Review of international grid codes for wind power integration: Diversity, technology and a case for global standard. Renew Sustain Energy Rev 16(6):3876–3890
[15] Technical regulation 3.2.5 for wind power plants with a power output greater than 11 kW. 55986/10, Transmission System Operator of Denmark, Fredericia, Denmark, 2010
[16] Requirements for offshore grid connections in the grid of TenneT TSO GmbH. TenneT, Arnhem, Netherlands, 2012
[17] ISO rules, Part 500—Facilities, Division 502—Technical requirements, Section 502.1—Wind aggregated generating facilities, Technical requirements. Alberta Electric System Operator (AESO), 2013
[18] Grid code version 4.0. EirGrid, Dublin, Ireland, 2012
[19] GB/T 19963—2011 Technical rule for connecting wind farm to power network (in Chinese)
[20] Hansen AD, Sørensen P, Iov F et al (2006) Centralised power control of wind farm with doubly fed induction generators. Renew Energy 31(7):935–951
[21] Moon SI, Pyo GC, Park JW (2010) Study on development of grid code and operation scheme of Jeju Island with high wind penetration. In: Proceedings of the 21st world energy congress, Montréal, Canada, 12–16 Sept 2010
[22] Howlader AM, Urasaki N, Yona A et al (2013) A review of output power smoothing methods for wind energy conversion systems. Renew Sustain Energy Rev 26:135–146
[23] Uehara A, Pratap A, Goya T et al (2011) A coordinated control method to smooth wind power fluctuations of a PMSG-based WECS. IEEE Trans Energy Convers 26(2):550–558
[24] Jauch C (2006) Stability and control of wind farms in power systems. Ph.D. Dissertation, Aalborg University, Aalborg, Denmark
[25] Domínguez T, de la Torre M, Juberías G et al (2008) Renewable energy supervision and real time production control in Spain. In: Proceedings of the international conference on renewable energies and power quality (ICREPQ'08), Santander, Spain, 12–14 Mar 2008, p 6
[26] Hansen AD, Hansen LH (2007) Market penetration of wind turbine concepts over the years. In: Proceedings of the European wind energy conference and exhibition (EWEC'07), Milan, Italy, 7–10 May 2007, p 6
[27] Akhmatov V (2005) Induction generators for wind power. Multi-Science Publishing Co Ltd, Brentwood
[28] Specifications for connecting wind farms to the transmission network. ELT 1999-411A, 2nd edn. ELTRA, Copenhagen, Denmark, 2000
[29] Abbey C, Joos G (2005) Effect of low voltage ride through (LVRT) characteristic on voltage stability. In: Proceedings of the 2005 IEEE Power Engineering Society general meeting (PES'10), vol 2, San Francisco, 12–16 Jun 2005, pp 1901–1907
[30] Arulampalam A, Ramtharan G, Jenkins N et al (2007) Trends in wind power technology and grid code requirements. In: Proceedings of the 2nd international conference on industrial and information systems (IIS'07), Peradeniya, Sri Lanka, 8–11 Aug 2007, pp 129–133

Author Biographies

Xiaoge LIU received his B.Eng. from Huazhong University of Science and Technology, Wuhan, China 2012. He is now an M.Phil. candidate with The Hong Kong Polytechnic University.

Zhao XU received his B.Eng., M.Eng., and Ph.D. degrees from Zhejiang University, China, in 1996, National University of Singapore, Singapore, in 2002, and The University of Queensland, Australia, in 2006, respectively. He is now with The Hong Kong Polytechnic University. He was previously an Associate Professor with Centre for Electric Power and Energy, Technical University of Denmark. His research interest includes demand side, grid integration of renewable energies and EVs, electricity market planning and management, and AI applications in power engineering. He is an Editor of *Electric Power Components and Systems* journal.

Kit Po WONG received the M.Sc, Ph.D., and higher doctorate D.Eng. degrees from the University of Manchester, Institute of Science and Technology, Manchester, UK, in 1972, 1974, and 2001, respectively. He is a Fellow of IEEE. Since 1974, he has been with the School of Electrical, Electronic and Computer Engineering, The University of Western Australia, where he is currently a Winthrop Professor. His current research interests include power system analysis, planning and operations, and smart grids. Prof. Wong received three Sir John Madsen Medals (1981, 1982, and 1988) from the Institution of Engineers Australia, the 1999 Outstanding Engineer Award from IEEE Power Chapter Western Australia, and the 2000 IEEE Third Millennium Award. He was General Chairman of IEEE/CSEE PowerCon2000 conference. He was an Editor-in-Chief of IEE Proceedings in Generation, Transmission and Distribution. Currently he is serving as Editor-in-Chief for IEEE Power Engineering Letters of the IEEE PES Transactions.

5

Operational reliability evaluation of restructured power systems with wind power penetration utilizing reliability network equivalent and time-sequential simulation approaches

Yi DING (✉), Lin CHENG, Yonghong ZHANG, Yusheng XUE

Abstract In the last two decades, the wind power generation has been rapidly and widely developed in many regions and countries for tackling the problems of environmental pollution and sustainability of energy supply. However, the high share of intermittent and fluctuating wind power production has also increased the burden of system operator for securing power system reliability during the operational phase. Moreover, the power system restructuring and deregulation have not only introduced the competition for reducing cost but also changed the strategy of reliability evaluation and management of power systems. The conventional long-term reliability evaluation techniques have been well developed, which have been more focused on planning and expansion rather than operation of power systems. This paper proposes a new technique for evaluating operational reliabilities of restructured power systems with high wind power penetration. The proposed technique is based on the combination of the reliability network equivalent and time-sequential simulation approaches. The operational reliability network equivalents are developed to represent reliability models of wind farms, conventional generation and reserve provides, fast reserve providers and transmission network in restructured power systems. A contingency management schema for real time operation considering its coupling with the day-ahead market is proposed. The time-sequential Monte Carlo simulation is used to model the chronological characteristics of corresponding reliability network equivalents. A simplified method is also developed in the simulation procedures for improving the computational efficiency. The proposed technique can be used to evaluate customers' reliabilities considering high penetration of wind power during the power system operation in the deregulated environment.

Keywords Operational reliability, Restructured power system, Wind power, Simulation

Y. DING, Y. ZHANG, School of Information and Control, Nanjing University of Information Science & Technology, Nanjing, China
(✉) e-mail: yding@elektro.dtu.dk
Y. DING, Department of Electrical Engineering, Technical University of Denmark, Lyngby, Denmark
L. CHENG, Department of Electrical Engineering, Tsinghua University, Beijing, China
Y. XUE, State Grid Electric Power Research Institute, Nanjing, China

1 Introduction

In recent years, the development and utilization of wind power generation have been rapidly expanding in many regions and countries for reducing reliance on conventional energy resources and reducing environmental pollutants. Wind power generation is a promising renewable energy resource, which can compete with conventional power generation in terms of abundance, accessibility and production cost. Wind energy will play an important role in the European Union's (EU) future energy plan [1]: For example, wind power will provide 50% of electricity production by 2025 [2] in Denmark. However, the fluctuation of wind velocity varying chronologically and random nature of failures of WTGs make the generation output of wind farm stochastic and totally different from that of the conventional generating units [3]. The high penetration of intermittent and fluctuating wind power production can therefore bring complexities for securing proper balancing between generation and demand and maintaining system reliabilities during the operational phase. The fluctuating wind power production can result in system imbalances

[4]: In Denmark, more than half of the system imbalances are caused by wind power fluctuation, which will be more frequent with the increasing penetration of wind power production in the future.

There are two categories of reliability evaluation methods for power systems with wind power generation including analytical techniques [3, 5, 6] and Monte Carlo simulation approaches [7, 8]. Direct mathematical methods such as universal generating functions [9, 10] are utilized by analytical techniques for evaluating system and customers' reliability indices for determined states. Monte Carlo simulation approaches is more flexible for considering the chronological characteristics of power system operation [7], hence they can provide more detailed and accurate information on the reliability indices of power system operation. Moreover Monte Carlo simulation approaches are more suitable when considering complex operational conditions or when the number of contingency events is large [11].

These research works mainly focus on the long term reliability evaluation based on the steady-state probabilities of system components. They are mainly utilized for planning and expansion of power systems considering high wind power penetration [9].

However, these methods can only provide a rough approximation of reliability indices in the operational phase because of high fluctuations of wind power generation. Moreover these developed methods are more concerned on the conventional integrated power systems. The restructuring of power system changes the basic reliability management strategies of system operation and planning [12]. Electric energy and reserve are traded in various electricity markets in restructured power systems. These markets are correlated with the real time operation of power systems, which may have significant impacts on system and customers' reliabilities in the operational stage. These changes make contingency management schema more complicated than that used in conventional integrated power systems.

This paper proposes a technique for evaluating operational reliabilities of restructured power systems with high wind power penetration. The proposed technique is based on the combination of the reliability network equivalent and time-sequential simulation approaches. The reliability network equivalents are developed to represent operational reliability models of conventional generation and reserve providers, wind farms, fast reserve providers and transmission network in restructured power systems. The time-sequential Monte Carlo simulation is used to model the chronological characteristics of corresponding reliability network equivalents. A contingency management schema for real time operation considering its coupling with the day-ahead market is proposed. A simplified method is developed in the simulation procedures for improving the computational efficiency. The proposed technique can be used to evaluate customers' reliabilities considering high penetration of wind power during the power system operation in the deregulated environment.

2 Reliability network equivalents of generation systems

2.1 Operational reliability equivalents of wind farms

Electric power generation from a wind farm is strongly dependent on the intermittent and fluctuating wind speed, which has great uncertainty due to the random nature of the weather. A well-known model used in the reliability evaluation of power systems with wind power generation is the Markov process model [3, 9, 13].

In the operational phase, the Markov process model can also be used to predict probabilities of future wind states, whose future development depends only on the present state and not on how the process arrived at that state [14]. The wind speed model is modelled as a continuous-time Markov chain illustrated in Fig. 1. As shown in Fig. 1, the wind speed $V^w(t)$ at any time t is a random variable taking values from the wind speed set $\{v_1, \ldots, v_{K^w}\}$.

The Markov chain model assumes that the state transition depends on its present state [11, 14]. In the operational phase, the probabilities of future wind states are also strongly dependent on transition rates among the present wind state and possible future wind states. For a given operation period, e.g., in the evening, the current wind speed is in the state j^w and $\sum_{s=1}^{K^w-j^w} \lambda_{j^w,j^w+s} > \sum_{s=1}^{j^w-1} \lambda_{j^w,j^w-s}$, it is more probable that wind speed will increase. Similarly if $\sum_{s=1}^{K^w-j^w} \lambda_{j^w,j^w+s} < \sum_{s=1}^{j^w-1} \lambda_{j^w,j^w-s}$, e.g., in the day, it shows that the probability of wind speed decrease is higher than the wind speed increase.

The following equation (1) is used to evaluate the power output $wp_l(t)$ of WTG l at time t corresponding to wind speed $V^w(t)$ [16]:

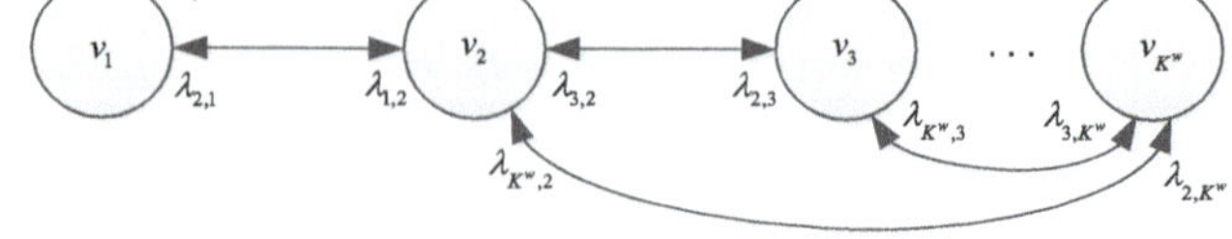

Fig. 1 State space diagram for wind speed model

$$wp_l(t) = \begin{cases} 0 & 0 \le V^w(t) \le V^w_{ci,l} \\ \left(A_l + B_l \times V^w(t) + C_l \times V^w(t)^2\right) \cdot P_{r,l} & V^w_{ci,l} \le V^w(t) \le V^w_{r,l} \\ P_{r,l} & V^w_{r,l} \le V^w(t) \le V^w_{co,l} \\ 0 & V^w(t) \ge V^w_{co,l} \end{cases} \quad (1)$$

where A_l, B_l and C_l are operational parameters of the WTG l presented in [16], respectively. The cut-in speed, the cut-out speed, the rated speed and the rated power of the WTG l are $V^w_{ci,l}$, $V^w_{co,l}$, $V^w_{r,l}$ and $P_{r,l}$, respectively. The power output at time t $wp_l(t)$ take values from the set $\{wp_{l,1}, \cdots, wp_{l,j^w}, \cdots, wp_{l,K^w}\}$. The power output of WTG l corresponding to wind state v_{j^w} is wp_{l,j^w}.

A wind farm usually consists of many WTGs and trades its generation in the day-ahead market, which usually has an hourly time resolution, e.g. the Nordic electricity market [17]. However, the hourly time resolution in the day-ahead market is a rough approximation of the generation of wind farms, which cannot handle the generation fluctuations and possible failures of WTGs during the operational hour. In real time operation, the power output of a wind farm is determined by the power output of each WTG for time t, which can be obtained as:

$$WP_i(t) = \sum_{l=1}^{n_i^w} wp_{li}(t) \quad (2)$$

where n_i^w is the number of WTGs in the wind farm at bus i.

Suppose the scheduled generated power of the wind farm at bus i for the operational hour h in the day-ahead market is $WP_i^{h,D}$. The difference between $WP_i(t)$ and $WP_i^{h,D}$ is the imbalance of the wind farm at time t because of wind power fluctuation, where $h \le t < h + 1$:

$$\Delta WP_i(t) = WP_i(t) - WP_i^{h,D} \quad (3)$$

If $\Delta WP_i(t) < 0$, it indicates that the power output of the wind farm at time t is less than the scheduled value in the day-ahead market, which needs to be compensated by up regulation. Similarly if $\Delta WP_i(t) > 0$, it indicates that the wind farm at time t can generate more power than the scheduled value in the day-ahead market, which can either be sold in the real-time market or spilled.

If only the stochastic behavior of wind speed is considered, $WP_i(t)$ is a random variable taking value from the set $\{WP_{i,1}, \cdots, WP_{i,j_i^w}, \cdots, WP_{i,K_i^w}\} = \{\sum_{l=1}^{n_i^w} wp_{l,1}, \cdots, \sum_{l=1}^{n_i^w} wp_{l,j^w}, \cdots, \sum_{l=1}^{n_i^w} wp_{l,K^w}\}$. The random failures of WTGs can also derate the power output of a wind farm. In this case, $WP_i(t)$ is the random variable taking the value from the set $\{\sum_{l=1}^{n_i^w} wp_{l,1}, \sum_{l=1}^{n_i^w-1} wp_{l,1}, \ldots, \sum_{l=1}^{n_i^w} wp_{l,j^w}, \sum_{l=1}^{n_i^w-1} wp_{l,j^w}, \ldots, \sum_{l=1}^{n_i^w} wp_{l,K^w}, \sum_{l=1}^{n_i^w-1} wp_{l,K^w}, \ldots, 0\}$ considering both the stochastic behavior of wind speed and random failures of WTGs. For a specific state of wind farm $J^{\hat{w}}$, the power output of the wind farm at bus i is

$$WP_{i,j^{\hat{w}}} = \sum_{l=1}^{n_i^w - n_{j^{\hat{w}}}^f} wp_{l,j^w} \quad (4)$$

Equation (4) indicates that wind speed $V^w(t)$ is in state j^w and there are $n^f_{j^{\hat{w}}}$ WTGs failed. The Markov model for representing the stochastic power output of the wind farm is represented in Fig. 2. The state transitions between non-adjacent states are not illustrated in Fig. 2 for the sake of clarity [13].

Reliability network equivalent techniques have been successfully used to represent reliability models of market participants in the restructured power systems [18–21]. The operational reliability model of a wind farm at bus i can be represented as an equivalent operational multi-state wind generation provider (EOWP). The characteristics of an EOWP depend on the power generation of the wind farm and its coupling with the day-ahead market such as the imbalance of the wind farm with the scheduled wind generation in the operational phase.

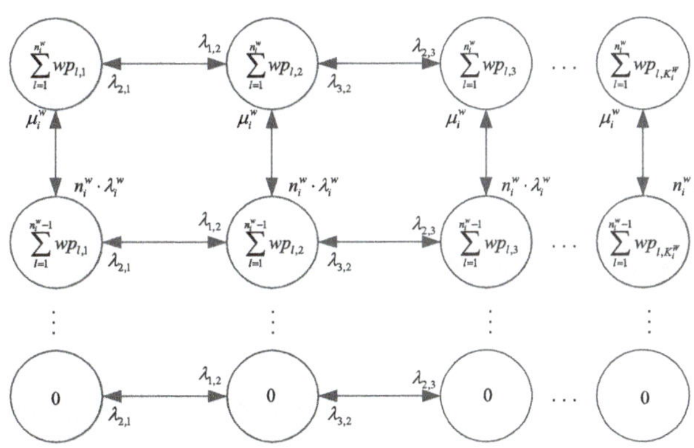

Fig. 2 Markov model the wind farm considering stochastic wind speed and random failures of WTGs

The following time-sequential simulation procedures are used to determine the characteristics of an EOWP.

Step 1: Suppose the initial state of the EOWP in operational phase is $j^{\hat{w}}$. Determine the corresponding power output $WP_{i,j^{\hat{w}}}$ based on (4).

Step 2: Generate a random number and convert the number to the time period of the state $D_{j^{\hat{w}}}$ following the corresponding exponential distribution

$$\left[\sum_{\substack{s^w=1\\s^w\neq j^w}}^{K^w}\lambda_{j^w,s^w}+\left(n_i^w-n_{j^{\hat{w}}}^f\right)\lambda_i^w+\mu_i^w\right]\cdot e^{-\left[\sum_{\substack{s^w=1\\s^w\neq j^w}}^{K^w}\lambda_{j^w,s^w}+\left(n_i^w-n_{j^{\hat{w}}}^f\right)\lambda_i^w+\mu_i^w\right]t}.$$

If the number of failed WTGs is zero in the state, the exponential distribution has become $\left[\sum_{\substack{s^w=1\\s^w\neq j^w}}^{K^w}\lambda_{j^w,s^w}+n_i^w\lambda_i^w\right]\cdot e^{-\left[\sum_{\substack{s^w=1\\s^w\neq j^w}}^{K^w}\lambda_{j^w,s^w}+n_i^w\lambda_i^w\right]t}$. The imbalance of the wind farm at time t can be evaluated as:

$$\Delta WP_{i,j^{\hat{w}}}(t)=WP_{i,j^{\hat{w}}}-WP_i^{h,D} \tag{5}$$

where $h\leq t<h+1$ and $t\leq D_{j^{\hat{w}}}$.

Step 3: Determine the next state of the wind farm $s^{\hat{w}}$ based on the system state transition sampling technique [22].

Step 4: Evaluate the corresponding power output $WP_{i,s^{\hat{w}}}$ and the time period of the state $D_{s^{\hat{w}}}$ as step 2. The imbalance of the wind farm for the state $D_{\widehat{sW}}$ at time t can also be calculated as (5).

Step 5: Repeat step 3 to step 4 till the sampled total duration has reached the studied operational period.

2.2 Operational reliability equivalents of conventional generation and reserve provides

In restructured power systems, a conventional generation and reserve provider usually consists of several large conventional generating units for trading electric energy and reserve in various forward and balancing markets [21]. The large generating units are economically dispatched in the normal operation and can provide balancing power during a contingency state. These generating units still play an important role for providing electricity and maintaining system and customers' reliabilities in existing power systems.

A conventional generating unit has both the characteristics of controllability and stochastic behavior. A conventional generating unit l with installed capacity IC_l has a scheduled power SP_l^h for the operational hour h, $SP_l^h<IC_l$, which usually is determined in the day-ahead market. The generating unit l usually also has an operating reserve margin, ΔP_l^h, e.g. primary reserve, which can be controlled and activated in real time for maintaining system reliability and security in the operational hour h. Let DC_l^h be the dispatchable generating capacity of the unit, in the operational hour h, where $DC_l^h=SP_l^h+\Delta P_l^h$ and $DC_l^h\leq IC_l$. Therefore, if the generating unit l is functioning well, it will be controlled for producing power within the limit of DC_l in the operational hour.

The stochastic behaviors of conventional generating units are caused by random failures. The random failures of generating units can reduce the available generating capacity completely or partially. The reliability model for a conventional generating unit can be represented as a binary-state model or a more complex multi-state model. Multi-state representations of generating units especially for large generators provide a more accurate and flexible tool in generating capacity adequacy assessment. A typical example is a coal fired unit with a nominal generating capacity of 576 MW used in real world [23], which is represented as a four-state reliability model as shown in Fig. 3. Assume that the scheduled power and primary reserve provided by the coal fired unit are 500 MW and 28 MW, respectively. Therefore the dispatchable generating capacity of the coal fired unit is 528 MW in the well functioning state, which indicates the maximum power which can be produced in real time. In the complete failure state, the available generating capacity is zero and therefore the generation output is also zero. In a derated state, e.g. state 3, the available generating capacity is only 482 MW, which indicates that the produced power of the coal fired unit will be reduced to 482 MW.

In general, a conventional generating unit l can have K_l^g states, $K_l^g\geq 2$. The evolution of unit l produces the stochastic process of the available generating capacity $GC_l(t)\in\{GC_{l,1},\cdots,GC_{l,j_l^g},\cdots,GC_{l,K_l^g}\}$. The available generating capacity for each state j_l^g, $j_l^g=1$, ···, K_l^g, is GC_{l,j_l^g}.

The dispatchable generating capacity of the generating unit l for each unit state j_l^g in the operational hour h can be evaluated as:

$$DC_{l,j_l^g}^h=\min\left\{\left(SP_l^h+\Delta P_l^h\right),GC_{l,j_l^g}\right\} \tag{6}$$

The generation and reserve provider at bus i can have $n_{j^{\hat{w}}}^f$ conventional generating units, which can be represented as a multi-state Markov model. The operational reliability model of a generation and reserve provider at bus i can be represented as an equivalent operational multi-state generation and reserve provider (EOGRP). The available generating capacity of the EOGRP is a random variable at

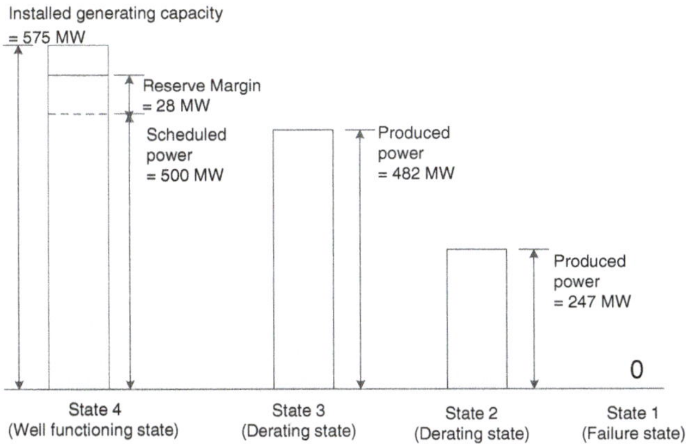

Fig. 3 The dispatchable generating capacities of the coal fired unit

time t taking values from $\{AG_{i,1}, \cdots, AG_{i,j^G}, \cdots, AG_{i,K^G}\}$. The available generating capacity of the EOGRP for each state j^G can be evaluated as:

$$AG_{i,j^G} = \sum_{l=1}^{n_i^g} DC^h_{l,j_l^g} \quad (7)$$

The EOGRP can trade its generation and operating reserve in the day-ahead energy and reserve markets, which are supposed as $GP^{h,D}$ and $GR^{h,D}$ for the operational hour h, respectively. The characteristics of an EOGRP are determined by the dispatchable generating capacity in real time considering the impact of generation and reserve scheduling in the day-ahead market.

The dispatchable generating capacity of the EOGRP for state j^G at time t can be evaluated as:

$$DG_{i,j^G}(t) = \min\left\{AG_{i,j_i^G}, \left(GP^{h,D} + GR^{h,D}\right)\right\} \quad (8)$$

where $h \le t < h + 1$ and $t \le D_{j^G}$.

If $GP^{h,D} \le DG_{i,j^G}(t) < \left(GP^{h,D} + GR^{h,D}\right)$, it indicates that the EOGRP cannot provide sufficient operating reserve and can satisfy the scheduled generation requirement for the operational hour h. If $DG_{i,j^G}(t) < GP^{h,D}$, it indicates that the EOGRP is even short of available capacity for satisfying the scheduled generation requirement for the operational hour h.

The following time-sequential simulation procedures are used to determine the characteristics of an EOGRP.

Step 1: Suppose the initial state of the EOGRP in operational phase is j^G. Determine the corresponding available generating capacity AG_{i,j^G} and dispatchable generating capacity $DG_{i,j^G}(t)$ based on (7) and (8), respectively.

Step 2: Generate a random number and convert the number to the time period of the state D_{j^G} following the corresponding exponential distribution.

$$\left[\sum_{\substack{s^G=1\\ s^G \ne j^G}}^{K^G} \lambda_{j^G,s^G}\right] \cdot e^{-\left[\sum_{\substack{s^G=1\\ s^G \ne j^G}}^{K^G} \lambda_{j^G,s^G}\right] t}.$$

Step 3: Determine the next state of the EOGRP s^G based on the system state transition sampling technique [22].

Step 4: Evaluate the corresponding available generating capacity AG_{i,s^G} and dispatchable generating capacity $DG_{i,s^G}(t)$, and the time period of the state D_{s^G} as step 2.

Step 5: Repeat step 3 to step 4 till the sampled total duration has reached the studied operational period.

2.3 Operational reliability equivalents of fast reserve providers

Besides the large online generating units, the rapid start-up generating units can start-up and synchronize with the system in a very short lead time. The rapid start-up generating units are usually utilized for providing additional operating reserve in contingency states.

Similar to the conventional online generating unit, the rapid start-up unit also has both the characteristics of controllability and stochastic behavior. Moreover, the frequent start-up can lead to extra starting stress of generating units.

If the unit starts up successfully and transits to the in-service state, the unit has the controllability for providing the necessary reserve. However, if the unit fails to start-up or meets a random failure during operation, it goes into the failure state, where the generation output of the unit is zero and the controllability of the unit is lost. The transitions among different states are also followed by the stochastic behaviors of the unit. The rapid start-up generating unit can be represented as a three-state model [24], which neglects repair in the short operational phase, as illustrated in Fig. 4.

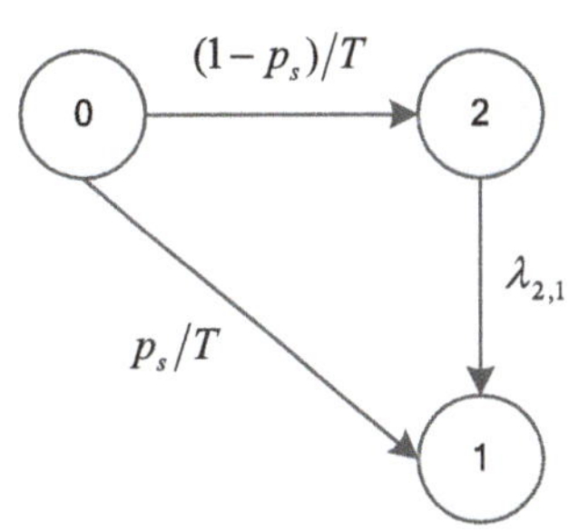

Fig. 4 State space diagram for the rapid start-up unit

Once committed, a unit initially in the ready-for-service state (state 0) can start up successfully to the in-service state (state 2). In the in-service state, the generating unit is controlled for the generation output within the limit of corresponding available capacity. Or the unit can fail to start-up and transit to the failure state (state 1). In the failure state, the generation output of the unit is zero and the controllability of the unit is lost. The transition rate between state 0 and state 2, and the transition rate between state 0 and state 1 are $(1 - p_s)/T$ and p_s/T, where p_s, T and $\lambda_{2,1}$ are the start-up failure probability, the average shut-down time between periods of commitment and failure rates from the in-service state to the failure state, respectively.

The fast reserve provider at bus i can have n_i^f rapid start-up generating units. The operational reliability model of a fast reserve provider at bus i can be represented as an equivalent operational multi-state fast reserve provider (EFRP). The available reserve capacity of the EFRP is a random variable at time t taking values from $\{AR_{i,1}, \cdots, AR_{i,j^f}, \cdots, AR_{i,K^f}\}$. The available reserve capacity of the EFRP for each state j^f can be evaluated as:

$$AR_{i,j^f} = \sum_{l=1}^{n_i^f} AG_{l,j_l^f} \tag{9}$$

where AG_{l,j_l^f} is the available generating capacity of the unit l for the corresponding state.

The EFRP can schedule operating reserve in the forward reserve market or through bilateral contract, which is supposed as $FR^{h,D}$ for the operational hour h. The characteristics of an EFRP are determined by the dispatchable reserve capacity in real time.

The dispatchable reserve capacity of the EFRP for state j^f at time t can be evaluated as:

$$DR_{i,j^f}(t) = \min\left\{AR_{i,j^f}, FR^{h,D}\right\} \tag{10}$$

where $h \leq t < h + 1$ and $t \leq D_{j^f}$.

The following time-sequential simulation procedures are used to determine the characteristics of an EFRP.

Step 1: Suppose the initial state of the EFRP in operational phase is j^f. Determine the corresponding available reserve capacity AR_{i,j^f} and dispatchable reserve capacity $DR_{i,j^f}(t)$ based on (9) and (10), respectively.

Step 2: Generate a random number and convert the number to the time period of the state D_{j^f} following the corresponding exponential distribution. $\left[\sum\limits_{\substack{s^f=1 \\ s^f \neq j^f}}^{K^f} \lambda_{j^f,s^f}\right] \cdot e^{-\left[\sum\limits_{\substack{s^f=1 \\ s^f \neq j^f}}^{K^f} \lambda_{j^f,s^f}\right] t}$.

Step 3: Determine the next state of the EFRP s^f based on the system state transition sampling technique [22].

Step 4: Evaluate the corresponding available reserve capacity AR_{i,s^f} and dispatchable reserve capacity $DR_{i,s^f}(t)$, and the time period of the state D_{s^f} as step 2.

Step 5: Repeat step 3 to step 4 till the sampled total duration has reached the studied operational period.

3 Contingency management schema for real time operation

If there is generation inadequacy for a contingency state, generation and reserve has to be re-dispatched and load may be curtailed to maintain the balance of system operation. The transmission network violation can also affect the electricity deliverability from the generation and reserve providers to bulk load points (BLP) [18]. The operational reliability model of the transmission network between the generation and reserve providers and a BLP can be represented as an equivalent operational multi-state transmission provider (EMTP).

For determining the possible load curtailment at each BLP in the contingency state, a contingency management schema is proposed as shown in Fig. 5, which is formulated as an optimal power flow (OPF) model. The day-ahead scheduling of wind power, the wind power fluctuation during real time operation, the dispatchable generating capacity and the dispatchable reserve capacity of the EOGRP and the EFRP considering their coupling with the forward electricity markets, and network constraints are included in the model.

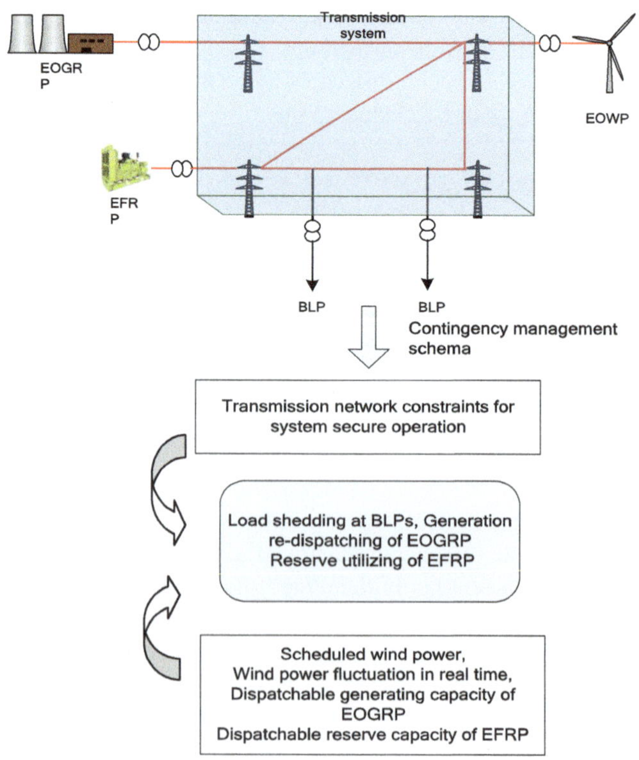

Fig. 5 Contingency management schema for real time operation

The objective of the OPF model is to minimize the total system load curtailment for the contingency state j considering network constraints and market scheduling and coupling during the real time operation.

For contingency state j, the objective function is:

$$\text{Min } f_j = \sum_{i=1}^{N} LC_{j_i}(t) \tag{11}$$

where $LC_{j_i}(t)$ is the load curtailment at BLP i for contingency state j at time t.

The objective function is subject to the following constraints for contingency state j:

DC power flow constraints:

$$\boldsymbol{B}_j\boldsymbol{\theta}_j(t) = \left(\boldsymbol{WP}^{h,D} + \Delta\,\boldsymbol{WP}_j(t)\right) + \boldsymbol{PG}_j(t) + \boldsymbol{PR}_j(t) + \boldsymbol{LC}_j(t) - \bar{\boldsymbol{D}}(t) \tag{12}$$

where $\boldsymbol{B}_j$ is the admittance matrix of the network; $\boldsymbol{\theta}_j(t)$ is phase angle vector of bus voltages at time t; $\boldsymbol{WP}^{h,D} = [WP_1^{h,D}, \cdots, WP_N^{h,D}]^{\mathrm{T}}$ is the vector of scheduled wind generations for the operational hour h in the day-ahead market; $\Delta\,\boldsymbol{WP}_j(t) = \left[\Delta\,WP_{1,\widehat{jW}}(t), \cdots, \Delta\,WP_{N,\widehat{jW}}(t)\right]^{\mathrm{T}}$ is the vector of wind generation imbalance at time t caused by wind power fluctuation; $\boldsymbol{PG}_j(t) = [PG_{1,j^G}(t), \cdots, PG_{N,j^G}(t)]^{\mathrm{T}}$ is the vector of power generations of EOGRPs at time t; $\boldsymbol{PR}_j(t) = [PR_{1,j^f}(t), \cdots, PR_{N,j^f}(t)]^{\mathrm{T}}$ is the vector of reserve dispatched of EFRPs at time t; $\boldsymbol{LC}_j(t) = [LC_{1,j}(t), \cdots, LC_{N,j}(t)]^{\mathrm{T}}$ is the vector of load curtailment at time t; $\bar{\boldsymbol{D}}(t) = [\bar{D}_1(t), \cdots, \bar{D}_N(t)]^{\mathrm{T}}$ represents the vector of the forecasted bus loads for the normal state at time t.

Equations (12) represents that the system has to be balanced with the wind power fluctuation from the day-ahead scheduling and system load after the generation re-dispatch of EOGRPs, fast reserve dispatch of EFRPs and load curtailments.

Load curtailment constraints:

$$0 \le LC_{i,j}(t) \le LC_{i,j}^{\max} \tag{13}$$

Generation output limits of EOGRPs:

$$0 \le PG_{i,j^G}(t) \le DG_{i,j^G}(t) \tag{14}$$

Fast reserve dispatch limits of EFRPs:

$$0 \le PR_{i,j^f}(t) \le DR_{i,j^f}(t) \tag{15}$$

Line flow constraints:

$$\left|\frac{1}{x_{j_{ik}}}\left(\theta_{j_i}(t) - \theta_{j_k}(t)\right)\right| \le \left|F_{ik}^{\max}\right| \tag{16}$$

where $\theta_{j_i}(t)$ is the phase angle of voltage at bus j at time t; $x_{j_{ik}}$ and $|F_{ik}^{\max}|$ are the reactance and maximum power flow of the line between buses i and k respectively.

4 Simulation procedures and reliability evaluation

The time-sequential Monte Carlo simulation can be extremely costly if the proposed contingency management schema is conducted for analyzing each sampled state by OPF. Therefore it is important to develop a simplified method for improving the computational efficiency of simulation procedures.

In this paper, it is assumed that the system is coherent: if a failed component has been repaired, the system performance would never be worse; conversely, if a working component has failed, the system performance would not be better [26]. The possible system states can be further split into reliable states, marginal successful states and failure states as shown in Fig. 6.

1) Reliable state: In this state, the generation system has adequate dispatchable capacity and transmission network is intact. The adequate dispatchable capacity indicates that the total dispatchable generating and reserve capacities from the wind farms, EOGRPs and the EFRPs are larger than the system demand plus a reserve margin:

$$\sum_{i=1}^{N}\left(\Delta\,WP_{i,j^W}(t) + WP_i^{h,D}\right) + \sum_{i=1}^{N} DG_{i,j^G}(t) + \sum_{i=1}^{N} DR_{i,j^f}(t) \ge \sum_{i=1}^{N} \bar{D}_i(t) + RM \tag{17}$$

where RM is the pre-defined reserve margin.

The adequate dispatchable capacity can guarantee the demand will not be shed because of insufficient generation capacity. The intact transmission network indicates that the system has sufficient transmission capacity for delivering generation to the demand. In this state, the contingency

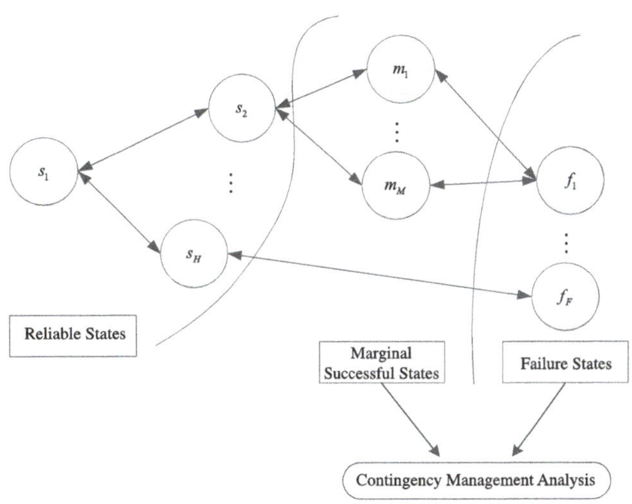

Fig. 6 Power system evolution among state space

management schema OPF is not necessary to be conducted because the system has sufficient margin for preventing load curtailment. The reliable state can be easily determined if (17) can be satisfied and the transmission network is intact.

2) Marginal successful state: As shown in Fig. 6, the system can transit from a reliable state to another state because of the failure of a system component. The failure of a transmission line or a generating unit may lead to possible network violations or inadequate dispatchable capacity, respectively. If two criterions of the reliable state cannot be satisfied, the contingency management schema will be implemented for determining the possible load curtailment at each BLP. In this case, if there is not any load curtailment, the power system is in the marginal successful state indicating that the system is still successful but may not have sufficient margin.
3) Failure state: In this state, there exists load curtailment caused by inadequate generation capacity or network violations, which is determined by the contingency management schema. The system can transit from the failure state to a reliable state or a marginal successful state if failed components have been repaired.

The sequential simulation procedure for evaluating reliability indices consists of the following steps.

Step1: Generate the state sequence of the EOWP at each bus for the studied operational period, e.g. 24 hours, utilizing the approach described in Section 2.A. From the sampled sequence, determine the state $j^{\hat{w}}$ of the EOWP at time t.

Step2: Generate the state sequence of the EOGRP at each bus for the studied operational period utilizing the approach described in Section 2.B. From the sampled sequence, determine the state $j^G$of the EOGRP at time t.

Step3: Generate the state sequence of the EFRP at each bus for the studied operational period utilizing the approach described in Section 2.C. From the sampled sequence, determine the state $j^f$of the EFRP at time t.

Step4: Similar procedures as Section 2 are used to generate the state sequence of the EMTP for the studied operational period. From the sampled sequence, determine the state j^L of the EMTP at time t.

Step5: Determine the total wind power output, dispatchable generating capacity and dispatchable reserve capacity of the system for time t, which can be evaluated as:

$$\sum_{i=1}^{N}\left(\Delta WP_{i,j^{\hat{w}}}(t)+WP_i^{h,D}\right) \text{and\$\$} \sum_{i=1}^{N} DG_{i,j^G}(t), \sum_{i=1}^{N} DR_{i,j^f}(t),$$

respectively. If

$$\sum_{i=1}^{N}\left(\Delta WP_{i,j^{\hat{w}}}(t)+WP_i^{h,D}\right)+\sum_{i=1}^{N} DG_{i,j^G}(t)+\sum_{i=1}^{N} DR_{i,j^f}(t) - TD(t) \geq RM$$

where RM is the reserve margin determined by the system operator and $TD(t)$ is the total system demand at time t, and the sampled EMTP state j^L is the best state indicating there is no network violation, the system is in the healthy state at time t. Otherwise, go to Step 6.

Step 6: The OPF model developed in (11)–(16) is used to evaluate the load curtailment at each BLP at time t. If there exists no load curtailment at each BLP, the system is in the marginal state at time t. Otherwise the system is in the failure state at time t.

Step 7: Go to step 1 if the confidence intervals are not satisfied, otherwise go to Step 8.

Step 8: Calculate the average reliability indices.

The customer and system reliability indices used in the operational phase are the loss of probability at BLP i at time t ($LOLP_i(t)$), the system loss of load expectation during the operational period T ($LOLE(T)$) and the system expected energy not supplied during the operational period T ($EENS(T)$).

These indices can be estimated using the following equations over NS sampling states:

$$LOLP_i(t) = \sum_{j=1}^{NS} X_{j,i}(t)/NS \tag{18}$$

where

$$X_{j,i}(t) = \begin{cases} 0 \text{ if there is no load curtailment at BLP } i \\ 1 \text{ if there is load curtailment at BLP } i \end{cases}$$

$$LOLE(T) = \int_0^T \left(\sum_{j=1}^{NS} X_j(t)/NS\right) \cdot dt \tag{19}$$

where $X_j(t) = \begin{cases} 0 \text{ if the system has no load curtailment} \\ 1 \text{ if the system has load curtailment} \end{cases}$

$$EENS(T) = \int_0^T \left(\sum_{j=1}^{NS}\sum_{i=1}^{N} LC_{j_i}(t)/NS\right) \cdot dt \tag{20}$$

where $LC_{j_i}(t)$ is the load curtailment at BLP i for state j at time t.

The $EENS$ coefficient of variation is used as the criterion for the convergence in the simulation:

$$\beta_{EENS} = \sqrt{V(EENS(T))}/EENS(T) \tag{21}$$

where $V(EENS(T))$ is the variance of $EENS(T)$.

The loss of load frequency (*LOLF*) is another index used in the reliability evaluation. Usually the *LOLF* is estimated by non-sequential simulation for long-term reliability evaluation rather than sequential simulation for the operational phase. Reference [27] proposed a method for evaluating the *LOLF* utilizing the state transition based sequential simulation. The method is based on the basic Markov assumption: system state durations follow corresponding exponential distributions. The *LOLF* for the operational period *T* can be assessed as:

$$LOLF(T) = \sum_{j=1}^{NS} X_j(T)/NS \qquad (22)$$

where

$$X_j(T) = \begin{cases} 0 & \text{if the system has no load curtailment} \\ 1/\mathrm{E}(\mathrm{D}_I) & \text{if the system has load curtailment} \end{cases} \qquad (23)$$

$\mathrm{E}(\mathrm{D}_I)$ is the expected total duration of interruption sequence for the operational period *T*, which can be evaluated as [27]:

$$\mathrm{E}(\mathrm{D}_I) = \sum_{j\in J} E(D_j) = \sum_{j\in J}\left(\frac{T}{\sum_j \lambda_j}\right) \qquad (24)$$

where state *j* is a failure state in the set of interruption sequence; λ_j is the transition rate between state *j* and any connected state [27].

The *LOLF* can provide useful information for system reliability studies. However, the corresponding evaluation is based on the basic Markov assumption including load models, which may not be satisfied in the short term (operational phase) reliability evaluation.

5 System studies

The IEEE-RTS [25] has been restructured to illustrate the proposed techniques. A large wind farm with total generation capacity of 600 MW has been installed in the system. The wind farm consists of 300 Vestas V-80 WTGs with 2 MW rated power [3]. The cut-in, rated, and cut-out wind speeds of a V-80 WTG are 4, 15 and 25 km/h, respectively. Markov model for the output power of a single V-80 WTG and corresponding transition rates are proposed in [3]. In [3], the wind speed series of a wind farm in the northern Iran region is utilized. When the wind speed is below the cut-in speed, the power output of the WTG is zero. When the wind speed is between the rated speed and the cut-out speed, the power output of the WTG is the rated power. When the wind speed is between the cut-in and the rated speed, the power output of the WTG is below the rated power depending on the wind speed. The power output of the WTG was split into 0, 0.5, 1, 1.5, and 2 MW steps in [3]. The transition rates in the Markov model is shown in Table 1, which is obtained from the statistical analysis of wind speed series [3].

Table 1 State transition rates of the WTG (occurrences/h)

	State 1	State 2	State 3	State 4	State 5
State 1		0.039	0.013	0.008	0.018
State 2	0.365		0.151	0.045	0.097
State 3	0.122	0.220		0.192	0.155
State 4	0.038	0.093	0.185		0.359
State 5	0.016	0.012	0.016	0.067	

The MTTF and MTTR of a WTG are assumed to be 3650 hours and 55 hours, respectively [12]. There are 6 EOGRPs in the system for providing electric energy and reserve. Each of the EOGRPs located at buses 15, 16, 18 and 23 owns one large 575-MW coal thermal generating unit, respectively. The EOGRPs located at buses 13 and 8 have three and one 197-MW oil thermal generating units, respectively. The 575-MW coal thermal generating units are utilized in real life [23], represented as the four-state Markov model. The four oil thermal generators are represented as binary Markov models [11].

The system also has two EFRPs located at buses 1 and 2, which are utilized for providing additional operating reserve. The EFRPs located at buses 1 and 2 have three and two 40-MW gas thermal generating units, which are used as rapid start-up units for providing fast start reserve. Three cases have been studied to analyze the reliabilities for the system and customers under different scenarios. The simulation codes were written in C language and running in a 2.67 GHz Fujitsu laptop.

5.1 Case 1

In the first case, it is assumed that the initial wind speed at the wind farm is 9.5 km/h and the corresponding power output of a WTG is 1 MW, half of the rated power. Suppose the scheduled generated power of the wind farm is 300 MW, which is about 10.5% of the total scheduled power at the beginning of the operating time. The total operating reserve provided by the large thermal generating units is 575 MW, which is equal to the capacity of the largest generator for satisfying the *N*-1 criterion. All generating units are in a good condition at the beginning of the operating time. The computational time for case 1 running in the 2.67 GHz Fujitsu laptop is 569.6 seconds.

Customers' *LOLP* at time *t* for representative load buses – bus 6 and bus 20 are shown in Fig. 7 and Fig. 8, respectively. It can be observed from Figures that the instant *LOLP* at each bus is a time variable rather than a constant value. The instant *LOLP* for bus 6 increases from 0.0215% at $t = 1$ h to 0.567% at $t = 24$ h. The instant *LOLP* for bus 20 increases from 0.0212% at $t = 1$ h to 0.541% at $t = 24$ h. In this case, Customers' reliability at bus 20 is a little higher than that at bus 6 because of a relatively strong transmission network connected with bus 20. Customers' risk is increasing during the operating period because of the wind power fluctuation and possible random failures of generating units. The system *EENS* and *LOLE* are 18.1207 MWh and 0.0883 hours, respectively.

5.2 Case 2

In the second case, we assume the initial wind speed at the wind farm is 20 km/h and WTGs are generating rated power at time $t = 0$. In this case, the scheduled generated power of the wind farm is doubled compared with that in case 1–600 MW, which is about 21.1% of the total scheduled power at the beginning of the operating time. Other conditions are the same as those in case 1: The total operating reserve provided by the large thermal generating units is still 575 MW for satisfying the *N*-1 criterion. All generating units are in a good condition at the beginning of the operating time. The computational time for case 2 running in the 2.67 GHz Fujitsu laptop is 817.9 seconds.

Customers' *LOLP* at time *t* for bus 6 and bus 20 are illustrated in Fig. 9 and Fig. 10, respectively. The instant *LOLP* for bus 6 increases from 1.8922% at $t = 1$ h to 39.22% at $t = 24$ h, which are relatively high values. The instant *LOLP* for bus 20 increases from 1.8922% at $t = 1$ h to 39.20% at $t = 24$ h, which almost has same values as those for bus 6. Therefore, lower customers' reliabilities are mainly caused by the increasing penetration of fluctuating wind power than the transmission network.

The system *EENS* and *LOLE* are 183.498 MWh and 5.8209 hours, which are about 10 times and 66 times than the values in case 1. Obviously only satisfying *N*-1 criterion of operating reserve cannot maintain reliability level of power system operation if high penetration of fluctuating wind power has been scheduled.

5.3 Case 3

In the third case, conditions are the same as those in case 2: the scheduled generated power of the wind farm and the operating reserve provided by the large thermal generating units are 600 MW and 575 MW, respectively. For increasing system and customers' reliabilities, the rapid start-up units owned by the EFRPs have been utilized for providing additional 200 MW operating reserve. The rapid start-up units are committed for operation at $t = 1$ h. The computational time for case 3 running in the 2.67 GHz Fujitsu laptop is 737.1 seconds.

Customers' *LOLP* for bus 6 and bus 20 at time *t* is illustrated in Fig. 11 and Fig. 12, respectively. As shown in Fig. 11, the commitment of rapid start-up units decreases

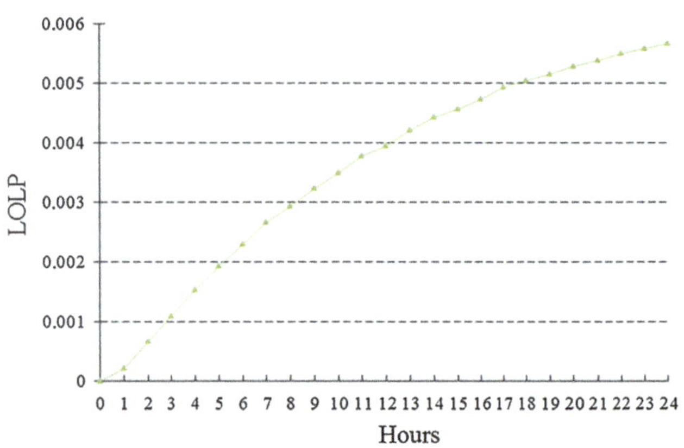

Fig. 7 Customers' instant LOLP at bus 6 for case 1

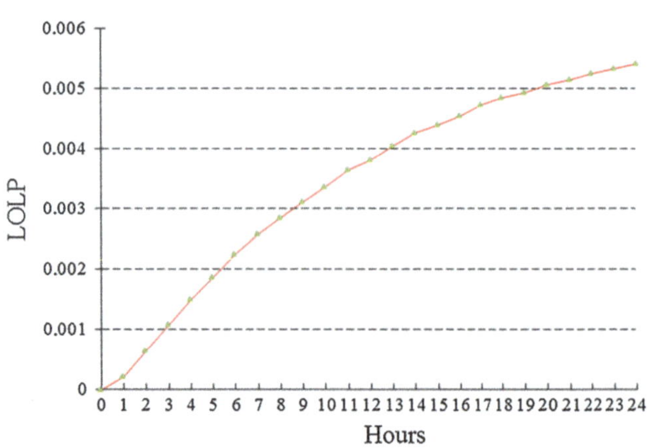

Fig. 8 Customers' instant LOLP at bus 20 for case 1

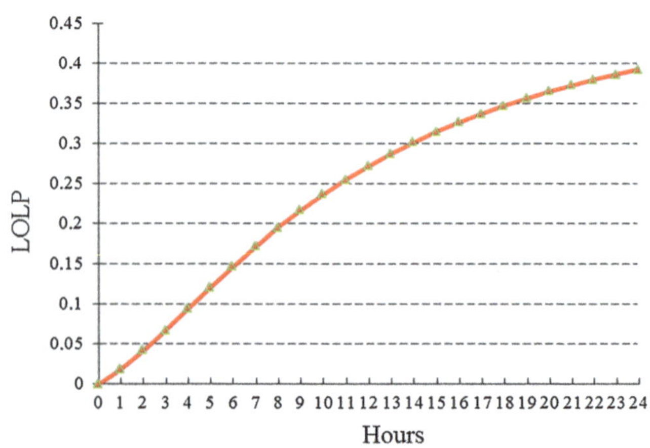

Fig. 9 Customers' instant LOLP at bus 6 for case 2

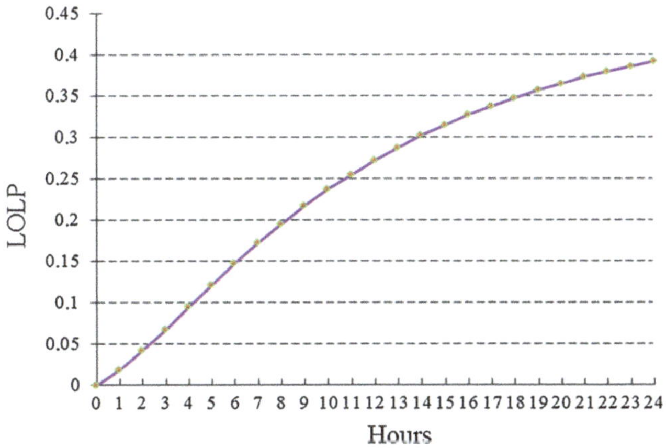

Fig. 10 Customers' instant LOLP at bus 20 for case 2

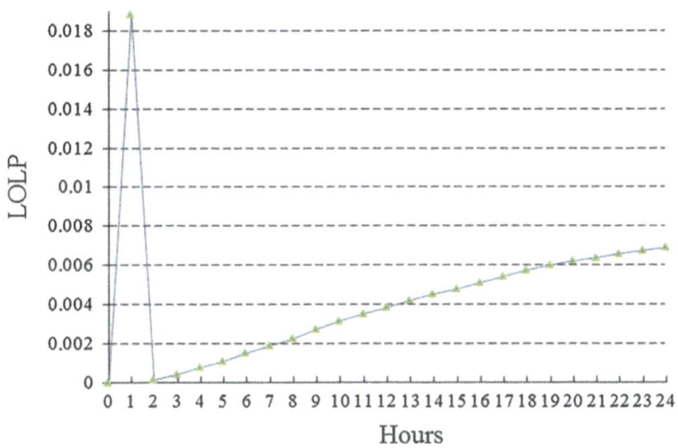

Fig. 11 Customers' instant LOLP at bus 6 for case 3

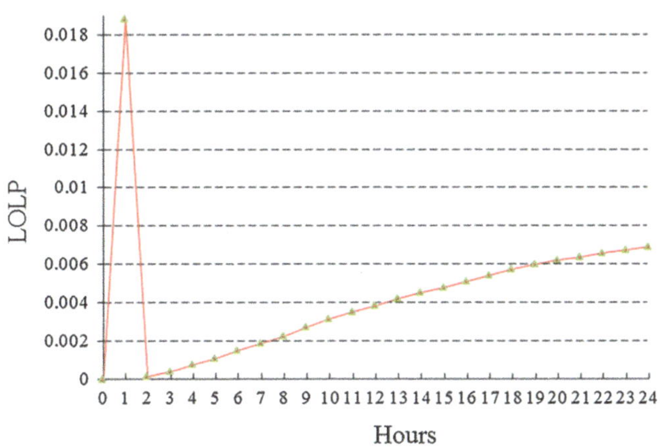

Fig. 12 Customers' instant LOLP at bus 20 for case 3

the *LOLP* for bus 6 from 1.881% at $t = 1$ h to 0.0189% at $t = 2$ h. Customers' *LOLP* for bus 6 at $t = 24$ h is 0.6895%, which decreases about 98.2% compared with that in case 2. The instant *LOLP* for bus 20 has the almost same pattern and values as those for bus 6. The system *EENS* and *LOLE* are 20.97 MWh and 0.0996 hours, which decreases about 88.6% and 98.3% compared with those in case 2.

6 Conclusions

In the last four decades, reliability evaluation techniques and reliability management strategies have been well developed and studied. These reliability evaluation techniques are more focused on planning and expansion of conventional power systems. However, the fast development and widely utilization of intermittent and fluctuating wind power generation have brought complexities for securing system balancing and maintaining system and customers' reliabilities during the operational phase. Moreover, in the last two decades, power systems have been restructured: electric energy and reserve are traded in different correlated markets. The restructuring of power systems has changed reliability management strategies fundamentally. Some existing and widely used criteria e.g., *N*-1 criterion may or may not be suitable for securing the reliable operation of power systems. In this paper, a new technique for assessing operational reliabilities of restructured power systems with high wind power penetration has been developed, which is based on reliability network equivalents and time-sequential simulation approaches. For reducing computational complexities, a simplified method has been developed and utilized in the simulation procedure. The proposed technique can be used to assistant system operator and market participants to assess their risks during system operation and make optimal decisions.

Acknowledgments This work was supported by Chinese National Natural Science Funds under Grant 51177078.

References

[1] Communication from the commission to the European parliament and the council, Renewable energy: Progressing towards the 2020 target. Commission staff working document. European Commission, Brussels, 2011

[2] Østergaard J, Foosnæs A, Xu Z et al (2009) Electric vehicles in power systems with 50% wind power penetration: the Danish case and the EDISON programme. In: Proceedings of the European conference on smart grids and mobility, Würzburg, 16–17 Jun 2009

[3] Dobakhshari AS, Fotuhi-Firuzabad M (2009) A reliability model of large wind farms for power system adequacy studies. IEEE Trans Energy Convers 24(3):792–801
[4] Ding Y, Pineda S, Nyeng P et al (2013) Real-time market concept architecture for EcoGrid EU—a prototype for European smart grids. IEEE Trans Smart Grid 4(4):2006–2016
[5] Karki R, Billinton R (2004) Cost-effective wind energy utilization for reliable power supply. IEEE Trans Energy Convers 19(2):435–440
[6] Karki R, Hu P, Billinton R (2006) A simplified wind power generation model for reliability evaluation. IEEE Trans Energy Convers 21(2):533–540
[7] Wang P, Billinton R (2001) Time sequential simulation technique for rural distribution system reliability cost / worth evaluation including wind generation as an alternative supply. IEE Proc Gener Transm Distrib 148(4):355–360
[8] Wangdee W, Billinton R (2006) Considering load-carrying capability and wind speed correlation of WECS in generation adequacy assessment. IEEE Trans Energy Convers 21(3):734–741
[9] Ding Y, Wang P, Goel L et al (2011) Long term reserve expansion of power systems with high wind power penetration using universal generating function methods. IEEE Trans Power Syst 26(2):766–774
[10] Lisnianski A, Ding Y (2009) Redundancy analysis for repairable multi-state system by using combined stochastic processes methods and universal generating function technique. Reliab Eng Syst Safe 94(11):1788–1795
[11] Billinton R, Allan RN (1996) Reliability evaluation of power systems. Plenum Press, New York
[12] Wang P, Ding Y, Goel L (2009) Reliability assessment of restructured power systems using optimal load shedding technique. IET Proc Gener Transm Distrib 3(7):628–640
[13] Salehi-Dobakhshari A, Fotuhi-Firuzabad M (2011) Integration of large-scale wind farm projects including system reliability analysis. IET Renew Power Gener 5(1):89–98
[14] Lisnianski A, Frenkel I (2010) Ding Y (2010) Multi-state system reliability analysis and optimization for engineers and industrial managers. Springer, London
[15] Leite AP, Borges CLT, Falcão DM (2006) Probabilistic wind farms generation model for reliability studies applied to Brazilian sites. IEEE Trans Power Syst 21(4):1493–1501
[16] Giorsetto P, Utsurogi KF (1983) Development of a new procedure for reliability modeling of wind turbine generators. IEEE Trans Power Appar Syst 102(1):134–143
[17] Nord Pool Spot (2014)
[18] Ding Y, Wang P, Goel L et al (2007) Reliability assessment of restructured power systems using reliability network equivalent and pseudo-sequential simulation techniques. Electr Power Syst Res 77(12):1665–1671
[19] Goel L, Viswanath PA, Wang P (2004) Monte Carlo simulation based reliability evaluation in a multi-bilateral contracts market. IEE Proc Gener Transm Distrib 151(6):728–734
[20] Wang P, Billinton R (2003) Reliability assessment of a restructured power system using reliability network equivalent techniques. IEE Proc Gener Transm Distrib 150(5):555–560
[21] Ding Y, Wang P (2006) Reliability and price risk assessment of a restructured power system with hybrid market structure. IEEE Trans Power Syst 21(1):108–116
[22] Billinton R, Li W (1993) A system state transition sampling method for composite system reliability evaluation. IEEE Trans Power Syst 8(3):761–770
[23] Lisnianski A, Elmakias D, Ben Haim H (2012) A multi-state Markov model for a short-term reliability analysis of a power generating unit. Reliab Eng Syst Safe 98(1):1–6
[24] Ortega-Vazquez MA, Kirschen DS (2008) Optimising the spinning reserve requirements considering failures to synchronise. IET Proc Gener Transm Distrib 2(5):655–665
[25] Reliability Test System Task Force of the Application of Probability Methods Subcommittee (1979) IEEE reliability test system. IEEE Trans Power Appar Syst 98(6):2047–2054
[26] Melo ACG, Pereira MV, Leite da Silva AM (1992) Frequency and duration calculations in composite generation and transmission reliability evaluation. IEEE Trans Power Syst 7(2): 469–476
[27] Leite da Silva AM, da Fonseca Manso L A, de Oliveira Mello J C et al (2000) Pseudo-chronological simulation for composite reliability analysis with time varying loads. IEEE Trans Power Syst 15(1):73–80

Yi DING received the B.Eng. degree from Shanghai Jiaotong University, China, and the Ph.D. degree from Nanyang Technological University (NTU), Singapore, both in electrical engineering. He is a Professor in the College of Electrical Engineering, Zhejiang University (ZJU), China. Before he Joined in ZJU, he was an Associate Professor in the Department of Electrical Engineering, Technical University of Denmark (DTU), Denmark. He also held research and teaching positions in University of Alberta, Canada and NTU. He was a Consultant as Energy Economist for Asian Development Bank in 2010. He is editorial member of international journals of Electric Power System Research and Journal of Modern Power Systems and Clean Energy. He is also a guest editor for the special section of IEEE Trans. on Power Systems. Dr. Ding is member of IEC working groups for micro-grid standards. His research areas include power system planning and reliability evaluation, smart grid and complex system risk assessment.

Lin CHENG was born in 1973. He received the B.S. degree in electrical engineering from Tianjin University, Tianjin, China, in 1996, and the Ph.D. degree in electrical engineering from Tsinghua University, Beijing, China, in 2001. Currently, he is an Associate Professor with the State Key Laboratory of Power Systems, Department of Electrical Engineering, Tsinghua University. His research interests include the areas of power system reliability and planning, power system dynamics and control based on wide-area measurement systems (WAMS).

Yonghong ZHANG received the Ph.D degree from Shanghai Jiaotong University, China in 2005. He had been a research fellow in the University of Alberta, Canada from 2007 to 2008. From 2009, he has been a professor in the Nanjing University of Information Science & Technology, China. His research areas include smart energy systems, artificial intelligence and information technology.

Yusheng XUE received his Ph.D. degree in Electrical Engineering from the University of Liege (Belgium) in 1987. He became a Member of Chinese Academy of Engineering in 1995. He is now the Honorary President of State Grid Electric Power Research Institute (SGEPRI), State Grid Corporation of China. His research interests include nonlinear stability, control and power system automation.

6

Reliability assessment in active distribution networks with detailed effects of PV systems

Zulu ESAU, Dilan JAYAWEERA (✉)

Abstract With the increased integration of photovoltaic (PV) power generation into active distribution networks, the operational challenges and their complexities are increased. Such networks need detailed characterization PV systems under multiple operating conditions to understand true impacts. This paper presents a new mathematical model for a PV system to capture detailed effects of PV systems. The proposed model incorporates state transitions of components in a PV system and captures effects of insolation variations. The paper performs a reliability assessment incorporating PV systems at resource locations. The results suggest that the reliability performance of an active distribution network at stressed operating conditions might be influenced by the high penetration of PV system. Reliability performance of distribution networks that are high penetrated with PV systems can be affected by cloudy effects resulting from insolation variations. The proposed model can be used to quantify the level of reduction in reliability, resulting with cloudy effects.

Keywords Distribution network planning, Insolation variation, Mathematical model, Strategic integration

1 Introduction

Traditional power distribution systems were principally passive because the networks only consumed power from the conventional generators through the transmission lines. However, the modern distribution networks have the ability to generate power through the use of distributed generators [1]. PV systems can be integrated into distribution networks as micro grids or standalone units, depend on the regulatory framework and constraints in the country.

A micro-grid is a cluster of distributed generators and loads operating under different conditions [2]. The micro-grids are typically designed for bi-mode of operations: the first mode is the grid connected mode and the other is the stand alone or islanded mode. When operating in the grid connected mode, the micro-grid interchangeably imports and exports power to the utility grid as per the operating criteria. The distribution networks, which are connected with embedded generators and active controls, are termed as an active distribution network (ADN).

There are different types of micro grids that are operated using different types of generating technologies. In the recent past years, the micro grids that are driven through PV systems evolved considerably [3]. Some of published literature reports that PV micro-grids improves the network reliability [4], however, such micro grids are not always economically viable. Some of the other published literature looked at economical deployment of micro-grids with improved reliability [5]. The PV micro-girds without energy storage mediums are capable of exporting power during the day and shut down during the night. Such situations can lead to bidirectional power flows and unwanted congestion of lines [6]. The rapid growth of PV micro-grids has been motivated by many factors including pollution reduction in the environment and reduced customer interruptions. This growth has brought many challenges and invoked significant researches [7]. The penetration of PV generation through micro-grids poses a number of challenges in the evaluation of reliability [8]. Integration of distributed generation also requires the capacity assessment of a distribution network incorporating uncertainties [9]. Distributed generation can be integrated into distribution networks by prioritizing different aspects, including their values [10].

Z. ESAU, D. JAYAWEERA, Department of Electrical and Computer Engineering, Curtin University, Perth, WA, Australia
(✉) e-mail: dilan.jayaweera@ieee.org

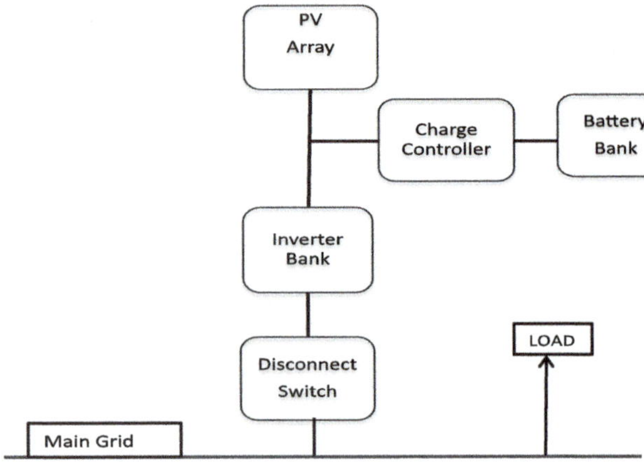

Fig. 1 Abstract model of a PV system

This paper proposes a new mathematical model that captures intermittent effects, uncertain operating conditions, and component based operation of PV systems. The model can be incorporated for the reliability planning of an active distribution network. The major benefit of the proposed model is its adaptability and flexibility to capture the detailed effects of varying insolation levels through stochastic transitions.

2 Proposed model for a PV system

A PV system can be comprised of solar power conversion components (PV arrays), the energy storage units (batteries) and the power conditioning units (inverters) [11]. Figure 1 shows an abstract model of a PV system.

2.1 Stochastic model for the PV array

The output power of the PV generation is constantly changing with changes in insolation levels during a day [11], which results in operational complications such as bidirectional power flows, voltage rise effects, and congestion. A PV array subsystem can be accurately represented by a three state model. One state for fully operational or UP state, one for fully non-operational or DOWN state, and another for partially operational or DERATED state [12].

2.2 Stochastic model for storage (battery bank)

Figure 2 shows the stochastic model for the storage facility where λ and μ terms give failure and repair rates correspond to the transitions, respectively. The subscripts b and c denote battery storage and charger controller respectively.

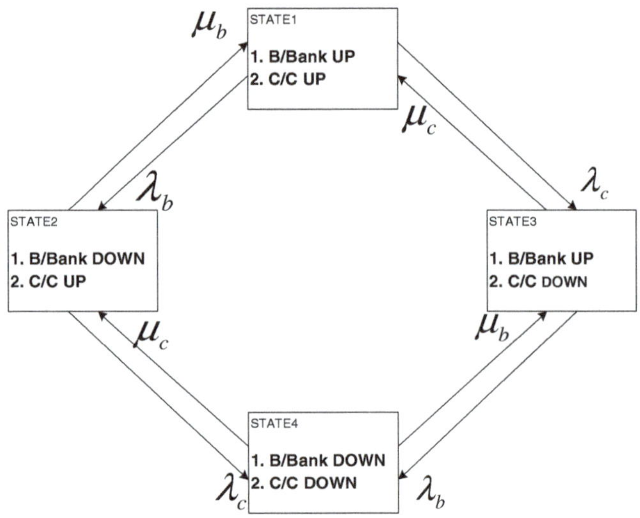

Fig. 2 Stochastic model for storage facility

The storage facility has two important components, which are the battery bank and the charge controller. The effect of other components of this subsystem can be considered negligible for the reliability studies. Figure 2 shows the Markov model for the battery bank (B/Bank) and charge controller (C/C) with four states. Both the battery bank and the charge controller have two operational states: the up or fully operational state and the down or the fully non-operational states.

Since two components are connected in series, the storage system is only operational if both the battery bank and the charge controller are fully operational. Therefore, the system will operate if, and only if, it is in STATE1 otherwise it will be in 'DOWN state'. Equation (1) gives the stochastic transitional probability matrix for the architecture shown in Fig. 2.

$$\boldsymbol{P}_{cb} = \begin{bmatrix} 1-\lambda_b-\lambda_c & \lambda_c & \lambda_b & - \\ \mu_c & 1-\lambda_b-\mu_c & - & \lambda_b \\ \mu_b & - & 1-\mu_b-\lambda_c & \lambda_c \\ - & \mu_b & \mu_c & 1-\mu_b-\mu_c \end{bmatrix}. \tag{1}$$

The state probability matrix shows the probability of the system residing in one of the states shown in Fig. 2. As a series configuration, the subsystem is only operational if both components are operable. Therefore, STATES2, 3 and 4 (being failure states) can be represented by absorbing states which results (2) as a truncated matrix.

$$\boldsymbol{Q} = [1-\lambda_b-\lambda_c]. \tag{2}$$

The mean time to failure (MTTF) for this subsystem can be calculated as per the steps given in (3)–(4).

$$t_{MTTF} = [\boldsymbol{I} - \boldsymbol{Q}]^{-1}, \tag{3}$$

where $\boldsymbol{I}$ is the identity matrix. Therefore,

$$t_{MTTF} = [1 - [1 - \lambda_b - \lambda_c]]^{-1} = [\lambda_b + \lambda_c]^{-1} = \frac{1}{\lambda_b + \lambda_c}. \tag{4}$$

Thus, the composite failure rate for this subsystem is determined using (5).

$$\lambda_{bc} = \frac{1}{t_{MTTF}} = \frac{1}{1/(\lambda_b + \lambda_c)} = \lambda_b + \lambda_c, \tag{5}$$

where λ_{bc} is the composite failure rate of the storage subsystem.

The mean time to repair (MTTR) is obtained by truncating the original stochastic probability matrix and eliminating the operational state as per (6)–(8). Thus, the composite mean time to repair for this subsystem is determined using (9).

$$t_{MTTR} = [\boldsymbol{I} - \boldsymbol{Q}_t]^{-1}, \tag{6}$$

where

$$\boldsymbol{Q}_t = \begin{bmatrix} 1 - \lambda_b - \mu_c & 0 & \lambda_b \\ 0 & 1 - \mu_b - \lambda_c & \lambda_c \\ \mu_b & \mu_c & 1 - \mu_b - \mu_c \end{bmatrix},$$

$$\boldsymbol{t}_{MTTR} = \left[\begin{bmatrix} 1 & 0 & 0 \\ 0 & 1 & 0 \\ 0 & 0 & 1 \end{bmatrix} - \begin{bmatrix} 1 - \lambda_b - \mu_c & 0 & \lambda_b \\ 0 & 1 - \mu_b - \lambda_c & \lambda_c \\ \mu_b & \mu_c & 1 - \mu_b - \mu_c \end{bmatrix}\right]^{-1}. \tag{7}$$

Equation (7) reduces to the following:

$$\boldsymbol{t}_{MTTR} = \begin{bmatrix} \lambda_b + \mu_c & 0 & -\lambda_b \\ 0 & \mu_b + \lambda_c & -\lambda_c \\ -\mu_b & -\mu_c & \mu_b + \mu_c \end{bmatrix}^{-1}, \tag{8}$$

$$t_{MTTR} = \frac{\lambda_b + \mu_b + \mu_c}{(\mu_b + \mu_c)^2}. \tag{9}$$

Then, the composite storage system repair rate is determined using (10).

$$\mu_{cb} = \frac{(\mu_b + \mu_c)^2}{(\lambda_b + \mu_b + \mu_c)}. \tag{10}$$

The battery bank and the charge controller can be modeled as a single component with a constant failure rate (λ_{bc}) and a constant repair rate (μ_{cb}).

2.3 Stochastic model for power conditioning unit

Figure 3 shows the Markov model for the power conditioning unit (PCU), where subscripts *inv* and *CB* give the inverter and disconnect switch respectively. The subsystem has an inverter and power disconnection switch (D/Switch) as the two most important components. Thus, its Markov model is similar to the storage subsystem. The resulting transitional probability matrix is given in (11).

$$\boldsymbol{P}_{inv} = \begin{bmatrix} 1 - \lambda_{inv} - \lambda_{CB} & \lambda_{CB} & \lambda_{inv} & - \\ \mu_{CB} & 1 - \lambda_{inv} - \mu_{CB} & - & \lambda_{inv} \\ \mu_{inv} & - & 1 - \mu_{inv} - \lambda_{CB} & \lambda_{CB} \\ - & \mu_{inv} & \mu_{CB} & 1 - \mu_{inv} - \mu_{CB} \end{bmatrix}. \tag{11}$$

From (11), the failure rate can be evaluated using (12).

$$\lambda_{invCB} = \frac{1}{t_{MTTR}} = \frac{1}{1/(\lambda_{inv} + \lambda_{CB})} = \lambda_{inv} + \lambda_{CB}, \tag{12}$$

and the repair rate is

$$\mu_{invCB} = \frac{(\mu_{inv} + \mu_{CB})^2}{\lambda_{inv} + \mu_{inv} + \mu_{CB}}. \tag{13}$$

2.4 Stochastic model for the entire PV system

Figure 4 shows the Markov model for the entire PV system. The composite failure and repair rates are defined as:

PV array $\lambda_{PV} = \lambda_1$, $\mu_{PV} = \mu_1$;
Battery bank/charge controller $\lambda_{cb} = \lambda_2$, $\mu_{cb} = \mu_2$;
Inverter/disconnect switch $\lambda_{invCB} = \lambda_3$, $\mu_{invCB} = \mu_3$.

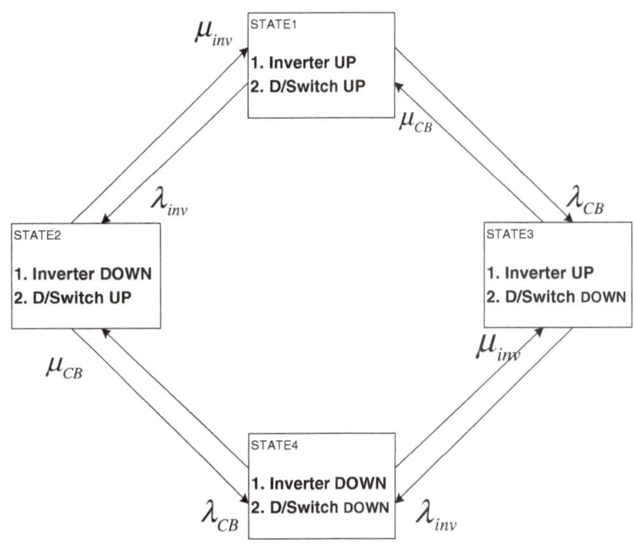

Fig. 3 Stochastic model for power conditioning unit (PCU)

The probability transitional matrix derived from the above diagram is shown in (14).

$$
P = \begin{bmatrix}
(1-\lambda_1-\lambda_2-\lambda_3) & \lambda_1 & \lambda_2 & \lambda_3 & 0 & 0 & 0 & 0 \\
\mu_1 & (1-\mu_1-\lambda_2-\lambda_3) & 0 & 0 & \lambda_2 & 0 & \lambda_3 & 0 \\
\mu_2 & 0 & (1-\mu_2-\lambda_1-\lambda_2) & 0 & \lambda_1 & \lambda_3 & 0 & 0 \\
\mu_3 & 0 & 0 & (1-\mu_3-\lambda_1-\lambda_2) & 0 & \lambda_2 & \lambda_1 & 0 \\
0 & \mu_2 & \mu_1 & 0 & (1-\mu_2-\mu_1-\lambda_3) & 0 & 0 & \lambda_3 \\
0 & 0 & \mu_3 & \mu_2 & 0 & (1-\mu_2-\mu_3-\lambda_1) & 0 & \lambda_2 \\
0 & \mu_3 & 0 & \mu_1 & 0 & 0 & (1-\mu_1-\mu_2-\lambda_2) & \lambda_1 \\
0 & 0 & 0 & 0 & \mu_3 & \mu_1 & \mu_2 & (1-\mu_1-\mu_2-\mu_3)
\end{bmatrix}. \tag{14}
$$

This matrix reduces to a truncated matrix involving the operational states of the system as in (15).

$$
Q_{PV} = \begin{bmatrix}
1-\lambda_1-\lambda_2-\lambda_3 & \lambda_1 & \lambda_2 \\
\mu_1 & 1-\mu_1-\lambda_2-\lambda_3 & 0 \\
\mu_2 & 0 & 1-\mu_2-\lambda_1-\lambda_2
\end{bmatrix}. \tag{15}
$$

Equation (15) is used for reliability analysis and to evaluate the reliability performance of an active distribution network.

The reliability assessment, which described in Sect. 3, incorporates de-rating of PV outputs into the calculations; however, it is not modelled within the Markov process of the PV system directly. The reason is that, unlike conventional power generating units, PVs are inherited with intermittent outputs and within the up-state of an intermittent output; the de-rating resulting through PV string failures can also be included. In this way, the size of the stochastic transitional matrix can be kept at a reasonable level. Alternatively, the Markov model given in Fig. 4 can also be extended to incorporate the de-rated state of the PV system; however; it will further complicate the stochastic transitional matrix, demanding an increased processing time for calculating the reliability indices. This is particularly the case with high penetration of dispersed PVs [13].

Stochastic generation output of the reliability calculation is modelled by using time series of intermittent power outputs, superimposed by de-rating effects of PV strings.

3 Calculation of reliability indices

The calculation of reliability indices involves the following steps.

Step 1: Model the distribution network base case.

Step 2: Define possible load demands or load forecast and growth scenarios.

Step 3: Characterise load points based on the type of customers (customer types include agricultural, domestic, commercial and industrial).

Step 4: Model component failures of buses, lines, cables, transformers, shunt devices, generators, protection/switch, and common mode as appropriate.

Step 5: Create the network condition with one or more simultaneous faults and specific load conditions.

Step 6: Establish the power balance with power flow analysis.

Step 7: Analyse the effects of failure of components through clearing faults by tripping of protection breakers or fuses, separating the faults by opening separating switches, restoring the power by closing normally open switches, alleviating overload by load transfer and load shedding, alleviating voltage constraint violations by load shedding at critical conditions. Thus, this step determines if system faults lead to load interruptions and if so, which loads are to be interrupted and how long. The loads can be shed in discrete blocks (e.g., 25%, 50%, 75%, and 100% of the base value), the minimum level considering priority, or the minimum level ignoring priority. The load shedding is attempted if the load transfer through switches of the distribution network is failed. The approach alleviates overload after power restoration with a minimum level of load shedding. It uses linear sensitivity indices to first select those loads with any contribution to overloading. A linear optimization is then used to find the best shedding option.

Step 8: Calculate load point reliability indices and then system level reliability indices. At first, the amount of load

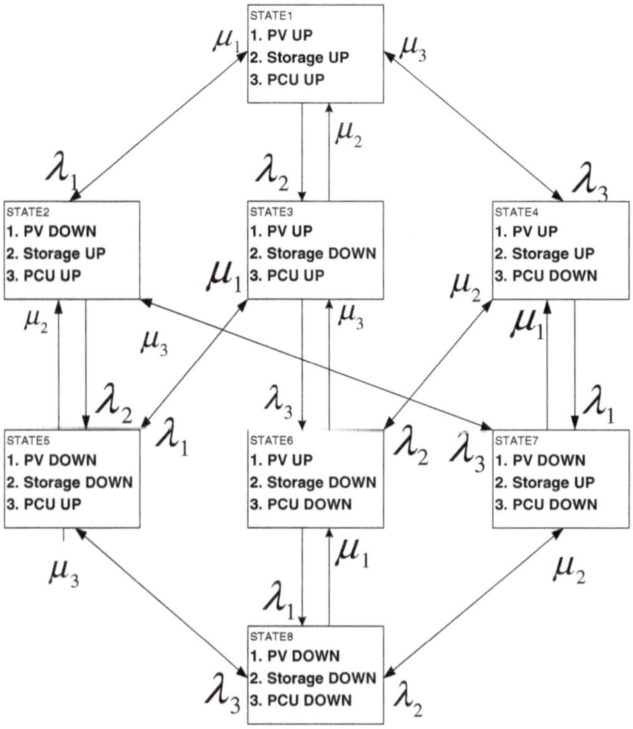

Fig. 4 Stochastic model for the PV system

shedding is multiplied by the duration of the system state to calculate the energy not supplied. The system energy not supplied for all possible system states is calculated after the entire system reliability assessment process is completed. CAIDI (customer average interruption duration index) and CAIFI (customer average interruption frequency index) are calculated taking into account duration of customers affected and respective frequencies. Similarly, SAIDI (system average interruption duration index) and SAIFI (system average interruption frequency index) are also calculated taking into account duration of customer interruptions of the system and the frequency of the system [14, 15].

4 Application of the proposed model

The proposed PV system model was applied to a modified RBTS (Roy Billinton Test System) model, shown in Fig. 5 [16]. This is a primary four radial feeders and secondary loop distribution network with a peak active power load of 20 MW. The network has a PV system maximum capacity of 5 MW. The reliability data of all the system equipment is obtained from the IEEE test system [17] and the RBTS test system [18–20]. However, the reliability data of the PV systems are obtained from [21]. Table 1 shows the definition of loads. The aggregate power production of each PV system and the reliability data is given in Table 2.

4.1 Case 1: varying load

The load is varied regardless of the integration of PV systems. It is varied from its base case level to 200% of the base case level. The results of SAIDI, SAIFI, and ENS (energy not supplied) are shown in Fig. 6a–c. Results suggest that the increase in load up to 40% does not affect the system reliability performance of this particular network considerably. However, after point A, a further increase in load results in a significant increase in the values of indices. This is due to the rapid increase in the total number of loads being shed off from the system coupled with a significant increase in the magnitude of the load.

In region L to K, there is a marked increase in ENS from 220 to 495 MWh/year. This represents a 125% increase in ENS for an increase of 100% in the load which is less than that obtained for the initial section. The increase in the load of the order of 175% of the base case load leads to overload system transformers and lines. This results in more loads being curtailed.

4.2 Case 2: varying PV generation output

The effect of hourly variation in the output of the PV generator on the network reliability is investigated and shown in Fig. 7a–d. Figure 7a shows active power output of the PV generators in the system in per unit on a 1 MVA base. The graph shows that how the power of the PV system varies during the day. The power production increases sharply to the maximum at 11 am and then drops steadily down to zero at around 9 pm.

Figure 7b shows the variation of SAIDI during a particular day with the same PV output power variation as shown in Fig.7a. It is seen that the SAIDI value improves from the initial value of 3.48 h, for the initial condition in which the PV output power is zero, to 2.8 h at point B when the outputs of the PV generators increase. This represents a steep improvement of 19.5% within 30 min of PV outputs increase from 0 to 0.15 pu.

SAIDI continues to improve at point B to point C with reduced steepness. At point C, the PV output power is the highest and thus the SAIDI is also at the maximum value. Compared to the initial value, SAIDI at this point is improved to 2.15 h representing a 38% improvement.

Between C and D, SAIDI remains substantially constant because of the outputs of the PV generators are almost constant. SAIDI increases gradually from D to E and

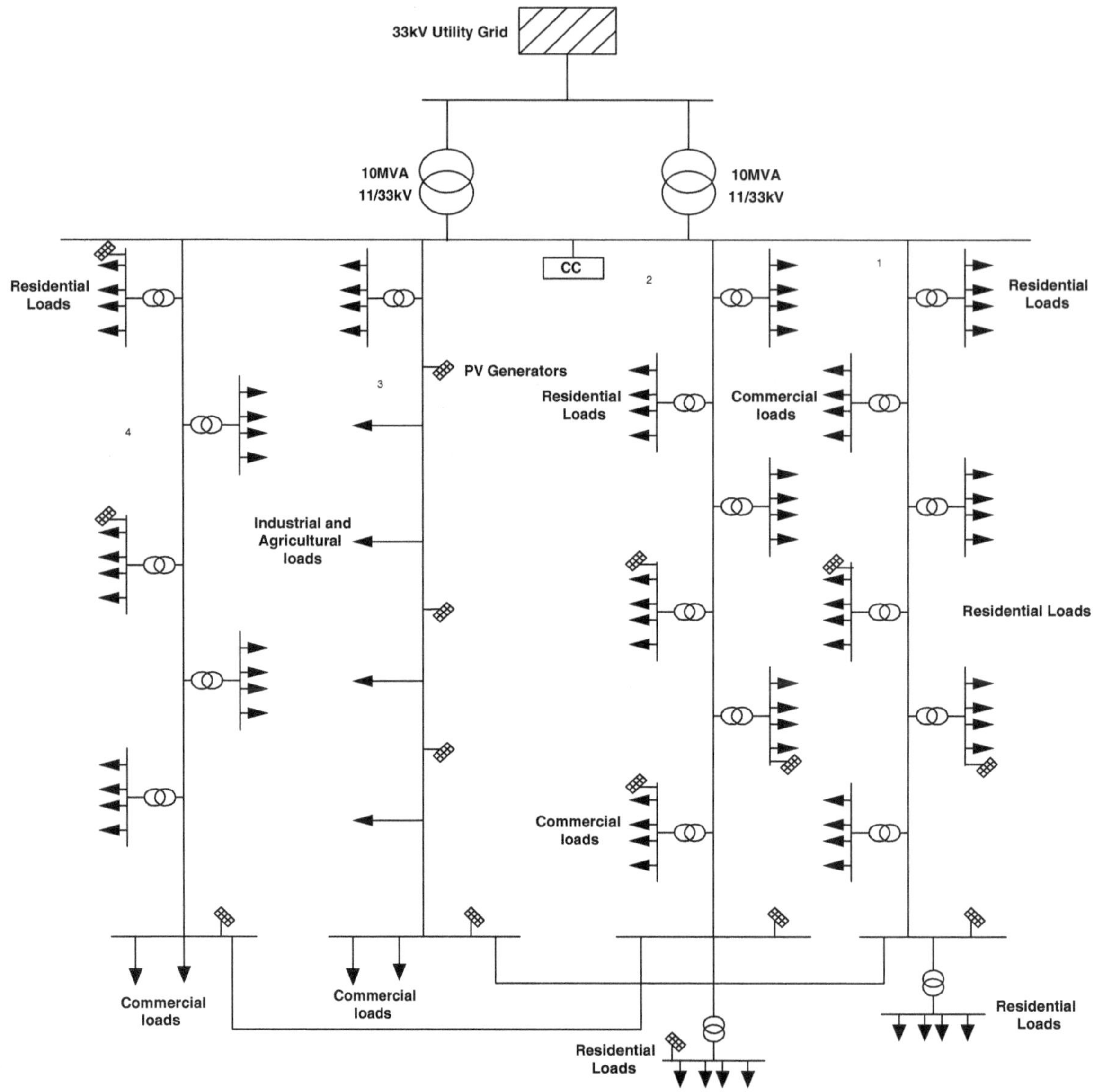

Fig. 5 Modified RBTS model

Table 1 Definition of loads

Load	Type	Number of customers	*P* (MW)	*Q* (Mvar)	Priority (%)	Shedding steps
Load1	Agricultural	100	0.69	0.025	80	10
Load2	Commercial	30	0.8	0.223	85	5
Load3	Commercial	50	1	0.215	85	5
Load4	Domestic	195	1.2	0.025	70	10
Load5	Domestic	200	1.2	0.025	70	10
Load6	Domestic	100	0.69	0.021	65	15
Load7	Industrial	25	5	0.981	95	5
Load8	Industrial	25	5	0.981	93	5
Load9	Industrial	25	5	0.981	95	5

Table 2 PV system data

Designation	Location	Capacity (MW)	Failure rate per year	Repair rate per year
PVG1	Feeder 1	1.5	0.1	18.25
PVG2	Feeder 2	1.5	0.1	18.25
PVG3	Feeder 3	1.0	0.1	18.25
PVG4	Feeder 4	1.0	0.1	18.25

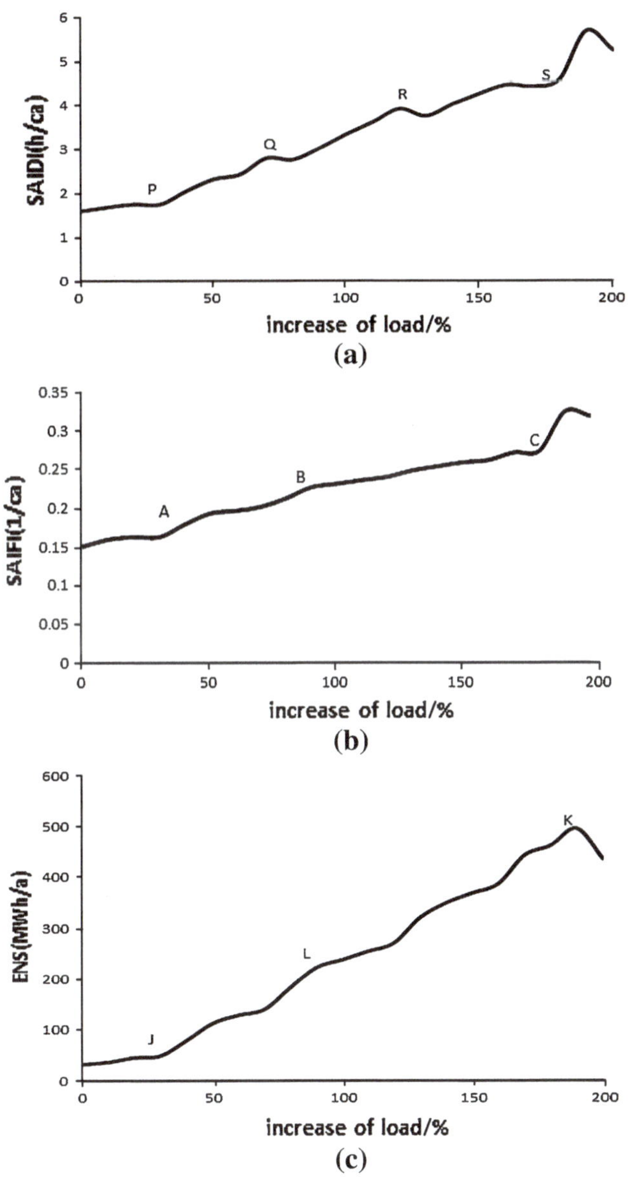

Fig. 6 Variation of index with load increases. **a** SAIDI variation with load increases, **b** SAIFI variation with load increases, **c** ENS variation with load increases

finally goes to the initial value at F when the insolation reduces to zero.

Figure 7c and d show the variation of CAIDI and ENS respectively with time of the day. The graphs are similar to that for SAIDI and as such follow the same trends as SAIDI.

4.3 Case 3: effect of cloud cover on reliability

Because of the power production capabilities of the PV generators are entirely dependent on the insolation in a particular area; the reliability of a distribution network with high PV penetration is heavily affected by insolation variations. The graph of Fig. 7a shows the output of the PV generators on a particular day, and it follows the insolation closely. One way in which the insolation is affected is by the clouds.

The results for the system under the influence of cloud cover are shown in Fig. 8a–c.

The entire results suggest that the reliability performance of an active distribution network can potentially be impacted by insolation variations, level of the stress of the network, and the level of penetration of PV generation. The proposed model is able to characterise the detailed transitions of states in a PV system that is a requirement for the assessment of true impacts in a distribution network.

5 Conclusions

A new mathematical model is proposed to assess the impacts on the reliability of an active distribution network with high penetration of PV power generation. The model, which is flexible, is able to capture intermittent outputs and uncertainties of PV power generating systems.

The study results suggest that PV systems impact less on the reliability performance of moderately loaded networks. In contrast, the highly stressed networks may have significant impacts from high penetration of PV systems. Cloudy effects resulting insolation variations can potentially impact the reliability performance in stressed networks that are high penetrated with PV power generation. The new model-based approach proposed in the paper can quantify the benefit thresholds in a distribution network with high penetration of PV power generation. Results also argue that PV systems can potentially improve the reliability of distribution networks if they are strategically integrated within thresholds.

The proposed model-based approach can be used as a benchmarking entity for the operational planning of a distribution network with high PV concentrations.

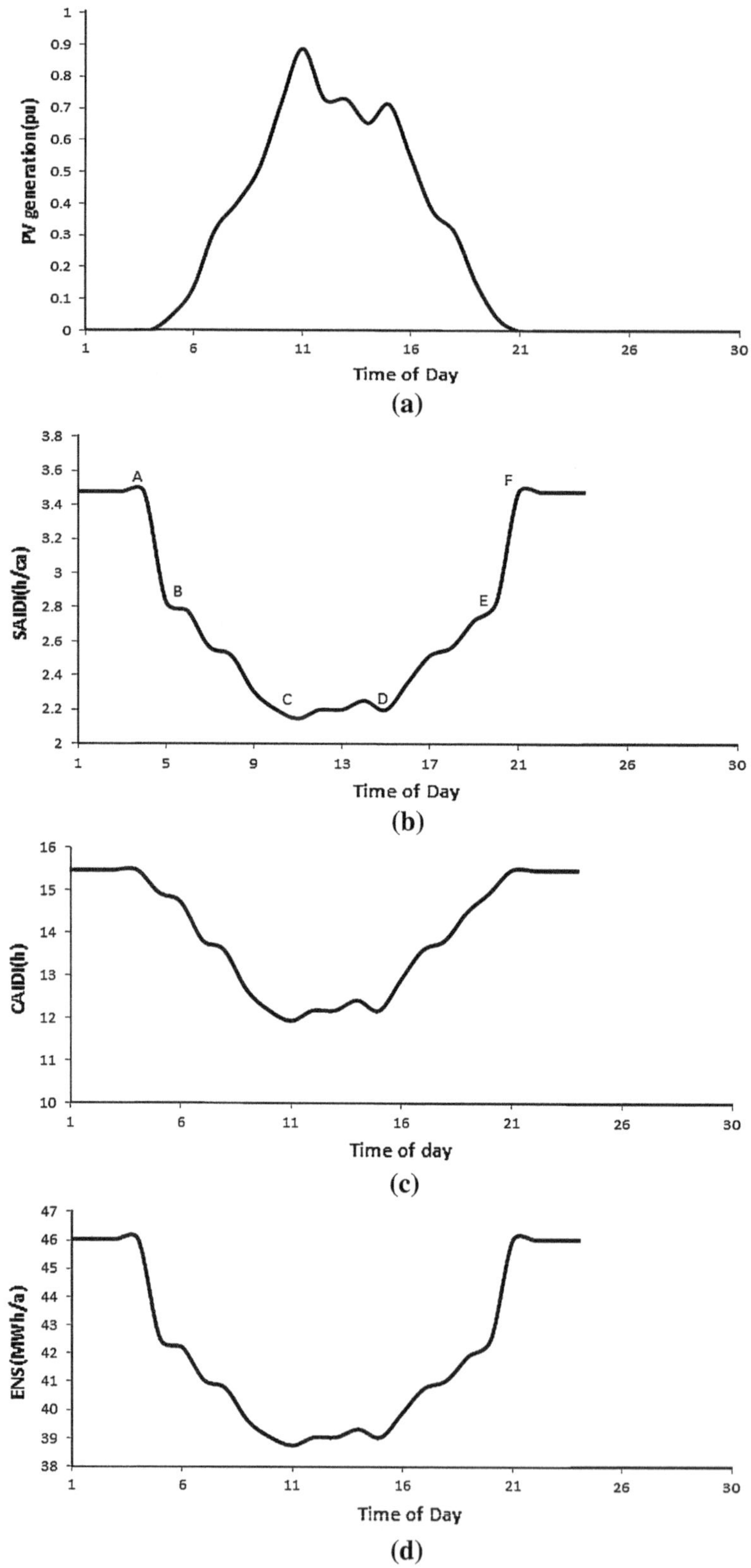

Fig. 7 Variation of power output and the indices of the PV system. **a** Power output variation of the PV system, **b** SAIDI against PV generation variation, **c** CAIDI against PV generation variation, **d** ENS against PV generation variation

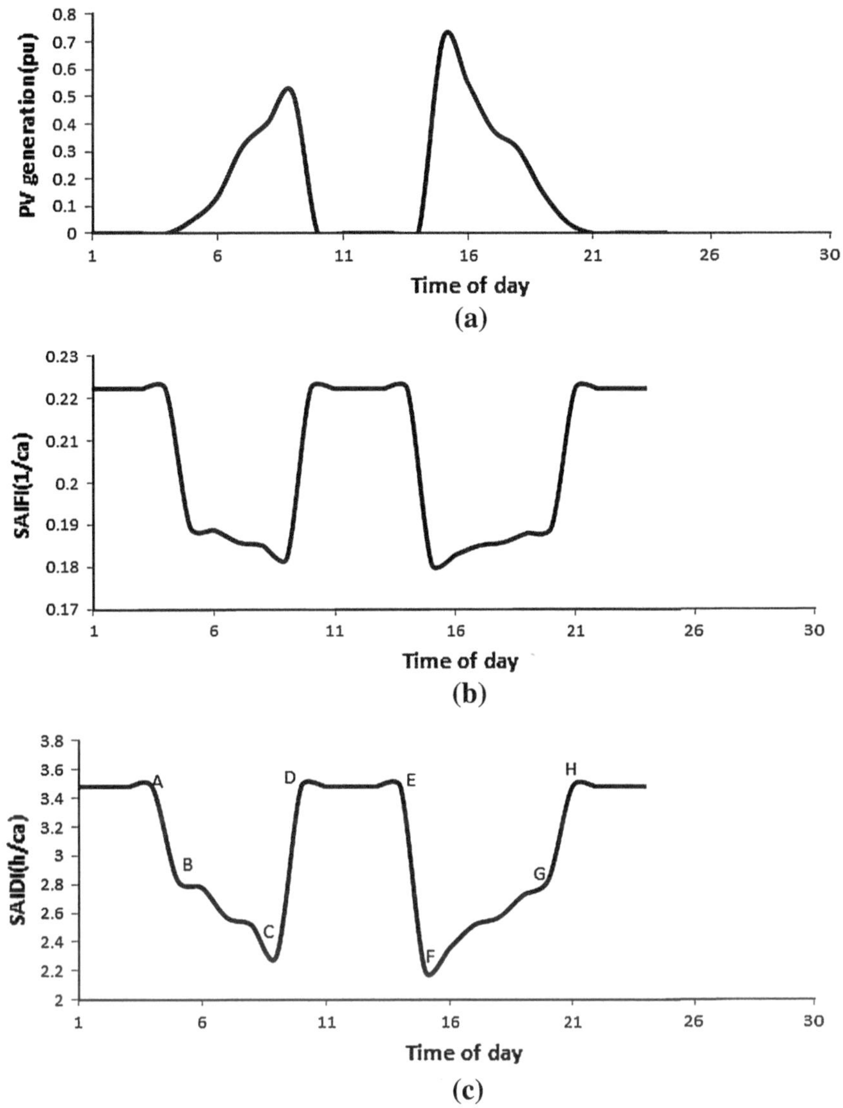

Fig. 8 Variation of PV generation and the indices on a cloudy day. **a** PV generation variation on a cloudy day, **b** SAIFI against PV generation variation on a cloudy day, **c** SAIDI against PV generation variation on a cloudy day

References

[1] Bie ZH, Zhang P, Li GF et al (2012) Reliability evaluation of active distribution systems including microgrids. IEEE Trans Power Syst 27(4):2341–2350
[2] Lasseter RH (2011) Smart distribution: coupled micro-grids. Proc IEEE 99(6):1074–1082
[3] Park J, Liang W, Choi J et al (2009) A probabilistic reliability evaluation of a power system including solar/photovoltaic cell generator. In: Proceedings of the 2009 IEEE power & energy society general meeting (PES'09), Calgary, Canada, 26–30 Jul 2009
[4] Zhang P, Li WY (2010) Boundary analysis of distribution reliability and economic assessment. IEEE Trans Power Syst 25(2):714–721
[5] Basu AK, Chowdhury S, Chowdhury SP (2010) Impact of strategic deployment of CHP-based DERs on microgrid reliability. IEEE Trans Power Deliv 25(3):1697–1705
[6] Bae IS, Kim JO (2008) Reliability evaluation of customers in a micro-grid. IEEE Trans Power Syst 23(3):1416–1422
[7] Sun Y, Bollen MHJ, Ault GW (2006) Probabilistic reliability evaluation for distribution systems with DER and microgrids. In: Proceedings of the international conference on probabilistic methods applied to power systems (PMAPS'06), Stockholm, Sweden, 11–15 Jun 2006
[8] Hlatshwayo M, Chowdhury S, Chowdhury SP, et al (2010) Impacts of DG penetration in the reliability of distribution systems. In: Proceedings of the 2010 international conference on power system technology (POWERCON'10), Hangzhou, China, 24–28 Oct 2010

[9] Jayaweera D, Islam S (2009) Probabilistic assessment of distribution network capacity for wind power generation integration. In: Proceedings of the Australasian universities power engineering conference (AUPEC'09), Adelaide, Australia, 27–30 Sept 2009
[10] Jayaweera D, Islam S (2010) Value based strategic integration of distributed generation. In: Proceedings of the Australasian universities power engineering conference (AUPEC'10), Christchurch, New Zealand, 5–8 Dec 2010
[11] Chowdhury S, Chowdhury SP, Crossley P (2009) Microgrids and active distribution networks. Institution of Engineering and Technology, Stevenage
[12] Billinton R, Ronald N (1983) Reliability evaluation of engineering systems—concepts and techniques, 2nd edn. Pitman Advanced Publishing, Boston
[13] Ding Y, Shen WX, Levitin G et al (2013) Economical evaluation of large-scale photovoltaic systems using universal generating function techniques. J Mod Power Syst Clean Energy 1(2):166–174
[14] DIgSILENT (2013) PowerFactory 15, tutorial. DIgSILENT GmbH, Gomaringen
[15] Billinton R, Ronald N (1996) Reliability evaluation of power systems, 2nd edn. Plenum Press, New York
[16] Billinton R, Jonnavinthula S (1996) A test system for teaching overall power system reliability assessment. IEEE Trans Power Syst 11(4):1670–1676
[17] Grigg C, Wong P, Albrecht P et al (1999) The IEEE reliability test system-1996: a report prepared by the reliability test system task force of the application of probability methods subcommittee. IEEE Trans Power Syst 14(3):1010–1020
[18] Billinton R, Kumar S, Chowdhury N et al (1989) A reliability test system for educational purposes-basic data. IEEE Trans Power Syst 4(3):1238–1244
[19] Billinton R, Kumar S, Chowdhury N et al (1990) A reliability test system for educational purposes-basic results. IEEE Trans Power Syst 5(1):319–325
[20] Allan RN, Billinton R, Sjarief I et al (1991) A reliability test system for educational purposes-basic distribution system data and results. IEEE Trans Power Syst 6(2):813–820
[21] Dhople SV, Davondi A, Chapman PL (2010). Integrating photovoltaic inverter reliability into energy yield estimation with Markov models. In: Proceedings of the IEEE 12th workshop on control and modeling for power electronics (COMPEL'10), Boulder, CO, USA, 28–30 Jun 2010

Zulu ESAU received the B.Eng. and M.Eng.Sc. degrees from Copperbelt University, Zambia and Curtin University, Australia in 2010 and 2013 respectively. His research interests include reliability of power systems, micro-grids, and integration of distributed generation into distribution networks.

Dilan JAYAWEERA received the M.Sc. and Ph.D. degrees in electrical power engineering from the University of Manchester Institute of Science and Technology (UMIST), Manchester, UK, in 2000 and 2003, respectively. He was with Imperial College, London and University of Strathclyde, Glasgow. Currently, he works at Curtin University, Australia as a Senior Lecturer. He is a senior member of the IEEE and a chartered engineer in the United Kingdom.

7

Optimal reactive power dispatch of wind power plant cluster considering static voltage stability for low-carbon power system

Shuo YANG (✉), Weisheng WANG, Chun LIU, Yuehui HUANG

Abstract The implementation of developing the wind power is an important way to achieve the low-carbon power system. However, the voltage stability issues caused by the random fluctuations of active power output and the irrational regulations of reactive power compensation equipment have become the prominent problems of the regions where large-scale wind power integrated. In view of these problems, this paper proposed an optimal reactive power dispatch (ORPD) strategy of wind power plants cluster (WPPC) considering static voltage stability for low-carbon power system. The control model of the ORPD strategy was built according to the wind power prediction, the present operation information and the historical operation information. By utilizing the automatic voltage control capability of wind power plants and central substations, the ORPD strategy can achieve differentiated management between the discrete devices and the dynamic devices of the WPPC. Simulation results of an actual WPPC in North China show that the ORPD strategy can improve the voltage control performance of the pilot nodes and coordinate the operation between discrete devices and the dynamic devices, thus maintaining the static voltage stability as well.

Keywords Low-carbon power system, Wind power plants cluster, Optimal reactive power dispatch, Wind power prediction, Static voltage stability

S. YANG, W. WANG, C. LIU, Y. HUANG, Renewable Energy Department, China Electric Power Research Institute, Beijing 100192, China
(✉) e-mail: yangshuo04cn@126.com

1 Introduction

Because of the worsening of global ecological problems, the implementation of reducing carbon emissions and achieving sustainable development has become the issue of common concern around the world. China is one of the major contributions to the recent emission surge, and the electricity sector is the largest single source of carbon emissions, emitting approximately 40% of China's CO_2 from fossil fuel combustion [1]. Therefore, the electricity sector development in China will be filled with great challenges in the low-carbon and clean era [2, 3].

The Chinese government has developed the renewable energy all over the country in order to achieve the low-carbon power system. Among various renewable energy sources, the wind power becomes the most rapidly growing one due to its rich resources and mature technology, and the wind power installed capacity in China has been more than 91.412 GW by the end of 2013 [4]. However, limited by the distribution of the wind resource, most of the large-scale wind power bases under construction or planning are located in the regions far away from load center, such as North China, Northeast China and Northwest China. These regions where large-scale wind power plant cluster (WPPC) integrated generally possess the weakly connected grid structure, and the voltage stability issues caused by the random fluctuations of active power output of the wind power plants have become the prominent problems in low-carbon energy generation systems [5–7].

In order to improve the voltage quality of the point of common coupling (PCC), many wind power plants are equipped with wide variety of reactive power regulation devices and their own auto-voltage control system, so they can supply active power and reactive power to the grid [8–11]. However, due to the non-simultaneous variation of the wind speeds of different wind power plants in the WPPC, the reactive power and voltage control strategy of each wind power plant, which only aims at the reactive power balance of the single PCC, cannot satisfy comprehensive voltage stability needs of the WPPC in whole. While the reality is that some wind power plants where the voltage stability margin (VSM) doesn't satisfy the security constraints have to limit their wind power output for ensuring the safe and stable operation of the WPPC. Therefore, the reactive power and voltage control of the WPPC have great significant for the large-scale effective utilization of this low-carbon energy.

In previous work, many scholars have investigated reactive power and voltage control strategy for wind power plants centralized region [12–17]. These studies mainly focus on the characteristics of reactive power compensation equipment in wind power plants and central substations, utilizing the operation information based on the time section that the voltage control cycle begin to decrease the power loss of WPPC [12–14] or to reduce the voltage deviation of the pilot nodes [15–17].

Actually, the WPPC is the generation system of intermittent low-carbon energy, and its reactive power dispatch has some particularities. On the one hand, it is much more difficult for the wind turbine generators (WTG) to maintain the active power output constant during the control cycle of reactive power dispatch than the conventional generators. Therefore, the reactive power dispatch strategy based on the operation information at one time section cannot guarantee the control performance during the whole control cycle. On the other hand, the discrete devices, such as shunt capacitor bank (SCB) and on load tap changer (OLTC), cannot smooth the rapid fluctuation of active power output due to the limit of their response time constant. Then, the frequent even irrational regulation of these discrete devices will induce the cascading trip-off of WTGs in the WPPC caused by the reactive power overcompensation or undercompensation.

This paper proposes an optimal reactive power dispatch (ORPD) strategy for the regions where large-scale WPPC integrated. In order to maintain sufficient VSM and restrict the frequent regulation of discrete devices of the WPPC during each control cycle, the control model for the strategy is established according to the wind power prediction, the present and the historical operation information. By implementing the differentiated management between the discrete devices and the dynamic devices, the strategy can achieve the coordination different reactive power regulation devices of the WPPC.

The remainder of this paper is organized as follows. The description of static voltage stability assessment of the WPPC is brought out in Section 2. Section 3 presents the technical aspects of the ORPD strategy. The mathematical model and solution algorithm are described in Section 4, and the simulation results are presented in Section 5. Finally, the conclusions are drawn in Section 6.

2 Static voltage stability assessment of WPPC

The static voltage stability is defined as an important subject in power systems and is considered in many planning and scheduling problems. In conventional power systems, the requirement of load is generally regarded as the driving force of voltage stability issue, and the static voltage stability can be established by the VSM index based on the P-V curve of the load and is expressed as [18, 19]:

$$\lambda = \frac{P_{\max} - P}{P} \tag{1}$$

where λ is the load factor; P is the active power of load in the actual operation point; $P_{\max}$ is the critical value of the load active power corresponding to the bifurcation point of the P-V curve which is calculated by a certain kind of ways to increase the load in the power system from the actual operation point.

However, the conventional VSM index expressed in (1) isn't suitable for the regions where large-scale wind power plants integrated due to the random fluctuation characteristic of wind power generation. On the one hand, the fluctuation of the wind power output is bidirectional, and the significant or sudden changes of the bus voltage profiles caused by both sides of the active power fluctuation will induce voltage instability to the WPPC. Therefore, the VSM index only considering the increase of the active power output is inadequate for the WPPC voltage stability analysis. On the other hand, the active power output of the wind power plant is sometimes close to zero, and this will lead to the infinite calculation result of the index expressed in (1).

Based on the above analysis, the up-fluctuation factor λ^{u} and down-fluctuation factor λ^{d} are introduced in this paper to estimate the VSM of the wind power plant in the WPPC, which can be defined as:

$$\lambda^{\mathrm{u}} = \frac{P^{\mathrm{U}} - P^{\mathrm{N}}}{P_{\mathrm{CAP}}} \geq \lambda^{\mathrm{ulim}} \tag{2}$$

$$\lambda^{\mathrm{d}} = \frac{P^{\mathrm{D}} - P^{\mathrm{N}}}{P_{\mathrm{CAP}}} \leq \lambda^{\mathrm{dlim}} \tag{3}$$

where P^{U} and P^{D} are the critical values of the active power output of the wind power plant in up-fluctuation state and down-fluctuation state, respectively; P^{N} is the active power output of the wind power plant in the actual state; P_{CAP} is the installed capability of the wind power plant; λ^{ulim} and λ^{dlim} are the preset thresholds of VSM for up-fluctuation state and down-fluctuation state respectively. It is obvious that $\lambda^{\mathrm{u}} \in [0, 1]$ and $\lambda^{\mathrm{d}} \in [-1, 0]$.

3 Technical aspects of ORPD strategy of WPPC

Random fluctuations of active power output and wide variety of reactive power regulation devices are some of the typical characteristics of the regions where large-scale wind power plants are integrated. In order to solve the voltage issues caused by these characteristics, the ORPD strategy of the WPPC is in accordance with the wind power prediction information, the present operation information and the historical operation information, and the automatic voltage control capability of wind power plants, and central substations is also taken into account. The overall framework of the ORPD strategy of WPPC is shown in Fig. 1. Specifically, the main technical aspects of the proposed ORPD strategy of WPPC include the utilization of the wind power prediction, the regulation restrict of the discrete devices and the differentiated management between different devices.

3.1 Utilization of wind power prediction

In order to avoid the significant or sudden changes of the bus voltage profiles caused by the active power fluctuation and enhance the static voltage stability of the wind power plants during $t \sim t+1$, the wind power prediction at $t+1$ is taken into account when the OPRD control scheme at t is made in the proposed strategy. According to the definitions of VSM in (2) and (3), by considering the wind power prediction error at $t+1$, the minimum VSM for up-fluctuation state and down-fluctuation state of wind power plant i at t is described in red and blue grid patterns, respectively, as shown in Fig. 1, which can be determined as:

$$\lambda_{i(t)}^{\text{ulim}} = \frac{\max\{P_{\mathrm{P}_i(t+1)} + \xi_{i(t+1)}^{\text{u}} P_{\mathrm{CAP}_i}, P_{\mathrm{W}_i(t)}\} - P_{\mathrm{W}_i(t)}}{P_{\mathrm{CAP}_i}} \tag{4}$$

$$\lambda_{i(t)}^{\text{dlim}} = \frac{\min\{P_{\mathrm{P}_i(t+1)} - \xi_{i(t+1)}^{\text{d}} P_{\mathrm{CAP}_i}, P_{\mathrm{W}_i(t)}\} - P_{\mathrm{W}_i(t)}}{P_{\mathrm{CAP}_i}} \tag{5}$$

where $P_{\mathrm{W}_i(t)}$ is the actual wind power output of wind power plant i at time section t; $P_{\mathrm{P}_i(t+1)}$ is the wind power prediction value of wind power plant i at $t+1$; $\xi_{i(t+1)}^{\text{u}}$ and $\xi_{i(t+1)}^{\text{d}}$ are the prediction error for up-fluctuation state and down-fluctuation state of wind power plant i at time section $t+1$ under some confidence level respectively.

In (4), the max{ } means that the up-fluctuation state wind power at t is no less than the current state value; while in (5), the min{ } means that the up-fluctuation state wind power should be no more than the current state value. In such a case, the up-fluctuation factor λ^{u} and down-fluctuation factor λ^{d} introduced in Section 2 can meet the rules, that is, $\lambda^{\text{u}} \in [0,1]$ and $\lambda^{\text{d}} \in [-1,0]$.

3.2 Regulation restriction of discrete devices

In the proposed strategy, the historical operation information is utilized to reduce the disturbances caused by the frequent or irrational regulation of the discrete reactive power regulation devices. Based on the operational position of SCBs or OLTCs at $t-1$, the regulating times of these discrete devices during $t \sim t+1$ are converted to the operational cost which is regarded as one of the control objectives.

3.3 Differentiated management between different devices

In order to prevent the inappropriate reactive power compensation due to the frequent or irrational regulation of the discrete reactive power regulation devices, this paper implements differentiated management between the discrete devices and the dynamic devices of the WPPC. The discrete devices in central substations and wind power plant stations can adopt the constant reactive power control method, responding the tap position command during the control cycle. The dynamic devices in wind power plants can adopt the voltage control method, thus maintaining the PCC voltage within the allowed deviation range of the reference value by utilizing the automatic voltage control capability of the wind power plant.

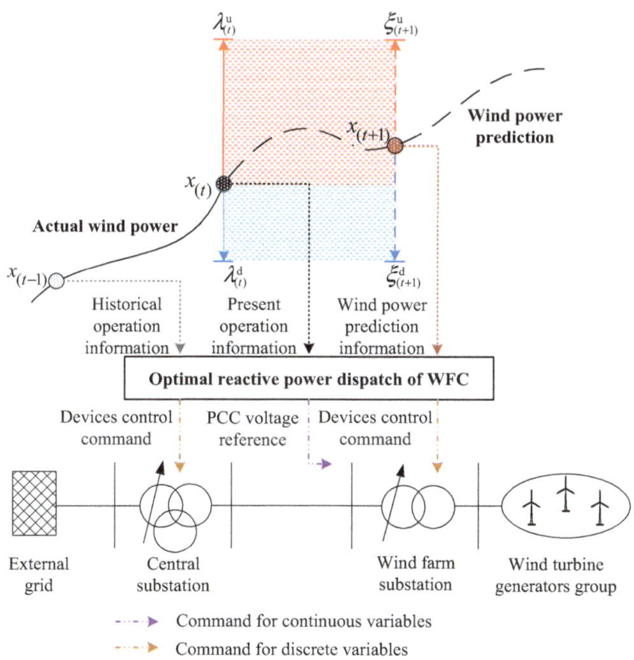

Fig. 1 Overall framework of the ORPD strategy of WPPC

4 Optimization formulations

4.1 Objective function

In this ORPD strategy, the total voltage deviation of pilot nodes of WPPC is defined as the deviation index, the total reactive power reserve of the dynamic devices of WPPC is defined as the reserve index, and the total regulating times of the discrete devices of WPPC are defined as the economic indexes. Therefore, the objective function can be expressed as:

$$\min f(\boldsymbol{x}) = \omega_v f_v(\boldsymbol{x}) + \omega_q f_q(\boldsymbol{x}) + \omega_n f_n(\boldsymbol{x}) \tag{6}$$

where $f_v(\boldsymbol{x})$, $f_q(\boldsymbol{x})$ and $f_n(\boldsymbol{x})$ are the deviation, the reserve and the economic indexes of the ORPD strategy, respectively; ω_v, ω_q and ω_n are the weight coefficients of the three indexes, respectively.

1) Deviation index

The deviation index in (6) can be expressed as:

$$f_v(\boldsymbol{x}) = \alpha\|\Delta \boldsymbol{V}_P\|^2 + \alpha^u\|\Delta \boldsymbol{V}_P^u\|^2 + \alpha^d\|\Delta \boldsymbol{V}_P^d\|^2 \tag{7}$$

where

$$\|\Delta \boldsymbol{V}_P\|^2 = \sum_{i=1}^{N_P} \Delta V_{P_i}^2 = \sum_{i=1}^{N_P} (V_{P_i} - V_{P_i}^{ref})^2 \tag{8}$$

$$\|\Delta \boldsymbol{V}_P^u\|^2 = \sum_{i=1}^{N_P} \Delta V_{P_i}^{u2} = \sum_{i=1}^{N_P} (V_{P_i}^u - V_{P_i}^{ref})^2 \tag{9}$$

$$\|\Delta \boldsymbol{V}_P^d\|^2 = \sum_{i=1}^{N_P} \Delta V_{P_i}^{d2} = \sum_{i=1}^{N_P} (V_{P_i}^d - V_{P_i}^{ref})^2 \tag{10}$$

In (7)–(10): $i \in \boldsymbol{N}_P$, $\boldsymbol{N}_P$ is the set of pilot node of WPPC; V_{P_i}, $V_{P_i}^{ref}$ are the actual and reference voltages of the pilot node i, respectively; $V_{P_i}^u$, $V_{P_i}^d$ are the voltages of pilot node i in up-fluctuation state and down- fluctuation state, respectively; α, α^u and α^d are the acceleration factors for the three different operation states, respectively.

2) Reserve index

The reserve index in (6) can be expressed as:

$$f_q(\boldsymbol{x}) = \|\Delta \boldsymbol{Q}_S\|^2 = \sum_{i=1}^{N_Q} \left(\frac{Q_{S_i}}{Q_{S_{imax}} - Q_{S_{imin}}}\right)^2 \tag{11}$$

where $i \in \boldsymbol{N}_Q$, $\boldsymbol{N}_Q$ is the set of dynamic reactive power devices of the WPPC; Q_{S_i}, $Q_{S_{imax}}$, and $Q_{S_{imin}}$ are the actual, the maximum and the minimum reactive power outputs of dynamic device i, respectively.

3) Economic index

The economic index in (6) can be expressed as:

$$f_n(\boldsymbol{x}) = \sum_{i=1}^{N_T} S_{T_i}\Delta u_{T_i} + \sum_{j=1}^{N_C} S_{C_j}\Delta u_{C_j} \tag{12}$$

where $i \in \boldsymbol{N}_T$, $j \in \boldsymbol{N}_C$, $\boldsymbol{N}_T$ and $\boldsymbol{N}_C$ are the sets of OLTCs and SCBs of the WPPC, respectively; Δu_{T_i} and S_{T_i} are the adjustment variation and the cost of unit variation of OLTC i, respectively; Δu_{C_j} and S_{C_j} are the adjustment variation and the cost of unit variation of SCB j, respectively.

4.2 Constraints of control strategy

The constraints of this control strategy include the power flow, the static voltage stability and the operating constraints.

1) Power flow constraint

The power flow constraints in the actual state, the up-fluctuation state and the down-fluctuation state can be expressed as:

$$\begin{cases} P_{W_i} - V_i \sum_{j=1}^{N_S} V_j(G_{ij}\cos\theta_{ij} + B_{ij}\sin\theta_{ij}) = 0 \\ Q_{W_i} - V_i \sum_{j=1}^{N_S} V_j(G_{ij}\sin\theta_{ij} - B_{ij}\cos\theta_{ij}) = 0 \end{cases} \tag{13}$$

$$\begin{cases} \lambda_i^u P_{CAP_i} + P_{W_i} - V_i^u \sum_{j=1}^{N_S} V_j^u(G_{ij}\cos\theta_{ij}^u + B_{ij}\sin\theta_{ij}^u) = 0 \\ Q_{W_i}^u - V_i^u \sum_{j=1}^{N_S} V_j^u(G_{ij}\sin\theta_{ij}^u - B_{ij}\cos\theta_{ij}^u) = 0 \end{cases} \tag{14}$$

$$\begin{cases} \lambda_i^d P_{CAP_i} + P_{W_i} - V_i^d \sum_{j=1}^{N_S} V_j^d(G_{ij}\cos\theta_{ij}^d + B_{ij}\sin\theta_{ij}^d) = 0 \\ Q_{W_i}^d - V_i^d \sum_{j=1}^{N_S} V_j^d(G_{ij}\sin\theta_{ij}^d - B_{ij}\cos\theta_{ij}^d) = 0 \end{cases} \tag{15}$$

where $i \in \boldsymbol{N}_S$, $\boldsymbol{N}_S$ is the set of total nodes of WPPC; P_{W_i} and Q_{W_i} are the active and reactive power injected at node i in normal state, respectively; P_{CAP_i} is the installed capability of the wind power plant connected at node i; $Q_{W_i}^u$ and $Q_{W_i}^d$ are the reactive power injected at node i in up-fluctuation state and down-fluctuation state, respectively; θ_{ij}, θ_{ij}^u and θ_{ij}^d are the phase angle differences between nodes i and j for the three different operational states, respectively;

G_{ij} and B_{ij} are the real and imaginary parts of the ijth elements of the admittance matrix, respectively.

2) Static voltage stability constraint

In order to maintain sufficient VSM in each control cycle, the static voltage stability constraints should be satisfied. They can be expressed as:

$$\lambda_i^{\mathrm{u}} \geq \lambda_i^{\mathrm{ulim}} \tag{16}$$

$$\lambda_i^{\mathrm{d}} \leq \lambda_i^{\mathrm{dlim}} \tag{17}$$

where λ_i^{u} and λ_i^{d} are the up-fluctuation and the down-fluctuation factors of wind power plant i, respectively; $\lambda_i^{\mathrm{ulim}}$ and $\lambda_i^{\mathrm{dlim}}$ are the thresholds of VSM of wind power plant i in the up-fluctuation state and the down-fluctuation state, respectively, which are described in (4) and (5).

3) Operating constraints

The operating constraints are the upper and lower bounds of the control, and the state variables which can be expressed as:

$$\begin{cases} V_{i\min} \leq V_i \leq V_{i\max} \\ V_{i\min} \leq V_i^{\mathrm{u}} \leq V_{i\max} \quad i \in \boldsymbol{N}_{\mathrm{S}} \\ V_{i\min} \leq V_i^{\mathrm{d}} \leq V_{i\max} \end{cases} \tag{18}$$

$$\begin{cases} Q_{\mathrm{S}_{i\min}} \leq Q_{\mathrm{S}_i} \leq Q_{\mathrm{S}_{i\max}} \\ Q_{\mathrm{S}_{i\min}} \leq Q_{\mathrm{S}_i}^{\mathrm{u}} \leq Q_{\mathrm{S}_{i\max}} \quad i \in \boldsymbol{N}_{\mathrm{Q}} \\ Q_{\mathrm{S}_{i\min}} \leq Q_{\mathrm{S}_i}^{\mathrm{d}} \leq Q_{\mathrm{S}_{i\max}} \end{cases} \tag{19}$$

$$T_{\mathrm{K}_{i\min}} \leq T_{\mathrm{K}_i} \leq T_{\mathrm{K}_{i\max}} \quad i \in \boldsymbol{N}_{\mathrm{T}} \tag{20}$$

$$C_{\mathrm{P}_{i\min}} \leq C_{\mathrm{P}_i} \leq C_{\mathrm{P}_{i\max}} \quad i \in \boldsymbol{N}_{\mathrm{C}} \tag{21}$$

where $Q_{\mathrm{S}_i}^{\mathrm{u}}$ and $Q_{\mathrm{S}_i}^{\mathrm{d}}$ are the reactive power outputs of dynamic device i in the up-fluctuation state and the down-fluctuation state respectively; T_{K_i},$T_{\mathrm{K}_{i\max}}$ and $T_{\mathrm{K}_{i\min}}$ are the actual, the maximum and the minimum tap positions of transformer i. C_{P_i}, $C_{\mathrm{P}_{i\max}}$ and $C_{\mathrm{P}_{i\min}}$ are the actual, the maximum and the minimum banks of capacitor i.

4.3 Solving algorithm and control flow

The analytic hierarchy process (AHP) is a method for analyzing the multi-objective problems both quantitatively and qualitatively [20]. In this paper, the AHP is utilized to decide the weight coefficients of different indexes in the objective function (6). The weight coefficients can be decided when formulating the power system operating mode. Also, they can be decided by the dispatchers according to their operating experience during the process of ORPD.

The interior point filter algorithm (IPFA) has been concerned for possessing faster calculation speed and good robustness [21–23]. It is based on a primal dual interior point algorithm with a filter set which can be updated after every iteration. The filter can replace merit functions, avoiding the update of penalty parameters associated with the penalization of the constraints on merit functions. And the filter will be augmented after every iterative calculation, ensuring that the iteration point x_{k+1} cannot return to the neighborhood of the previous iteration point x_k. If the iteration point x_{k+1} does not satisfy the filter rule, the backtracking line-search procedure will be used to decrease the trial step sizes unless the iteration point x_{k+1} does not meet the filter rule. Over all, these procedures ensure that the algorithm cannot cycle between two points that alternatively decrease the constraint violation and the barrier objective function. In this paper, the mathematical model of WPPC ORPD strategy is solved by the IPFA. By using the IPFA, the discrete variables are adjusted continuously firstly, then the calculation results of discrete variables are fixed on the nearest rounding value. Finally, the optimization problem is calculated again and the discrete optimal solution will be found later.

The control flow chart of the proposed ORPD strategy for one control cycle is shown in Fig. 2. The large deviation between the preset prediction error and the real prediction

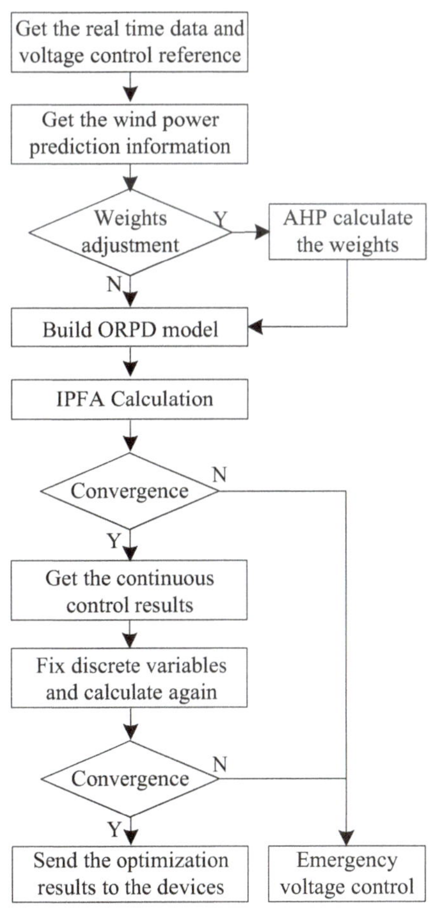

Fig. 2 Flow chart of the ORPD strategy for one control cycle

error will lead to inappropriate VSM thresholds and the optimization problem will be infeasible with the inappropriate static voltage stability constraints. This means that there is not enough excess reactive power in the WPPC to meet the request of the active power output at $t + 1$. In this situation, the WPPC ORPD system will give up this ORPD control module and turn to the emergency voltage control module whose control objective is to find the maximum active power out of the WPPC with the current reactive power reserve. Due to space limitations, the emergency voltage control module is not discussed in this paper.

5 Case studies

An actual WPPC system in North China is used to verify the validity of the proposed strategy. The structure of the simulation system is shown in Fig. 3. There are 14 wind power plants connected to 2 central substations, and the installed capability of the WPPC is 1,780.5 MW. The statistical information of installed capability and reactive power equipment in different regions is shown in Table 1.

Considering the influences of non-simultaneous variation on the wind speeds of different wind power plants, two different wind power output curves are used in the simulation. The wind power prediction and actual values of the wind power plants in different regions for 24 h are shown in Fig. 4, where the time resolution of the prediction value is 15 min and the sampling period of the actual value is 1 min. According to the historical statistics, the probability that the wind power prediction error of the wind power plants in the WPPC exceeds 15% is less than 2%.

In this simulation, two different ORPD modes of WPPC are utilized to compare the control performance:

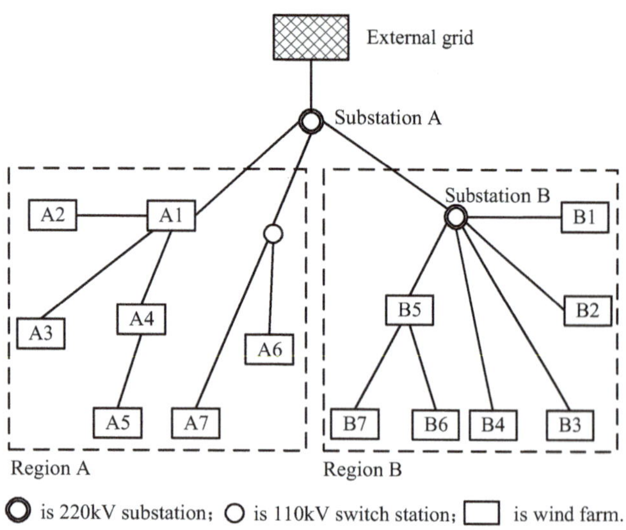

Fig. 3 Structure of the simulation system

1) In control mode I, the prediction and the historical operation information is ignored, and the control objective is only the voltage deviation of the pilot nodes. All the reactive power devices can adopt the constant reactive power control method and the control cycle is 15 min.
2) In control mode II, the proposed ORPD strategy in this paper is used. The thresholds of the VSM are set by the prediction error at 98% confidence level. The control cycle is also 15 min.

5.1 Control performance of voltage deviation of pilot nodes

The two pilot nodes of the WPPC are the 220 kV bus of the central substations A and B. The voltage control performance of the two pilot nodes in different control modes are shown in Fig. 5, and the statistical results of the voltage control performance are given in Table 2.

As indicated in Fig. 5 and Table 2, in control mode I, although the voltage control of the pilot nodes is accurate at the time section when ORPD control cycle begin, the voltage deviation of the pilot nodes cannot be guaranteed during the whole control cycle due to the fluctuation of the wind power. In control mode II, by utilizing the wind power prediction, the up-fluctuation and down-fluctuation states can be considered and therefore the voltage deviation of the pilot nodes can be guaranteed during the whole ORPD control cycle.

5.2 Control performance of coordination between different reactive power devices

The statistical results of the reactive power device regulation in different control modes are given in Table 3. The total reactive power consumption results of the discrete and the dynamic devices for 24 h are shown in Fig. 6.

As can be seen from Table 3 and Fig. 6, considering the reserve index of the dynamic devices and the economic index of the discrete devices in the control objectives utilizing the historical operation information, the regulating times of the discrete devices are decreased and the dynamic reactive power reserve is increased in control mode II. Moreover, by implementing the differentiated management between the discrete devices and the dynamic devices of the WPPC, the discrete devices can undertake the basis reactive power support and the dynamic devices can compensate the real-time shortfall of the reactive power caused by the wind power fluctuation in control mode II.

5.3 Control performance of static voltage stability

The VSM of the WPPC by the ORPD strategy in control mode II is shown in Fig. 7, and the corresponding

Table 1 Statistics of installed capability and reactive power equipment

Region	Installed capability (MW)	Discrete device capability (Mvar)	Dynamic device capability (Mvar)
A	793.5	−60 to 340.8	±103.83
B	987	−60 to 330	±209.7

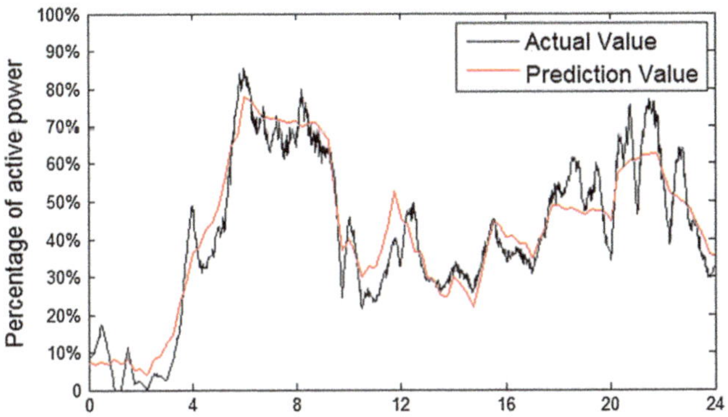

(**a**) Wind power of the wind power plants in region A

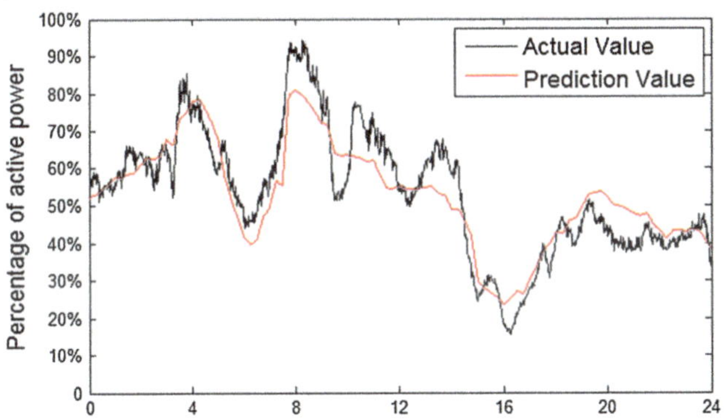

(**b**) Wind power of the wind power plants in region B

Fig. 4 Prediction and actual values of wind power for wind power plants in different regions

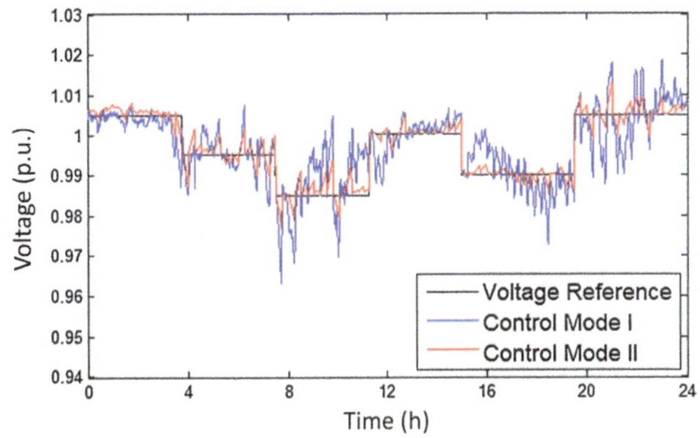

(**a**) Control performance of 220 kV bus of substation A

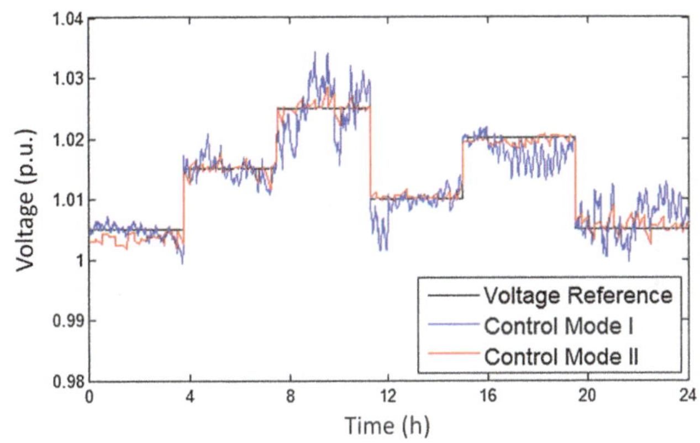

(**b**) Control performance of 220 kV bus of substation B

Fig. 5 Voltage reference and voltage control performance of the pilot nodes in different control modes

Table 2 Maximum voltage deviation of pilot nodes (kV)

Control mode	Pilot node A	Pilot node B
Mode I	4.782	2.233
Mode II	1.883	0.845

permitted fluctuation values of wind power of WPPC in control mode II are shown in Fig. 8. As indicated in Fig. 7 and Fig. 8, the WPPC can maintain sufficient VSM in each control cycle by setting the static voltage stability constraints utilizing wind power prediction information in control mode II.

6 Conclusions

The implementation of reactive power and voltage control of WPPC have great significant for the large-scale effective utilization of wind power. This paper proposes an ORPD strategy of WPPC considering static voltage stability. Through theoretical analysis and simulation verification, the following conclusions are obtained.

1) By utilizing the wind power prediction in the ORPD strategy, the voltage deviation of the pilot nodes is improved and the WPPC can maintain sufficient VSM in each control cycle. Therefore, the WPPC can operate in a more security mode during the period when the wind power is in heavy-load or large-fluctuation conditions.
2) The reserve index of the dynamic devices and the economic index of the discrete devices are taken into

Table 3 Statistical results of reactive power devices

Control mode	Discrete device regulating times	Dynamic devices reactive power consumption (Mvar)
Mode I	283	2.925×10^5
Mode II	68	2.267×10^5

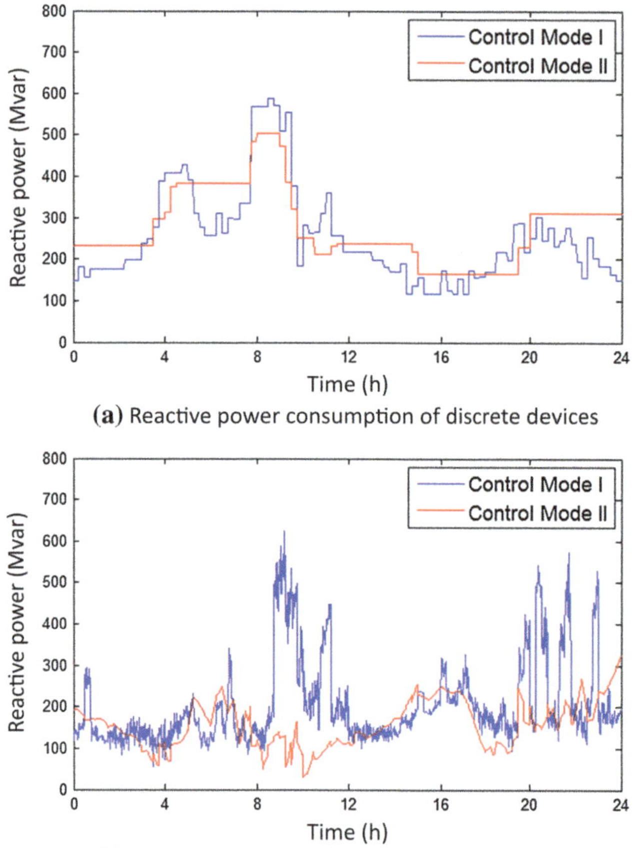

(a) Reactive power consumption of discrete devices

(b) Reactive power consumption of dynamic devices

Fig. 6 Total reactive power consumption of the WPPC

consideration on the control objectives utilizing the historical operation information. By doing this, the regulating times of the discrete devices are decreased, and meanwhile the dynamic reactive power reserve is increased.

3) By utilizing the automatic voltage control capability of the wind power plants and central substations, the differentiated management between the reactive power devices of the WPPC is implemented. Therefore, the discrete devices can undertake the basis reactive power support, while the dynamic devices can compensate the real-time shortfall of the reactive power caused by the wind power fluctuation. thus, the coordination between different devices of the WPPC is achieved.

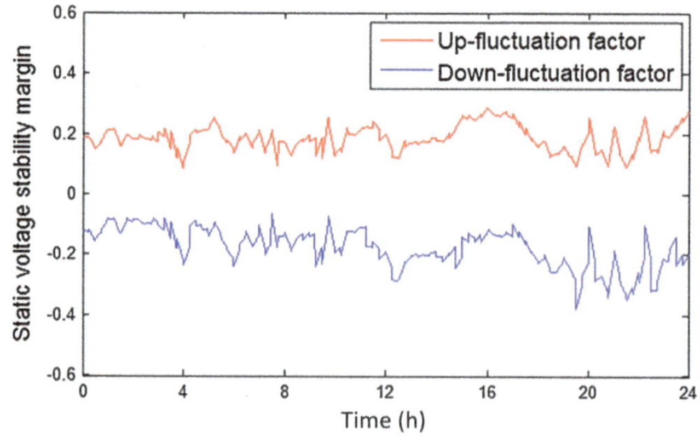

Fig. 7 VSM of the WPPC in control mode II

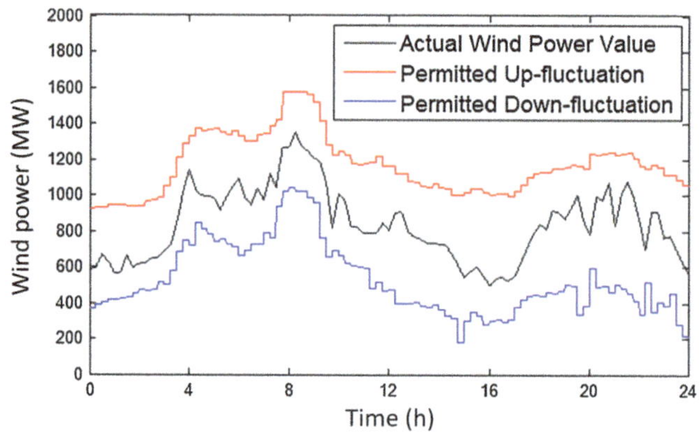

Fig. 8 Actual and permitted fluctuation values of wind power of WPPC in control mode II

Acknowledgment This work was supported by the National Natural Science Foundation of China (No. 51207145), the Science and Technology Project of State Grid Corporation of China (No. NY71-14-035).

References

[1] National Bureau of Statistics of China. China Statistics Yearbook (2010) Accessed 7 Feb 2011

[2] Chen QX, Kang CQ, Ming H et al (2014) Assessing the low-carbon effects of inter-regional energy delivery in China's electricity sector. Renew Sustain Energy Rev 32:671–683

[3] Chen QX, Kang CQ, Xia Q (2010) Power generation expansion planning model towards low-carbon economy and its application in China. IEEE Trans Power Syst 25(2):1117–1125

[4] Li JF, Cai FB, Qiao LM et al (2013) China wind power development report 2013. China Enviromental Science Press, Beijing

[5] Vittal E, O'Malley M, Keane A (2010) A steady-state voltage stability analysis of power systems with high penetrations of wind. IEEE Trans Power Syst 25(1):433–442

[6] Yang S, Wang WS, Liu C et al (2014) Coordinative strategy of reactive power and voltage control for improving static voltage stability of wind power centralized system. Power Syst Technol 38(5):1250–1256
[7] Hossain MJ, Hemanshu RP, Mahmud MA et al (2012) Investigation of the impacts of large-scale wind power penetration on the angle and voltage stability of power systems. IEEE Syst J 6(1):76–84
[8] Moursi ME, Joos G, Abbey C (2008) A secondary voltage control strategy for transmission level interconnection of wind generation. IEEE Trans Power Electron 23(3):1178–1190
[9] Pappala VS, Nakawiro W, Erlich I (2010) Predictive optimal control of wind farm reactive sources. In: 2010 IEEE/PES transmission and distribution conference and exhibition, New Orleans, USA
[10] Erlich I, Nakawiro W, Martinez M (2011) Optimal dispatch of reactive sources in wind farms. In: 2011 IEEE power and energy society general meeting, Detroit, USA
[11] Ullah NR, Bhattacharya K, Thiringer T (2009) Wind farms as reactive power ancillary service providers—technical and economic issues. IEEE Trans Energy Convers 24(3):661–672
[12] Mònica AP, Agustí E, Samuel GA (2014) Droop control for loss minimization in HVDC multi-terminal transmission systems for large offshore wind farms. Electr Power Syst Res 112:48–55
[13] Sáiz-Marín E, Lobato E, Egido I (2014) Optimal voltage control by wind farms using data mining techniques. Renew Power Gener 8(2):141–150
[14] Zhang XG, Pan WM, Liu YC et al (2012) Improved grid voltage control strategy for wind farms with DFIGs connected to distribution networks. J Power Electron 12(3):495–502
[15] Chen HF, Zhang YW, Min Y et al (2013) Graded voltage control strategy for clustering wind farms based on doubly fed induction generators. Autom Electr Power Syst 37(4):7–13
[16] Ahmidi A, Guillaud X, Besanger Y et al (2012) A multilevel approach for optimal participating of wind farms at reactive power balancing in transmission power system. IEEE Syst J 6(2):260–269
[17] Kumar VSS, Reddy KK, Thukaram D (2014) Coordination of reactive power in grid-connected wind farms for voltage stability enhancement. IEEE Trans Power Syst 29(5):2381–2390
[18] Kunder P (1994) Power system stability and control. McGraw-Hill, New York
[19] Abed AM (1999) WSCC voltage stability criteria, under voltage load shedding strategy, and reactive power reserve monitoring methodology. Proc IEEE PES Summer Meet 1:191–197
[20] Gustavo S, Vladimiro M (2009) A two-stage planning and control model toward economically adapted power distribution systems using analytical hierarchy processes and fuzzy optimization. Int J Electr Power Energy Syst 31(6):277–284
[21] Wachter A, Bieglery LT (2001) Line search filter methods for nonlinear programming: motivation and global convergence. IBM T. J. Watson Research Center, Yorktown Heights
[22] Ulbrich M, Ulbrich S, Vicente L (2004) A globally convergent primal-dual interior point filter method for nonconvex nonlinear programming. Math Program 100(2):379–410
[23] Córcoles J, González MA, Rubio J (2009) Multi-objective optimization of real and coupled antenna array excitations via primal dual interior point filter method from spherical mode expansions. IEEE Trans Antennas Propag 57(1):110–121

Shuo YANG received the B.S. degree in electrical engineering from Wuhan University, China in 2008 and M.S. degree in Power Automation Department, China Electric Power Research Institute, Beijing, China in 2011. He is now pursuing his Ph.D degree in Renewable Energy Department, China Electric Power Research Institute, Beijing, China. His research interest is power system operation with renewable energy.

Weisheng WANG received his B.S. degree and M.S. degree at Xi'an University of Technology, Ph.D degree at Xi'an Jiaotong University, in 1990, 1993 and 1996, respectively, all in Electrical Engineering. His research areas are renewable energy generation and its grid integration.

Chun LIU received his B.S. degree at North China Electric Power University, in July 1990, in Environment Engineering. His research areas are renewable energy generation and its grid integration.

Yuehui HUANG received her B.S. and M.S. degree from Xi'an Jiaotong University in 2002 and 2005 respectively. In 2008, she got her PhD degree from The Hong Kong Polytechnic University in the department of Electronic and Information Engineering. Her research interests include renewable generation integration and its dispatch and operation technology.

8

Variable speed wind turbine for maximum power capture using adaptive fuzzy integral sliding mode control

Saravanakumar RAJENDRAN, Debashisha JENA (✉)

Abstract This paper presents a nonlinear control approach to variable speed wind turbine (VSWT) with a wind speed estimator. The dynamics of the wind turbine (WT) is derived from single mass model. In this work, a modified Newton Raphson estimator has been considered for exact estimation of effective wind speed. The main objective of this work is to extract maximum energy from the wind at below rated wind speed while reducing drive train oscillation. In order to achieve the above objectives, VSWT should operate close to the optimal power coefficient. The generator torque is considered as the control input to achieve maximum energy capture. From the literature, it is clear that existing linear and nonlinear control techniques suffer from poor tracking of WT dynamics, increased power loss and complex control law. In addition, they are not robust with respect to input disturbances. In order to overcome the above drawbacks, adaptive fuzzy integral sliding mode control (AFISMC) is proposed for VSWT control. The proposed controller is tested with different types of disturbances and compared with other nonlinear controllers such as sliding mode control and integral sliding mode control. The result shows the better performance of AFISMC and its robustness to input disturbances. In this paper, the discontinuity in integral sliding mode controller is smoothed by using hyperbolic tangent function, and the sliding gain is adapted using a fuzzy technique which makes the controller more robust.

S. RAJENDRAN, D. JENA, National Institute of Technology Karnataka, Surathkal, India
(✉) e-mail: bapu4002@gmail.com

Keywords Variable speed wind turbine, Integral sliding mode controller (ISMC), Sliding mode control (SMC), Adaptive fuzzy integral sliding mode control (AFISMC)

1 Introduction

Because of the power crises and environmental issues, renewable energy sources play a vital role in energy market. Among all renewable energy sources, wind energy is one of the rapidly growing energy technology and has its own benefits such as pollution free, clean and environmental friendly. In recent years, due to the advance in drive technology and grid interconnection control, the production of wind power is increased. Generally, WT has two different types: fixed speed WT (FSWT) and variable speed WT (VSWT). The main advantages of VSWT over FSWT is that, the annual power production is increased by 5%–10%, and reduction in mechanical stress and power fluctuations [1, 2]. Generally, wind speed is classified into two types: below and above rated wind speed. Accordingly, the WT control is classified into two types: torque control and pitch control. At below rated wind speed, the main objective of the WT is, to maximize the wind energy capture from the wind by rotating the WT rotor at optimal rotor speed, which is derived from effective wind speed. At above rated speed, the main objective of the WT is to control the pitch angles which correspond to the reference power. In literature, some of the authors have discussed the control of WT with the assumption that the effective wind speed is available. In [3], a fuzzy controller was used to maximize the power capture, improve the efficiency, and the controller was found to be more robust to the wind gust and oscillatory torque. In [4], control algorithm based on fuzzy logic control (FLC) tracks the maximum power by controlling the WT rotor speed without estimating the effective wind speed. In WT control, several literatures

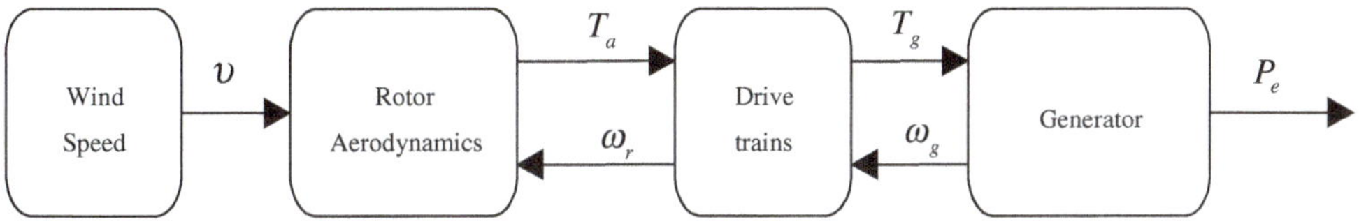

Fig. 1 Schematic of WT

reported either to estimate or to calculate the effective wind speed with WT control. In [5], the rotor speed and aerodynamic torque are estimated by the input and state based estimation with the known pitch angle, the effective wind speed is calculated by the inversion of the static aerodynamic model. In [6–8], Kalman filter is used to estimate rotor speed and aerodynamic torque, and finally the effective wind speed is calculated using Newton Raphson. For the single mass model given in [6, 7] and two mass model given in [8], nonlinear controllers such as nonlinear static state feedback with estimator and nonlinear dynamic state feedback with estimator (NDSFE) are used to control the WT at below rated wind speed. For both the controllers, the wind speed is estimated using Newton Raphson. In [9], calculation of effective wind speed is achieved by the particle filter, and FLC is used to control the WT at below rated wind speed. In [10–12], the SMC based controllers are applied to the WT without estimating the effective wind speed. References [10, 11] discussed higher order sliding mode control (HSMC) of WT at below and above rated speed and concluded that HSMC is more robust with respect to parameter uncertainty of the WT. In [12, 15], conventional SMC based control with adaptive sliding gain is used to control the WT where the sliding gain is varied by an adaptation algorithm.

The objective of this work is to maximize the energy capture from the wind with reduced oscillation in drive train. Modified Newton Raphson (MNR) is used to estimate the effective wind speed. A comparison of WT efficiency, with respect to maximum power capture, reduced transient load on drive train, and robustness to input disturbances is done by using different control algorithms. It is observed that, AFISMC is achieving the above objectives with more robustness to different types of input disturbances. The results are validated for different wind speed profile.

2 WT model

A WT is a device which converts the kinetic energy of the wind into electric energy. Simulation complexity of the WT purely depends on the type of control objectives. In case of WT modeling, complex simulators are required to verify the dynamic response of multiple components and aerodynamic loading. Generally, dynamic loads and interaction of large components are verified by the aero elastic simulator. For designing a WT controller, instead of going with complex simulator, the design objective can be achieved by using simplified mathematical model. In this work, WT model is described by the set of nonlinear ordinary differential equation with limited degree of freedom. This paper describes the control law for a simplified mathematical model with the objective of optimal power capture at below rated wind speed and reduced oscillation of the drive train. The proposed controller is tested with different wind profiles in the presence of different types of input disturbances.

Generally, VSWT system consists of the following components: aerodynamics, drive trains, and generator. The schematic of WT is shown in Fig. 1.

Equation (1) gives the nonlinear expression for aerodynamic power capture by the rotor

$$P_a = \frac{1}{2}\rho\pi R^2 C_P(\lambda,\beta)\upsilon^3 \tag{1}$$

where R is the radius; ρ is the air density; ω_r is the rotor speed (rad/s); C_P is the power coefficient of the WT; and υ is the wind speed (m/s).

From (1), it is clear that the aerodynamic power (P_a) is directly proportional to the cube of the wind speed. The power coefficient C_P is the function of blade pitch angle (β) and tip speed ratio (λ). The tip speed ratio is defined as ratio between linear tip speed and wind speed.

$$\lambda = \frac{\omega_r R}{\upsilon} \tag{2}$$

Generally, wind speed is stochastic with respect to time. Because this tip speed ratio gets affected, it will lead to the variation in power coefficient. The relationship between aerodynamic torque (T_a) and the aerodynamic power is given in (3).

$$P_a = T_a\omega_r \tag{3}$$

$$T_a = \frac{1}{2}\rho\pi R^3 C_q(\lambda,\beta)\upsilon^2 \tag{4}$$

where C_q is the torque coefficient given as

$$C_q(\lambda,\beta) = \frac{C_P(\lambda,\beta)}{\lambda} \tag{5}$$

Substituting (5) in (4), we get

Table 1 Coefficients values

$a_0 = 0.1667$	$a_3 = -0.01617$
$a_1 = -0.2558$	$a_4 = 0.00095$
$a_2 = 0.115$	$a_5 = -2.05 \times 10^{-5}$

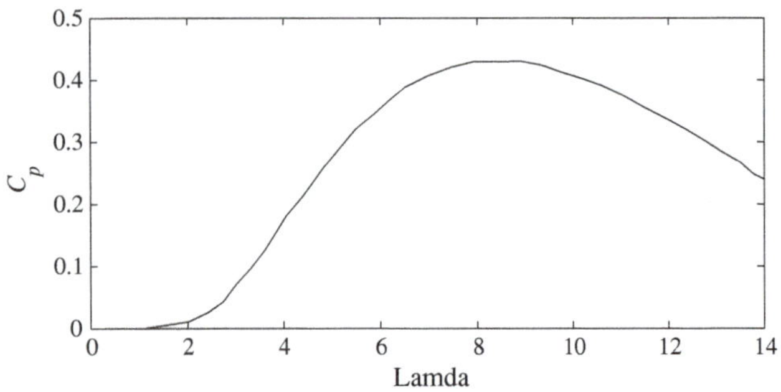

Fig. 2 C_p vs λ curve

$$T_a = \frac{1}{2}\rho\pi R^3 \frac{C_P(\lambda, \beta)}{\lambda} \upsilon^2 \tag{6}$$

In above equation, the nonlinear term is C_p which can be approximated by the 5th order polynomial given in (7).

$$C_P(\lambda) = \sum_{n=0}^{5} a_n\lambda^n = a_0 + \lambda a_1 + \lambda^2 a_2 + \lambda^3 a_3 + \lambda^4 a_4 + \lambda^5 a_5 \tag{7}$$

where a_0 to a_5 are the WT power coefficient. The values of approximated coefficients are given in Table 1. Figure 2 shows the C_p versus λ curve.

Figure 3 shows the two mass model of the WT. Equation (8) represents dynamics of the rotor speed ω_r with rotor inertia J_r driven by the aerodynamic torque (T_a).

$$J_r\dot{\omega}_r = T_a - T_{ls} - K_r\omega_r \tag{8}$$

Breaking torque acting on the rotor is low speed shaft torque (T_{ls}) which can be derived by using stiffness and damping factor of the low speed shaft given in (9).

$$T_{ls} = B_{ls}(\theta_r - \theta_{ls}) + K_{ls}(\omega_r - \omega_{ls}) \tag{9}$$

Equation (10) represents dynamics of the generator speed ω_g with generator inertia J_g driven by the high speed shaft torque (T_{hs}) and braking electromagnetic torque (T_{em}).

$$J_g\dot{\omega}_g = T_{hs} - K_g\omega_g - T_{em} \tag{10}$$

Gearbox ratio is defined as

$$n_g = \frac{T_{ls}}{T_{hs}} = \frac{\omega_g}{\omega_{ls}} \tag{11}$$

Transforming the generator side dynamics into the low speed shaft side, we will get

$$n_g^2 J_g\dot{\omega}_g = T_{ls} - n_g K_g\omega_g - n_g T_{em} \tag{12}$$

If a perfectly rigid low-speed shaft is assumed, the dynamics of the rotor characteristics of a single mass WT model can be expressed by a first order differential equation given as

$$J_t\dot{\omega}_r = T_a - T_g - K_t\omega_r \tag{13}$$

where

$$J_t = J_r + n_g^2 J_g \tag{14}$$

$$K_t = K_r + n_g^2 K_g \tag{15}$$

$$T_g = n_g T_{em} \tag{16}$$

3 Control objectives

Generally, WT is classified into two types: fixed and variable speed WT. Variable speed WT has more advanced and flexible operation than fixed speed WT. Operating regions in variable speed WT are divided in to three types. Figure 4 shows the various operating region in variable speed WT. Region 1 represents the wind speed below the cut in wind speed. Region 2 represents the wind speed between cut in and cut out. In this region, the main objective is to maximize the energy capture from the wind with reduced oscillation on the drive train.

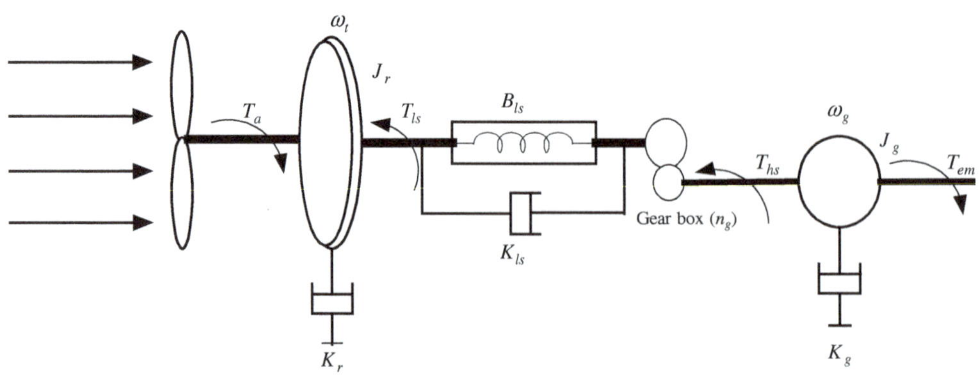

Fig. 3 Two mass model of the WT [8]

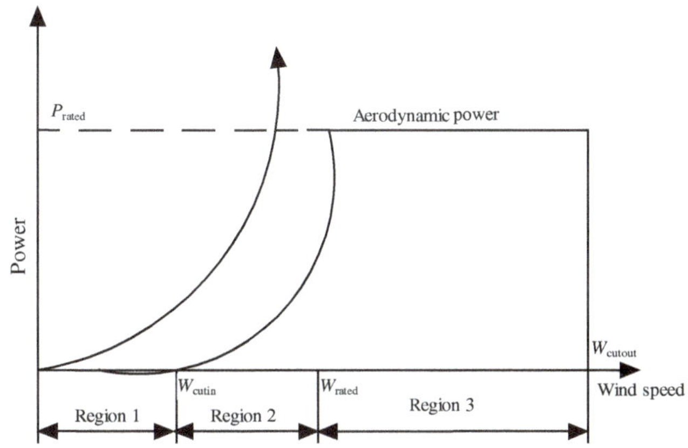

Fig. 4 Power operating region of wind turbines

Region 3 describes the wind speed above the cut out speed. In this region, pitch controller is used to maintain the WT at its rated power.

To achieve the above objective (Region 2), the blade pitch angle (β_{opt}) and tip speed ratio (λ_{opt}) are set to be its optimal value. In order to achieve the optimal tip speed ratio, the rotor speed must be adjusted to the reference/optimal rotor speed (ω_{ropt}) by adjusting the generator torque (T_g). Equation (17) defines the reference/optimal rotor speed (ω_{ropt}).

$$\omega_{ropt} = \frac{\lambda_{opt} \upsilon}{R} \tag{17}$$

Figure 5 shows the WT control scheme. It is clear that WT has two control loops: inner and outer loop. The inner control loop consists of electrical generator with power converters. The outer loop having the aero turbine control which gives the reference to the inner loop is shown in Fig. 5. In this paper, we made an assumption that, the inner loop is well controlled.

3.1 Wind speed estimator

The aero dynamic torque is approximated with 5th order polynomial given in (7). It is assumed that rotor speed measurement is available. Estimation of effective wind speed depends on aerodynamic torque and rotor speed with the pitch angle at optimal value.

$$F(\upsilon) = T_a - \frac{1}{2}\rho\pi R^3 \frac{C_P(\lambda)}{\lambda}\upsilon^2 \tag{18}$$

The MNR algorithm is used to solve the effective wind speed (υ) from (18). This equation has unique solution at below rated region. With known υ, the optimal rotor speed ω_{ropt} is calculated by using (17).

3.2 Sliding mode control

To achieve the maximum power at below rated wind speed, sliding mode based torque control is proposed in [12]. The main objective of this controller is to track the reference rotor speed ω_{ref} for maximum power extraction. In conventional sliding mode control, sliding surface generally depends on error, and derivative of the error signal is given in (19).

$$S(t) = \left(\lambda + \frac{d}{dt}\right)^{n-1} e(t) \tag{19}$$

where λ is the positive constant and n is the order of the uncontrolled system.

For speed control, a sliding surface is defined as

$$S(t) = \omega_r(t) - \omega_{ref}(t) \tag{20}$$

The reference rotor speed has defined in (17). Taking the time derivative of (20), we get

$$\dot{S}(t) = \dot{\omega}_r(t) - \dot{\omega}_{ref}(t) \tag{21}$$

Substituting $\dot{\omega}_r$ from (13) into (21), we get

$$\dot{S} = \frac{1}{J_t}T_a - \frac{K_t}{J_t}\omega_r - \frac{1}{J_t}T_g - \dot{\omega}_{ref} \tag{22}$$

Stability of SMC can be evaluated by using Lyapunov candidate function given in (23):

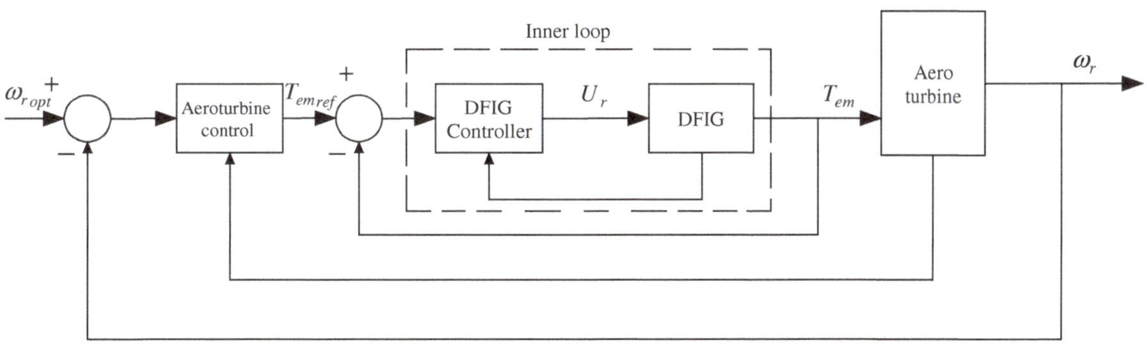

Fig. 5 WT control scheme

$$V = \frac{1}{2}S^2 \quad (23)$$

Taking the time derivative of the above equation, then

$$\dot{V} = S\dot{S} = S\left[\frac{1}{J_t}T_a - \frac{K_t}{J_t}\omega_r - \frac{1}{J_t}T_g - \dot{\omega}_{ref}\right] \quad (24)$$

if $\dot{V}$ is negative definite

$$\frac{1}{J_t}T_a - \frac{K_t}{J_t}\omega_r - \frac{1}{J_t}T_g - \dot{\omega}_{ref}\left\{\begin{array}{ll} <0 & \text{for } S>0 \\ >0 & \text{for } S<0 \end{array}\right\} \quad (25)$$

Stability of the controller is guaranteed if the torque control law satisfies (26).

$$T_g\left\{\begin{array}{ll} <T_a - K_t\omega_r - J_t\dot{\omega}_{ref} & \text{for } S>0 \\ >T_a - K_t\omega_r - J_t\dot{\omega}_{ref} & \text{for } S<0 \end{array}\right\} \quad (26)$$

Generally, the SMC have two parts: equivalent control U_{eq} and switching control U_{sw}. Combining these two controls for minimizing the tracking error, the control input can be expressed as

$$U(t) = U_{eq}(t) + U_{sw}(t) \quad (27)$$

Finally, the torque control structure is given in (28) as

$$T_g = T_a - K_t\omega_r - J_t\dot{\omega}_{ref} + J_tk\,\text{sign}(S) \quad (28)$$

The major drawback in the signum function is that, it has the discontinued value between +1 and −1 due to which it introduces the chattering phenomenon. So the signum function is changed by a smooth hyperbolic tangent (tanh) function with boundary layer (φ).

$$T_g = T_a - K_t\omega_r - J_t\dot{\omega}_{ref} + J_tk\tanh\left(\frac{S}{\varphi}\right) \quad (29)$$

where k is the sliding gain.

3.3 Integral fuzzy sliding mode control

To improve the sliding surface and overcome the steady state error, the integral action is included in the sliding surface [14]. An integral sliding surface is defined as

$$S(t) = \left(\lambda + \frac{d}{dt}\right)^{n-1} e(t) + k_i\int_0^\infty e(t)\,dt \quad (30)$$

where k_i is the integral gain.

For first order $n = 1$, then the sliding surface modified as

$$S(t) = e(t) + k_i\int_0^\infty e(t)\,dt \quad (31)$$

The major objective of the controller is that, the tracking error $e(t)$ should converge to zero. The stability of the controller is determined by using the Lyapunov candidate function given in (23) with $V(0) = 0$ and $V(t) > 0$ for $S \neq 0$.

By taking the time derivative of the (31), we get

$$\dot{S}(t) = \dot{e}(t) + k_ie(t) \quad (32)$$

$$\dot{V} = \dot{S}S = S\left[\frac{1}{J_t}T_a - \frac{K_t}{J_t}\omega_r - \frac{1}{J_t}T_g - \dot{\omega}_{ref} + k_ie(t)\right] \quad (33)$$

Generally, the SMC have two parts: equivalent control U_{eq} and switching control U_{sw}. By combining these two controls, tracking error can be minimized, and we will get the total control as

$$U(t) = U_{eq} + U_{sw} \quad (34)$$

The Switching control is defined in two ways

$$U_{sw} = k\,\text{sign}\left(\frac{S}{\varphi}\right) \quad (35)$$

or

$$U_{sw} = k\tanh\left(\frac{S}{\varphi}\right) \quad (36)$$

Finally, the torque control structure is given as

$$T_g = T_a - K_t\omega_r - J_t\dot{\omega}_{ref} + J_tk_ie(t) + J_tk\,\tanh\left(\frac{S}{\varphi}\right) \quad (37)$$

3.4 Adaptive fuzzy integral sliding mode control (AFISMC)

In order to accommodate the input disturbances, the fixed gain k in ISMC has been replaced by the variable gain which is achieved by AFISMC. The main aim of the AFISMC controller is to achieve robustness with respect to disturbances. In Fuzzy SMC, variable boundary layer (φ) is a function of S and $\dot{S}(t)$ [13]. In this problem, we need to avoid more tracking errors, in the presence of disturbance in the control input like generator torque (T_g). Fuzzy logic (FL) is used to improve the performance of the controller as well as the system. FL is used to automatically adjust the variable gain based on sliding surface and derivative of the sliding surface. The input to the fuzzy controller is sliding surface and derivative of the sliding surface, and the output is variable gain. Equation (38) defines the control input for AFISMC.

$$T_g = T_a - K_t\omega_r - J_t\dot{\omega}_{ref} + J_tk_ie(t) + J_tk_{fuzzy}\,\tanh\left(\frac{S}{\varphi}\right) \quad (38)$$

A variable sliding gain layer is introduced to smooth out the discontinuity which ensures reduced chattering effect. The condition $S = 0$ indicates the tracking error is zero, and the switching control is also zero. But during simulation, it may not possible to achieve the tracking

Table 2 Fuzzy rules S, $\dot{S}(t)$and k_{fuzzy}

U(t)	$\dot{S}(t)$				
S(t)	NB	NS	Z	PS	PB
NB	Z	NS	PS	NS	NS
NS	NS	PS	NS	NS	Z
Z	NB	NS	NB	PB	NS
PS	PS	PB	Z	NS	NS
PB	PS	PB	NS	PB	NS

error to zero because of high tracking dynamics in the reference signal. So the varying fuzzy sliding gain is introduced in the ISMC control.

The triangular membership function is used for both inputs and output fuzzy variables. The inputs are error and derivative of the error which is varying between the {−0.4, 0.4} and {−0.04, 0.04} respectively. The output variable is sliding gain which is varying between {0.006, 0.2}. The fuzzy variables are defined in the rule base as {NS (Negative Small), NB (Negative Big), Z (Zero), PS (Positive Small) and PB (Positive Big)}. From the conventional SMC and ISMC, the knowledge of the sliding gain is obtained. With this knowledge, the fuzzy rules in Table 2 are initially derived by trial and error method. After obtained the rule base, the simulation is carried out and it is tuned appropriately as per the control objectives. Finally, the derived rule base is validated for different disturbances and different mean wind speed conditions. From the results, it is found that the rule base is optimal for achieving the given control objectives. A typical fuzzy control rule of the proposed AFISMC is expressed as: $R^{(i)}$ ::IF $S(t)$ is H_1^i and $\dot{S}(t)$ is H_2^i THEN k_{fuzzy} is M^i where H_1^i and H_2^i are the labels of the input fuzzy sets and M^i is the labels of the output fuzzy sets, $i = 1,\ldots, p$ represents the number of IF-THEN fuzzy rules. Table 2 explains the linguistic fuzzy rules for AFISMC. The fuzzy control surface of the output is shown in Fig. 6.

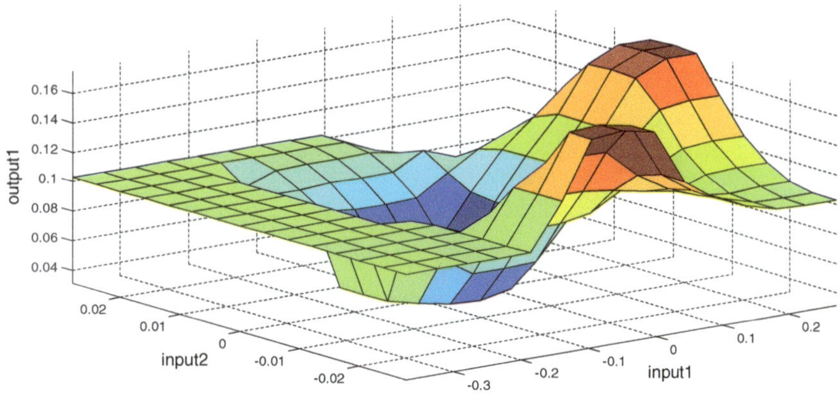

Fig. 6 3D plot of S, S and k_{fuzzy}

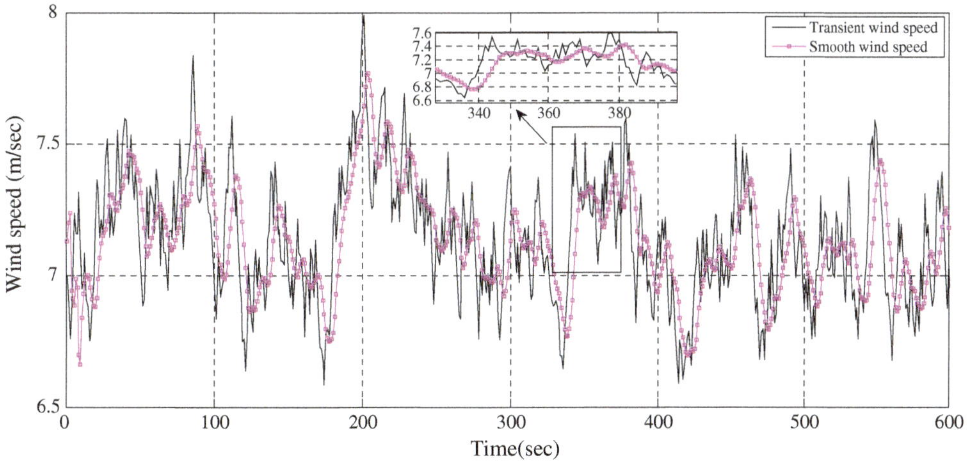

Fig. 7 Test wind speed profiles with a mean of 7 m/s

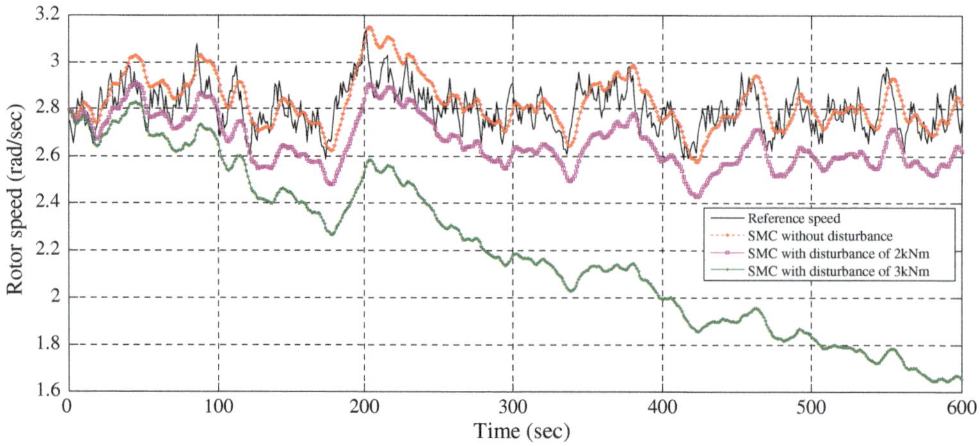

Fig. 8 Comparison of rotor speed with different disturbance for SMC (transient wind speed)

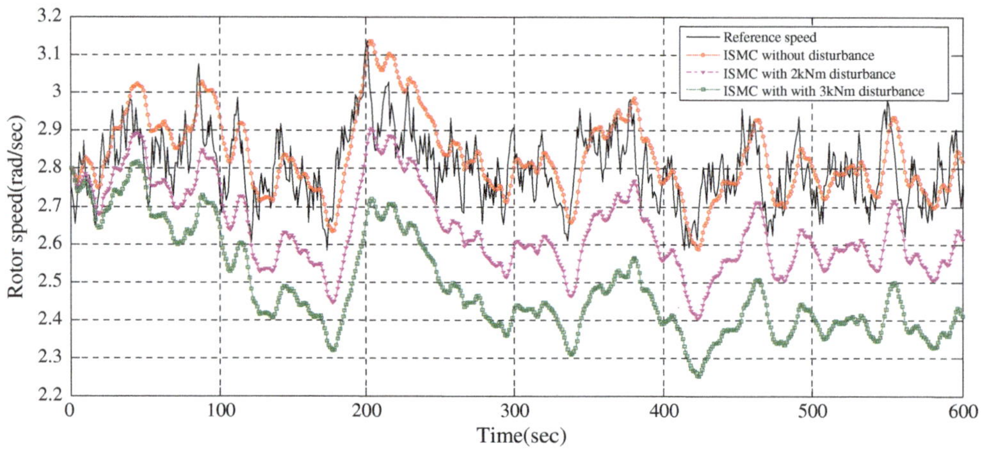

Fig. 9 Comparison of rotor speed with different disturbance for ISMC (transient wind speed)

4 Results and discussion

Figure 7 shows the test wind profile of the WT with mean wind speed of 7 m/s. Generally, wind speed consists of two components: mean and turbulent component. From this figure, it is clear that two different wind speeds are used with different turbulence intensity. Both the wind speeds are having 10 min wind data with the standard deviation (STD) of 0.25 m/s (transient wind speed) and 0.19 m/s (smooth wind speed). In order to analyze the robustness of the proposed AFISMC, three different input disturbances such as constant, sinusoidal and random disturbance of maximum magnitude 10 kNm has been considered in this work.

Figures 8 and 9 show the rotor speed comparisons for SMC and ISMC considering both with and without input disturbances. From these figures, it is clear that without any input disturbance, the obtained rotor speed for SMC and ISMC are almost following the reference rotor speed. As shown in the Figs. 8 and 9, an increase in constant input disturbance level from 2 to 3 kNm results a significant increase in tracking error due to which the rotor is unable to track the reference speed. This leads to more power loss and reduced electrical efficiency. The performances of the controllers are analyzed with respect to the following:

1) Aero dynamic and electrical efficiency
2) Maximum and standard deviation (STD) of control input
3) Controllers are tested with different wind profile.
4) Controllers are tested in presence of different level and types of input torque disturbance.

Figure 10 shows the rotor speed comparison for AFISMC with different level of constant input disturbances ranging from 2 to 7 kNm. Figures 11 and 12 shows the comparison of rotor speed and generator torque for AFISMC with different types of input disturbances such as constant, sinusoidal and random disturbances of magnitude 10 kNm. From Figs. 10 to 12, it is observed that the

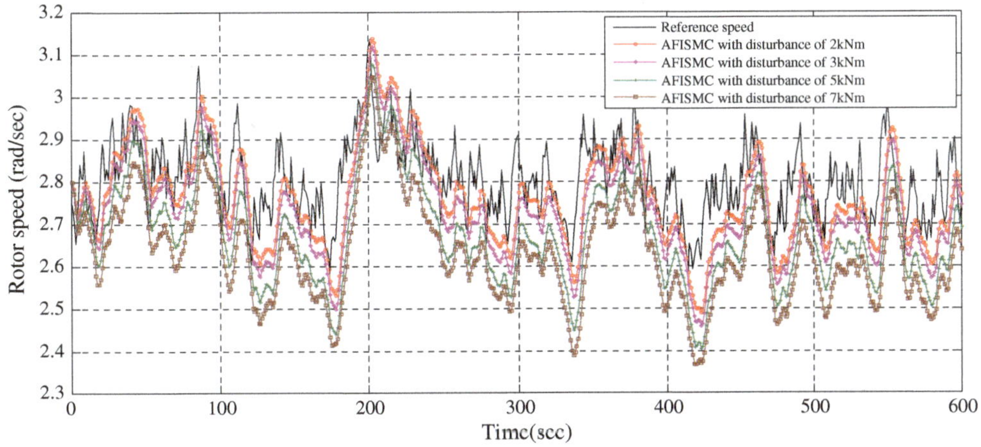

Fig. 10 Rotor speed comparisons for AFISMC with different level of constant disturbance (transient wind speed)

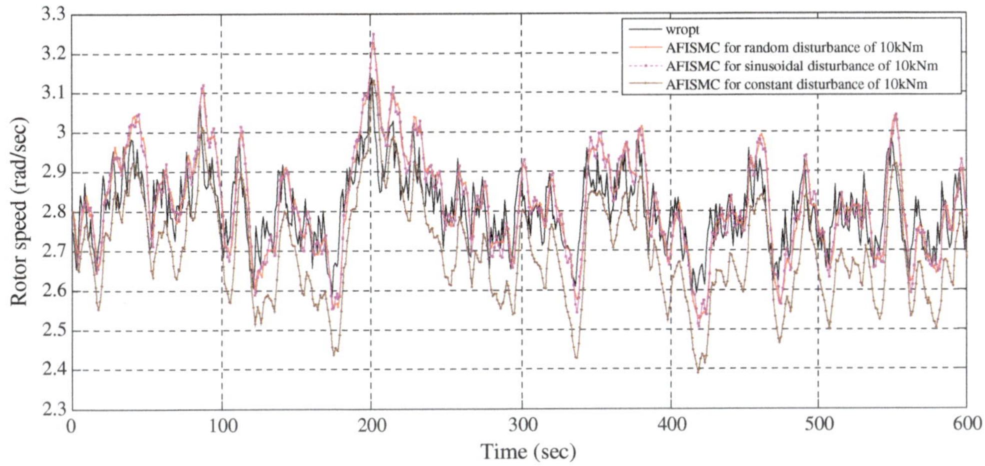

Fig. 11 Comparison of rotor speed with different disturbance for AFISMC (transient wind speed)

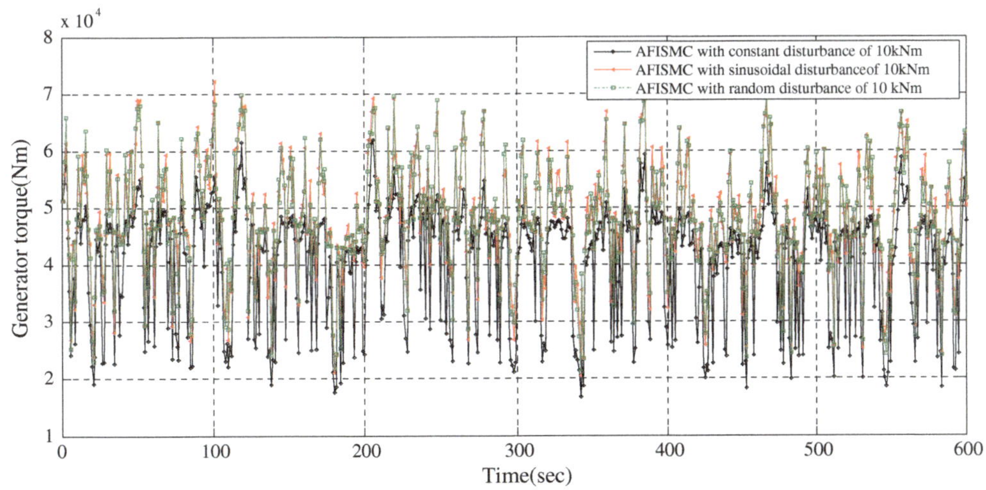

Fig. 12 Comparison of generator torque with different disturbance for AFISMC (transient wind speed)

Table 3 Performance analysis of SMC with and without disturbances (wind speed 7 m/s)

SMC with and without disturbances	SMC without disturbance	SMC with disturbance of 2 kNm	SMC with disturbance of 3 kNm
Max(T_g) kNm	51.92	55.20	61.44
Electrical efficiency (%)	91.10	89.50	81.17
Aerodynamic efficiency (%)	93.23	94.86	85.82
STD (T_g) kNm	1.928	1.705	3.232

Table 4 Performance analysis of ISMC with and without disturbances (wind speed 7 m/s)

ISMC with and without disturbances	ISMC without disturbance	ISMC with disturbance of 2 kNm	ISMC with disturbance of 3 kNm
Max(T_g) kNm	51.58	55.26	57.44
Electrical efficiency (%)	91.07	89.56	87.86
Aerodynamic efficiency (%)	93.21	94.71	94.05
STD (T_g) kNm	1.606	1.744	2.064

Table 5 Performance analysis of AFISMC with and without disturbances (7 m/s wind profile)

AFISMC with different types of disturbances (7 m/s wind speed)	AFISMC with constant disturbance of 10 kNm	AFISMC with sinusoidal disturbance of 10 kNm	AFISMC with random disturbance of 10 kNm
Max(T_g) kNm	61.69	72.03	70.97
Electrical efficiency (%)	73.77	91.25	91.38
Aerodynamic efficiency (%)	94.56	93.44	93.45
STD (T_g) kNm	10.257	9.484	9.414

performance of the WT with the proposed adaptive fuzzy based ISMC is able to track the maximum power delivery point in the presence of different types as well as the different level of input disturbances. Table 3 shows the performance of SMC as per the above mentioned criteria. From the results, it is clear that with an increase of disturbance level, the STD of the generated torque increases, which ensures more oscillation on the drive train. Also, it is found that the electrical and aerodynamic efficiency decreases with the increase in disturbance level, which introduces more power loss. Analyzing the results given in Table 3 and Fig. 8, it is concluded that conventional SMC is not robust to a disturbance level of more than 3 kNm.

Table 6 Performance analysis of AFISMC with and without disturbances (8 m/s wind profile)

AFISMC with different types of disturbances (8 m/s wind speed)	AFISMC with constant disturbance of 10 kNm	AFISMC with sinusoidal disturbance of 10 kNm	AFISMC with random disturbance of 10 kNm
Max(T_g) kNm	71.56	67.93	80.00
Electrical efficiency (%)	75.59	89.31	89.39
Aerodynamic efficiency (%)	93.27	91.88	91.88
STD (T_g) kNm	9.289	8.301	7.987

Table 7 Performance analysis of AFISMC with and without disturbances (8.5 m/s wind profile)

AFISMC with different types of disturbances (8.5 m/s wind speed)	AFISMC with constant disturbance of 10 kNm	AFISMC with sinusoidal disturbance of 10 kNm	AFISMC with random disturbance of 10 kNm
Max(T_g) kNm	76.69	86.87	86.26
Electrical efficiency (%)	76.93	89.53	89.58
Aerodynamic efficiency (%)	93.25	92.04	92.04
STD (T_g) kNm	8.875	7.891	7.487

Table 8 Analysis of electrical efficiency with respect to different level of disturbance for a wind speed of 7 m/s

AFISMC with disturbance	AFISMC with 2 kNm disturbance	AFISMC with 3 kNm disturbance	AFISMC with 5 kNm disturbance	AFISMC with 7 kNm disturbance
Electrical efficiency (%)	88.25	86.82	83.83	80.89
Mean square error	0.008	0.011	0.023	0.037

Table 4 shows the performance of ISMC in terms of efficiency and STD of input torque. This table ensures that an increase in disturbance level introduces more power loss and more drive train oscillation. It is observed that for ISMC, the WT is not able to track the reference rotor speed beyond the disturbance level of 3 kNm. For a disturbance level of 3 kNm, the electrical efficiency decreases to 87.86% for ISMC and 81.17% for SMC, which indicates that SMC and ISMC controllers are not robust with respect to disturbance of more than 3 kNm. Generally, WT system disturbance is not predictable and the controller should accommodate the disturbance with maximum power

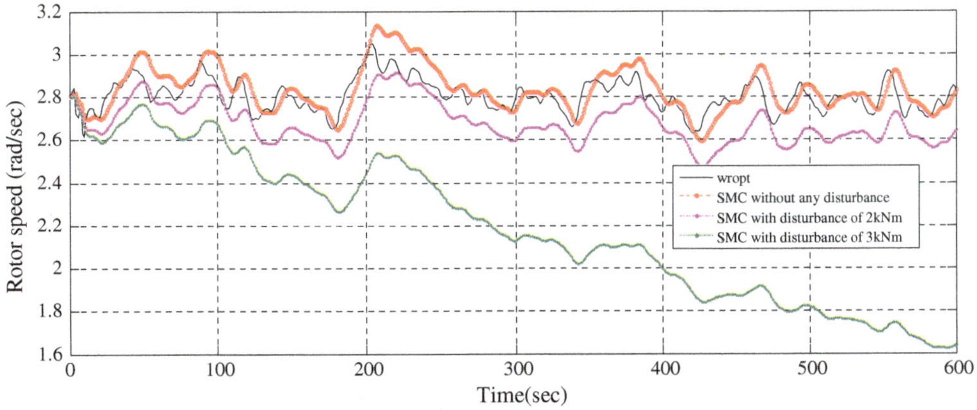

Fig. 13 Comparison of rotor speed with different disturbance for SMC (smooth wind speed)

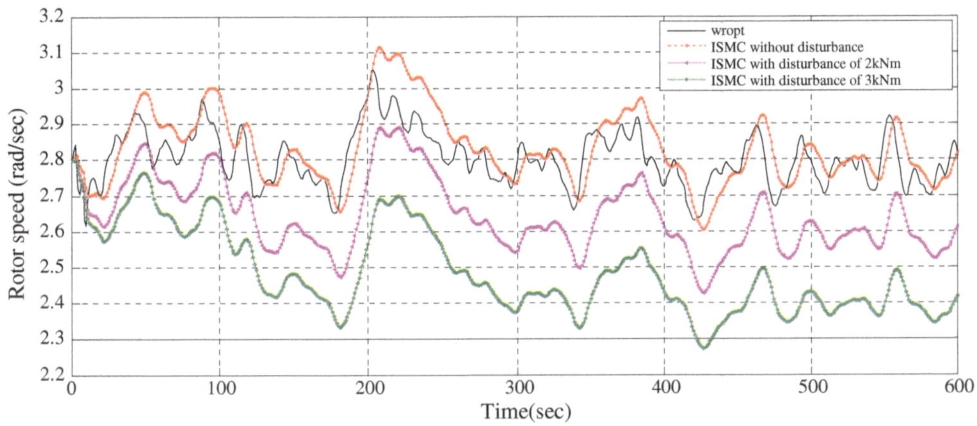

Fig. 14 Comparison of rotor speed with different disturbance for ISMC (smooth wind speed)

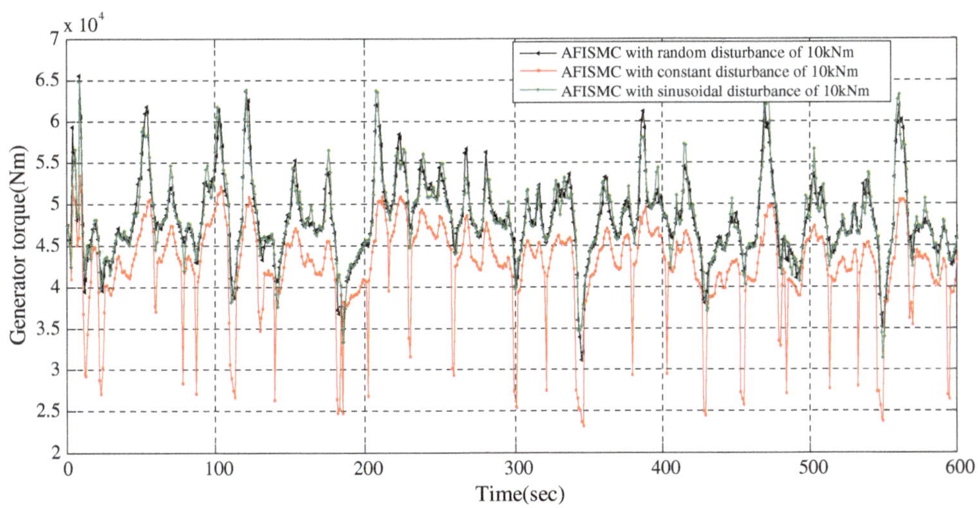

Fig. 15 Comparison of generator torque with different disturbance for AFISMC (smooth wind speed)

Table 9 Performance analysis of SMC with and without disturbances for smooth wind speed of 7 m/s

SMC with and without disturbances (smooth wind speed)	SMC without disturbance	SMC with disturbance of 2 kNm	SMC with disturbance of 3 kNm
Max(T_g) kNm	51.82	53.29	60.97
Electrical efficiency (%)	91.14	89.40	80.64
Aerodynamic efficiency (%)	93.34	94.73	85.23
STD (T_g) kNm	1.77	1.423	3.114

Table 10 Performance analysis of ISMC with and without disturbances for smooth wind speed of 7 m/s

ISMC with and without disturbances (smooth wind speed)	ISMC without disturbance	ISMC with disturbance of 2 kNm	ISMC with disturbance of 3 kNm
Max(T_g) kNm	51.12	54.11	56.91
Electrical efficiency (%)	91.15	89.61	87.94
Aerodynamic efficiency (%)	93.34	94.82	94.18
STD (T_g) kNm	1.388	1.482	1.865

Table 11 Performance analysis of AFISMC with disturbances for smooth wind speed of 7 m/s

AFISMC with different types of disturbances (smooth wind speed)	AFISMC with constant disturbance of 10 kNm	AFISMC with sinusoidal disturbance of 10 kNm	AFISMC with random disturbance of 10 kNm
Max(T_g) kNm	53.37	64.81	65.53
Electrical efficiency (%)	76.20	91.31	91.37
Aerodynamic efficiency (%)	94.79	93.53	93.53
STD (T_g) kNm	5.612	4.814	4.962

capture and reduced oscillation. Table 5 shows the performance of AFISMC with an objective of maximum energy extraction from the wind and reduced mechanical stress on the drive train. Different types of disturbances are given to the AFISMC controller: random, sinusoidal and constant disturbances with the magnitude of 10 kNm. From this table, it is clear that in the presence of 10 kNm disturbance, the AFISMC is able track the reference rotor speed, whereas for SMC and ISMC, the WT is not able to track the reference for a disturbance level more than 3 kNm.

The electrical and aerodynamic efficiency are almost found to be the same for sinusoidal and random disturbances, but with constant disturbance, the electrical efficiency is 18% less than the other two disturbances. From Tables 3, 4, and 5, it is concluded that SMC and ISMC are not robust to disturbances more than 3 kNm, but AFISMC is able to track the optimal rotor speed with different types of disturbances of magnitude 10 kNm.

In order to analyze the performance of the controllers critically, two different wind profiles are tested for simulation. The results for the mean wind speed of 8 and 8.5 m/s are given in Tables 6 and 7 respectively. Results given in the table can be analyzed in the same way as discussed in the previous sections. Finally, it is concluded that with increase in wind speed, there is an increase in electrical and aerodynamic efficiency, at the same time the oscillation in the drive train is reduced. Table 8 shows that, the increase in disturbance level decreases the efficiency of AFISMC. It can be observed that even though the efficiency decreases from 88.25% to 80.89%, for a constant disturbance range of 2 to 7 kNm, the WT with AFISMC control is able to track the reference speed with minimum tracking error (mean square error) and maximum efficiency. In comparison with SMC and ISMC, the tracking error of AFISMC was found to be minimum with improved electrical efficiency.

Figures 13 and 14 show the rotor speed comparisons for SMC and ISMC with and without input disturbances, where a smooth variation in wind is considered. From these figures, it is clear that without any disturbance, the obtained rotor speed for SMC and ISMC are almost following the reference rotor speed. As shown in the Figs. 13 and 14, with an increase in input disturbance level from 2 to 3 kNm, there is a significant increase in tracking error due to which the rotor is unable to track the reference speed.

Figure 15 shows the comparisons of generator torque for AFISMC with different disturbances. Tables 9, 10, and 11 gives the performance of SMC, ISMC and the proposed AFISMC respectively for smooth wind speed of 7 m/s. The performance of the controller can be analyzed in the same way of analysis done for transient wind speed profile of 7 m/s. The only difference between two wind speed profiles is that, for a smooth variation of wind speed, the efficiency of the WT increases, at the same time, the STD of the input torque reduces. This indicates a smooth variation in generated torque for a smooth varying wind speed.

5 Conclusions

Extracting maximum power at below rated wind speed is an important component of efficiently harnessing wind power. Selection of the right control strategy is therefore important to ensure that the system performs optimally. The objective is to design robust controllers that maximize the energy extracted from the wind while reducing transient loads on the drive train. To achieve the above objective, nonlinear controllers such as SMC and ISMC are adapted initially to control a single mass model of the WT of 600 kW machine. Those controllers are found to give weak performance particularly in the presence of input torque disturbance of more than 3 kNm. In this work, a nonlinear AFISMC with MNR wind speed estimator has been developed. The proposed controller is a combination of fuzzy technique, and the ISMC where the discontinuity in ISMC is smoothed by using hyperbolic tangent function, and the sliding gain is adapted using a fuzzy technique which makes the controller more robust. Compared to SMC and ISMC, the proposed AFISMC is found to be more robust in achieving the above objectives in the presence of input torque disturbance varying from 2 to 10 kNm. To prove the efficiency of the proposed AFISMC control, the simulation has been conducted for different wind speed profiles with different turbulence component.

References

[1] Carlin PW, Laxson S, Muljadi EB (2010) The history and state of the art of variable-speed wind turbine technology. Wind Energy 6(2):129–159

[2] Wang Q, Chang L (2004) An intelligent maximum power extraction algorithm for inverter-based variable speed wind turbine systems. IEEE Trans Power Electron 19(5):1242–1249

[3] Simoes MG, Bose BK, Spiegel RJ (1997) Fuzzy logic based intelligent control of a variable speed cage machine wind generation system. IEEE Trans Power Electron 12(1):87–95

[4] Eltamaly AM, Farh HM (2013) Maximum power extraction from wind energy system based on fuzzy logic control. Electro Power Syst 97(19):144–150

[5] Østergaard KZ, Brath P, Stoustrup J (2007) Estimation of effective wind speed. J Phys Conf Ser 75:1–9

[6] Boukhezzar B, Siguerdidjane H (2005) Nonlinear control of variable speed wind turbines without wind speed measurement. In: Proceedings of the IEEE conference on decision and control (CDE-ECC '05), 12–15 Dec 2005, pp 3456–3461

[7] Boukhezzar B, Siguerdidjane H, Hand MM (2006) Nonlinear control of variable-speed wind turbines for generator torque limiting and power optimization. J Sol Energy Eng 128(4):516–530

[8] Boukhezzar B, Siguerdidjane H (2011) Nonlinear control of a variable-speed wind turbine using a two-mass model. IEEE Trans Energy Convers 26(1):149–162

[9] Ćirić I, Ćojbašić Z, Vlastimir N et al (2011) Hybrid fuzzy control strategies for variable speed wind turbines. Facta Universitatis 10:205–217

[10] Beltran B, Ahmed-ali T, Bensouzid MEH (2008) Sliding mode power control of variable speed wind energy conversion systems. IEEE Trans Energy Conv 23(2):551–558

[11] Beltran B, Ahmed-ali T, BeNSouzid MEH (2009) High-order sliding-mode control of variable speed wind turbines. IEEE Trans Ind Electron 56(9):3314–3321

[12] Merabet A, Beguenane R, Thongam Jogendra ST et al (2011) Adaptive sliding mode speed control for wind turbine systems. In: Proceedings of the 37th Annual Conference on IEEE Industrial Electronics Society (IECON '11), 7–10 Nov 2011, pp 2461–2466

[13] Ben-Tzvi P, Bai S, Zhou Q et al (2011) Fuzzy sliding mode control of rigid-flexible multibody systems with bounded inputs. J Dyn Syst Meas Control 133(6):061012

[14] Eker I, Akinal SA (2005) Sliding mode control with integral action and experimental application to an electromechnical system. In: Proceedings of the international conference on computational intelligence methods and application (ICSC '05), Turkey, pp 1–6

[15] Boukezzar B, M'Saad M (2008) Robust Sliding mode control of a DFIG variable speed wind turbine for power production optimization. In: Proceedings of the 16th Mediterranean conference on control and automation, Ajaccia, France, 25–27 June, pp 795–800

Saravanakumar RAJENDRAN received the Master degree in Electrical Engineering from Anna University, Coimbatore Region, India, in 2010. Currently, he is pursuing the Ph.D. degree in the Department of Electrical & Electronics Engineering, National Institute of Technology Karnataka, Surathkal, India.

Debashisha JENA received his Bachelor degree of Electrical Engineering degree from University College of Engineering, Burla, India, in 1996. He received Master degree of Engineering in Power System in 2004 and Ph.D. degree in Control System Engineering from the Department of Electrical Engineering, National Institute of Technology, Rourkela, India in 2010. He was awarded a GSEP fellowship in 2009 from Canada for research on neuro control of complex dynamic systems. Currently, he is an Assistant Professor in the Department of Electrical & Electronics Engineering in the National Institute of Technology Karnataka, Surathkal, Mangalore, India. His research interests include evolutionary computation, system identification and neuro-evolutionary computation, and its application to distributed generation.

New inverter topology for ground current suppression in transformerless photovoltaic system application

Xiaoqiang GUO (✉), Huaibao WANG, Zhigang LU, Baocheng WANG

Abstract An interesting inverter topology is proposed in this paper. It is similar to the typical three-phase full bridge inverter from the topology point of view, but smartly designed for the ground current reduction in single-phase photovoltaic (PV) inverter applications. Theoretical analysis is conducted to clarify the operation mechanism of the proposed topology. Performance evaluation is carried out to verify the effectiveness of the proposed topology for the ground current suppression.

Keywords Ground current, Grid-connected inverter, Photovoltaic (PV) system

1 Introduction

Photovoltaic (PV) power is a clean and renewable alternative form of energy, which becomes increasingly attractive in recent years. According to the latest national survey report of PV power applications [1], in 2011, 2.5 GW PV systems has been installed in China, reached a total installed capacity of 3.3 GW, which represents 5% of the global total installed capacity. Fig. 1 shows annual and cumulated PV installation in China.

In order to improve the efficiency and reduce the cost of the PV power system, the transformerless inverter topology is a good choice, because it does not need heavy and costly transformer. However, the ground current may arise because there is no galvanic isolation between the inverter and grid. The ground current results in electromagnetic interference and safety issues [2]. Therefore, the suppression of ground current is a necessity for the transformerless PV inverter [3]. Many transformerless inverter topologies have been reported in literature. Most of them are derived from the single-phase full bridge inverter, such as H5 [4] and H6 [5], which has the potential risk of the bridge shoot through. It may lead to the inverter switch damage and reduce the system reliability. In order to solve the problem, a new inverter topology is proposed in this paper. It is similar to the typical three-phase full bridge inverter from the topology point of view, but smartly designed for the ground current suppression in single-phase PV inverter applications. The theoretical analysis and performance evaluation are carried out to demonstrate the feasibility of the proposed topology.

2 Proposed topology

Fig. 2 depicts the schematic of the proposed topology. It is interesting to note that the proposed topology is analogy to the typical three-phase full bridge inverter topology [6], but smartly modified for the ground current reduction. Furthermore, it has the advantages that the bridge shoot through is enabled without any switch damage, which increase the system reliability. Following we will present the operation mechanism of the propose topology.

In general, the ground current results from the high-frequency voltage components across the parasite capacitor between the PV system and the ground. Therefore, in this paper, the basic idea for the ground current suppression is to keep the parasite capacitor voltage free of high-frequency components.

Fig. 3 shows the system operation in the positive half cycle. When S_1 and S_6 are on, the parasite capacitor voltage in Fig. 3a is equal to the voltage across the inductor L_2, which is a half the input voltage and grid voltage. On the

X. GUO, H. WANG, Z. LU, B. WANG, Yanshan University, Qinhuangdao 066004, China
(✉) e-mail: gxq@ysu.edu

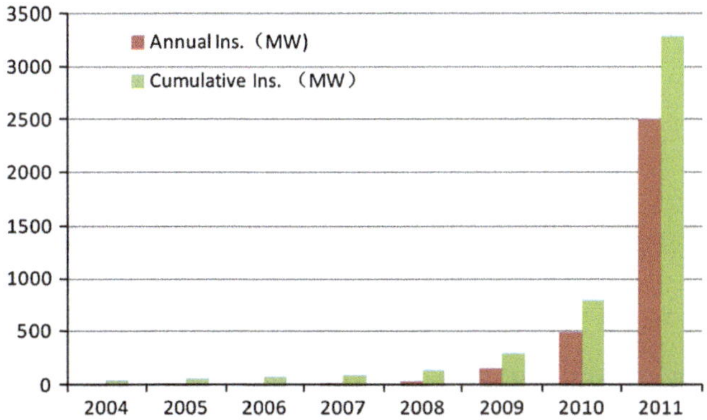

Fig. 1 Annual and cumulated PV installation in China [1]

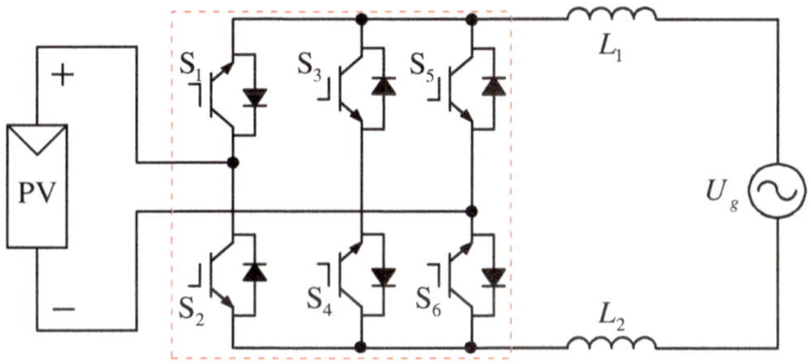

Fig. 2 Proposed topology

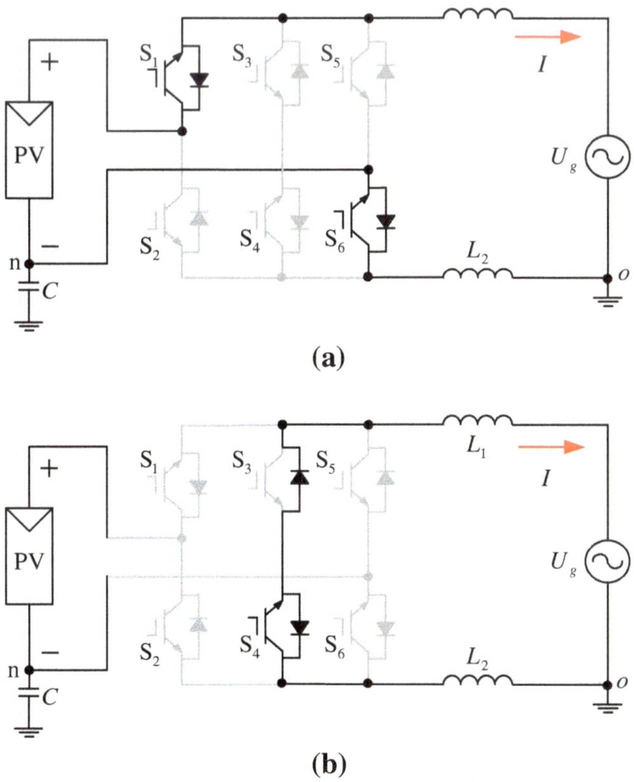

Fig. 3 Operation in the positive half cycle

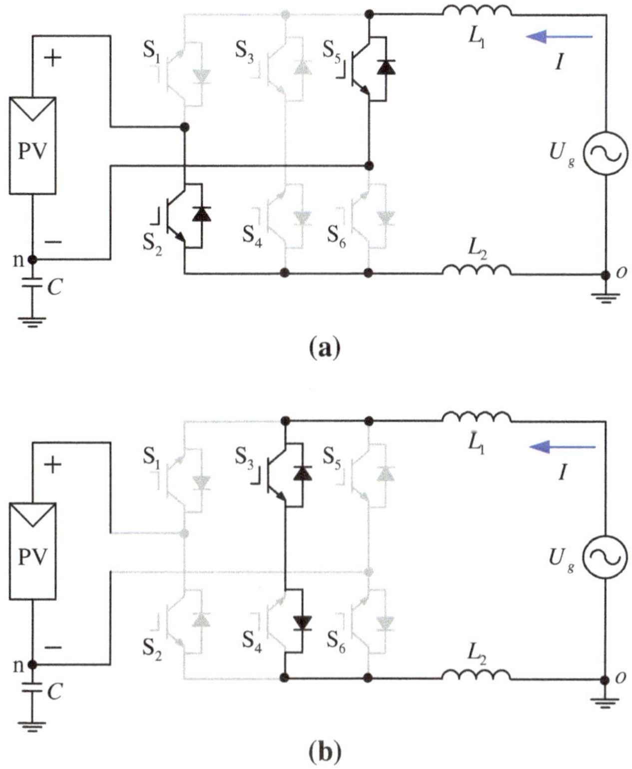

Fig. 4 Operation in the negative half cycle

other hand, When S_4 and S_3 diode are on, the parasite capacitor voltage in Fig. 3b is equal to the voltage across the S_6 and inductor L_2, where the voltage drop of S_6 is a half the input voltage and the voltage drop of inductor L_2 is a half the grid voltage.

Fig. 4 shows the system operation in the negative half cycle. When either S_2/S_5 (See Fig. 4a) or S_3/S_4 Diode (See Fig. 4b) are on, the parasite capacitor voltages are equal to a half the input voltage and grid voltage, which is exactly the same as one in the positive half cycle.

In summary, the parasite capacitor voltage in the entire cycle is free of high frequency components. Therefore, the ground current can be reduced.

Another consideration that should be noted is the effect of the switch parasitic capacitance on the ground current. An interesting analysis and solution has been reported in [7]. In practice, however, these parasitic capacitances can not be accurately predicted [8], and may be time-varying as the operating temperature and operating state change [9]. Therefore, further investigation is needed to deal with it, which is the subject of our future research.

3 Performance evaluation

As we know, the typical H-bridge inverter with bipolar modulation or unipolar modulation is not suitable for the

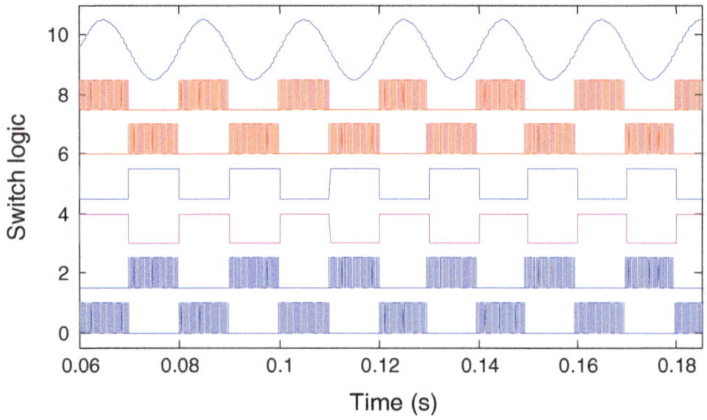

Fig. 5 Switch logic, from bottom to top: S_6, S_5, S_4, S_3, S_2 and S_1

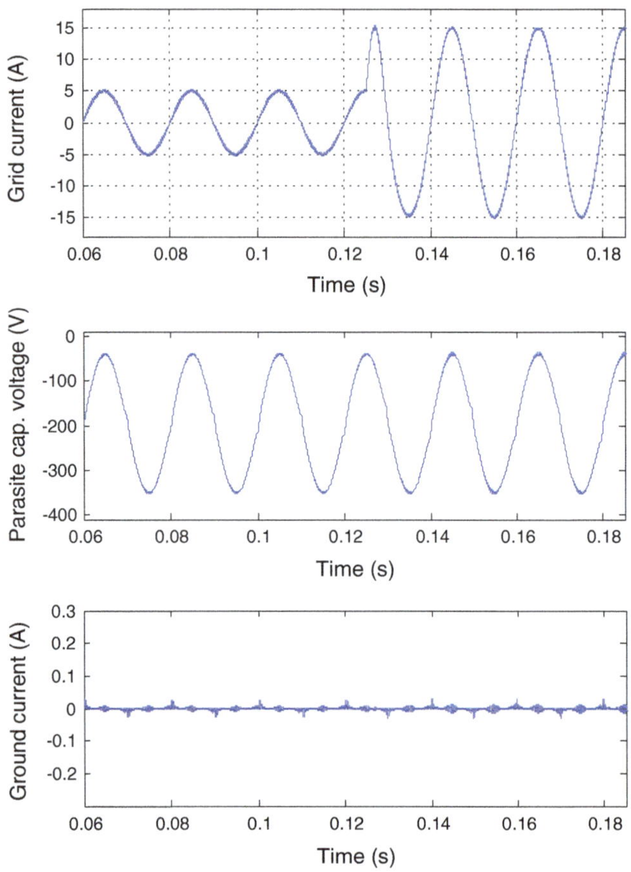

Fig. 6 Performance evaluation results

transformerless PV system application due to its low efficiency or high ground current [10], and the extensive evaluation results have been reported in [2, 3]. Therefore, it is not duplicated here. In order to verify the effectiveness of the proposed solution, following will present the performance evaluation of the proposed topology. System parameters are listed as follows: input voltage is 390 V; grid voltage is 220 V/50 Hz; inverter switching frequency is 16 kHz; filter $L_1 = L_2 = 5$ mH; ground parasitic capacitor is 25 nF. The evaluation results are presented as follows. From Fig. 5, it can be observed that only two switches operate in high frequency mode at any instance, leaving four switches in the fundamental frequency mode, which greatly reduces the switching losses and increase the system efficiency.

From Fig. 6, it can be observed that the current waveform is sinusoidal. At 0.12 s, the current steps from 5 A to 15 A. It can be observed that the grid current has good tracking capability and dynamic response. On the other hand, the parasite capacitor voltage is free of high-frequency components. The ground current is much less than 300 mA, and it meets the international standard of VDE 0126-1-1. It should be noted that this paper mainly focused on the ground current reduction in single-phase PV systems. For three-phase PV systems, some possible solution has been reported in [11], which is beyond the scope of this paper.

4 Conclusions

This paper has presented an interesting inverter topology for the single-phase transformerless PV systems. It can achieve the ground current reduction by keeping the parasite capacitor voltage free of high frequency components. It is expected that the proposed topology is attractive and promising for the ground current reduction in transformerless grid-connected PV applications.

Acknowledgment This work was supported by the National Natural Science Foundation of China (51307149), China Postdoctoral Science Foundation (2014M551050), and Specialized Research Fund for the Doctoral Program of Higher Education (20131333120016).

References

[1] Xu HH, Charlie D, Wang SC et al (2012) National survey report of PV power applications in China 2011. IEA PVPS Task 1
[2] Lopez O, Freijedo FD, Yepes AG et al (2010) Eliminating ground current in a transformerless photovoltaic application. IEEE Trans Energy Convers 25(1):140–147
[3] Gubia E, Sanchis P, Ursua A et al (2007) Ground currents in single-phase transformerless photovoltaic systems. Prog Photovolt Res Appl 15(7):629–650
[4] Xiao HF, Xie SJ, Chen Y et al (2011) An optimized transformerless photovoltaic grid-connected inverter. IEEE Trans Ind Electron 58(5):1887–1895
[5] Yu WS, Lai JS, Qian H et al (2011) High-efficiency MOSFET inverter with H6-type configuration for photovoltaic nonisolated AC-module applications. IEEE Trans Power Electron 26(4): 1253–1260

[6] Guo XQ, Wu WY (2010) Improved current regulation of three-phase grid connected voltage source inverter for distributed generation systems. IET Renew Power Gener 4(2):101–115
[7] Yang B, Li WH, Gu WJ et al (2012) Improved transformerless inverter with common-mode leakage current elimination for a photovoltaic grid-connected power system. IEEE Trans Power Electron 27(2):752–762
[8] Xiao HF, Xie SJ (2010) Leakage current analytical model and application in single-phase transformerless photovoltaic grid-connected inverter. IEEE Trans Electromagn Compat 52(4):902–913
[9] Oh KS (2001) IGBT basics 1. Application note 9016, Fairchild Semiconductor, San Jose, CA, USA
[10] Kerekes T, Teodorescu R, Rodrıguez P et al (2011) A new high-efficiency single-phase transformerless PV inverter topology. IEEE Trans Ind Electron 58(1):184–191
[11] Guo XQ, Cavalcanti MC, Farias AM et al (2013) Single-carrier modulation for neutral-point-clamped inverters in three-phase transformerless photovoltaic systems. IEEE Trans Power Electron 28(6):2635–2637

Xiaoqiang GUO (M′ 10) received the Ph.D. degrees in electrical engineering from Yanshan University, China, in 2009. He is an associate professor with Yanshan University, where he is engaged in research associated with power electronics, industrial electronics, etc. His current research interests include power electronics applications to distributed power generation systems and smartgrids.

Huaibao WANG is an assistant researcher at Department of Electrical Engineering, Yanshan University. His research interests include renewable energy systems.

Zhigang LU is a professor at Department of Electrical Engineering, Yanshan University. His research interests include power system operation and renewable energy systems.

Baocheng WANG is a professor at Department of Electrical Engineering, Yanshan University. His research interests include renewable energy systems.

10

Key technologies and the implementation of wind, PV and storage co-generation monitoring system

Xianliang TENG (✉), Zonghe GAO, Yingyuan ZHANG, Hua HUANG, Lili LI, Tingting LIANG

Abstract The coordinated control of multiple-sources including wind, photovoltaic (PV) and storage brings new challenges to traditional dispatch and control technologies. This paper firstly introduces a framework of wind, PV and storage co-generation monitoring system. Then, key technologies of co-generation monitoring system including day-ahead optimal dispatching, active power coordinated control and reactive power and voltage control are proposed. The framework and the techniques described in this paper have been applied in the National Wind, Photovoltaic, Storage and Transmission Demonstration Project of China, and their validity have been tested and verified.

Keywords Wind, photovoltaic (PV) and storage, Co-generation monitoring system, Day-ahead optimal dispatching, Active power coordinated control, Reactive and voltage control, Coordination control

1 Introduction

Because of the increasing energy crisis and environmental pollution, energy saving and emission reduction are of vital importance for social development. Large-scale utilization and centralized integration of renewable energy in China becomes an inevitable trend. However, wind and solar energy are much different from conventional resources, their inherent characteristics of stochastic volatility and intermittency bring massive challenges to the dispatch and operation of power systems [1–3]. In experimental research field, large-scale energy storage technologies have drawn much attention. They are expected to be effective ways to increase the penetration of renewable energy power generation [4–7].

The National Wind, Photovoltaic, Storage and Transmission Demonstration Project has been constructed by the State Grid Corporation. The project is located in Zhangbei county, Hebei province, which aims to increase the penetration of large-scale renewable energy generation.

At present, some achievements and application practices have been realized in renewable energy generation monitoring. Reference [8, 9] analyzes the technical requirements about wind power integrating into the grid and introduces a scheme of wind farm monitoring system which aims to make wind farm a predictable, adjustable and high-quality power supply. Reference [10] presents a structure of PV generation monitoring system for data processing and analysis, which also makes remote control available. However, the system does not meet the technical requirements of PV power generation with grid-friendly interaction [3]. Besides configuration, some key technologies in grid connection and data monitoring for a small-scale wind-solar complementary generation system are described in reference [11]. However, coordination control of wind power and PV power is not involved. Reference [12] proposes a solution of operation and control for hybrid energy storage system. Wind power fluctuations can be stabilized by two controllers which are interconnected with ultra-capacitor storage and battery storage respectively. But the battery storage cannot participate in the process of power regulation more flexibly. Reference [13] discusses the general principles of control strategy for wind, PV and storage co-generation system. But a complete set of implementation scheme is not put forward.

X. TENG, Z. GAO, Y. ZHANG, H. HUANG, L. LI, NARI Group Corporation, Nanjing 211102, China
(✉) e-mail: tengxianliang@sgepri.sgcc.com.cn
T. LIANG, Zhangjiakou Wind and Solar Power Energy Demonstration Station Co. Ltd, Hebei, China

Reference [14, 15] explains the operation and management technologies for large-scale wind power dispatch and control in provincial dispatching center. Control schemes of AGC and AVC are discussed. And the framework of on-line power and voltage control to wind farm are put forward at the same time.

In general, application of wind PV and storage co-generation monitoring system is still concerned with the theoretical research and simulation phase. There are few engineering applications and operating experiences. It is expected to do further research in models and strategies of multiple-source complementary coordination control.

Based on the National Wind, PV, Storage and Transmission Demonstration Project, this paper mainly studies the coordinated operation mechanism and focuses on three key dispatching technologies discussed in the following. In addition, the proposed technical schemes have been tested and verified in the practical project.

1) Day-ahead optimal dispatching: An optimization model aiming to maximize the benefits of the co-generation system is given in this paper. Meanwhile, penalties for multi-point coupling fluctuations of generating curves are considered and are added to the optimal object.
2) Active power coordinated control: An active power coordinated control strategy is proposed to satisfy different operation requirements including co-generation output fluctuation smoothening, planned generation output tracking and frequency regulation participating. Taken co-generation and single plant into account, this paper designs various combinational control modes, which can be flexibly configured.
3) Reactive power and voltage control: The general ideas for coordinated control are presented as follows. Wind turbines, PV inverters and storage devices are given priority during the regulating process. Static var generator (SVG) is expected to respond as quickly as possible. Capacitor switching acts as a supplementary method. Cooperative and coordinated control of reactive power resources can be realized in multiple modes, such as continuous or discrete mode, high speed or low speed mode and voltage command or reactive power command mode.

2 Framework of the wind, PV and storage co-generation system

The National Wind, PV, Storage and Transmission Demonstration Project plans to construct 500 MW wind farms, a 100 MW PV power station and a corresponding

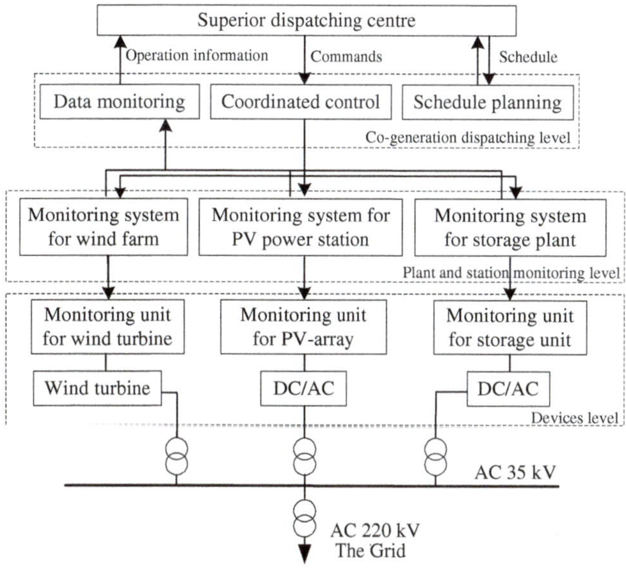

Fig. 1 Framework of the wind, PV and storage co-generation system

energy storage plant. The first-phase project was completed and put into operation on December 25, 2011. Construction of the first-phase project includes 100 MW wind farms, a 40 MW PV power station, a 20 MW energy storage plant and a 220 kV smart substation. At present, the project is the largest comprehensive renewable energy utilization platform worldwide, aiming to make the renewable energy involved predictable, controllable and dispatchable. Besides, the grid capability for large-scale renewable energy integrating can be effectively improved.

The general framework of the demonstration project consists of three levels as shown in Fig. 1. They are co-generation dispatching level, plant and station monitoring level, and devices level, among which, co-generation dispatching level is a bridge connecting the co-generation system to the superior dispatching centre. On one hand, this level designs open and coincident models of the co-generation system and provides operation analysis support for superior dispatching centre. On the other hand, as a controlling centre of the whole demonstration project, this level carries on intelligent analysis and decision-making, and the coordinated control among wind farms, PV power station and storage plant is realized. *Plant and station monitoring level* is used to monitor each plant and upload real-time operation information of every device. This level also receives control commands sent by co-generation dispatching level and distributes to control units in devices level simultaneously.

Wind, PV and storage co-generation panoramic monitoring system is developed and applied in co-generation dispatching level. The monitoring system offers a variety of advanced applications which mainly include Day-ahead optimal dispatching, active power coordinated control,

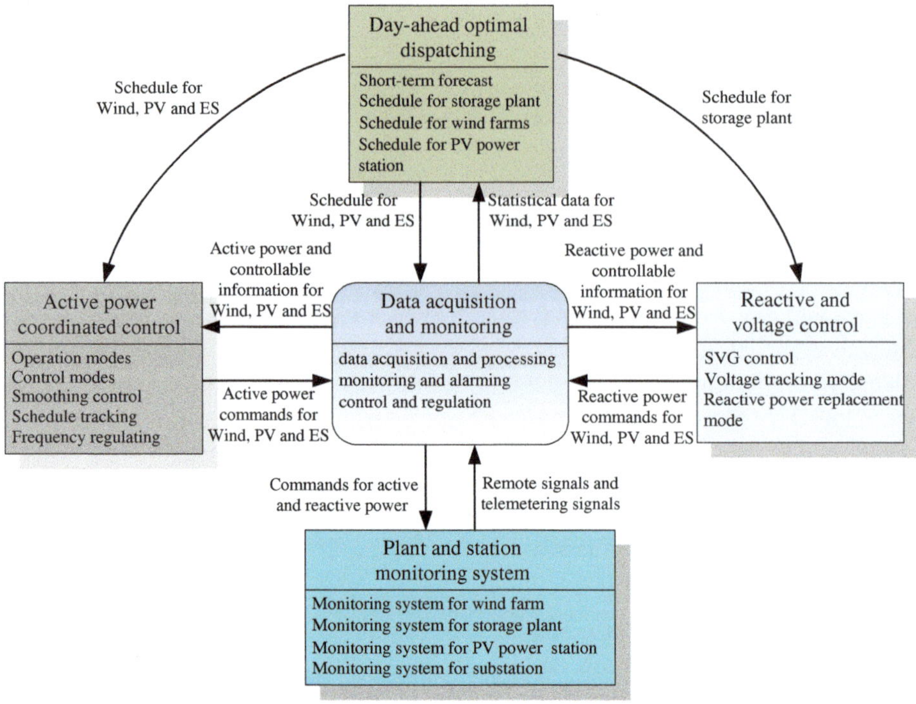

Fig. 2 Logical relationship among each applications of the wind, PV and storage co-generation panorama monitoring system

reactive power and voltage control, etc. Logical relationship among those applications is shown in Fig. 2.

3 Key technologies and their application

3.1 Day-ahead optimal dispatching

There are two key requirements of wind, PV and storage co-generation system in the day-ahead optimal dispatching. One is to keep generation curve smooth, which means to smooth out the fluctuations caused by the intermittent energy generation using bidirectional characteristics of storage devices. The other is to increase the integration capacity of intermittent energy. Based on short-term prediction of renewable energy generation, reasonable schedules are drawn up and expected to ensure maximum integration capacity of renewable energy.

3.1.1 Problem formulation

Smoothing model could be achieved in two ways: constrained or objective. The latter model is adopted here.

1) Objective function
   The objective of co-generation schedule is to maximize the deviation between integration revenue of the co-generation system and penalty cost of power fluctuation:

$$\max f = \sum_{t=1}^{T} [C_t P_t - \Delta S_t] \tag{1}$$

where T is the interval numbers in a schedule period; P_t is the total co-generation schedule over time interval t; C_t is the generation revenue per unit; ΔS_t is the penalty cost of power fluctuation. The objective is a superposition of generation revenue and penalty cost. Since generation revenue can be used to measure the integration benefits of co-generation directly, the penalty cost has also been given certain economic significance. The penalty cost is the mechanical wear of thermal units, caused by the regulation of power output to meet the requirement of co-generation schedule.
In order to ensure that the generation revenue and penalty cost are in the same dimension, the penalty cost is normalized to the dimension of generation revenue. The formulation is

$$\Delta S_t = \lambda C_t \Delta_t \tag{2}$$

where Δ_t is the power fluctuation of co-generation schedule over time interval t and obtained by formulating a smoothing model for co-generation schedule. $\lambda \in (0, 1)$ is the offset scale factor of generation revenue per unit caused by power fluctuation. Thus, when co-generation schedule fluctuates, the total revenue will be reduced after superimposed on the penalty cost and will in turn suppress the fluctuation of the schedule.

2) Constraints
The main constraints are listed as follows.

a) System power balance

$$\sum_{w=1}^{N_w} P_{w,t} + \sum_{v=1}^{N_v} P_{v,t} + P_{c,t} + P_{d,t} = P_t \qquad (3)$$

where N_W is the number of wind farms; $P_{w,t}$ is the power output of the wind farm w over interval t; N_V is the number of PV power stations; $P_{v,t}$ is the power output of the PV power station v over interval t; $P_{c,t}$ is the power of storage plant in charging mode over interval t; and $P_{d,t}$ is the power of storage plant in discharging mode over interval t.

b) Wind generation power output

$$0 \leq P_{w,t} \leq P_{0,w,t} \qquad (4)$$

where $P_{0,w,t}$ is the forecasting power output of the wind farm w over interval t.

c) PV generation power output

$$0 \leq P_{v,t} \leq P_{0,v,t} \qquad (5)$$

where $P_{0,v,t}$ is the forecasting power output of the PV power station v over interval t.

d) Inverter power output of storage plant in charging and discharging mode

$$-P_{inv} \leq P_{c,t} \leq 0 \qquad (6)$$

$$0 \leq P_{d,t} \leq P_{inv} \qquad (7)$$

where P_{inv} is the allowed inverter power output of storage plant.

e) Capacity constraints of storage plant
Charging energy for each interval:

$$E_{c,t} = \alpha P_{c,t} dt \qquad (8)$$

Discharging energy for each interval:

$$E_{d,t} = \beta P_{d,t} dt \qquad (9)$$

where α is the efficiency coefficient in charging mode; β is the efficiency coefficient in discharging mode.
The stored energy of storage plant at the end of each interval is

$$E_t = E_{t-1} + E_{c,t} - E_{d,t} \qquad (10)$$

$$E_0 = E_{ini} \qquad (11)$$

where E_{ini} is the initial stored energy of storage plant.
Capacity limits:

$$E_{\min} \leq E_t \leq E_{\max} \qquad (12)$$

where $E_{\max}$ and $E_{\min}$ are the upper and lower capacity limit of storage plant respectively.

3) Smooth modeling
For co-generation system, its modeling solution should consider multi-period coupling changes to avoid sharp swings of total power output. A power fluctuation penalty is added into the objective function. There are two methods for modeling of the penalty: three-point coupling smoothing model and two-point coupling smoothing model.

a) Three-point coupling smoothing model
This model considers the changes of the co-generation power output during three adjacent periods. And the mathematical model is

$$\Delta_t = |(P_{t+1} - P_t) - (P_t - P_{t-1})| \qquad (13)$$

It's mainly used for reducing the frequency of fluctuations.

b) Two-point coupling smoothing model
This model considers the changes of the co-generation power output during two adjacent periods. And the mathematical model is

$$\Delta_t = |P_t - P_{t-1}| \qquad (14)$$

This model is mainly used for reducing the magnitude of fluctuations.

Linear programming (LP) algorithm has been widely applied in practical systems because of its advantage of fast computational speed and reliable convergence. After linearization of optimization model, this paper solves the optimal problem by adopting a commercial package CPLEX. CPLEX is a sophisticated and computationally efficient solution that can handle thousands of variables.

3.1.2 Case study

Based on measured data of the demonstration project, a study case is constructed and a series of numerical analysis is carried out. The case includes two wind farms with installation capacity of 100 MW, a PV power station with installation capacity of 40 MW, and a storage power plant with capacity of 20 MW. The co-generation system is connected to the grid via a 220 kV substation. On the basis of daily forecasting data of wind power and PV power, day-ahead generation scheduling is optimized and drawn up in 96 periods. The time period is 15 min.

Related information of the storage plant is described as follows. Upper and lower capacity limits are 50 MWh and 5 MWh respectively. Initial stored energy is 10 MWh. Allowed maximum inverter power output is 50 MW. The efficiency coefficient in charging mode is 0.8, while the efficiency coefficient in discharging mode is 1.

Predictive power outputs of wind farms and PV power station are shown in Fig. 3.

Fig. 4 shows the accumulated forecasting power of wind farms and PV power station. The optimized total power of the co-generation system is drawn as well. The charging and discharging result of the storage plant is shown in Fig. 5. Stored energy of the storage plant is described in Fig. 6.

The above results show that the optimized fluctuation of accumulated power of wind farms and PV power station is restrained by the storage plant. The power output of the co-generation system keeps at a certain constant value. The curve of total power output is comparatively smooth and ideal.

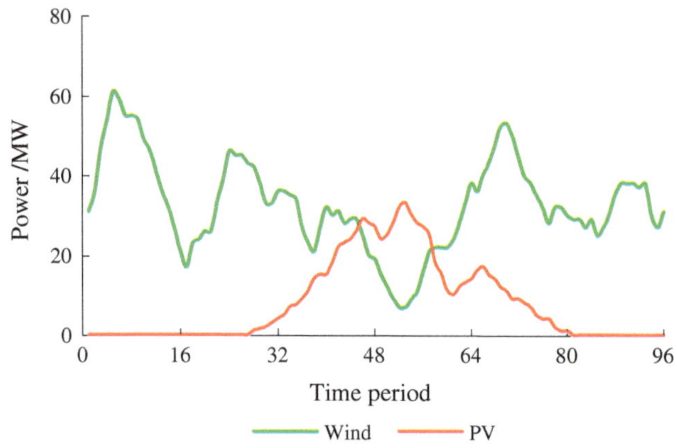

Fig. 3 Forecast of renewable energy generation

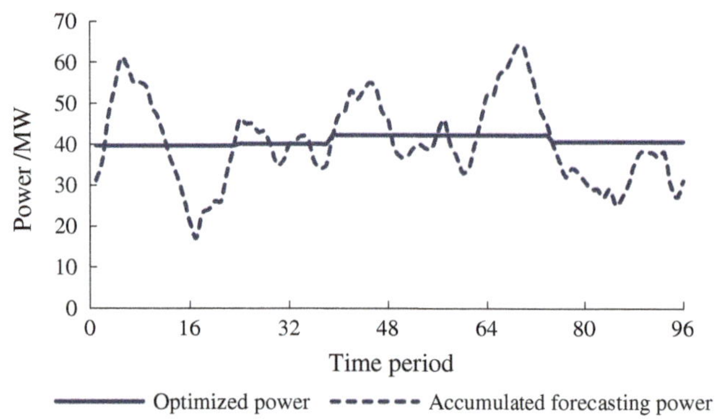

Fig. 4 Optimized results of the co-generation system

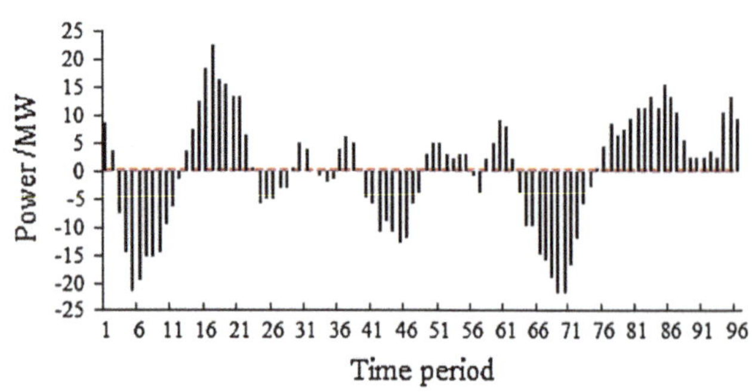

Fig. 5 Optimized charging and discharging result of the storage plant

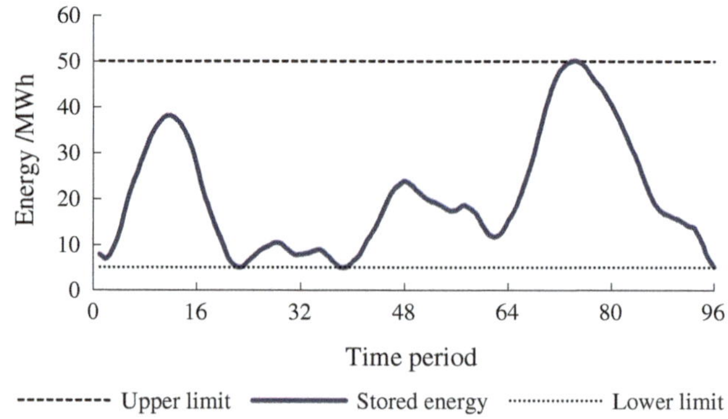

Fig. 6 Stored energy of the storage plant

Fig. 6 shows that stored energy of the storage plant reaches its lower limit at period 23 and 39, and it reaches the upper limit at period 74. These limitations lead to the variations of the co-generation system power output.

3.2 Active power coordinated control

Active power coordinated control is utilized for further promoting the grid-friendly characteristic of the co-generation system.

3.2.1 Overall scheme

Real-time operation of the co-generation system involves various operating configurations. Its control objection may keep changing according to different power regulation demands. They all raise the difficulty of dispatching. As a result, this part presents two fundamental control modes: plant and station control mode, co-generation control mode.

Regarding diverse features and control requirements, control modes for wind farms, PV power station and storage plant are designed to cope with different scenarios. These control modes include output maximization mode, schedule tracking mode, output limit mode, reserve mode and automation mode.

Co-generation control mode is used to describe the global run mode of the co-generation system. In this paper, a kind of co-generation operation with coordinated multi-objective control is realized, the power generation resources are all combined properly. For each operation mode, corresponding control modes are listed in Table 1.

3.2.2 Control strategy

The control objects are determined by the operation mode, while the control objectives are decided by the

Table 1 List of operation and control modes

Operation modes	Control modes			
Wind-PV-ES	Smoothing	Tracking AGC attaching superior dispatching centre	Tracking schedule	Tracking a setting object
Wind-ES	Smoothing	Tracking schedule	Tracking a setting object	
PV-ES				
Wind-PV	Free power generation	Tracking schedule	Tracking a setting object	
Wind alone				
PV alone				
Storage alone	Frequency regulating	Tracking schedule	Tracking a setting object	

control mode. In accordance with features among objective functions, control modes can also be distinguished as smoothing control mode and tracking control mode.

1) Smoothing control mode
In smoothing control mode, the storage plant is used to smooth the fluctuations of power output. The fluctuation is possibly caused by wind power, PV generation, or wind and PV co-generation. For the storage plant, a state of charge (SOC)-feedback, which is built upon traditional one-order low pass filter to avoid excessive charge-discharge, is added to the control loop. The control block diagram is interpreted in Fig. 7.
where P_{wp} is the power input of wind and PV co-generation; P_{wpb} is the power output of wind, PV and storage co-generation; P_{b0} is the charge-discharge power which is obtained by one-order low pass filter; P_b is the revised charge-discharge power which is gained by accounting SOC; T is the time constant of one-order low pass filter; B is the compensation coefficient; K is the feedback gain; and O_{offset} is the offset compensation and used to guarantee the storage devices to be in the optimum area.
2) Tracking control mode
Wind, PV and storage co-generation system can provide steady power output as planned. But the premise is the coordination with the storage plant and real-time bias compensation. This part puts forward a

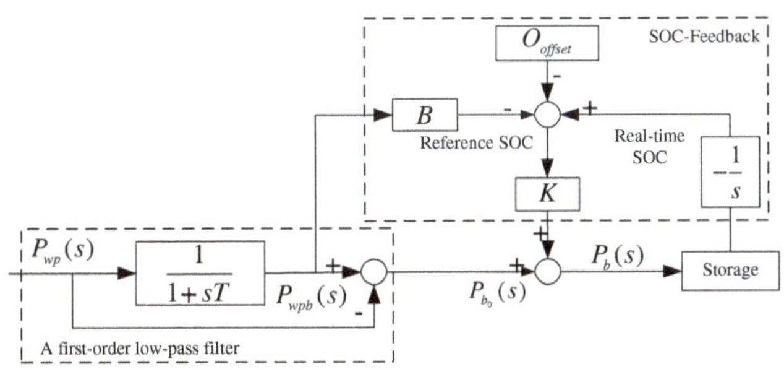

Fig. 7 Block diagram of the SOC-feedback control

kind of tracking control strategy. The principle of the strategy is to bundle wind energy generation and PV generation, then decouple the control of storage energy. The intermittent energy generation plants and the storage plant have relatively independent control target.
In order to maximize the utilization of wind and PV resources, wind farm and PV should produce maximum power if the storage plant has sufficient storage space. Thus, for wind and PV co-generation, the objective function can be written as

$$P_{wp\_des}(t) = P_{ref}(t) + \Delta P_{b\max}(t) \tag{15}$$

where $P_{wp\_des}(t)$ is the ideal objective of wind and PV co-generation at period t; $P_{ref}(t)$ is the tracked reference objective at period t; and $\Delta P_{b\max}(t)$ is the allowed maximal charged power of the storage plant at period t.
After decoupling, the real-time output objective of the storage plant is the deviation between real-time co-generation power and reference objective. The ideal adjustment can be written as

$$\Delta P_b(t) = P_{ref}(t) - P_{wpb}(t) \tag{16}$$

where $\Delta P_b(t)$ is the ideal adjustment of the storage plant at period t; and $P_{wpb}(t)$ is the output of wind and PV co-generation at period t.

3.2.3 Application

1) The effect of smoothing control mode
In Wind-PV-ES operation mode, power output with smoothing control mode (within an hour) is shown in Fig. 8.
The grid-connected technology specification of wind farm and PV power station has clarified the maximum variation of power fluctuations, but the degree of fluctuations is not evaluated. To quantify the smoothing effect of the storage plant, a volatility index α is introduced.

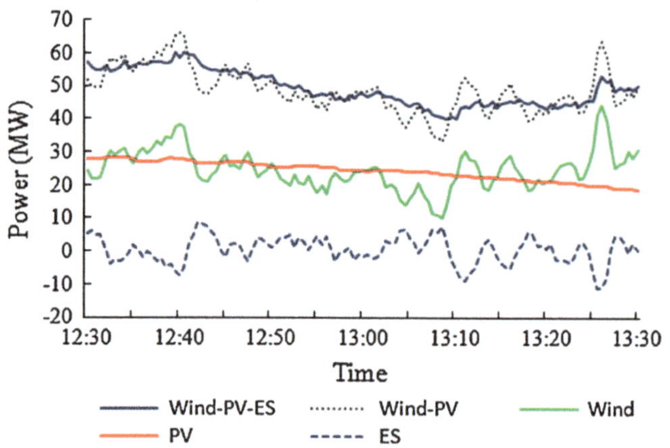

Fig. 8 Power output of Wind-PV-ES co-generation in smoothing mode

$$\alpha = \frac{P_{\max,1\min} - P_{\min,1\min}}{P_{avg,1\min}} \times 100\% \tag{17}$$

where $P_{\max,1\min}$ is the upper limit of power output within 1 min; $P_{\min,1\min}$ is the lower limit of power output within 1 min; $P_{avg,1\min}$ is the average power output within 1 min.

Fig. 9 draws the volatility index α of Wind-PV-ES co-generation power output before and after smoothing. It can be found that the volatility index of Wind-PV co-generation power output is generally around 10%. After smoothing, the average volatility index decreases from 7.97% to 4.39%. It shows that the volatility index is notably improved.

2) The effect of tracking control mode

In Wind-PV-ES operation mode, the effect of tracking control mode (within an hour) is shown in Fig. 10.

In Wind-PV-ES operation mode, control precision within an hour in tracking control mode is listed in Table 2.

As is shown above, in case the regulation reserve of the storage plant is sufficient, control errors within an hour are all below 10%. The proportion of the points with a

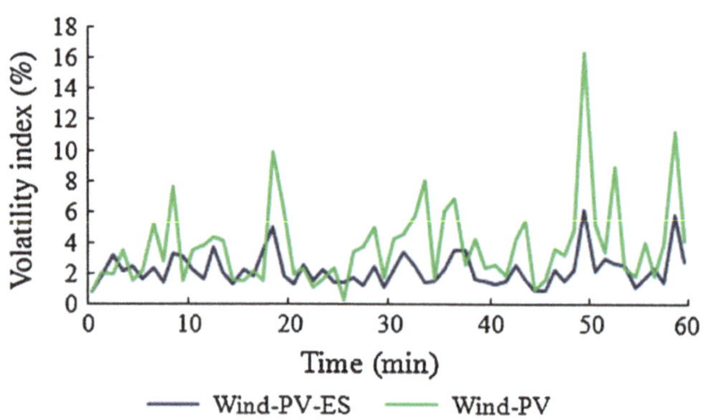

Fig. 9 Comparison of volatility

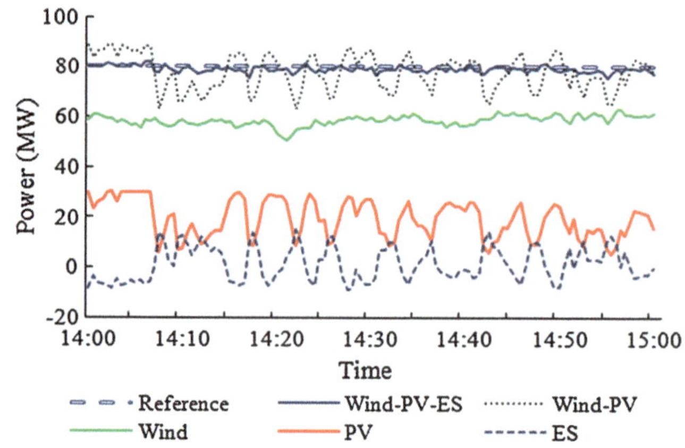

Fig. 10 Power output of Wind-PV-ES co-generation in tracking mode

Table 2 Control precision in tracking mode

	Control error/η		
	<10%	<5%	<3%
Percent	100%	98.35%	92.56%

control error percent below 5% is 98.35%. The proportion comes to 92.56% with a control error below 3%. This verifies the effectiveness of the control. The power output characteristic of the Wind-PV-ES co-generation system can be similar to that of the conventional power sources.

3.3 Reactive power and voltage control

Each reactive power regulating device has its difference in regulating performance in the co-generation system. So a major point of this part is to research the coordinated control among various reactive power sources.

3.3.1 Overall scheme

According to the hierarchical voltage regulation principle [16], the structure of reactive power and voltage control can be divided into two levels in this system.

1) Primary control: In this level, SVG attempts to compensate voltage fluctuation caused by rapid variations of renewable energy. It is expected to maintain the constancy of the voltage of 35 kV bus. This cycle will be in milliseconds.
2) Secondary control: The object is to keep the voltage of the injection point at the desired reference values. It is realized through coordinated control among SVGs, wind turbines, inverters and capacitors. The control

cycle will be set up from 30 s to 5 min.

SVG is expensive and limited by its installed capacity, which makes it a precious reactive power resource. Thus, the general ideas for coordinated control are as follows. Wind turbines, PV inverters and storage devices are given priority during regulating process. SVG is expected to respond as soon as possible. Capacitor switching acts as a supplementary method. Taking full advantage of reactive power regulation capability of renewable energy generation, we can own maximum dynamic reactive power reserve of SVG and carry out minimum number of capacitors action.

3.3.2 Control strategy

Reactive power and voltage control can be divided into two steps: statistical calculation for regulatory demand and assignment among multi-sources as illustrated in Fig. 11.

Assignment strategy among multi-sources is the core link throughout this process. Reactive power demand is distributed among those reactive power sources in reasonable orders. There are two different implementations that include voltage tracking mode and reactive power replacement mode. The latter one is expected to keep maximum dynamic reactive power capability of SVG and provide constant total reactive power.

Based on calculation results, adjustment target for each feeder will be determined according to equal power factor principle.

3.3.3 Application

1) The effect of voltage tracking mode

Fig. 12 shows the effect of voltage tracking control. Take the following case as an example. The reference voltage was changed from 230 kV to 225 kV at 14:31:10 on December 23, 2011. The dead zone of command was set at 1 kV. At this point, real-time voltage was 230.5 kV, with a 5.5 kV deviation from reference voltage, which obviously exceeded the dead zone. Hence the co-generation monitoring system started to control immediately. At 14:34:10, real-time voltage was adjusted to 230.5 kV, with a 0.5 kV deviation from reference voltage which met the control requirements. Meanwhile, reactive power of the injection point was varied from −18 MVar to −30 MVar.

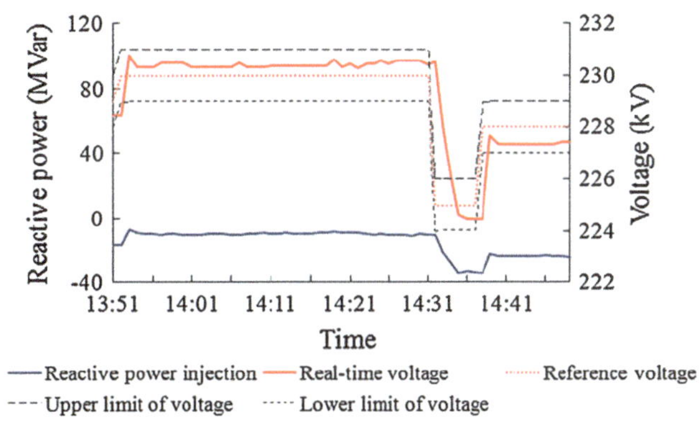

Fig. 12 Effect of voltage tracking mode

Fig. 13 shows the control sequence of all devices in the same section. Voltage of the 35 kV bus, which connects with SVG, dropped gradually from 37.2 kV to 35.7 kV. Reactive power of Xiaodongliang wind farm varied from 3.8 MVar to −3 MVar. And the voltage of PV power station decreased from 5 MVar to 0 MVar.

It can be concluded that the above-mentioned regulating process can rapidly track the voltage of the injection point at 220 kV bus. And the control effect is satisfying. The rate of regulation is greater than 1 kV/min.

2) The effect of reactive power replacement mode

While active power is stable, co-generation monitoring system will start to adjust low-speed reactive power resources to substitute for capacity of SVG, thus it increases the dynamic reactive power reserves. The capability to smooth the rapid voltage fluctuations will be reserved consequently. Fig. 14 shows the control sequence of all devices on Dec.19, 2011.

During that period, active power of the co-generation system was relatively steady. The voltage of the 35 kV bus remained around 37 kV. Reactive power of the SVG

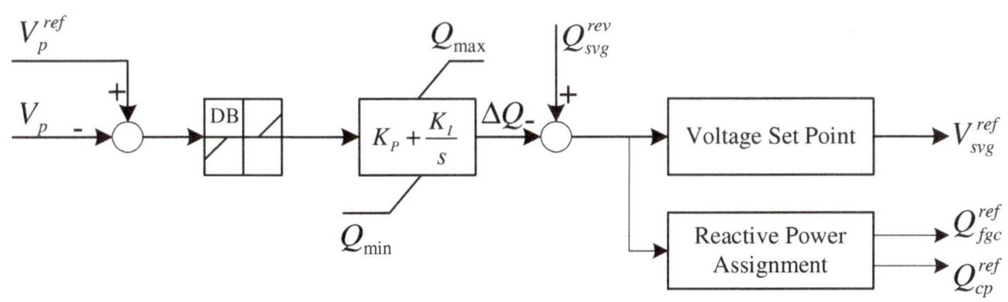

Fig. 11 Schematic diagram of reactive power and voltage control

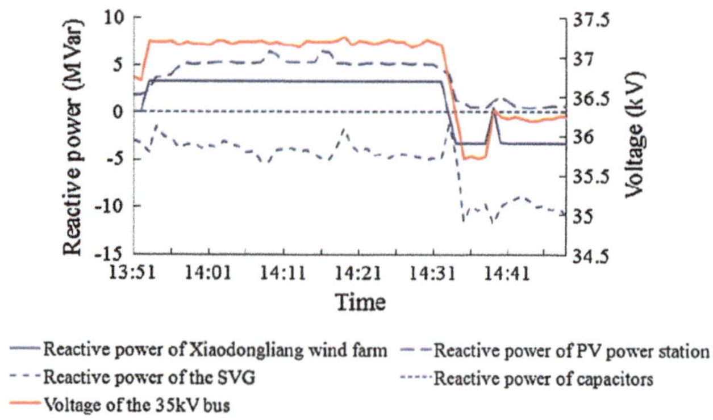

Fig. 13 Control sequence of reactive power sources for voltage control

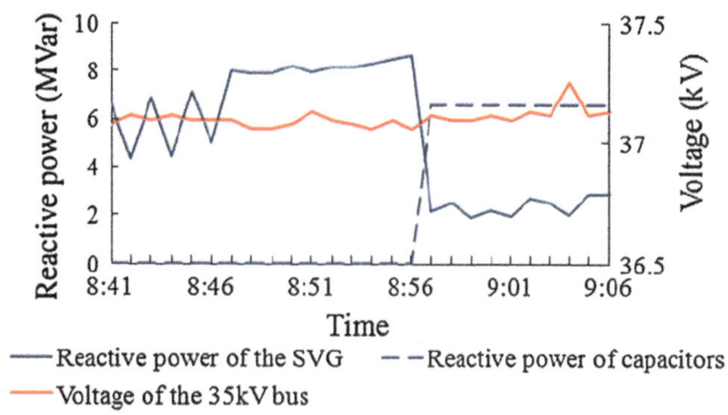

Fig. 14 Control sequence of reactive power sources under reactive power replacement mode

kept around 8 MVar before 8:56. At 8:56, capacitors were ordered to put into regulation, reactive power of the SVG started to adjust and gradually dropped to 2 MVar. In other words, dynamic reactive power capability of SVG was improved.

4 Conclusions

Wind, PV and storage co-generation monitoring system has been implemented in the demonstration project. With several months of trial operation, the project is stable and reliable. Complementary advantages of each participant are made to the most. Power output of the co-generation system is grid-friendly and is similar to conventional power sources. Regulation characteristic of grid-connected voltage is improved. Grid capability to accept renewable energy is effectively improved.

Both theory and practice prove that it is feasible to make use of renewable energy resources through wind, PV and storage co-generation mode. This will be one of the developing directions for utilization of renewable energy resources and a demonstrative case for large-scale renewable energy power integration.

References

[1] Wang CS, Wang SX (2008) Study on some key problems related to distributed generation systems. Automat Electr Power Syst 32(20):1–4 (in Chinese)

[2] Sun YZ, Wu J, Li GJ (2007) Influence research of wind power generation on power systems. Power Syst Technol 31(20):55–62

[3] Zhao ZM, Lei Y, He FB et al (2011) Overview of large-scale grid-connected photovoltaic power plants. Automat Electr Power Syst 35(12):101–107 (in Chinese)

[4] Barton JP, Infield DG (2004) Energy storage and its use with intermittent renewable energy. IEEE Trans Energy Convers 19(2):441–448

[5] Tan CW, Green TC, Hernandez-Aramburo CA(2007) A stochastic simulation of battery sizing for demand shifting and uninterruptible power supply facility. In:Proceedings of the IEEE Power Electronics Specialists Conference(PESC'07), Orlando, FL, USA, 17–21 June 2007, pp 2607–2613

[6] Bhargava B, Dishaw G (1998) Application of an energy source power system stabilizer on the 10 MW battery energy storage system at Chino substation. IEEE Trans Power Syst 13(1):145–151

[7] Billinton RB (2006) Reliability considerations in the utilization of wind energy, solar energy and energy storage in electric power systems. In: Proceedings of the International Conference on Probabilistic Methods Applied to Power Systems (PMAPS'06), Stockholm, Sweden, 11–15 June 2006, 6 pp

[8] Huang SZ, Dai ZQ, Ma XP (2010) Design and realization of computer integrated monitoring system of wind farm. Jiangsu Electr Eng 29(1):38–40 (in Chinese)

[9] Qiao Y, Lu ZX, Xu F et al (2011) Study on the integrated monitoring & control platform of wind farms. Power Syst Protect Contr 39(6):117–123 (in Chinese)

[10] Wang CF, Li R, Liu HR et al (2011) The design and implementation of photovoltaic power generation monitor control system. Telecommun Electr Power Syst 32(6):53–57 (in Chinese)

[11] Zhao M, Lu JZ, Zhou RJ et al (2011) The small-scale wind-solar hybrid grid-connected generation system. Hunan Electr Power 31(1):10–13 (in Chinese)

[12] Yu P, Zhou W, Sun H et al (2011) Hybrid energy storage system and control system design for wind power balancing. Proc CSEE 31(17):127–133 (in Chinese)

[13] Li JX, Zhang JC, Zhou Y (2011) Energy management strategy of wind-PV-ES hybrid power system. East China Electr Power 39(12):2026–2028 (in Chinese)

[14] Feng LM, Fan GY, Zheng TY et al (2011) Design of wind power dispatch automation system in Jilin power grid. Automat Electr Power Syst 35(11):39–43 (in Chinese)

[15] Bai YX, Fang DZ, Zhu CS (2011) Design and application of wind farms dispatching and management technical support system in Inner Mongolia power grid. Automat Electr Power Syst 35(7):86–90 (in Chinese)

[16] Vu H, Pruvot P, Launay C et al (1996) An improved voltage control on large-scale power system. IEEE Trans Power Syst 11(3):1295–1303

Xianliang TENG received the B.E. degree in 2000 from Zhengzhou University of Technology and M.E. degree in 2011 from Southeast University. Currently, he is working at NARI Group Corporation, China. His work has been in power system dispatch and control, renewable energy generation and its integration.

Zonghe GAO received the M.E. degree in 1989 from State Grid Electric Power Research Institute (SGEPRI), Nanjing, China. Currently, he is the Managing Director of Nanjing Power Grid Dispatching & Control Technology Branch Company, NARI Group Corporation. His work has been in the power system dispatch and control.

Yingyuan ZHANG received the Ph.D degree in 2011 from Hefei University of Technology, Anhui, China. Currently, she is working at NARI Group Corporation, China. Her work has been in the power system dispatch and control.

Hua HUANG received the M.E. degree in 2001 from Huazhong University of Science and Technology, Wuhan, China. Currently, he is working at NARI Group Corporation, China. His work has been in the power system dispatch and control, especially voltage control and reactive power optimization.

Lili LI received the M.E. degree in 2011 from State Grid Electric Power Research Institute (SGEPRI), Nanjing, China. Currently, he is working at NARI Group Corporation. His work has been in the power system operation and optimization.

Tingting LIANG received the B.E. degree in 2004 from Northeast electric power University, Jilin, China. Currently, she is working at Zhangjiakou Wind and Solar Power Energy Demonstration Station Co. Ltd. Her work has been in the renewable energy generation and its integration.

11

Renewable energy based microgrid system sizing and energy management for green buildings

Y. V. PAVAN KUMAR (✉), Ravikumar BHIMASINGU

Abstract The objective of this paper is to model a hybrid power system for buildings, which is technically feasible and economically optimal. With a view to promote renewable energy sources, photovoltaics and wind turbines are integrated with the grid connected building. The system is modeled and the optimal system configuration is estimated with the help of hybrid optimization model for electric renewables (HOMER). The logic is illustrated with a case study based on the practical data of a building located in southern India. This building is associated with 3.4 MWh/day priority load (peak load as 422 kW), as well as 3.3 MWh/day deferrable load (peak load as 500 kW). Sensitivity analysis is performed to deal with uncertainties such as the increase in electricity consumption and grid tariff, environmental changes, etc. From the simulation result, it is observed that the designed system is cost effective and environment friendly, which leads to 6.18 % annual cost savings and reduces CO_2 emissions by 38.3 %. Sensitivity results indicate that the system is optimal and adaptable in a certain range of unanticipated variances with respect to best estimated value. Finally, an energy management strategy is developed for the optimal system to ensure reliable power during contingency and disturbances. The green and hybrid power system designed can be adaptable to any critical and large consumers of urban buildings.

Y. V. PAVAN KUMAR, R. BHIMASINGU, Department of Electrical Engineering, Indian Institute of Technology Hyderabad (IITH), Hyderabad, India
(✉) e-mail: ee14resch01008@iith.ac.in
R. BHIMASINGU
e-mail: ravikumar@iith.ac.in

Keywords Renewable energy sources (RES), Hybrid power systems (HPS), Photovoltaic (PV), Wind turbine (WT), Low-carbon electricity, Energy management, Total net present cost (TNPC)

1 Introduction

India, a dwelling place for 1.21 billion people, which is over 17 % of the world's population, has a seemingly unquenchable thirst for energy. One blaring result of its meteoric growth is the widening gap between the energy demand and the supply; therefore, the government is paying increased focus to bridge the gap by capacity addition. In chorus, attention is being paid to that the growth should be in a sustainable manner while addressing the climatic change concerns. In 2008, with the announcement of the National Action Plan on Climate Change (NAPCC), there is a marked shift in policy to diversify the energy mix to the lower carbon intensity [1]. Besides, low-carbon generation systems bring competition in electricity markets and are in turn improves the country's sustenance and economy [2]. Nowadays, the country is promoting renewable energy to augment the total power supply and to meet the rural needs either by augmenting grid supply or by off-grid supply.

The statistics show that approximately 38 % of the total electric energy production is being consumed by the industrial sector, which is playing a crucial role in the economic growth of the country. With India emerging as an IT/BPO hub, erratic power supply hampers the business process, which ultimately reflects on the country's economy. Besides, cross subsidies, higher tariff, power outages, load shedding and inconsistent quality of power supplied by utilities are other factors that adversely affect such industries. With the demand far outstripping the electricity generation in the country, many industries are now looking at alternatives based on the guidelines and roadmaps given by various forums and government bodies for clean energy generation [3, 4]. The solution as envisioned by many organizations is to set up captive power plants for electricity generation or diesel generator set as a back-up, which is again not a clean energy.

In this framework, the hybrid and green power system uses renewable energy resources that are available at the site of use and are used to produce electricity in and around the premises of the building in an economic way. An energy management strategy is developed, which ensures uninterruptable power supply to all the priority loads connected to the building. Optimization result shows that the system is cost effective and reliable with less carbon emission. Sensitivity analysis shows that the system performance is nearly optimal in all scenarios.

1.1 Literature review

Below literature indicates some of the related works carried out in the design of low-carbon electricity systems. Various studies were done on microgrid formation for a particular combination of distributed units in a location. This paper deals with the formation of microgrid with various combinations of distributed resources to achieve a sustainable design for the low-carbon power system. Further, it compares all the combinations in terms of cost and carbon emissions to select best sized microgrid for a specific case study. Finally, optimum energy usage is achieved by the design of energy management strategy.

Reference [5] identified and suggested some insights for the people live in Germany and USA to choose and develop low-carbon electrical power generation systems. It also presented the advantages of low-carbon initiatives and crisis with the nuclear based generation systems briefly. Reference [6] presented the design of low-carbon optimum evaluation model by using multi-objective function to improve the efficiency of carbon footprint calculations. This considers dealer manufacturing capacity, green-house gas values, product components transport modes, design phase costs, as well as the product decisions made by enterprises. Reference [7] presented the scenarios and consideration of combined nuclear and renewable based energy systems design in Japan in 2030 to adhere the country's policies of CO_2 emission reduction. The safe and clean electricity was analyzed for "smart control" on both demand and supply sides. Reference [8] presented whole-systems method to appreciate the role of grid-scale electricity storage by concurrently optimizing investment into network and storage capacity, new generation. This minimizes system operation cost by considering system's reserve and security constraints. Reference [9] presented an integrated methodology that considered demand responses and renewable generation as inputs for the distribution system planning to make the system as a low-carbon sustainable system. This method optimized the allocation of renewable/non-renewable units and smart metering units simultaneously to minimize total carbon emissions cost. Similar integrated model for low-carbon power generation in China was presented in [10]. Reference [11] presented a comprehensive model to study low-carbon power system dispatches. The operating mechanisms and characteristics were analyzed by considering feasible power output limits, rates, and the carbon emissions. Reference [12] presented a generation output model to evaluate the power dispatch problems under low-carbon and economy systems development. In [13–16], similar analysis was carried out for the design of low-carbon electricity systems and study on their impacts on the energy crunches for different cases.

Reference [17] presented the design of hierarchical control for microgrids. Energy management system was designed to achieve a reasonable balance between generation and battery usage. However, the system was developed only for island mode but not for grid tied mode. Reference [18] presented the dynamic behavior of a mixed system consisting of a wind farm and a diesel group supplying a load, under different disturbances. Reference [19] presented two droop control strategy for distributed generation in interconnected grids. Reference [20] presented a simulation model to investigate specific hybrid architecture, which is based on the use of DC bus, DC/AC and DC/DC converters for power sources. This increases number of converters usage and leads to high cost and poor quality. Reference [21] had done an optimal cost analysis of hybrid renewable system using hybrid optimization model for electric renewables (HOMER). The real time cost analysis and system optimization was done based on the load profile, solar radiation and wind speed. Reference [22] and [23] described the integration of wind power into traditional diesel-only power system, which reduced greenhouse gases and thereby global warming. The wind-diesel power plant was designed, and the operational and economic analysis were compared with traditional diesel system using HOMER. In [24], the distributed energy resource system was designed and simulated. The cost analysis was done by calculating net present value. Reference [25] presented microgrid with micro-controller based energy management unit. Hydro generators, PV panels, and diesel generators were integrated to study the generation patterns under various ecological conditions. Reference [26] presented a voltage and frequency control strategy for active island microgrids design. There are few literatures available on the study of economic and technical feasibility of microgrids and hybrid systems in [27–33]. However, most of these were designed for standalone systems of remote areas, where it is either difficult or impossible to extend the grid service. And these studies considered the loads in a nonspecific manner. This paper focuses on the grid-connected HPS developed for a practical case study.

2 Hybrid power based microgrid description

Microgrids are formed by the hybrid interconnection of various units such as AC and DC energy sources, storage, AC and DC loads, AC/DC, DC/DC, or DC/AC converters. The selection of any architecture for a location depends on the geographical, economical, and technical factors at that location. The architectural segregations are as follows.

1) In the view of operation: stand-alone/off-grid/islanding mode and grid connected mode architectures.
2) In the view of integration: centralized DC bus (Fig. 1), centralized AC bus (Fig. 2), distributed AC bus (Fig. 3), hybrid coupled (AC–DC) bus (Fig. 4) architectures.

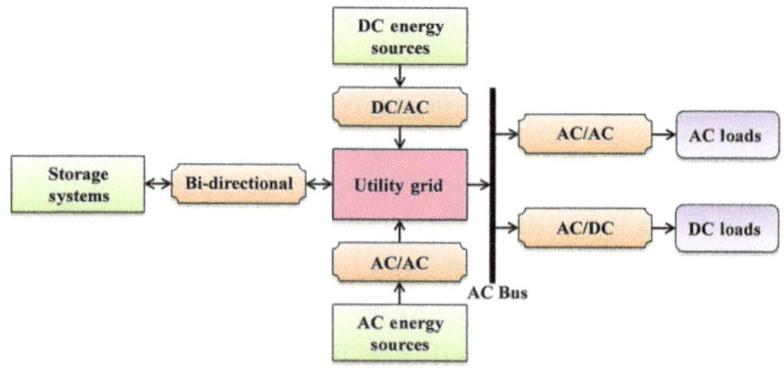

Fig. 3 Distributed AC bus architecture of microgrid

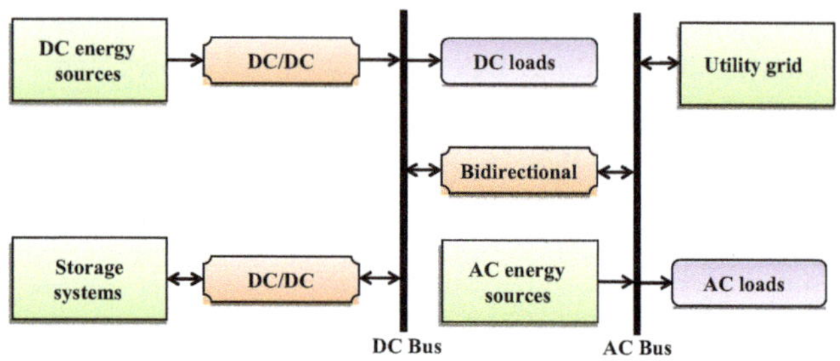

Fig. 4 Hybrid coupled (AC–DC) bus architecture of microgrid

3 Case study description

In this case study, the considered building is an organization located at Hyderabad, India. The area enclosed by the building is about 125000 sq ft. The existing architecture of the building is that it is connected to the utility grid that serves the entire load connected. A diesel generator set was also installed to meet certain portion of the load at the time of grid outage. The new proposed architecture of the system is that the electric energy produced from the PV and WT augments the grid supply to meet the demand. A converter is connected to convert the DC power produced by the PV array to AC, as the entire load served is the AC type. A 396 V battery bank is also connected to the DC bus, which can store energy and be used as a backup energy source. The system is modeled, and the best configuration of hybrid system is estimated by using HOMER [34, 35]. The profiles of load demands and energy resources are given as follows.

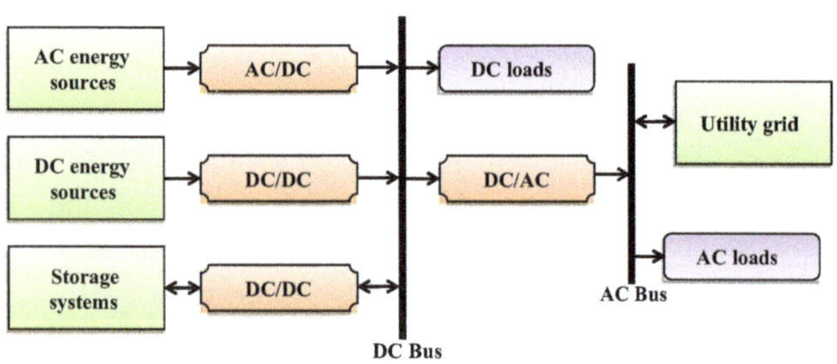

Fig. 1 Centralized DC bus architecture of microgrid

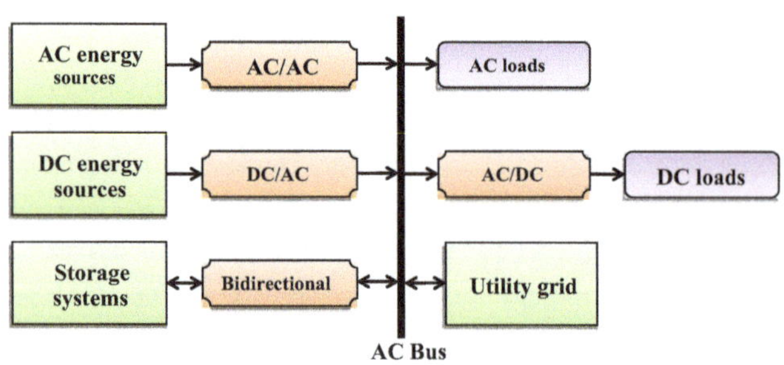

Fig. 2 Centralized AC bus architecture of microgrid

3.1 Load profile

The analysis does not perceive the entire load connected to the building. Moreover, it categorizes the load in the most smart and reasonable way. The type of load that is connected to this building and other such buildings are heating, ventilation, and air conditioning (HVAC) load, uninterrupted power supply (UPS) load, air handling unit (AHU) load, lighting load and other utility loads. These loads are further classified as shown in Table 1.

3.1.1 Priority load

These loads are electrical demand that the power system must serve at a specific time. In any business or industry,

Table 1 Categorized breakdown of building loads

Load category	Particulars	Total energy consumption/%
Priority load	UPS	32
	Lighting	10
	Elevators	7
	Others	2
Deferrable load	HVAC	31
	Pump Load	8
	AHU	6
	Others	4

computers are the integral part, thus all the UPS load is considered as a priority load. Other loads associated with this category are lighting load and some utility loads like elevators. As these loads varies according to time of the day and also do not attain the same value for weekdays and weekends, a separate survey for this category of loads is needed. The variation of this load is recorded for a year (separately for weekday and weekend), to make the data more realistic. An average 5 % day-to-day and hour-to - hour randomness are added to the practical data such that load profile of each day is unique. Figs. 5 and 6 illustrates a typical profile of the priority load.

3.1.2 Deferrable load

Deferrable loads can be classified as an electrical demand that can be met anytime within a definite time interval. Typical example of this load type is the chiller load, AHU load, water pump load, battery charging station, etc. Certain amount of energy is stored in such type of devices and is generally switched on intermediately. For example, after the chiller unit and the pump associated with it are charged to its full capacity, it can serve the building for certain period without demanding electrical input. This type of load is often advantageous as one can track the level of energy that is stored in this load type, and can put excess renewable energy into this as long as energy level is above zero. Once the energy level drops to zero, these loads can be treated as a priority load and can be served immediately. The typical monthly variation of such load type is shown in Fig. 7.

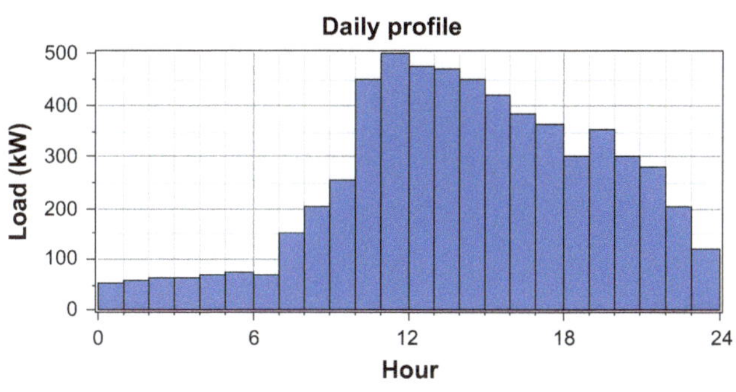

Fig. 5 Hourly profile of priority load in a weekday

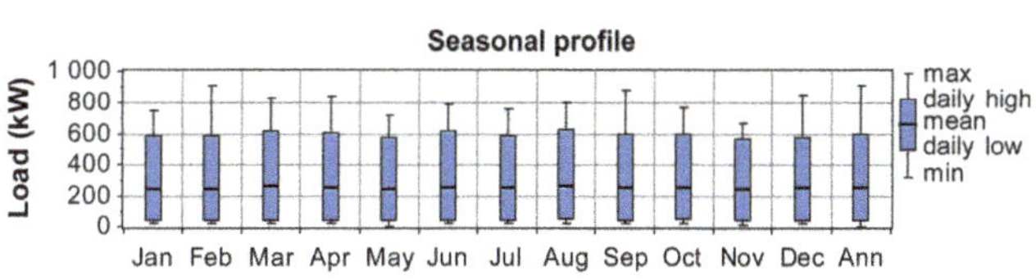

Fig. 6 Monthly profile of priority load in a year

3.2 Solar and wind resources

The considered building is located at a latitude of 17.37 (North), longitude of 78.47 (East), and an altitude of 536 m. It experiences a combination of tropical wet and dry climate [36]. The place is blessed with around 5.34 kWh/m^2 yearly average daily solar radiation and more than 6 kWh/m^2 maximum radiation in March, April and May. The radiation data can be obtained from the database of NASA [37]. The monthly average of solar radiation and clearness index is shown in Fig. 8. Recorded data shows that the monthly wind speed in a year is 4.141 m/s. Fig. 9 illustrates the wind speed variation throughout the year [38]. From the solar and wind resource data, it is clear that

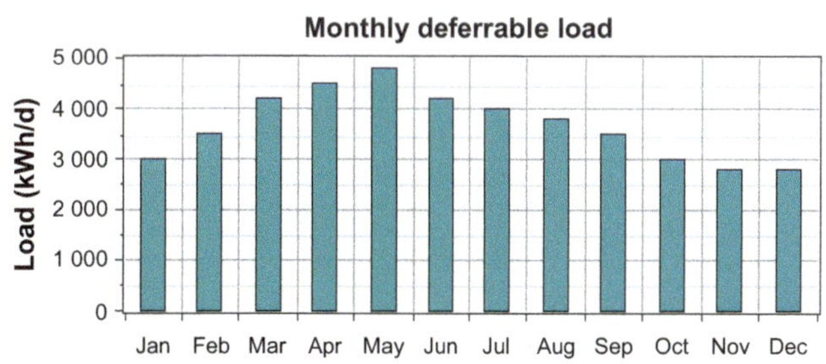

Fig. 7 Monthly profile of deferrable load in a year

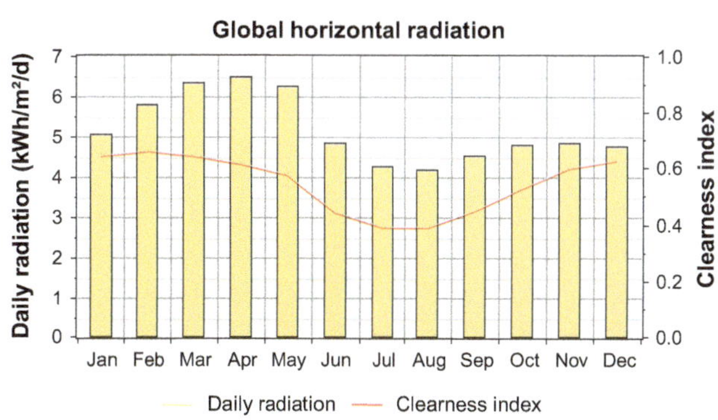

Fig. 8 Profile of solar radiation and clearness index in a year

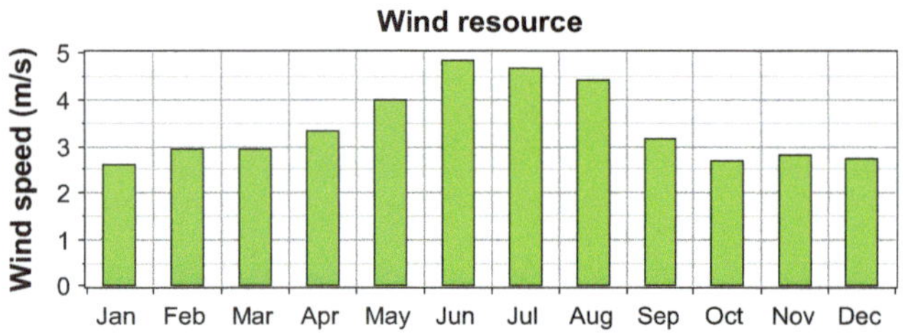

Fig. 9 Monthly profile of wind resource in a year

June, July, and August are experiencing less solar radiation, but the wind speed compensates this effect and makes the total natural resource available throughout the year.

4 Hybrid power system modeling and simulation

4.1 Introduction to HOMER and system simulation

HOMER is a micro-power optimization tool, which is developed by the U.S. National Renewable Energy Laboratory (NREL). It models the physical behavior of a power system and calculates its lifecycle cost or the total net present cost (TNPC), which includes installation and operating cost of the system over its life span. The modeler provides the solar and wind resources data, load profile of the building to be served by the HPS, components to be considered and their cost as input to the HOMER. The HOMER then performs an hourly energy balance calculation for each configuration for a year. After simulating all the possible configurations, the infeasible configurations are discarded and the feasible solutions are ranked according to the TNPC. It also allows the user to perform the sensitivity analysis on various inputs to access the effects of uncertainty on which the system has no control. Fig. 10 illustrates the relationship among simulation, optimization and sensitivity analysis. The optimization oval encloses the simulation oval to represent the fact that a single optimization consists of multiple simulations. Similarly, the sensitivity analysis oval encompasses optimization oval, representing that a single sensitivity analysis consists of multiple optimizations.

4.1.1 Simulation

Microgrid architectures with different combinations of distributed resources are developed for the performance analysis in each hour of the year to determine its technical feasibility and life-cycle cost. Energy balance calculations are performed based on the system architecture consisting of different combinations of the components. This paper chooses PV, WT, diesel generators, battery, and converter as the elements of HPS and the best feasible configuration is determined that can adequately serve the load.

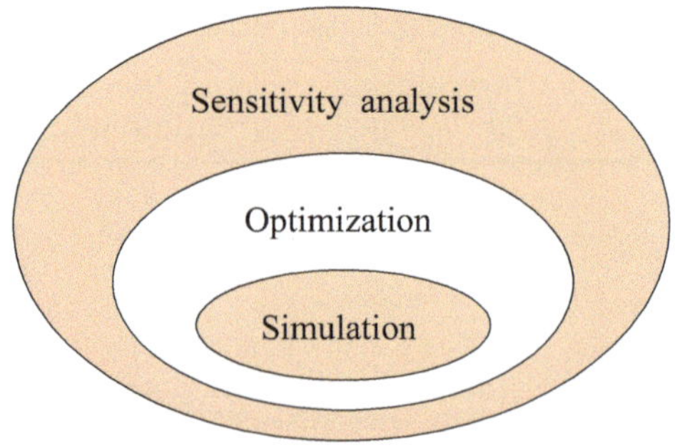

Fig. 10 Relationship among simulation, optimization, sensitivity

4.1.2 Optimization process

Power system should be appropriately designed in terms of economy, reliability, and ecological measures subjected to various physical and operational constraints of the location. The optimal system configuration can be decided with the mixture of components that the system should comprise the quantity of each component and the dispatch strategy the system should use. The overall objective is to identify a configuration which has low TNPC. Therefore, different system configurations and technical constraints are simulated and formed various combinations of system components. Finally, a list of configurations are sorted out and compared. Best feasible combinations are extracted and the least feasible ones are omitted. The overall optimization process is shown in Fig. 11. The objective function of the optimization problem is given by (1).

$$\min(TNPC) = \min\left(\sum TNPC(C_i)\right) \tag{1}$$

where TNPC can be calculated by (2).

$$TNPC(C_i) = N_i \times \left(CptlC_i + (RplC_i \times M_i) + \frac{OMC_i}{CRF}\right) \tag{2}$$

Overall system annualized cost (C_i) in general form is calculated as the sum of costs of all components and is given by (3). This cost consists annualized costs of the components procurement, replacement, and operations including fuel maintenance. Hence, the total TNPC value is calculated as the sum of TNPC of each component.

$$C_i = C_{PV} + C_{WT} + C_{DG} + C_{CONV} + C_{BAT} \tag{3}$$

$$CRF = \frac{ir(1+ir)^R}{(1+ir)^R - 1} \tag{4}$$

where C_{PV}, C_{WT}, C_{DG}, C_{BAT}, C_{CONV} are the costs of PV, WT, diesel generator, battery, and converter; i the index

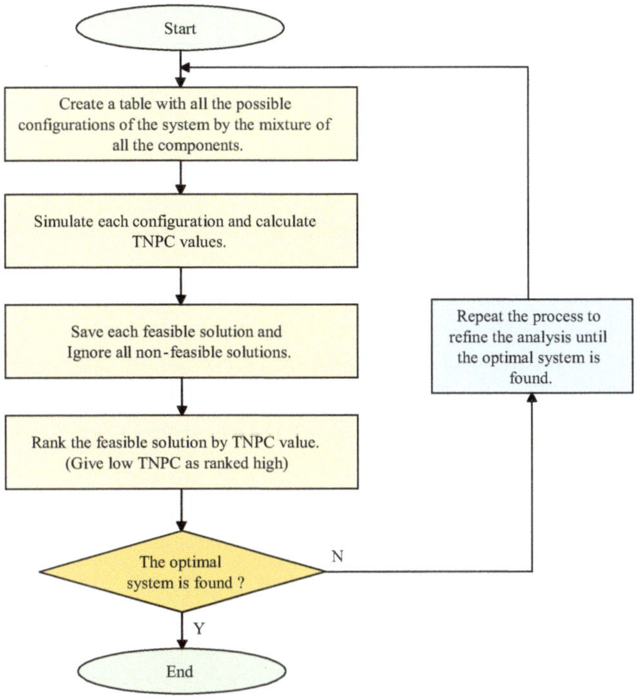

Fig. 11 Flow chart for the optimization process

of PV, WT, battery, converter, and diesel generator; N_i the total number of system components; $CptlC_i$ the capital cost of a component; $RplC_i$ the replacement cost of a component; M_i the single payment present worth of a component; OMC_i the operation and maintenance cost of a component; CRF the capital recovery factor; R the project lifetime; and ir the annual interest rate for the project life time.

4.1.3 Sensitivity analysis

As the TNPC varies based on sensitivity variables chosen, in this analysis, multiple optimizations are performed to measure the effects of the uncertainties such as wind speed, solar radiations, grid tariffs, and fuel prices on the system performance. Hence, the HPS configuration has to be chosen to tolerate all these situations.

4.2 Microgrid system modeling

The microgrid system is formed as an interconnection of various distributed units such as PV, WT, utility grid, diesel generator, converter, and battery that are modeled as shown below. It is important to know the technical and economic detail of these units precisely to get an accurate simulation result. Besides, decisions on system control parameters, constraints, and economic inputs have an equal effect on the simulation results.

4.2.1 PV panel modeling

The model of the PV array is developed that it produces DC electricity in direct proportion to the global solar radiation. It is independent of temperature and voltage to which it is exposed. Output power is given by (5).

$$P_{PV} = f_{PV} Y_{PV} \frac{I_T}{I_S} \tag{5}$$

where f_{PV} is PV derating factor; Y_{PV} the rated capacity of the PV array; I_T the global solar radiation incident on the surface of the PV array; and I_S is 1 kW/m^2 which is the standard amount of radiation used to rate the PV capacity.

The range of the PV array rating is permitted to vary between 125 kW and 275 kW to choose the optimal size. The capital cost of the PV panel is 3500 \$/kW and after a reasonable percentage of discount on bulk amount, the replacement cost is considered as 3400 \$/kW. The subsidy provided for using solar energy was also discounted in the given capital cost. Operation and maintenance cost for the PV array is considered as 0 \$. The lifetime of the selected PV array is 25 years with a derating factor of 85 %. Two-axis tracking system is established for the PV panel.

4.2.2 WT power unit modeling

Wind turbine converts kinetic energy of the wind into AC electricity according to a particular power curve. Each hour, power output of the wind turbine is calculated in a four-step process. Firstly, it determines the average wind speed for the hour at the anemometer height by referring to the wind resource data. Secondly, it calculates the corresponding wind speed at the turbine's hub height using either the logarithmic law or the power law. Thirdly, it refers to the turbine's power curve to calculate its power output at that wind speed assuming standard air density. Fourthly, it multiplies output power by air density ratio.

The rating of each wind turbine considered is 50 kW, AC. Range of 2 to 5 numbers of such turbines is taken as decision variable. The cost of each wind turbine is 181035 \$. The subsidy provided to set up wind firm is deducted in the given capital cost. Operation and maintenance cost of the wind turbine is 4320 \$/year/turbine and the total operation and maintenance cost is reduced as the number of turbines increases. The lifetime of the turbine is 25 years and the height of the turbine hub is 40 m.

4.2.3 Battery system modeling

Batteries are used to store excess power in the HPS, and to operate when the system has deficit power. Its life depends on how many times the battery is charge/ discharge per day. Battery bank is formed for 396 V DC bus. The rating of each battery is 12 V, 200 Ah, 2.4 kWh. A string of 33 such batteries are connected in series to deliver the power at a desired voltage level. The cost of each battery is 475 $ and the replacement cost is 430 $. The battery lifetime is 10 years. Value of 0, 1 and 2 strings was entered as decision variable.

4.2.4 Diesel generator modeling

A diesel generator consumes fuel to produce electricity, and possibly heat as a by-product. The fuel curve is assumed as a straight line with a y-intercept and uses (6) for the generator's fuel consumption. Diesel generators are added to the microgrid system to cover the peak load. Moreover, they can supply the electric demand when the RES system is producing insufficient power.

$$F = F_0 Y_{gen} + F_1 P_{gen} \tag{6}$$

where F_0 the fuel curve intercept coefficient; F_1 the fuel curve slope; Y_{gen} the rated capacity of the generator; and P_{gen} the electrical output of the generator.

Three diesel generator sets are installed in the existing system (one with a capacity of 380 kW and the second, third sets with a capacity of 1055 kW each). The capital cost is not considered as these are already installed but operation and maintenance cost of 0.002 $/hour was taken into account. Cost of the diesel is 0.9 $/liter. The lifetime of diesel generator set is 15000 operating hour and the minimum load factor is 30 %. In the decision variable, '0' value was also included to obtain the most optimal unit.

4.2.5 Converter modeling

The rating of the converter (working as both rectifier and inverter) is chosen according to the total PV panel output, as the total PV panel output is converted in to AC by the converter. Hence, the size considered for converter is 125 kW to 275 kW, with a cost of 110 $/kW. The replacement cost is 100 $/kW. The lifetime of converter is 20 years with an efficiency of 95 % in both the directions.

4.2.6 Utility grid

In this model, the building is connected to the utility grid, and the building produces maximum demand of 600 kW on the utility grid. The tariff of the electricity charged by the utility is 0.065 $/kWh of energy consumption and demand rate of 5 $/kW/month [39]. The interconnection charge paid to the utility is 8000 $.

4.2.7 System control and constraints

The dispatch strategy selected for this system is "load following" type. The simulation time step is kept at one hour [40] and is performed with the following constraints.

1) The operating reserve is set at 7 % of the total energy demand and 20 % of each solar and wind power output.
2) The minimum renewable fraction (RF) is set at 35 %, i.e. at any moment the minimum energy from RES is 35 % of the total energy consumed at that moment.
3) Maximum annual capacity shortage is set at 0 %.

4.2.8 System economics

The lifetime of the project is considered for 25 years at an annual interest rate of 6 % throughout. Also, it is supposed that there is no capacity shortage penalty. The overall microgrid system model is shown in Fig. 12.

5 Energy management strategy

While simulating the proposed system, it was assumed that the grid is supplying rated energy, and RES produces nominal power throughout the project lifecycle, which is not obvious. Again, while perceiving the grid in Indian context, the grid outage and load shedding occur in an

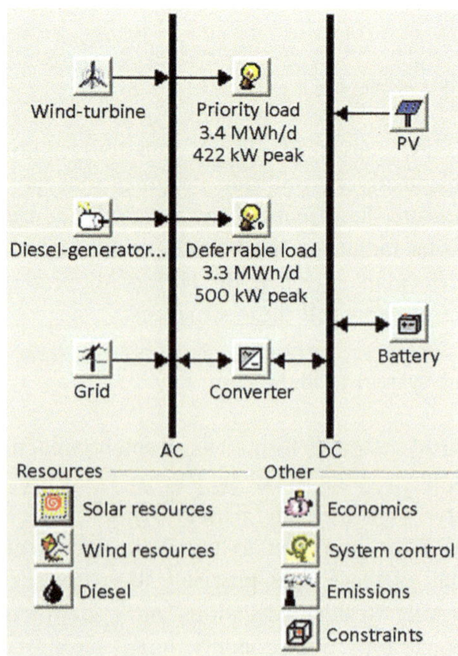

Fig. 12 Microgrid system model

unpredictable manner. In this case, the microgrid designer faces a major challenge as to how reliability can manage and coordinate with the energy system. In simple term, managing the energy system means, how smartly use the available energy from all the sources to meet the load reliably [41, 42]. A simplified energy management method is presented here by considering four operation states.

5.1 Normal operating condition

In this state, it is assumed that the grid is supplying required energy and the RES are producing energy as per the nominal rating. In this case, just like the "load following" strategy, the priority loads are served by the grid and the deferrable loads are served by the RES. Excess renewable energy is used to charge the battery.

5.2 Grid outage

This state occurs during load shedding or an occurrence of fault in the transmission or distribution lines so that the utility is unable to serve the load. Simulation result shows that the renewable sources in the system produce 38 % of total energy in a year with minimum renewable fraction of 35.6 %. Thus in this state, the energy produced from the renewable source is fed to the most important priority load (i.e. UPS load which is 32 % of total energy consumption at any time). The excess renewable energy can be fed to other priority loads in the order of lighting, elevator, etc.

5.3 Unavailability of renewable energy sources

Sometimes due to constant poor solar radiation and wind speed less than the cut-in speed of the turbine, the renewable energy sources do not respond. During such period, the grid energy can be used to serve the entire priority load and a fraction of important deferrable load.

5.4 Total blackout

In this state, it is assumed that, the unavailability of the RES due to climatic condition and grid outage occurs at the same time. This can be the worst case and may rarely occur. In such scenario, the system is also capable to meet the priority load. Fig. 13 shows the frequency of state of charge (SOC) of the battery. It is clear that around 57 % time in a year, the battery is in 100 % SOC. Besides this, the building installed the diesel generator set which can be synchronized on demand. Battery and diesel generator set can serve the priority load and can also be extended up to certain deferrable load.

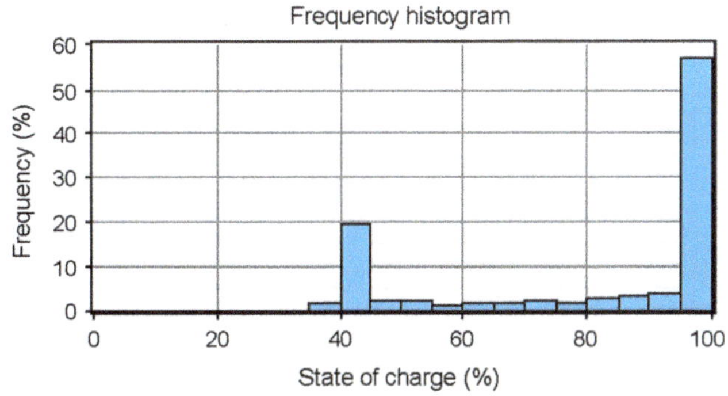

Fig. 13 Frequency of SOC of battery bank over year

6 Simulation results

The microgrid architecture is simulated in various aspects based on the variations in load profile, component cost and size, system constraints, control strategy, and system economics. The cumulative simulation results can be segregated as follows.

1) Energy management results
2) Optimization results
3) Sensitivity analysis results

6.1 Energy management results

Energy management results indicate the energy availability and transactions between microgrid and utility grid as shown in Figs. 14, 15, 16, 17, 18, 19 and 20 respectively.

6.2 Optimization results

HOMER considers all the possible combinations that are available in the search space to form the microgrid and

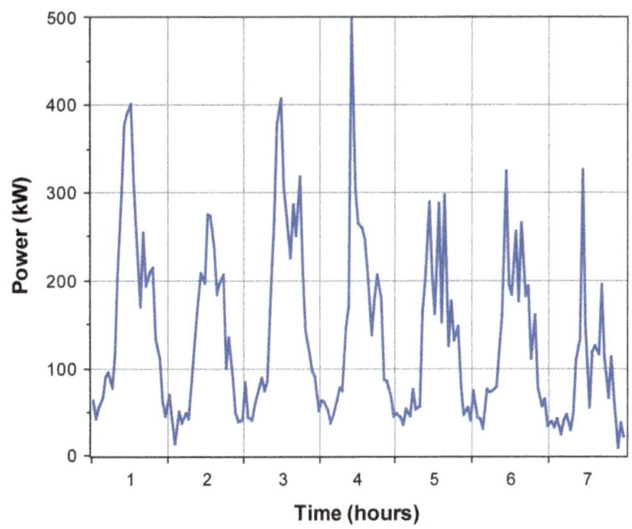

Fig. 14 Hourly load variation of the building

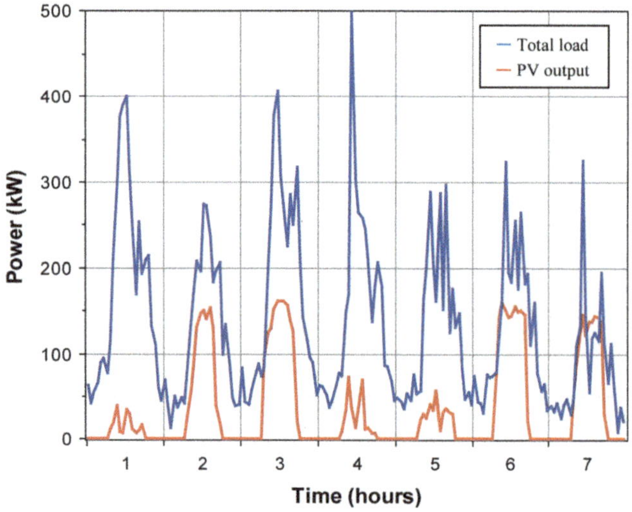

Fig. 15 Hourly load variation (Vs.) PV system output

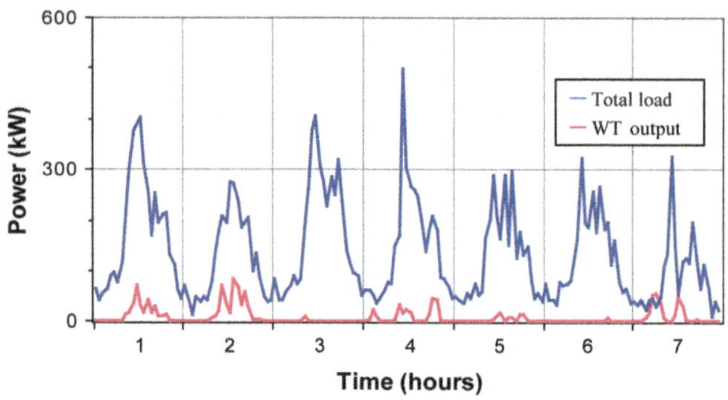

Fig. 16 Hourly load variation (Vs.) WT system output

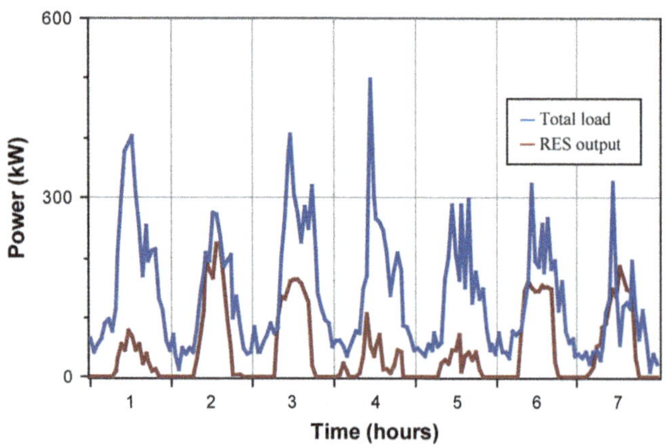

Fig. 17 Hourly load variation (Vs.) total RES system output

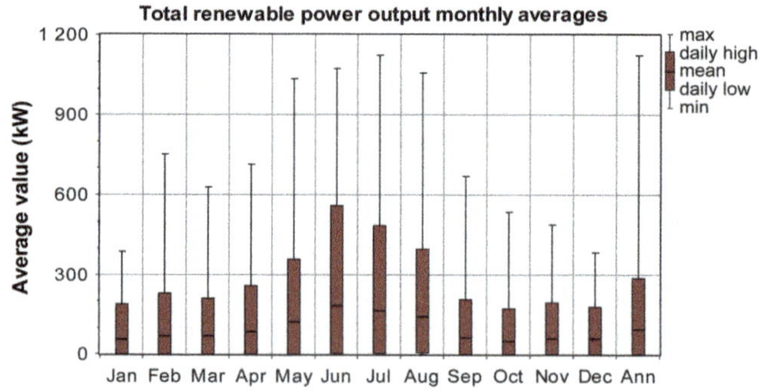

Fig. 18 Total RES power generation monthly wise in a year

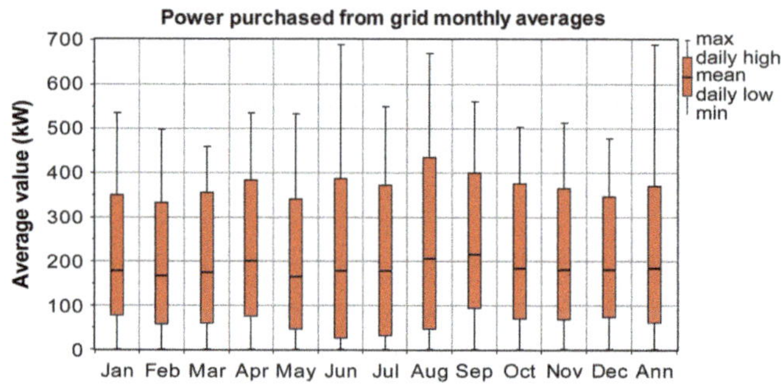

Fig. 19 Power purchased from grid monthly wise in a year

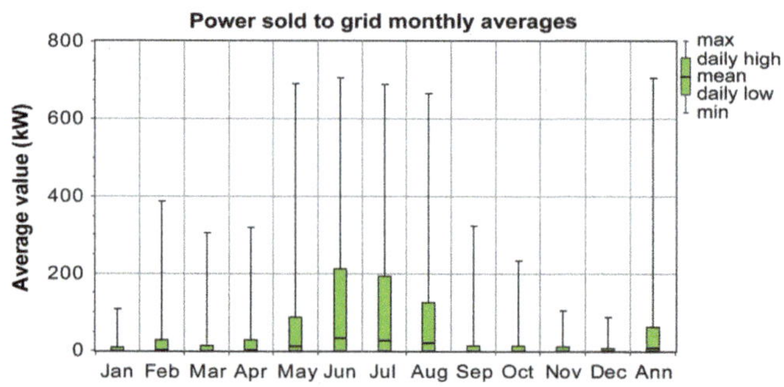

Fig. 20 Power sold to grid monthly wise in a year

checks the feasibility of each system configuration by imposing all the user specified constraints. TNPC for all the feasible system is calculated and sorted accordingly. Table 2 ranks the list of categorized optimization results for the microgrid. From the optimization result, it is clear that the most economic system configuration consists of the 180 kW PV array, three WTs, one battery string, a 160 kW converter and a grid connection having the demand of 600 kW. The TNPC for this configuration is 2878236 $.

The breakdown of TNPC is as follows: 968589 $ capital cost; 27248 $ replacement cost; 1885545 $ operation and maintenance cost; 0 $ fuel cost; and 3146 $ salvage value. This system has levelized 0.092 $/kWh cost of energy and produced 35.6 % minimum renewable energy. The considered 'zero' capital cost for diesel generator set does not impact on TNPC, as the optimal configuration does not include diesel generator set. Fig. 21 presents the electricity produced from different sources in a year.

Table 2 Optimization result of proposed microgrid system

PV (kW)	WT	DG (kW)	Battery	Converter (kW)	Grid (kW)	Initial capital	Operating cost ($/year)	Total NPC	COE ($/kWh)	Ren. fraction
180	3		33	160	600	968589	149385	2878236	0.092	0.36
170	3	380	33	170	600	947967	158841	2978492	0.096	0.35
180	3	380		170	600	956752	159508	2995795	0.096	0.36

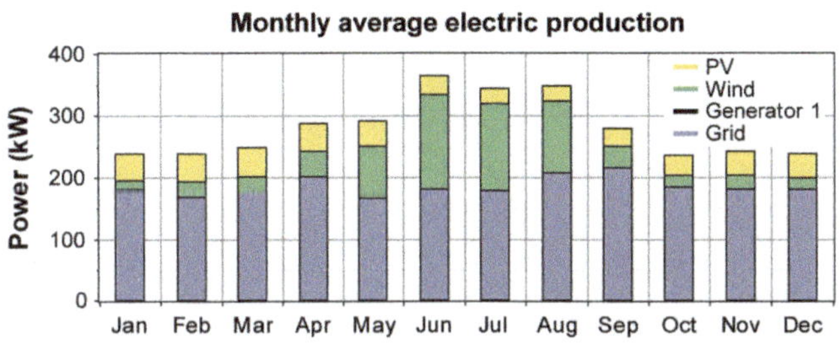

Fig. 21 Energy distribution from different sources in a year

The system has unmet load of only 1167 kWh/year which is 0.05 % of the total load. The total annualized cost is calculated [43]. In comparison, the grid-only connected system, that has a total annualized cost of 240000 $ and microgrid system, that has total annualized cost of 225155 $ produces savings of 14845 $ per year. Table 3 shows a comparison between the existing grid-only connected system and the proposed microgrid system.

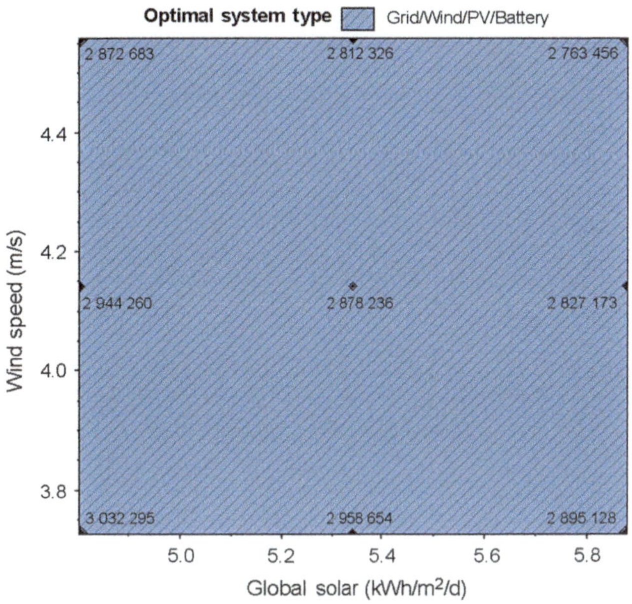

Fig. 22 Sensitivity for solar radiation and wind speed variation

6.3 Sensitivity analysis results

As there are several sensitivity variables on which the user has no control, several sensitivity analyses were also performed on the optimal system configuration to guarantee that the proposed system configuration is optimal, robust and adaptable in unanticipated variances with respect to best estimated value. Some examples of such variable are environment changes like solar radiation and wind speed variation, variation in the total electricity consumed, grid tariff and demand rate variation, etc.

6.3.1 Effect of solar radiation and wind speed variations

The solar radiation and wind speed is allowed to vary from 90 % to 110 % of its measured value. For all the values in this range, simulation is done and the results indicate that the same system configuration is optimal for this variation. The sensitivity result and TNPC variations are shown in Figs. 22 and 23 respectively.

6.3.2 Effect of electricity consumption variations

As the connected load or total electric consumption in the building will not remain constant throughout the lifetime of the project, it is also considered as one of the sensitivity variable. Both priority and deferrable loads are varied from 100 % to 105 % of the present value. For this variation, the sensitivity result shows that the same system type is optimal. It is also observed that priority load variation is more influential to TNPC than deferred load. Fig. 24 illustrates the variation of TNPC with load.

Table 3 Comparison between existing system and microgrid

Load category	Grid-only connected system	Hybrid power based microgrid	Savings (%)
Annualized cost	240000 $	225155 $	6.18
CO_2 emission	2090577 kg/year	1288131 kg/year	38.3

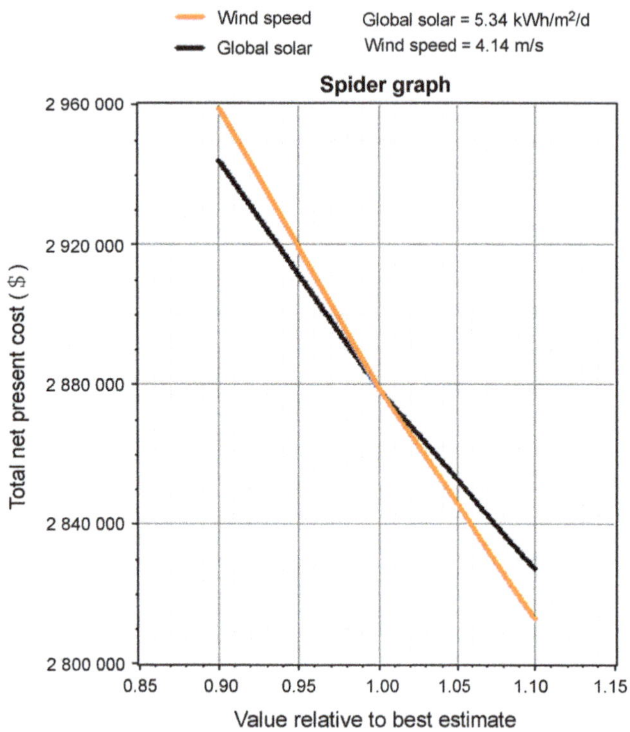

Fig. 23 Variation of TNPC with solar radiation and wind speed

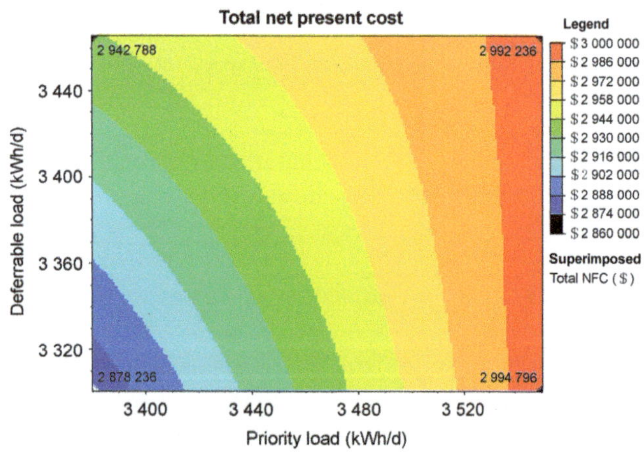

Fig. 24 Surface plot for variation of TNPC with load

6.3.3 Effect of utility grid tariff variations

As grid tariff keeps increasing to influence customer behavior, it is taken as another sensitivity variable for the evaluation. In this case, the grid tariff is allowed to vary up to 0.085 $/kWh and demand rate is also varied from 5 to 6 $/kW/month. The sensitivity result for grid tariff and demand rate variation shows that beyond grid tariff of 0.07 $/kWh, battery bank is discarded instead, the capacity

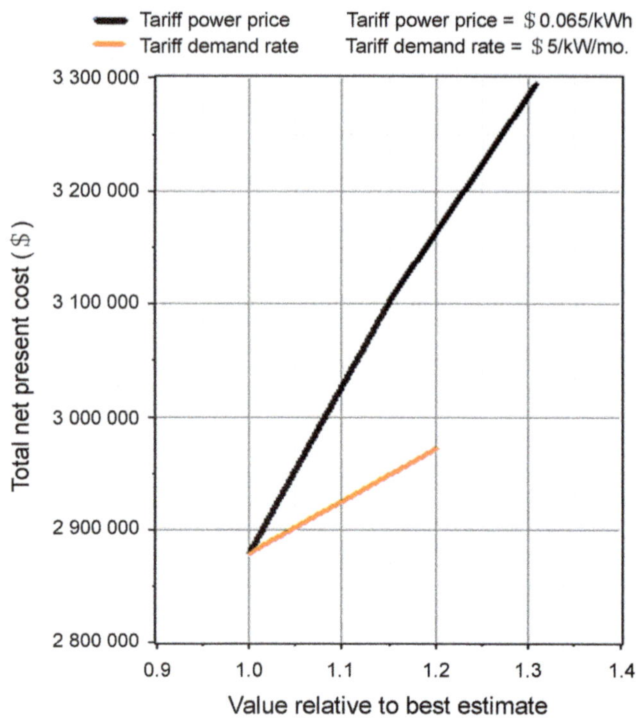

Fig. 25 Variation of TNPC with grid tariff rate and demand rate

of the PV is increased. Fig. 25 shows that the TNPC is more sensitive to grid power price to that of demand rate.

6.3.4 Effect of variation of minimum renewable fraction and shortage capacity fraction

It is considered that the proposed system must able to produce 35 % minimum renewable energy of the total electric energy consumed, which is a user specified constraints and depends on system design. This value can be varied as per the user choice and causes remarkable impact on the TNPC. Similarly, one more constraint imposed during the simulation process is maximum annual capacity shortage. This parameter is set at '0' as the building cannot compromise with reliability or interruption in energy supply. With this assumption, the system configuration is estimated that the system is able to supply peak load, which even occurs for a short interval.

However, practically if the peak load for a very short period can be avoided, then the system configuration and NPC may be other than as above. Thus, a sensitivity analysis for variation of minimum renewable fraction from 35 % to 40 %, and maximum annual capacity shortage vary from 0 % to 1 % is performed. It is observed from the simulation result that, for this variation, beyond minimum capacity shortage fraction of 0.37 %, battery is discarded from the system. TNPC variation is shown in Fig. 26.

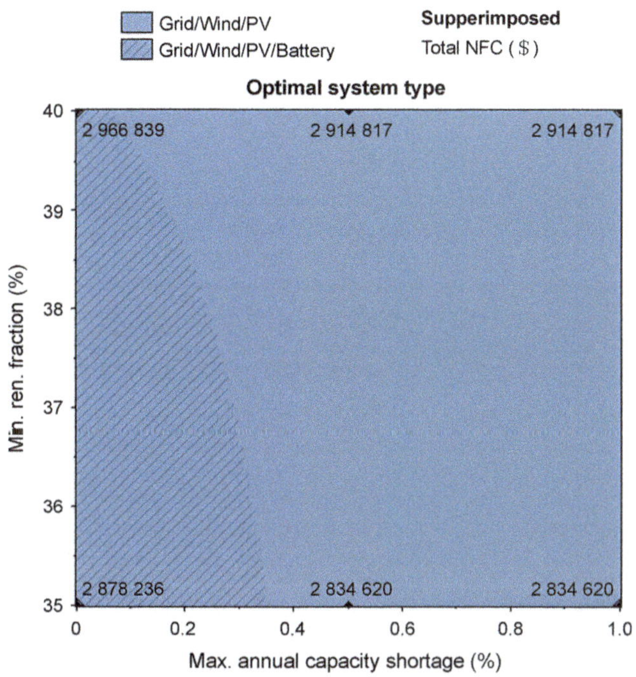

Fig. 26 Sensitivity for minimum renewable fraction variations

7 Conclusion

Hence, the design aspects of renewable energy based microgrid are discussed in this paper. The study is conducted for an urban building having 3.4 MWh/day priority load along with a 3.3 MWh/day deferrable load. The cumulative merits of this system are as follows.

1) The most economic system configuration is achieved for the combination of 600 kW utility grid contribution, 180 kW PV contribution, and 3 units WT contribution (50 kW for each, 150 kW in total), 33 batteries (1 string) and a 160 kW converter.
2) The levelized cost of energy of such system is achieved as 0.092 $/kWh which is less than the grid-only connected system.
3) Besides, the system is environment friendly and beneficial for sustainable development which leads to 6.18 % annual cost savings and reduces CO_2 emissions by 38.3 %.
4) From sensitivity results, it was observed that the system work satisfactorily in a range of varying scenarios such as solar radiation, wind speed and increase in energy consumption.
5) The energy management strategy developed ensures uninterruptable and reliable supply for the priority load for smooth running of the business.

References

[1] National Action Plan on Climate Change, Ministry of Environment, Forest, and Climate Change, Government of India.

[2] Thomas S (2014) Are low-carbon generation and competitive electricity markets compatible? Evidence from the UK electricity market reforms. In: Brunnengräber A, Di Nucci MR (eds) Im Hürdenlauf zur Energiewende. Springer, Berlin, pp 153–164

[3] Nelson D (2014) Roadmap to a low carbon electricity system in the US and Europe. Climate Policy Initiative, San Francisco

[4] Smart grid vision and roadmap for India. India Ministry of Power (2013)

[5] Scheer D, Konrad W, Scheel O (2013) Public evaluation of electricity technologies and future low-carbon portfolios in Germany and the USA. Energ Sustain Soc 3:1–13

[6] Kuo TC, Chen HM, Liu J et al (2014) Applying multi-objective planning in low-carbon product design. Int J Precis Eng Manuf 15(2):241–249

[7] Zhang Q, Tezuka T, Mclellan BC et al (2012) Scenario analysis of low-carbon smart electricity systems in Japan in 2030. In: Yao T (ed) Zero-carbon energy Kyoto 2011, Part 1. Springer, Berlin, pp 33–44

[8] Pudjianto D, Aunedi M, Djapic P et al (2014) Whole-systems assessment of the value of energy storage in low-carbon electricity systems. IEEE Trans Smart Grid 5(2):1098–1109

[9] Zeng B, Zhang JH, Yang X et al (2014) Integrated planning for transition to low-carbon distribution system with renewable energy generation and demand response. IEEE Trans Power Syst 29(3):1153–1165

[10] Chen Q, Kang CQ, Xia Q et al (2010) Power generation expansion planning model towards low-carbon economy and its application in China. IEEE Trans Power Syst 25(2):1117–1125

[11] Ji Z, Kang CQ, Chen QX et al (2013) Low-carbon power system dispatch incorporating carbon capture power plants. IEEE Trans Power Syst 28(4):4615–4623

[12] Lu SY, Lou SH, Wu YW et al (2013) Power system economic dispatch under low-carbon economy with carbon capture plants considered. IET Gener Transm Distrib 7(9):991–1001

[13] Turconi R, Tonini D, Nielsen CFB et al (2014) Environmental impacts of future low-carbon electricity systems: detailed life cycle assessment of a Danish case study. Appl Energy 132: 66–73

[14] Roelicha K, Dawson DA, Purnell P et al (2014) Assessing the dynamic material criticality of infrastructure transitions: a case of low carbon electricity. Appl Energy 123:378–386

[15] Walmsley MRW, Walmsley TG, Atkins MJ et al (2014) Minimising carbon emissions and energy expended for electricity generation in New Zealand through to 2050. Appl Energy 135:656–665

[16] Zhang Q, Mclellan BC, Utama NA et al (2011) A methodology for designing future zero-carbon electricity systems with smart grid and its application to Kansai area, Japan. In: Design for innovative value towards a sustainable society: Proceedings of the 7th international symposium on environmentally conscious design and inverse manufacturing (EcoDesign'11), Kyoto, Japan, 30 Nov–2 Dec 2011, pp 50–54

[17] Jin C, Wang P, Xiao JF et al (2014) Implementation of hierarchical control in DC microgrids. IEEE Trans Ind Electron 61(8):4032–4042
[18] Cristea C, Lopes JP, Eremia M et al (2007) The control of isolated power systems with wind generation. In: Proceedings of the 2007 IEEE Lausanne power technology conference, Lausanne, 1–5 Jul 2007, pp 567–572
[19] Mohd A, Ortjohann E, Sinsukthavorn W et al (2009) Supervisory control and energy management of an inverter-based modular smart grid. In: Proceedings of the IEEE PES power systems conference and exposition (PSCE'09), Seattle, 15–18 Mar 2009, 6 pp
[20] Omari O, Ortjohann E, Mohd A et al (2007) An online control strategy for DC coupled hybrid power systems. In: Proceedings of the 2007 IEEE power engineering society general meeting, Tampa, 24–28 Jun 2007, 8 pp
[21] Johnson NG, Glassmire JW, Lilienthal PD (2012) Comparing power system architectures for domestic lighting in isolated rural villages with HOMER. In: Proceedings of the 2012 IEEE global humanitarian technology conference (GHTC'12), Seattle, 21–24 Oct 2012, pp 375–380
[22] Anwari M, Rashid MIM, Muhyiddin HTM et al (2012) An evaluation of hybrid wind/diesel energy potential in Pemanggil Island Malaysia. In: Proceedings of the 2012 international conference on power engineering and renewable energy (ICPERE'12), Bali, 3–5 Jul 2012, 5 pp
[23] Yadav DK, Girimaji SP, Bhatti TS (2012) Optimal hybrid power system design using HOMER. In: Proceedings of the IEEE 5th India international conference on power electronics (IICPE'12), Delhi, 6–8 Dec 2012, 6 pp
[24] Kansara BU, Parekh BR (2011) Modeling and simulation of distributed generation system using HOMER. In: Proceedings of the 2011 IEEE international conference on recent advancements in electrical, electronics and control engineering (ICONRAEeCE'11), Sivakasi, 15–17 Dec 2011, pp 328–332
[25] Sureshkumar U, Manoharan PS, Ramalakshmi APS (2012) Economic cost analysis of hybrid renewable energy system using HOMER. In: Proceedings of the 2012 international conference on advances in engineering, science and management (ICAESM'12), Nagapattinam, 30–31 Mar 2012, pp 94–99
[26] De Brabandere K, Vanthournout K, Driesen J, et al (2007) Control of microgrids. In: Proceedings of the 2007 IEEE power engineering society general meeting, Tampa, 24–28 Jun 2007, 7 pp
[27] Alam MS, Gao DW (2007) Modeling and analysis of a wind/PV/fuel cell hybrid power system in HOMER. In: Proceedings of the 2nd IEEE conference on industrial electronics and applications (ICIEA'07), Harbin, 23–25 May 2007, pp 1594–1599
[28] Rohani A, Mazlumi K, Kord H (2010) Modeling of a hybrid power system for economic analysis and environmental impact in HOMER. In: Proceedings of the 18th Iranian conference on electrical engineering (ICEE'10), Isfahan, 11–13 May 2010, pp 819–823
[29] Konig O, Hametner C, Prochart G et al (2014) Battery emulation for power-HIL using local model networks and robust impedance control. IEEE Trans Ind Electron 61(2):943–955
[30] Reddy YJ, Kumar YVP, Raju KP et al (2012) Retrofitted hybrid power system design with renewable energy sources for buildings. IEEE Trans Smart Grid 3(4):2174–2187
[31] Nehrir MH, Wang C, Strunz K et al (2011) A review of hybrid renewable/alternative energy systems for electric power generation: configurations, control, and applications. IEEE Trans Sustain Energ 2(4):392–403
[32] Shaahid SM, El-Amin I (2009) Techno-economic evaluation of off-grid hybrid photovoltaic-diesel-battery power systems for rural electrification in Saudi Arabia—a way forward for sustainable development. Renew Sustain Energy Rev 13(3):625–633
[33] Dihrab SS, Sopian K (2010) Electricity generation of hybrid PV/wind systems in Iraq. Renew Energy 35(6):1303–1307
[34] Microgrid Power System Design Services Using HOMER.
[35] Felix A, Simoes MG (2006) Integration of alternative sources of energy. IEEE Press, Hoboken
[36] Hyderabad planet.
[37] Atmospheric science data center (2013)
[38] The mean monthly wind speed in Hyderabad, India.
[39] Southern Power, Distribution Company of T.S.Limited (2014)
[40] Barley CD, Winn CB (1996) Optimal dispatch strategy in remote hybrid power systems. Sol Energy 58(4–6):165–179
[41] Jalivand A, Kord H, Rohini A (2009) Design, control & power management for a hybrid PV/WG/FC system for standalone applications. In: Proceedings of the 14th electrical power distribution conference (EPDC'09), Kerman
[42] Wang CS, Nehrir MH (2008) Power management of a standalone wind/photovoltaic/fuel-cell energy system. IEEE Trans Energy Convers 23(3):957–967
[43] Nelson DB, Nehrir MH, Wang C (2006) Unit sizing and cost analysis of stand-alone hybrid wind/PV/fuel cell power generation system. Renew Energy 31(10):1641–1656

Y. V. PAVAN KUMAR (S'14) received B.Tech in Electrical and Electronics Engineering in 2007 from JNTUH, India; M.Tech in Instrumentation and Control Systems in 2011 from JNTUK, India; and currently pursuing Ph.D in Electrical Engineering in the area of Smart Microgrids at Indian Institute of Technology Hyderabad (IITH), India. Prior to this, he has worked as a design engineer for Control Systems, Power Systems, and Engineering Test Systems in the departments of Aerospace engineering and Energy Research & Development center of Honeywell Technology Solutions Lab, India for 3 years. He also has 2.5 years of teaching experience in Electrical and Electronics Engineering for undergraduate and postgraduate students. He has published 42 papers in International Journals and Conferences, 2 Books, 3 Honeywell innovation disclosures. He has guided 6 M.Tech theses and 3 B.Tech theses. His research area includes Microgrids, Smartgrids, Advanced Control systems, and Artificial Intelligence. He is a student member of IEEE Power & Energy Society, IEEE Communications Society, and IEEE Smart Grid Community.

Ravikumar BHIMASINGU obtained his Ph.D and M.Sc (Engg.) both in the Department of Electrical Engineering, Indian Institute of Science (IISc), Bangalore, India in the year 2009 and 2004 respectively. He obtained his B.Tech degree from Nagarjuna University, India in the year 2002. After completion of his Ph.D, he worked as a Senior Research Associate and Junior Research Associate both at the Department of Electrical Engineering, IISc. Later he joined Global R&D Center, Crompton Greaves Ltd., Mumbai, India as a Senior Executive-Technology and served for 3 years from 2010 to 2013. Currently, he is working as an Assistant Professor in the department of Electrical Engineering at Indian Institute of Technology Hyderabad (IITH), India. His research interests include AI techniques applications to power systems security improvement, computer-aided power system analysis and modeling, and distribution system automation. He has authored 7 National and International Journal publications and 35 National and International Conference publications.

12

Low-carbon benefits analysis of energy-intensive industrial demand response resources for ancillary services

Mingtao YAO (✉), Zhaoguang HU, Ning ZHANG, Wei DUAN, Jian ZHANG

Abstract As a major CO_2 emission source, the low-carbon development of the power sector requires the sector's own efforts and the cooperation with other industries, especially in the context of rapid development of renewable generation technologies. The industrial demand response resources (IDRR) will be helpful to improve wind power penetration and bring low-carbon benefits if they are utilized to provide ancillary services (AS) for the power system. In this paper, demand response (DR) characteristics of industrial users are firstly analyzed according to their production process and electricity consumption distribution. In order to have an in-depth study of the response mechanism of industrial loads to provide AS, cement and aluminum smelter are selected as two typical IDRR, and the AS type they provided and response mechanism are analyzed. Based on the data of these two industries in certain provinces of China, low-carbon benefits considering IDRR to provide AS are analyzed.

Keywords Demand response (DR), Ancillary services (AS), Energy-intensive industrial users, Low-carbon benefits

1 Introduction

As a large developing country, China is facing the pressure of promoting economic growth while saving energy resources as well as protecting the environment. Given that China is the world's largest carbon emitter [1], the government has announced to reduce its carbon emission intensity by 40%–45% by 2020 compared to that in 2005. As the important link in the chain of energy industry, the power sector should actively explore low-carbon development ways that are conducive to long-term interests. And power sector decarbonization is an important component of the low-carbon development of energy-intensive industries.

Renewable energy technologies have been considered as a key factor for promoting the power sector decarbonization [2]. The Renewable Energy Law promulgated by the government was put into effect on January 1, 2006, according to which, grid enterprises are required to buy all the power produced by renewable energy generators [3]. As a result, China's renewable generation has experienced a rapid development in terms of both installed capacity and generation in past few years, but the problem of wind power penetration has become more serious and aroused continuous concern [4]. Generally, wind feed-in is unresponsive to system demand and is associated with a high level of uncertainty due to the volatile nature of wind [5]. It is different from the majority of thermal generation, which provides not only schedulable energy but also vital control services necessary to maintain the integrity of the power system [6]. With existing dispatching practices, different kinds of AS are provided by thermal generation, which has slow response speed and restricted ramp rates. Consequently, the thermal generation cannot operate with the optimum efficiency and causes significant carbon emissions and energy wastage. Under the background of power sector decarbonization, the real-time AS is increasingly needed. If the AS is all provided by thermal generation running part-loaded, it will not only reduce the efficiency of system operation, but also significantly undermine the

M. YAO, N. ZHANG, W. DUAN, J. ZHANG, School of Electrical Engineering, Beijing Jiaotong University, Beijing 100044, China
(✉) e-mail: yaomingtao@bjtu.edu.cn
Z. HU, State Grid Energy Research Institute, Beijing 102209, China

system ability for absorbing intermittent renewable output [7], which will result in increasing pressure of energy conservation and emission reduction.

From the perspective of energy-intensive industries, the government regulation is mainly adopted in Chinese AS fields and the main measure is power rationing. This kind of load management mode is simple and extensive, thus it has limited effects on low-carbon development and industrial structure adjustment. The proportion of iron and steel, building materials, colored metals and other energy-intensive industries in Chinese power consumption is relatively high. Exploring the potential of energy-intensive industries to provide AS can not only relieve the energy wastage brought about by disordered development, but also supplement the shortage of reserve resources caused by the rapid development of renewable energy.

Smart grid realizes the real-time interaction between power sector and other sectors on the basis of modern technologies of information, communication, sensor and control. It enables DR to another interaction resource with considerable potential apart from the thermal generation [8]. With the development and improvement of demand side management technology, users can response to remote control or price signal quickly and automatically, and have a technical advantage in terms of speed and response accuracy. Likewise, they have advantages in response costs and the feasibility in providing AS from the technical perspective [9]. Therefore, it is a new topic in the AS field to research the DR actively participating AS and explore the mechanism of operation design.

This paper is organized as follows: Section 2 describes several kinds of AS and analyses the characteristics and low-carbon benefit of industrial DR resources. In section 3, by choosing cement and aluminum smelter as two typical IDRR, the response mechanism of these two industries is established considering their own production process. In section 4, two different scenarios, in which the DR potential of industrial users is fulfilled or not, are developed and low-carbon benefits are analyzed based on the industry data of certain province in China. Finally, some conclusions are shown in section 5.

2 IDRR for AS ancillary services

2.1 Ancillary services

Although AS include both real and reactive power applications, and their classifications are quite different according to different research view. This paper focuses only on real power AS and concerns main contents as follows [10, 11].

1) Regulation: It requires continuous response to short-term load fluctuation (including regulation up/down). Some loads may be capable of supplying regulation considering their fundamental capabilities.
2) Spinning reserve: It responses to unplanned load increase or decrease (mainly provided by the unit in the running state).
3) Non-spinning reserve: It responses to sudden but infrequent disruptions (mainly provided by the off-grid unit that can start up at any time without maintenance plan). Generally, some responsive loads are allowed to supply non-spinning reserve.

The three types of AS can also be distinguished through their physical parameters. Fig. 1 plots the response speed and duration on the horizontal axis and advance notice on the vertical. For example, if load i wants to provide non-spinning reserve, it must satisfy the following standards:

$$\begin{cases} T_{\text{an}\_i} \leq t_3 \\ T_{\text{rs}\_i} \leq t_1 \\ T_{\text{rd}\_i} \geq t_2 - t_1 \end{cases} \tag{1}$$

where T_{an_i} is the advance notice time of load i; T_{rs_i} the response speed of load i; T_{rd_i} the single response duration of load i; t_1 the slowest acceptable response speed after load received the signal; t_3 the maximum acceptable advance notice time; and $t_2 - t_1$ the shortest acceptable response duration.

For different DR resources, there are differences in advance notice time. Generally, industrial users may require 2 to 4 hours of advanced notice to prepare. Other load types, such as commercial buildings, may not need it in general [12]. Longer response duration is more useful for grid operators, which could be applied to most periods. Response speed is a strict parameter and there is no difference for all users.

2.2 Industrial demand response resources

In recent decades, especially after 2002, the proportion of industrial electricity consumption has maintained a sustained upward trend [13]. Considering electricity price

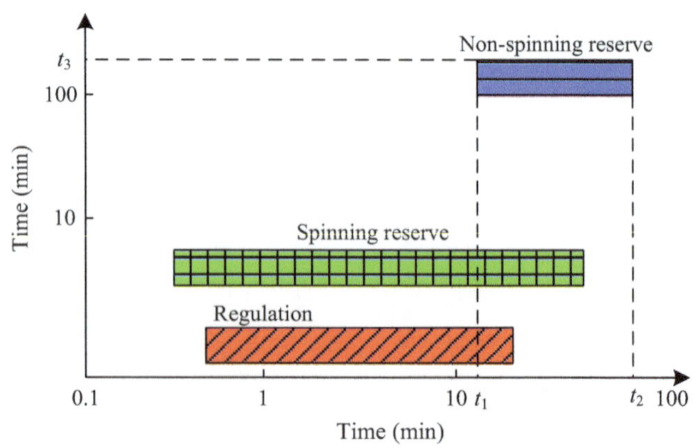

Fig. 1 Advance notice, response speed and duration of AS

as the main cost of these energy-intensive industries, power sector should participate more in building low-carbon development mode and promoting industrial upgrading. A specific DR program for these industries is necessary in the long term.

DR has been traditionally defined as a set of time dependent activities that reduce, shed or shift electricity use to improve electric grid reliability and manage costs, which mainly relies on customers' motivation for reducing electricity tariff [14]. For AS, technical requirements are more challenging in terms of speed and accuracy when compared with traditional DR. Specifically, DR for AS is needed year-round and not just for peak hours [15]. Different works have been presented in the past in order to evaluate DR for AS in different sectors. Lawrence Berkeley National Laboratory has carried out a lot of research about DR for AS under the framework of smart grid and verified a series of power loads which have the ability to provide AS by experiment and demonstration project, such as lighting, air conditioning, refrigeration, data center and water pumping, etc [16–21]. Reference [22] examined the characteristics of cement plants and their ability to shed or shift load to participate in DR. Reference [23] identified that six hours of load shifting could be achieved without adversely affecting production at a South African cement plant. Reference [24, 25] investigated the available opportunities of one aluminum smelter factory in AS market and discussed the regulation service that electrolytic aluminum enterprises could provide. Reference [26] analyzed the flexibility potential of customers in one meat industry based on the management of the most energy consuming process in this type of segment: cooling production and distribution, which is one of the most representative sectors in Spain. Reference [27] briefly estimated the technical and economic potential of energy-intensive industries to provide demand side management in electricity and balance markets through 2030, including chloralkali process, mechanical wood pulp production, aluminum electrolysis, electric arc furnace, cement mills. However, all the aforementioned analyses are limited in that they are case studies for a single industrial plant or sector, and they have not investigated the economic influence and low-carbon benefit of IDRR for AS through day-ahead optimizing dispatching.

In recent years, public attention on DR has been directed mostly to the residential sector due to the rapid development of smart meters and smart homes. But industry sector has a much larger consumption footprint and peak load contribution, and in some respects is well advanced in the implementation of smart grid technologies [28]. It is noteworthy that industrial users' loads have greater potential for optimization, and they show strong interest in the management of power use for price benefits.

According to the industry category, DR resources have different kinds of types. As shown in Fig. 2, unlike thermal generation, the ability of DR to provide AS is definitely affected by a serious of external variables, especially for commerce and inhabitant users. However, the majority of industrial users are flow industry and the production must keep flowing. Compared with commerce and inhabitant users, IDRR have the following advantages.

1) Large capacity: Industrial users are large consumers. Peak loads can reach hundreds of megawatt, and annual consumptions can reach hundreds of millions of kilowatt-hour. However, other power consumers require additional mechanisms, such as aggregator, because their power level is negligible.
2) High load rate: The difference of industrial users' peak and off-peak load usually keep in a low level, which can fit well with the output fluctuation of wind farm and benefit for power system stability.
3) Climate insensitive: A factory's electricity consumption bears little relationship to the cycle of the seasons.
4) Easy to manage: Modern factories are highly automatic and can meet the requirements of providing AS by a simple modification.

2.3 Low-carbon benefit of industrial users to provide AS

The source of low carbon benefits from smart DR can be concluded as follows.

1) Reduce the frequent ramp up/down of the unit. In current situation, thermal generation undertakes the primary mission to provide AS, which will result in huge accounts of ramp up/down costs and extra carbon emissions. Some units must operate in uneconomic range in case a unit or a transmission line unexpectedly fails. If controlled responsive loads could provide AS in the situation of meeting criteria, especially for some industrial users which could perfectly follow AS

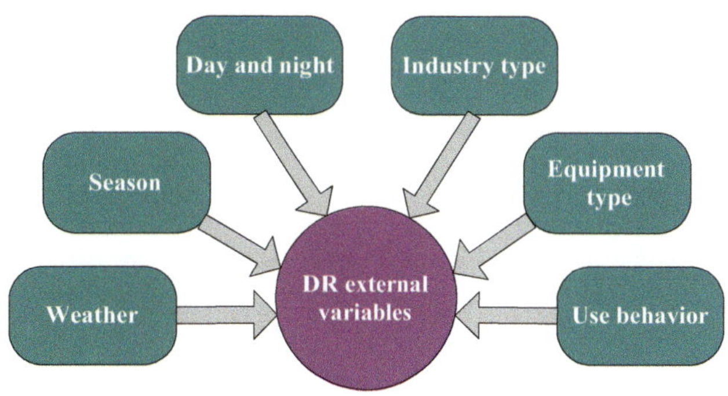

Fig. 2 DR external variables

commands, it will reduce the power system requirements for AS and lead to a large cut-in carbon emissions.

2) Improve the penetration of renewable energy resources. Under the traditional dispatching mode, we choose to adjust thermal power output to deal with the randomicity of wind farm. If there are more responsive loads, which could adjust their production process to fit the wind output characteristics, the construction requirement of thermal power for providing AS will be reduced significantly. This will also improve the penetration of wind power and make low-carbon benefits.

3 Response mechanism of typical industrial users for AS

3.1 Cement

The proportion of electricity bill in the total cost of cement enterprise is about 30%. In recent years, the production pressure of cement enterprise has been decreasing as a result of economic situation and industrial restructuring. The flexibility of participating in DR is improved. During the production process, just the kiln must keep running. The operating time of raw mill, coal mill and cement mill can be adjusted as needed. And these adjustable devices account for 60%–70% of the total electricity consumption of cement production. Generally, the production process and electricity consumption distribution of cement enterprise are shown in Fig. 3.

With the correlation between coal mill and kiln during production, it is not suitable to frequently start and stop coal mill [23]. Thus, raw mill and cement mill are the optimal non-spinning reserve resources of cement enterprise. The storage amount at various points in the cement process has the minimum storage capacity during the producing process, which determines the maximum response duration of all types of devices. Using raw mill as an example, on account of the good uniformity of raw meal quality, the remaining storage of raw meal should not be too low. When the storage is lower than a certain proportion, it will lead to the reduction of product quality if the raw mill participants in AS.

It is assumed that the load curve of mill j usually has a flat shape $P_{\text{base\_}j}$ as shown in Fig. 4. In the case of providing AS, the daily load curve of mill j can be expressed as $P_{\text{ns\_}j}$. For simplicity, we suppose the production decision satisfies the following conditions:

$$E_{\text{ns\_}j} - (E_{1\_j} + E_{2\_j}) = 0 \tag{2}$$

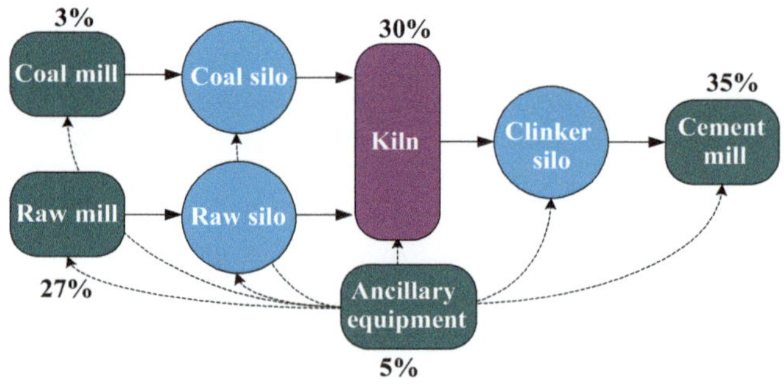

Fig. 3 Production process and electricity consumption distribution of cement enterprise [30]

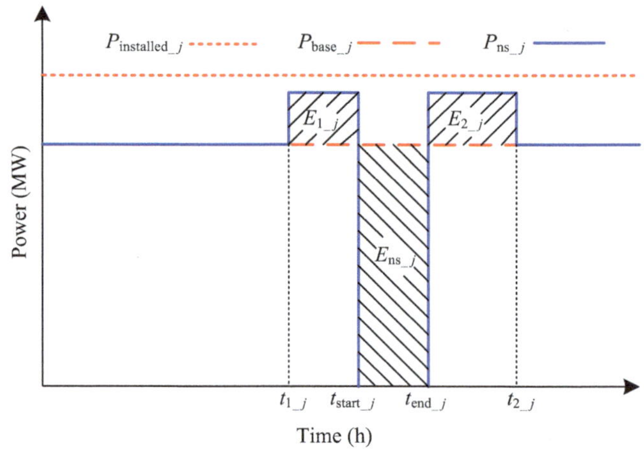

Fig. 4 The daily load curve of cement providing non-spinning reserve

where $E_{\text{ns\_}j}$ is the energy reduced during providing non-spinning reserve, and E_{1_j} and E_{2_j} are additional energy consumed before and after the non-spinning reserve respectively, which are used to maintain the relevant silo's input and output stability. It should be noted that the beginning and ending time of E_{1_j} and E_{2_j} could be adjusted according to power system's requirement without affecting the normal production of the premise.

Table 1 shows the technical parameters of production equipment in cement enterprise that is summarized based on different cement factories in China [29]. The flexibility of every process is expressed in the term of percentage, representing the proportion of one adjustable process in the installed capacity of cement enterprise.

Based on the above analyses, it is concluded that the large cement enterprises could provide greater flexibility because the storage space of raw silo and clinker silo are large enough to last for days. Therefore, responding to an AS event should have no impact on plant operation costs and productivity.

Table 1 Response parameters of various cement production equipments

	Flexibility (%)	Duration (h)	Times
Raw mill	16–20	1–3	1–2
Coal mill			
Cement mill	20–24	1–3	1–2
Kiln			

3.2 Aluminum smelter

Electrolytic aluminum is a key industry for energy saving. For a typical aluminum smelter, electric power accounts for 30% to 40% of the overall cost of producing primary aluminum. Compared with cement enterprise, it has fairly simple production process. Aluminum smelting pot is the core equipment in aluminum production and consumes the vast majority of energy. Thus the analysis of DR potential in an aluminum smelter is mainly focused on the pot line.

Aluminum smelting pot, which is electronically controlled, could potentially follow automatic generation control commands accurately [24]. Adjusting the incoming voltage could control the power consumption of each pot. Thermal balance is very important for the pot operation. It is maintained by controlling the energy input to the pot. Existing smelting plants can provide AS in two ways.

1) Adjust input voltage. As long as the smelting process is stable, reducing or increasing the input voltage of pot, thereby reducing or increasing electricity consumption, may achieve fast response. This action can be precisely achieved within seconds.
2) Turn the entire pot line off. A pot can provide AS by turning the entire pot off, thereby reducing the electricity consumption by a larger amount. In that case, the duration of an interruption is more critical and can be sustained only for short periods, depending upon the specific plant limitations.

In this paper, we just choose the way of adjusting input voltage, which was proved to be feasible [25].

As shown in Fig. 5, P_r and P_{base} represent the daily load curve of a pot with and without providing AS respectively. Regulation up and down could be regarded as two services to power system, but we assume that the up and down adjusting range should be consistent for convenience. In a dispatching cycle, regulation up and down are required to be equal, as in (3).

$$\sum E_{ru} - \sum E_{rd} = 0 \tag{3}$$

where E_{ru} is the energy consumed for providing regulation up, and E_{rd} is the energy consumed for providing regulation down.

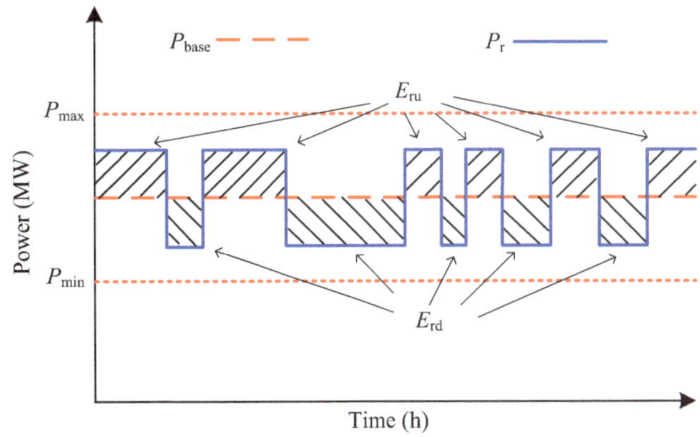

Fig. 5 The daily load curve of aluminum smelter providing regulation service

Table 2 The response parameters of aluminum smelting pot

	Flexibility (%)	Duration (h)	Times
Aluminum smelting pot	≤3	≤12	Continuously

Table 2 summarizes the response parameters of aluminum smelting pot. The flexibility of aluminum smelting pot is defined as the proportion of adjustable load in one pot. The dispatching cycle is one day so that the duration of provide regulation up or down must be no more than 12 hours.

As discussed above, regulation up and regulation down are required to be equal to reduce the impacts on product quality, process and productivity. Compared with small aluminum smelter enterprises, large aluminum smelter enterprises that have self-generation power plants will have obvious advantages in regulation response flexibility.

4 Scenario results

On basis of the analytical treatise above, we use data of cement and electrolytic aluminum industries in certain province of China to analysis the low-carbon benefits of IDRR. The basic data of these two industries is shown in Table 3. Due to the difficult of obtaining complete load data, we choose the daily average load of typical days in different seasons (April 18, July 26, September 12 and December 19) as a reference.

The implementation of IDRR to provide AS is similar to power plant capacity in electricity production, which could be regarded as the substitution of thermal generation. Table 4 and 5 summarize the DR potential of these two energy-intensive industrial users.

Table 3 The basic data of cement and aluminum smelter industries in 2012

	Production (million tons)	Load level (MW)			
		Spring	Summer	Autumn	Winter
Cement	15386	3084	3080	2543	2884
Electrolytic aluminum	197	3264	3335	3276	3201

Table 4 Overview of the potential of cement for AS

	Process	Capacity (MW)	AS type	Utilization level (%)
Cement	Raw mill	522	Non-spinning	75
	Cement mill	638	Non-spinning	80

Table 5 Overview of the potential of aluminum smelter for AS

	Process	Capacity (MW)	AS type	Utilization level (%)
Aluminum smelter	Aluminum smelting pot	98	Regulation up/down	95

In order to preliminary estimate the low-carbon benefits of two typical energy-intensive industrial users, two scenarios are set as follows.

1) Scenario 1: The AS of power systems required is still provided by thermal generation, regardless of IDRR. We suppose that the obligation of different units is fixed and their CO_2 emission parameters are shown in Table 6.
2) Scenario 2: The IDRR are admitted into the AS market and the thermal generators keep running smoothly in the case of existing available IDRR. In scenario 2, two parameters, transforming factor τ and penetration factor λ, are introduced to simulate the potential change of energy-intensive industrial users for AS.

As shown in Fig. 6, transforming factor reflects the process of integration of energy-intensive industries, while penetration factor represents the acceptability of providing AS for power system. By comparing two scenarios, the low-carbon benefits of cement and electrolytic aluminum industries for AS are obtained in certain province of China from 2012 to 2020, as shown in Table 7.

Thus, it can be seen that IDRR can bring significant low-carbon benefits. But it is hard to be widely applied in most factories due to the acceptance of present period. Considering the increasing penetration of wind power and the potential of energy-intensive industrial users for AS, some traditional ideas should be gradually changed. Loads, especially industrial loads, could not focus on maintaining a flat curve on the premise of safety production. Flexible loads represent a significant and large resource for power sector.

Table 6 CO_2 emission intensity of different generators

	AS type	CO_2 emission intensity (kg/kWh)
Large generators	Regulation up/down	0.80
Small generators	Non-spinning	1.05

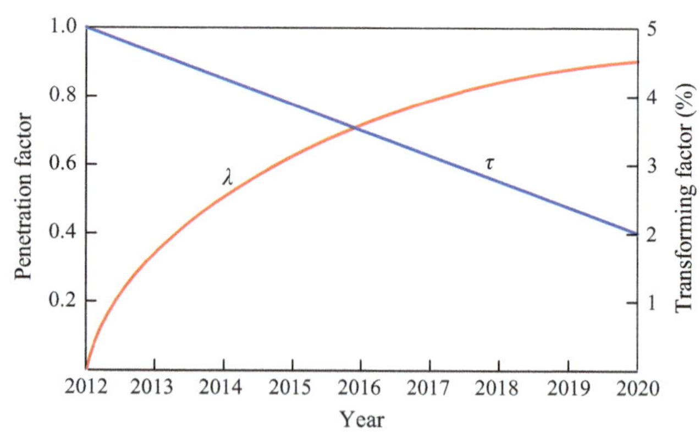

Fig. 6 Transforming factor and penetration factor

For the sake of economic interests, it is difficult to take the initiative to change the existing production mode for most enterprises. A relatively efficient incentive mechanism is essential. The existing response models that loads could participate in are mainly designed for peak load period, and there is still large gap between it and AS in terms of engagement standards and supervisor mode. Through the analysis of the response mechanism of cement

Table 7 CO_2 reduction volume

Year	CO_2 reduction volume (t)
2013	269.8×10^6
2014	354.2×10^6
2015	425.7×10^6
2016	465.5×10^6
2017	497.4×10^6
2018	516.0×10^6
2019	528.9×10^6
2020	555.3×10^6

and aluminum smelter enterprises, the incentive mechanisms of AS are put forward as follows.

1) Grid-industrial users: Grid could consider industrial users that have the ability to providing AS as power plants. Industrial users will get revenue according to real situation's requirements but not before they had been evaluated their qualifications. Besides, government regulators should give their additional subsidy for the low-carbon benefits they provide.
2) Generation-industrial users: On the basis of the geographical distribution and power supply business contact, all kinds of industrial users could form an AS alliance with related thermal generation separately. Then, industrial users could share the AS pressure and profits with generation. In this model, they also could earn additional subsidy for the low-carbon benefits.

5 Conclusions

This paper focuses on exploring the low-carbon benefits of energy-intensive industrial DR resources for AS. Cement and aluminum smelter are chosen as two typical IDRR for in-depth analysis. A preliminary calculation is carried out based on the response mechanism of cement and electrolytic aluminum. Then, the potential of CO_2 emission reduction of these two types of industrial users in certain province of China is revealed from 2012 to 2020. Due to the lack of related data, this paper could only present preliminary estimate. With the future development and improvement of DR technique, more attentions will be paid to industrial DR application.

Acknowledgments This work is sponsored by State Grid Science and Technology Program (XMERI2010KJKF001421) and the Fundamental Research Funds for Central Universities (2012YJS111). The authors gratefully acknowledge the contributions of Yichuan Power Group Corporation and ShenHuo Group Ltd. in Henan province, which are well-known enterprises engaging aluminum-processing in China.

References

[1] Bao C, Fang CL (2013) Geographical and environmental perspectives for the sustainable development of renewable energy in urbanizing China. Renew Sustain Energy Rev 27:464–474
[2] Chen QX, Kang CQ, Xia Q et al (2011) Preliminary exploration on low-carbon technology roadmap of China's power sector. Energy 36(3):1500–1512
[3] Renewable energy law of the People's Republic of China (2005) (in Chinese)
[4] He YX, Xia T, Liu ZY et al (2013) Evaluation of the capability of accepting large-scale wind power in China. Renew Sustain Energy Rev 19:509–516
[5] Holttinen H, Lemström B, Meibom P et al (2007) Design and operation of power systems with large amounts of wind power. State-of-the-art report. VTT Technical Research Centre of Finland, Espoo, Finland
[6] Aunedi M, Kountouriotis PA, Calderon JEO et al (2013) Economic and environmental benefits of dynamic demand in providing frequency regulation. IEEE Trans Smart Grid 4(4):2036–2048
[7] US Department of Energy (2011) Load participation in ancillary services. US Department of Energy, Washington, DC, USA
[8] Zhang Q, Wang XF, Fu M et al (2009) Smart grid from the perspective of demand response. Automat Electr Power Syst 33(17):49–55 (in Chinese)
[9] Kiliccote S, Sporborg P, Sheikh I et al (2010) Integrating renewable resources in California and the role of automated demand response. LBNL-4189E. Lawrence Berkeley National Laboratory, Berkeley, CA, USA
[10] Wang JX, Wang XF, Wu Y (2005) Operating reserve model in the power markets. IEEE Trans Power Syst 20(1):223–229
[11] Aghaei J, Alizadeh MI (2013) Demand response in smart electricity grids equipped with renewable energy sources: a review. Renew Sustain Energy Rev 18:64–72
[12] Watson DS, Matson N, Page J et al (2012) Fast automated demand response to enable the integration of renewable resources. LBNL-5555E. Lawrence Berkeley National Laboratory, Berkeley, CA, USA
[13] State Grid of Energy Research Institute (2013) The analysis report of energy conservation and power saving in China. State Grid of Energy Research Institute, Beijing, China (in Chinese)
[14] Zeng M (2013) Power demand side response theory and its application in electricity market. China Electric Power Press, Beijing, China (in Chinese)
[15] Ma O, Alkadi N, Coppers P et al (2013) Demand response for ancillary services. IEEE Trans Smart Grid 4(4):1988–1995
[16] Kiliccote S, Price PN, Piette MA et al (2010) Field testing of automated demand response for integration of renewable resources in California's ancillary services market for regulation products. LBNL-5557E. Lawrence Berkeley National Laboratory, Berkeley, CA, USA
[17] Ganti V, Ghatikar G (2012) Smart grid as a driver for energy-intensive industries: a data center case study. LBNL-6104E. Lawrence Berkeley National Laboratory, Berkeley, CA, USA
[18] Daniel O, Goli S, Faulkner D et al (2012) Opportunities for automated demand response in wastewater treatment facilities in

California—Southeast Water Pollution Control Plant case study. LBNL-6056E. Lawrence Berkeley National Laboratory, Berkeley, CA, USA
[19] Rubinstein FM, Li XL, Watson DS (2010) Using dimmable lighting for regulation capacity and non-spinning reserves in the ancillary services market: a feasibility study. LBNL-4190E. Lawrence Berkeley National Laboratory, Berkeley, CA, USA
[20] Han JQ, Piette MA (2010) Solutions for summer electric power shortages: demand response and its application in air conditioning and refrigerating systems. LBNL-63806. Lawrence Berkeley National Laboratory, Berkeley, CA, USA
[21] Kiliccote S, Piette MA, Dudley JH et al (2009) Open automated demand response for small commercial buildings. LBNL-2195E. Lawrence Berkeley National Laboratory, Berkeley, CA, USA
[22] Olsen D, Goli S, Faulkner D et al (2010) Opportunities for energy efficiency and demand response in the California cement industry. LBNL-4849E. Lawrence Berkeley National Laboratory, Berkeley, CA, USA
[23] Lidbetter RT (2013) Load-shifting opportunities for typical cement plants. J Energy Southern Africa 24(1):35–45
[24] Todd D, Caufield M, Helms B et al (2009) Providing reliability services through demand response: a preliminary evaluation of the demand response capabilities of Alcoa Inc. ORNL/TM-2008/233. Federal Energy Regulatory Commission, Washington, DC, USA
[25] Federal Energy Regulatory Commission (2010) Alcoa demand response innovation. Federal Energy Regulatory Commission, Washington, DC, USA
[26] Alcazar-Ortega M, Alvarez-Bel C, Escrivá-Escrivá G et al (2012) Evaluation and assessment of demand response potential applied to the meat industry. Appl Energy 92:84–91
[27] Paulus M, Borggrefe F (2011) The potential of demand-side management in energy-intensive industries for electricity markets in Germany. Appl Energy 88(2):432–441
[28] Samad T, Kiliccote S (2012) Smart grid technologies and applications for the industrial sector. Comput Chem Eng 47:76–84
[29] Jin RR (1993) An introduction to the process design of cement plant. Wuhan University of Technology Press, Wuhan, China (in Chinese)
[30] Madlool MA, Saidur R, Hossain MS (2011) A critical review on energy use and savings in the cement industries. Renew Sustain Energy Rev 15(4):2042–2060

Mingtao YAO received his B.S. degree in electrical engineering from Beijing Jiaotong University, China, in 2010. Now, he is a Ph.D. student in Beijing Jiaotong University. His research interests include power supply and demand, demand response and efficiency power plant.

Zhaoguang HU received his Ph.D. degree in China Electric Power Research Institute, Beijing, China, in 1989. He is the vice president and chief energy specialist at State Grid Energy Research Institute. His main research interests include low-carbon energy, energy economy, energy efficiency and demand side management, and policy study on the above areas.

Ning ZHANG received his B.S. degree in electrical engineering from Beijing Jiaotong University, China, in 2012. Now, he is a Ph.D. student in Beijing Jiaotong University. His research interests include power planning and demand response.

Wei DUAN received his B.S. degree in electrical engineering from Beijing Jiaotong University, China, in 2009. Now, he is a Ph.D. student in Beijing Jiaotong University. His research interests include power planning and efficiency power plant.

Jian ZHANG received his B.S. degree in electrical engineering from Beijing Jiaotong University, China, in 2011. Now, he is a Ph.D. student in Beijing Jiaotong University. His research interests include power supply and demand.

13

A preventive control strategy for static voltage stability based on an efficient power plant model of electric vehicles

Mingshen WANG, Yunfei MU (✉), Hongjie JIA, Jianzhong WU, Xiuping YAO, Xiaodan YU, Janaka EKANAYAKE

Abstract With the increasing integration of wind farms and electric vehicles (EVs) in power systems, voltage stability is becoming more and more serious. Based on vehicle-to-grid (V2G), an efficient power plant model of EVs (E-EPP) was developed to estimate EV charging load with available corresponding response capacity under different charging strategies. A preventive control strategy based on E-EPP was proposed to maintain the static voltage stability margin (VSM) of power system above a predefined security level. Two control modes were used including the disconnection of EV charging load ('V1G' mode) and the discharge of stored battery energy back to power grid ('V2G' mode). A modified IEEE 14-bus system with high penetration of wind power and EVs was used to verify the effectiveness of preventive control strategy. Simulation results showed that the proposed strategy can not only improve the static voltage stability of power system with considerable wind generation, but also guarantee the travelling comfort for EV owners.

Keywords Electric vehicle (EV), Vehicle-to-grid (V2G), Efficient power plant (EPP), Preventive control, Static voltage stability

M. WANG, Y. MU, H. JIA, X. YU, Key Laboratory of Smart Grid of Ministry of Education, Tianjin University, Tianjin 300072, China
(✉) e-mail: yunfeimu@tju.edu.cn
J. WU, J. EKANAYAKE, School of Engineering, Institute of Energy, Cardiff University, Wales CF10 3BN, UK
X. YAO, State Grid Xinjiang Electric Power Company, Urumqi 830002, China

1 Introduction

The development of "low-carbon economy" is drawing more and more attention around the world considering the environment and energy. As a consequence, a number of countries have taken specific initiatives to decarbonize their electrical power system and transport sectors [1–4]. In UK, it is anticipated that a large proportion of renewable energy is wind energy. There may be up to 30 GW wind generation of 100 GW total generation capacity serving around 60 GW load by 2020 [5, 6]. In order to realize the target of reducing emissions from domestic transport sector by 14 % by 2020, the UK government has supported EV trials with the anticipation that EVs will play a major role in future transport sector [7–9].

According to the large scale penetration of renewable energy generation from intermittent resources, especially wind power, it is fundamental that the electric system is able to appropriately compensate the effects on the variability and randomness of wind energy. Among all the issues caused by the intermittency of wind power, voltage stability is a main constraint for the further development of low-carbon power system [10, 11].

A lot of studies have been carried out to analyze the voltage stability integrated with wind generation. The impact of wind power with high penetration on the voltage stability of power system was analyzed in [12, 13]. A strategy was proposed in [14] to coordinate the reactive power of variable-speed wind generators and other reactive controllers for static voltage stability enhancement which may limit the output of wind power. Battery energy storage system (BESS) was used for regulating the wind power variation to improve the voltage stability of power system [15]. However, as BESS is expensive, its economic benefits for wide deployment are still in question. Although the application of static synchronous compensator can improve the voltage stability of power system with wind farms, large scale adoption of these kinds of devices is not economical and practical until now [16–18].

With the support from modern communication and control system, EVs can not only act as a rapid response load of power system, but also support the security operation of power system as a new kind of energy storage system [19]. A real-time scheduling method of EV charging load was proposed to increase voltage security margin in a low-voltage distribution system, but the energy stored in EV batteries was not fully utilized [20]. In [21], EVs were used to provide the voltage support of power system with the integration of photovoltaic power generation, which demonstrates the feasibility of using EVs for voltage control with intermittent renewable energy. However, these studies are unable to fully consider the transportation behaviors of EV owners, which lead to a spatial and temporal distribution of EV charging load and available response capability.

In this paper, an efficient power plant of EVs (E-EPP) was developed to determine the maximum/minimum available response capacity of EVs from a spatial and temporal perspective, considering the travelling comfort of EV owners. A preventive control strategy based on E-EPP was developed to fully utilize the vehicle-to-grid (V2G) capability of EV aggregation along a day, which can not only effectively improve the static voltage stability, but also reduce the emission of carbon dioxide by supporting the integration of wind farms in a power system. The rest of paper is organized as follows. The framework of preventive control strategy is introduced in Sect. 2. Section 3 develops E-EPP model considering the travelling behaviors of EV owners. The preventive control strategy is given in Sect. 4. In Sect. 5, simulations are carried out on a modified IEEE 14-bus system integrated with the high penetration of EVs and wind generation to verify the effectiveness of proposed preventive control strategy. The reduction of carbon dioxide emission is calculated. Conclusions are summarized in Sect. 6.

2 Framework of integrating EVs for preventive control

2.1 Traditional preventive control for static voltage stability

In conventional power systems, when load demand approaches or even exceeds its transfer limit, voltage instability/collapse happens at the heavy-duty point. With the integration of large wind farms, the transmission mode of power system may be changed on the receiving end. The intermittent output may cause transmission congestion problems. Besides, while sending out active power, wind farms usually absorb considerable reactive power from power system [22]. However, wind power will cause significant impact on voltage stability.

In order to analyze the voltage stability, $\boldsymbol{S}_{\mathrm{GL}} = \boldsymbol{S}_{\mathrm{G}} \cup \boldsymbol{S}_{\mathrm{L}}$ is used as the power injection vector; $\boldsymbol{S}_{\mathrm{G}} = (\boldsymbol{P}_{\mathrm{Gc}} \cup \boldsymbol{Q}_{\mathrm{Gc}}) \cup (\boldsymbol{P}_{\mathrm{Gw}} \cup \boldsymbol{Q}_{\mathrm{Gw}})$ is the corresponding power injection vector of conventional generators and wind farms; $\boldsymbol{S}_{\mathrm{L}}=\boldsymbol{P}_{\mathrm{L}} \cup \boldsymbol{Q}_{\mathrm{L}}$ is the power injector of loads; $\boldsymbol{P}_{\mathrm{Gc}}$ and $\boldsymbol{Q}_{\mathrm{Gc}}$ are the active and reactive power injection vectors of conventional generators, respectively; $\boldsymbol{P}_{\mathrm{Gw}}$ and $\boldsymbol{Q}_{\mathrm{Gw}}$ are the active and reactive power injection vectors of wind farms, respectively. The operating status $\boldsymbol{x}$ is determined by $\boldsymbol{S}_{\mathrm{GL}}$ as described as

$$\begin{cases} \boldsymbol{f}(\boldsymbol{x}, \boldsymbol{S}_{\mathrm{GL}}) = \boldsymbol{0} \\ \boldsymbol{g}(\boldsymbol{x}, \boldsymbol{S}_{\mathrm{GL}}) \leq \boldsymbol{0} \end{cases} \tag{1}$$

where $\boldsymbol{f}$ is the load flow equation; $\boldsymbol{g}$ is the system operating constrain equation. If $\boldsymbol{x}$ yielded by (1) also satisfies (2), the system is said to be static voltage stability. While, if $\boldsymbol{x}$ satisfies (1) as well as (3), the system is at the critical point of static voltage stability.

$$\det(\boldsymbol{f}_x) \neq 0 \tag{2}$$

$$\det(\boldsymbol{f}_x) = 0 \tag{3}$$

where $\boldsymbol{f}_x$ is the Jacobian matrix of load flow equation. λ is introduced to evaluate static voltage stability margin (VSM), which can be determined by continuous power flow (CPF) according to the direction of load growth and generation dispatch mode [23, 24]. In order to prevent voltage stability, retaining large enough λ under different contingencies is necessary.

Preventive control is a main means to maintain power system operating above a minimum voltage stability margin (λ_{cr}) under the increased uncertainty brought on by the integration of intermittent renewable generation, such as wind power. Usually, the traditional preventive control of power system for voltage control is divided into three stages, where if the previous stage is effective, the latter stage can be avoided.

Stage I: Adjusting the generator terminal voltage, shunts, and on-load tap changer.
Stage II: Considering the re-dispatching generator active power.
Stage III: Dispatching the load shedding.

The arrangement of above three stages aims to guarantee an uninterruptible power supply to end-users while taking the cost of control into account. Therefore, load-shedding is traditionally regarded as the last means of avoiding voltage collapse. However, this situation has been significantly changed as the V2G capability from EVs was introduced. Active participation based on the voluntary of EV owners will show lower control cost and higher efficiency, which can be used as an efficient power plant to participate in stage II for voltage control. In this paper, a preventive control strategy for static voltage stability using EVs is investigated. It is assumed that the capacity of stage I has been ran out, thereby only the capability of E-EPP provides static voltage control requirements.

2.2 Framework of preventive control using E-EPP

The framework of preventive control strategy based on E-EPP is shown in Fig. 1.

Power system dispatch center (PSDC) provides a valuable system-level metric for the minute-to-minute decision to ensure service consistency. One role of PSDC is to obtain the static voltage stability margin (λ) associated with the current operating point (distance to voltage collapse), and compute credible contingencies with respect to predefined load directions. CPF is used to compute the VSM of power system [25]. Also, the preventive control strategy is implemented in PSDC for static voltage control of power system.

E-EPP is an aggregate model of a large scale of geographically dispersed EVs connected to power system with V2G power electronic interface. Although the available response power capacity of a single EV is relatively small, the aggregate capacity of all EVs managed by E-EPP is considerable. E-EPP can serve as an intermediary between EVs and PSDC. On one hand, E-EPP estimates EV charging load with the corresponding available response capacity under different charging strategies. On the other hand, E-EPP is responsible for managing EVs charging/discharging process of voltage control considering the travelling comfort level of EV owners.

When a voltage contingency happens, E-EPP receives the control signals from PSDC. According to the preventive control strategy, part of EV charging loads are shed from power system ('V1G') or even discharge the stored energy back to grids ('V2G'). Battery state-of-charge (SOC) is used as an indicator for E-EPP to select EV candidates for responding the control signals given by preventive control strategy considering the travelling preference of EV owners. The battery SOC information is obtained via smart meter (SM) and charging point management system (CPM). Then the SOC information flow is transmitted to E-EPP control center for preventive control.

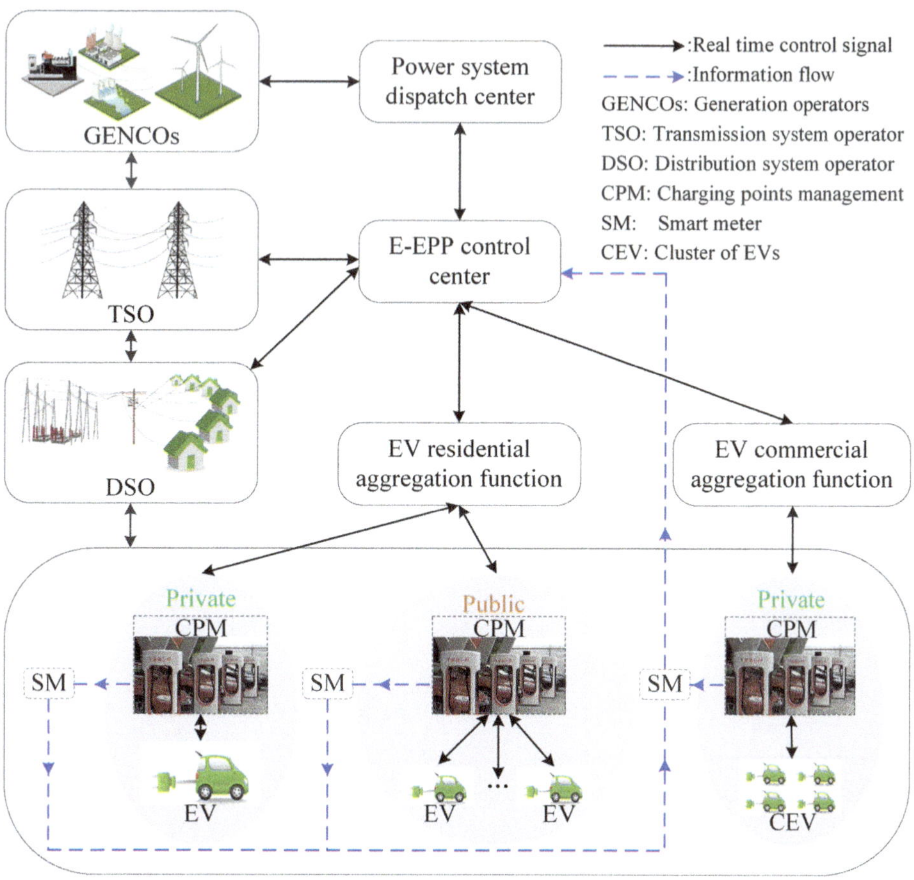

Fig. 1 Integration of EVs for preventive control based on E-EPP

3 Efficient power plant model of aggregate EVs

3.1 A generic battery model for EVs

A generic battery model (GBM) is firstly developed to describe the charging/discharging characteristics of various EV batteries. As shown in Fig. 2, GBM is a simple controlled voltage source in series with a constant resistance. The model uses only SOC as a state variable in order to accurately reproduce the manufacturers' curves for four major types of batteries (lead-acid, lithium-ion, nickel-cadmium and nickel-metal-hybrid batteries), which are commonly considered as the most promising EV batteries in the future EV market [26].

The battery terminal voltage V_{batt} is described as

$$V_{\mathrm{batt}} = E_0 - \frac{K}{K_{\mathrm{SOC}}} + A\exp\left(-B\int_0^t I\mathrm{d}t\right) - RI \tag{4}$$

$$K_{\mathrm{SOC}} = 1 - \int_0^t I(t)\mathrm{d}t/Q \tag{5}$$

where K_{SOC} is the value of SOC; E_0 is the constant battery voltage; K is the polarisation voltage; Q is the battery capacity; I is the charging current; A is the exponential zone amplitude; B is the exponential zone time constant inverse; R is the internal resistance.

Assuming a constant charging current I to simplify the integral part, the charging power P is shown as

$$P = E_0I - \frac{KI}{K_{\mathrm{SOC}}} + AI\exp(-BIt) - RI^2 \tag{6}$$

There are only a few parameters in the proposed GBM. Once the parameters of battery type and capacity are determined, SOC and charging power can be obtained in real time by using this GBM.

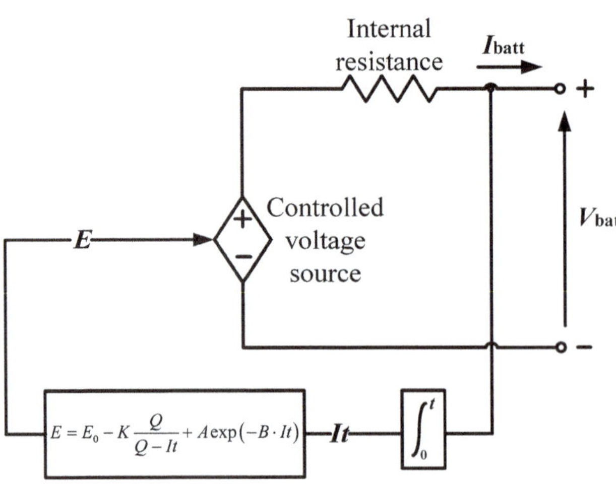

Fig. 2 The generic battery model of EV batteries

3.2 Formulation of E-EPP

Considering the travelling behaviors of EV owners, E-EPP is established to evaluate the response capacity of EVs along a day for voltage control. Several uncertainties are considered in this formulation, such as battery type, capacity, energy consumption, daily travelling distance, travelling time and the minimum desired battery SOC for travelling requirement.

3.2.1 EV classification

In the formulation of E-EPP, EVs are classified into the following two groups.

Based on the use of transportation, EVs are classified into three types, i.e., home-based-work (HBW), home-based-other (HBO) and non-home-based (NHB). Their proportions in UK are 61 %, 30 % and 9 %, respectively [27]. This classification is used to obtain EV daily travelling distance and time.

According to vehicle types, EVs are classified into four types, i.e., L7e, M1, N1 and N2, and their proportions are 1.49 %, 87.51 %, 10 % and 1 %, respectively [28]. This classification is used to obtain the battery type, capacity and energy consumption per kilometer of a single EV.

3.2.2 Battery type, capacity and energy consumption

Based on the market survey data of various battery types (B_t) in UK, the proportions of four promising EV batteries, i.e., lead-acid batteries, lithium-ion batteries, nickel–cadmium batteries and nickel-metal-hybrid batteries, are 20 %, 50 %, 10 % and 20 %, respectively [28]. According to the type of EVs, the distributions of battery capacities (Q_v) are shown in Table 1 [29], and the parameters in the distributions are further defined by (7) of Gamma distribution and (8) of normal distribution.

$$g(Q_v;\alpha,\beta) = \frac{1}{\beta^{\alpha}\Gamma(\alpha)}Q_v^{\alpha-1}\exp\left(-\frac{Q_v}{\beta}\right) \tag{7}$$

$$f(Q_v;\mu,\sigma) = \frac{1}{\sqrt{2\pi}\sigma}\exp\left(-\frac{(Q_v-\mu)^2}{2\sigma^2}\right) \tag{8}$$

Meanwhile, the distributions of EV energy consumption (C_e) are shown in Table 2 [28].

Table 1 Distributions of EV battery capacities

Type	Distribution	Parameters	Min (kWh)	Max (kWh)
L7e	Gamma	$\alpha = 10.8$; $\beta = 0.8$	3.0	15.0
M1	Gamma	$\alpha = 4.5$; $\beta = 6.7$	10.0	72.0
N1	Normal	$\mu = 23.0$; $\sigma = 9.5$	9.6	40.0
N2	Normal	$\mu = 85.3$; $\sigma = 28.0$	51.0	120.0

3.2.3 Daily travelling distance and time

Based on the use of transportation (HBW, HBO and NHB), daily travelling distance d follows a normal distribution, which is expressed as

$$f(d;\mu_d,\sigma_d)=\frac{1}{\sqrt{2\pi}\sigma_d}\exp\left(-\frac{(d-\mu_d)^2}{2\sigma_d^2}\right) \tag{9}$$

where μ_d is the mean daily travelling distance; σ_d is the standard deviation. For HBW and HBO vehicles, $\mu_d = 35.9$ km, $\sigma_d = 19.6$ km. While for NHB vehicles, $\mu_d = 87.1$ km, and $\sigma_d = 24.5$ km [27].

Daily travelling time has close relationship with the transportation behaviors of EV owners. The distributions of starting travelling time (t_s) and finishing travelling time (t_f) are shown in Fig. 3 [29].

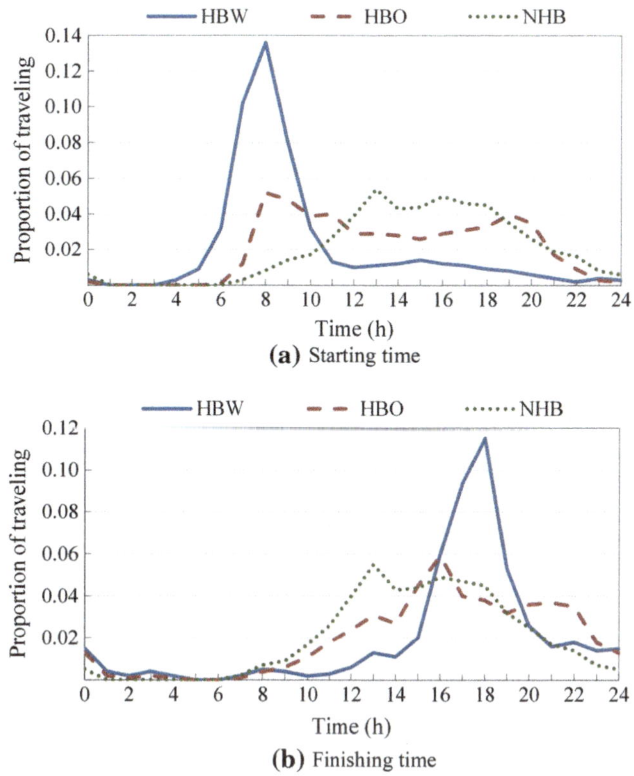

Fig. 3 EV travelling time distributions of HBW, HBO and NHB

3.2.4 Formulation of E-EPP

1) Determine B_t, C_e and Q_v of a single EV

For a single EV, a Monte Carlo simulation (MCS) method is used to obtain B_t based on the proportion of previous four EV batteries, also C_e and Q_v depending on the distributions in Tables 1 and 2, respectively.

2) Obtain the parameters in GBM

It is assumed that EV battery charging process is reversible. The power sent back to grid by an EV is assumed as a constant while providing voltage control, and it is equal to the power absorbed by the same EV before controlling. Once B_t and Q_v were determined in step 1, the other parameters in GBM are deduced from the discharging curves supplied by manufacturers [26]. Then the charging process consisting of charging power (P) and SOC is obtained via GBM.

3) Determine d and charging starting time

The daily travelling distance is obtained by (9) with a MCS process. The charging time t_{sc} of an EV battery is determined by daily transportation behavior and the charging strategies. Three charging strategies are considered in the formulation of E-EPP.

Table 2 Distributions of energy consumption per kilometer

C_e (kWh/km)	L7e	M1	N1	N2
0.05–0.10	0.34	0.05	0.00	0.00
0.10–0.15	0.58	0.35	0.29	0.00
0.15–0.20	0.08	0.45	0.14	0.00
0.20–0.25	0.00	0.15	0.57	0.00
0.45–0.50	0.00	0.00	0.00	0.50
0.50–0.55	0.00	0.00	0.00	0.25
0.80–0.85	0.00	0.00	0.00	0.25

a. Dumb Charging:

In dumb charging, it is assumed that all EVs start to charge as soon as their daily trips are finished. Therefore, t_{sc} is equal to t_f which is determined by the distributions shown in Fig. 3b.

b. Smart Charging:

It is envisaged that there will be an active management system based on two hierarchical control structures by an Aggregator and the system operators (TSO/DSO), respectively. Furthermore, it is assumed that EV charging is controlled according to the Aggregator's market negotiations or the need of system operators. Smart charging is described by (10) with μ_{sc} (1:00) and σ_{sc} (5 h) [29]. Compared with "dumb" charging, the model of "smart" charging represents the shift of EV charging load from the system peak demand time to the valley hours.

$$f(t_{sc};\mu_{sc},\sigma_{sc})=\frac{1}{\sqrt{2\pi}\sigma_{sc}}\exp\left(-\frac{(t_{sc}-\mu_{sc})^2}{2\sigma_{sc}^2}\right) \tag{10}$$

c. Hybrid Charging:

Hybrid charging is a charging strategy with 50 % of the EVs adopting dumb charging and the remaining 50 % following smart charging.

4) Determine the initial SOC when an EV starts to charge

Assuming SOC drops linearly with the travelling distance [27], K_{SOC_0} is determined as

$$K_{SOC_0} = \left(\delta - \frac{d}{d_t}\right) \times 100\,\% \tag{11}$$

where K_{SOC_0} is the value of SOC_0; d_t is the maximum travelling distance and $d_t = Q_v/C_e$; δ is SOC of a single EV before travelling, and it varies uniformly within [0.8, 0.9] to maintain the lifetime of a battery [28].

5) Determine the response state of an EV

Assuming a minimum desired SOC (K_{SOC_e}) (for satisfying the travelling requirements of EV owners) is 60 %. For an individual EV of EV_i, the charging state at time t ($\eta_{i,t}$) is determined by (12). If $\eta_{i,t}$ is 0, the charging EV_i is not controllable. If $\eta_{i,t}$ is 1, the charging EV_i can be selected to charge, stop charging ('V1G') or discharging ('V2G'). While if $\eta_{i,t}$ is 2, the idle EV_i can be selected to discharge of stored battery energy back to power grid ('V2G').

$$\eta_{i,t} = \begin{cases} 0 & 0 < K_{SOC_{i,t}} \leq K_{SOC_e} \\ 1 & K_{SOC_e} < K_{SOC_{i,t}} \leq \delta_i \\ 2 & \delta_i \leq K_{SOC_{i,t}} < 100\,\% \end{cases} \tag{12}$$

6) Based on the MCS process, steps 1–5 are repeated n times for n EVs. The real-time charging power ($P_{E\text{-}EPP,t}$) is obtain by (13). The upper available response boundary ($P_{upper,t}$) and lower available response boundary ($P_{lower,t}$) of E-EPP are given in (14) and (15), respectively.

$$P_{E-EPP,t} = \sum_{i=1}^{l_t} P_{i,t} + \sum_{j=1}^{s_t} P_{j,t} \tag{13}$$

$$P_{upper,t} = \sum_{i=1}^{l_t} P_{i,t} + \sum_{j=1}^{m_t} P_{j,t} + \sum_{k=1}^{n_t} P_{k,t} \tag{14}$$

$$P_{lower,t} = \sum_{i=1}^{l_t} P_{i,t} - \sum_{j=1}^{m_t} P_{j,t} - \sum_{k=1}^{n_t} P_{k,t} \tag{15}$$

where $P_{i,t}$ is the charging/discharging power of EV_i at time t; l_t is the number of EVs when $\eta_{i,t}$ is 0; m_t is the number of EVs when $\eta_{i,t}$ is 1; s_t is the number of charging EVs among the m_t EVs (s_t is obtained by the EV charging state information of E-EPP control center); n_t is the number of EVs when $\eta_{i,t}$ is 2.

4 Preventive control strategy based on E-EPP

In this section, a preventive control strategy for static voltage stability is proposed based on the E-EPP. During a voltage contingency, the available capacity of the E-EPP is fully utilized for voltage control considering the travelling comfort constraints of EV owners. Active Participation Factor (APF) is used in the preventive control to determine the optimum bus to response to the voltage event.

4.1 Bus response order based on APF

In this preventive control strategy, APF is used to determine the optimum load bus response order of power system for voltage control, as depicted in (16) [25].

$$\begin{bmatrix} \Delta \boldsymbol{P} \\ \Delta \boldsymbol{Q} \end{bmatrix} = \begin{bmatrix} \boldsymbol{J}_{P\theta} & \boldsymbol{J}_{PV} \\ \boldsymbol{J}_{Q\theta} & \boldsymbol{J}_{QV} \end{bmatrix} \begin{bmatrix} \Delta \boldsymbol{\theta} \\ \Delta \boldsymbol{V} \end{bmatrix} \tag{16}$$

where $\Delta \boldsymbol{P}$ is the active power variations of the buses; $\Delta \boldsymbol{Q}$ is the reactive power variations of the buses; $\Delta \boldsymbol{\theta}$ is the voltage angle variations of the buses; $\Delta \boldsymbol{V}$ is the voltage magnitude variations of the buses; $\boldsymbol{J}_{P\theta}$, $\boldsymbol{J}_{PV}$, $\boldsymbol{J}_{Q\theta}$ and $\boldsymbol{J}_{QV}$ are Jacobian sub-matrices representing the sensitivities of active and reactive power to voltage angle and magnitude, respectively.

During the CPF process, the Jacobian matrix $\boldsymbol{J}$ is obtained at the critical operating point of the system as

$$J_{2NPQ+NPV} = \boldsymbol{\Phi \Lambda \Gamma} \tag{17}$$

The first (NPV+NPQ) elements of the right and left eigenvector associated to the critical eigenvalue of $\boldsymbol{J}$ are defined as APF of all buses in the system. In this paper, EVs charging at load bus with the largest APF will be firstly selected to respond to the voltage events in the preventive control strategy.

As the operation status of power system (such as the load level) fluctuates with time, the load bus response order is changed dynamically.

4.2 Preventive control strategy based on E-EPP

The SOC is a key indicator to select the satisfied EVs for static voltage stability response in the preventive control strategy. Considering the minimum desired SOC for travelling of EV owners, EVs with high SOC level will participate in the control strategy firstly. The successive steps of the preventive control strategy are listed as follow.

1) Based on the CPF process, compute the VSM (λ_t) of power system in real time.

2) If $\lambda_t \geq \lambda_{cr}$, voltage is stable, calculate λ for next time. If $\lambda_t < \lambda_{cr}$, the voltage enters an emergency state, preventive control from EVs is started, go to next step.

3) Obtain the APFs, and queue the load buses by the APF values from high to low. According to the E-EPP model, evaluate the available response capacity of EVs at each load bus under 'V1G' and 'V2G' modes, respectively.

4) If there are existing available 'V1G' EVs for preventive control, go to step 5. Otherwise, go to step 6 for

'V2G'. If EVs for 'V1G' or 'V2G' are unavailable, some emergency methods (such as load shedding) are taken for voltage stability control.

The SOC threshold l ($l_{\min} \le l \le l_{\max}$) is used to select satisfying EVs with $K_{\mathrm{SOC}} \ge l$ to respond to voltage events, and $l_{\max}$ ($l_{\min}$) is the upper (lower) boundary, where $l_{\min}$ is equal to $K_{\mathrm{SOC_e}}$ for satisfying the travelling demand of EV owners.

5) Among all the load buses having connected EVs with $K_{\mathrm{SOC}} \ge l$, determine the bus with the highest APF, and the satisfying EVs connected to this load bus stop charging for preventive control.

If system still exists charging EVs with $K_{\mathrm{SOC}} > l$, go to step 1. Otherwise, $l = l - \Delta l \ge l_{\min}$, where $\Delta l = \alpha(\lambda_{\mathrm{cr}} - \lambda_t)$ (α is a control coefficient).

6) Among all the load buses having connected EVs with $K_{\mathrm{SOC}} \ge l$, determine the bus with the highest APF, and the satisfying EVs connected to this load bus discharge stored energy back to grid for preventive control.

If system still exists EVs with $K_{\mathrm{SOC}} \ge l$, go to step 1. Otherwise, $l = l - \Delta l \ge l_{\min}$, where $\Delta l = \alpha\ (\lambda_{\mathrm{cr}} - \lambda_t)$.

Taking a system with five load buses (N1, N2, N3, N4 and N5) as an example, and assuming the relative APF value at time t is N1 > N2 > N3 > N4 > N5, the 'V1G' or 'V2G' response order with a constant $\Delta l = 5\ \%$ is shown in Fig. 4.

If all the EVs for 'V1G' or 'V2G' service recharge simultaneously after the voltage event, there will be a sudden drop on the system VSM. Therefore, the gradual recharging strategy based on temporal-spatial distribution is also proposed. When the VSM is higher than λ_{high}, the EVs will connect to power system for recharging based on the opposite order of preventive control until the VSM is lower than λ_{low}, where $\lambda_{\mathrm{high}} > \lambda_{\mathrm{low}} > \lambda_{\mathrm{cr}}$. Taking the same system as an example, the recharging order after preventive control is shown in Fig. 5. As a result, the EV with lower SOC level can be recharged firstly, which can improve the travelling comfort level.

5 Case study

In this section, a modified IEEE 14-bus system with wind farms (GW3, GW6 and GW8) is used to illustrate the preventive control strategy, as shown in Fig. 6.

The system is divided into three function zones: residential, commercial and industrial zones. Figure 7 gives the load profiles during one day in different zones, which is provided by UKGDS [30].

It is assumed that the wind farms GW3, GW6 and GW8 are composed of doubly fed generators. A typical output of wind power during one day in UK is shown in Fig. 8 [30].

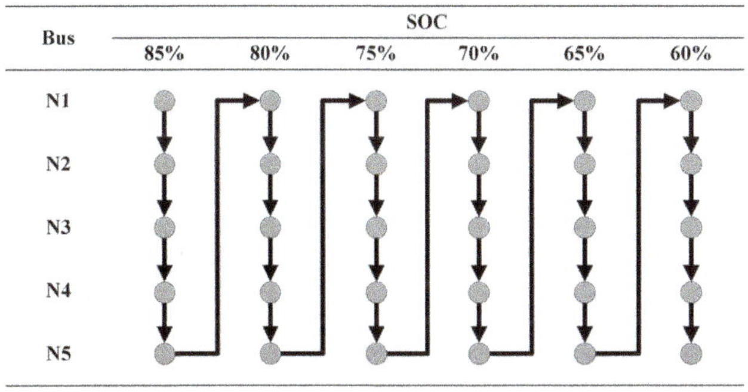

Fig. 4 Response order of 'V1G' or 'V2G' in the preventive control

In this modified IEEE 14-bus system, the peak load is 397.05 MW, which is 0.58 % of the UK peak electric demand [31]. The EV numbers in different zones are shown in Table 3, which also account for 0.58 % of all EVs in UK [32].

Assuming EV charging facilities have been set up, the EV number at each load bus depends on the proportion of its initial load (without EV charging loads). To illustrate the available capacity of aggregate EVs, the E-EPPs under three charging strategies at bus 10 along a day are shown in Fig. 9.

Comparing dumb charging with smart charging, it is clear that the peak charging load is shifted from 18:00 to 01:00. Hybrid charging is a compromised strategy between dumb and smart charging. To compare and analyze these three charging strategies, the VSM (λ) is shown in Fig. 10. Under dumb charging, the peak charging load is in coincidence with the initial peak load, and it is easy to see that the system VSM is lower than λ_{cr} from 17:00 to 23:00.

To verify the effectiveness of the preventive control strategy, it is defined that $\lambda_{\mathrm{cr}} = 2.5$, $\lambda_{\mathrm{high}} = 3.5$, $\lambda_{\mathrm{low}} = 3.3$, $l_{\min} = 60\ \%$, $l_{\max} = 85\ \%$ and $\alpha = 0.12$. Taking dumb charging as an example, the VSM results of a typical day before and after preventive control are shown in Fig. 11.

From 8:00 to 11:00, a line outage happens on line 6–11, and the VSM of power system decreases below λ_{cr} suddenly. As shown in Fig. 9a, the available response capacity

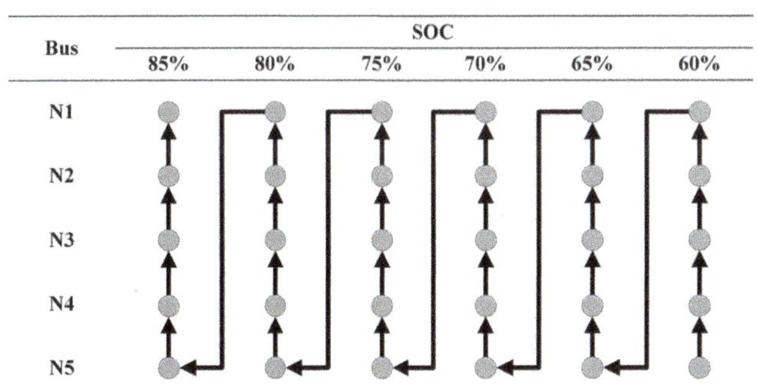

Fig. 5 Recharging order in the preventive control

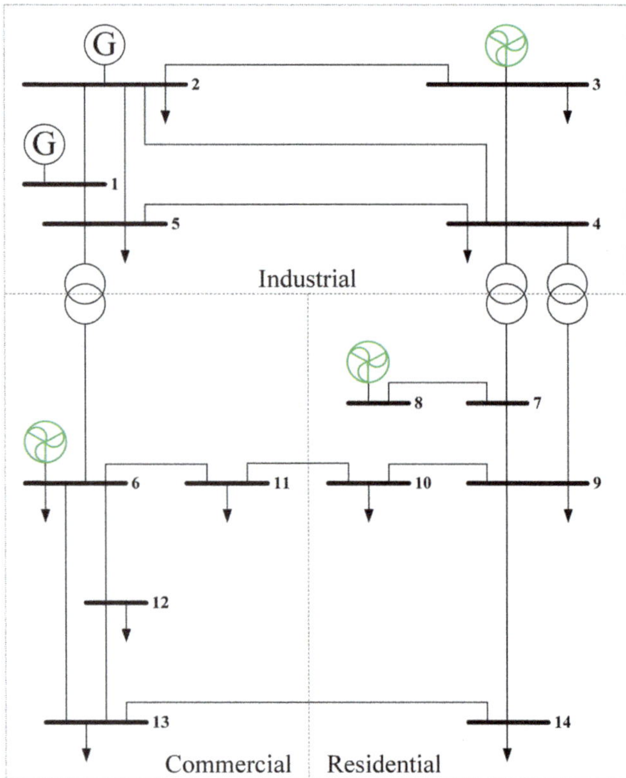

Fig. 6 The modified IEEE-14 bus system

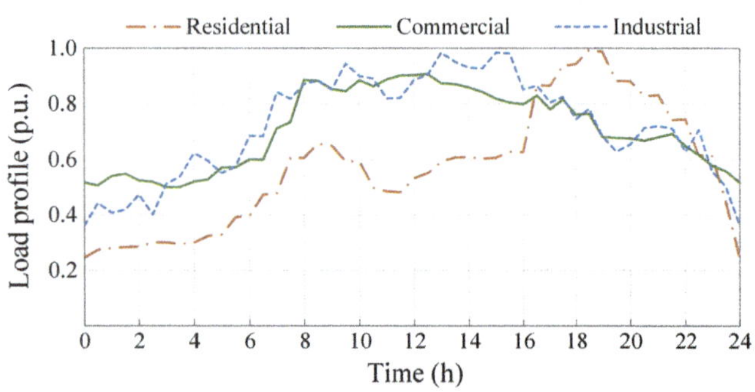

Fig. 7 Load profile of different zones for one day

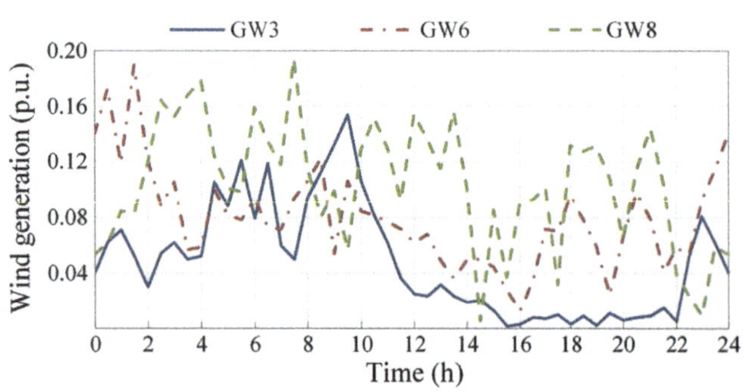

Fig. 8 Wind generation for one day

Table 3 EV numbers for different function zones

Zones	HBW	HBO	NHB	All EV groups
Residential	100650	49500	0	150150
Commercial	0	0	2500	2500
Industrial	0	0	12400	12400
Total	100650	49500	14900	165050

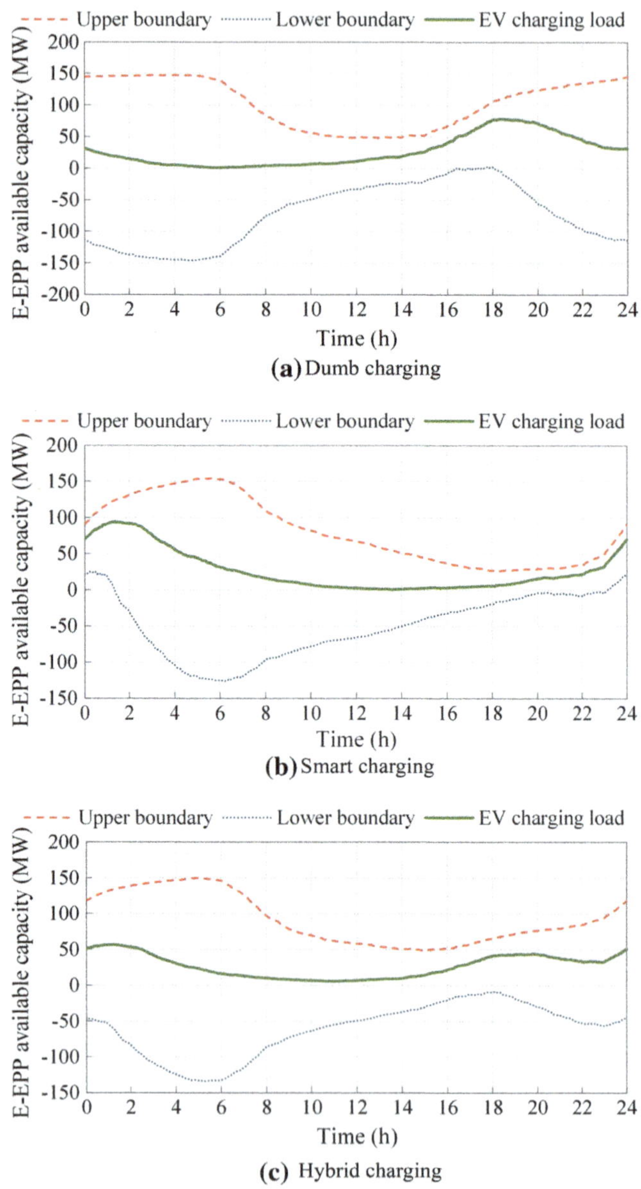

Fig. 9 The E-EPP with available response capacity along a day at bus 10

for 'V1G' is rather small during this period. But there are EVs which have completed their charging process, and still connecting to the power system. These EVs are available for 'V2G' control.

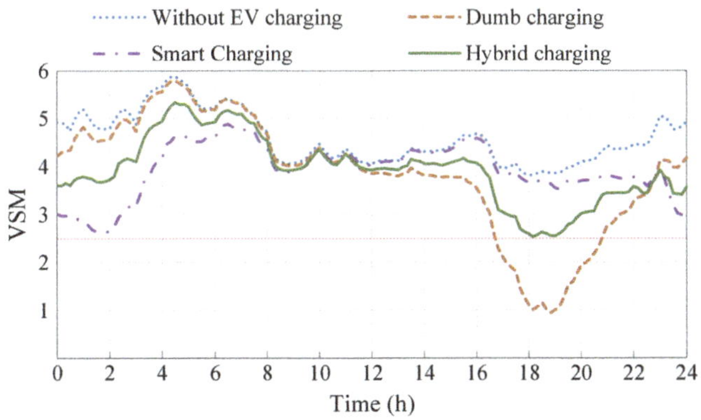

Fig. 10 The VSM along a day under three charging strategies

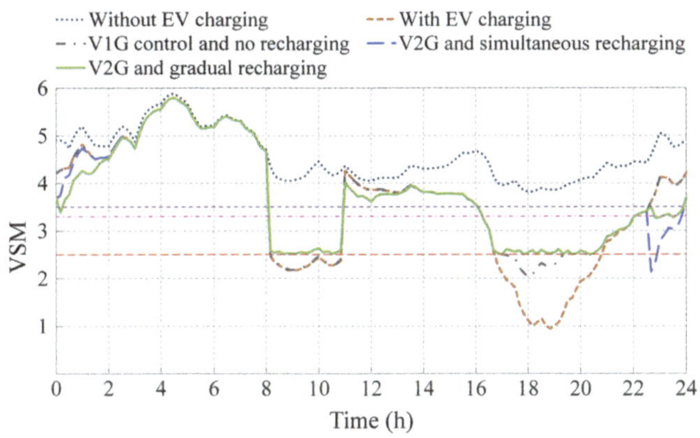

Fig. 11 The preventive control results with dumb charging

While from 17:00 to 21:00, because of a loss of the wind generation and the peak charging load of EVs, the system VSM falls below λ_{cr}. Although the charging load of EVs during this period is large, only 'V1G' control cannot restore the VSM higher than λ_{cr}, as the curve of 'V1G' control with no recharging shown in Fig. 11. 'V2G' is further needed for voltage stability control of power system compared with the curve of 'V2G' and gradual recharging in Fig. 11 during this period.

If all the EVs for 'V1G' and 'V2G' service recharge simultaneously, there will be a sudden drop of the system VSM at 22:30 as shown in Fig. 11. Under the gradual recharging strategy based on special-temporal distribution, this recharging strategy will ensure smaller fluctuations on voltage as the profile of 'V2G' and gradual recharging shown in Fig. 11 during 22:00 to 00:00 and 00:00 to 02:00.

Then to further introduce the process of preventive control strategy, the E-EPP response profile at bus 10 after control is shown in Fig. 12.

As depicted from the above results, using EVs for the preventive control of power system can significantly relieve the static voltage stability issue caused by the wind farms, which can support the integration of wind farms and

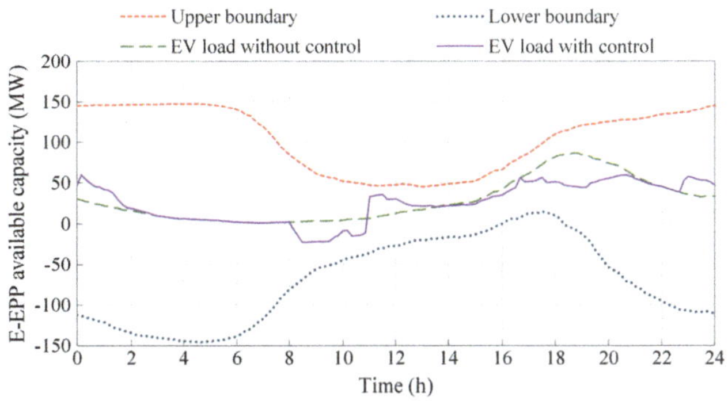

Fig. 12 E-EPP after preventive control

low-carbon economy. For example, according to the carbon dioxide emission factor of thermal power ($k_{CO_2} = 0.9109$ kg/(kWh)) and the calculation method proposed in [33], the emission reduction of carbon dioxide coming from the wind farms at bus 3, 6 and 8 is 506.55 t for a typical day. And this will be of significant importance to the low-carbon economy.

6 Conclusions

In this paper, a preventive control strategy from EVs based on E-EPP is proposed to maintain static voltage stability of power system under the V2G concept. The following conclusions are drawn:

1) EVs have great potential to support the voltage control of power system. The developed preventive control strategy can fully utilize the V2G capability of the EV aggregation along a day, which can effectively improve static voltage stability of power system. However, the contribution from EVs has an obvious temporal distribution along a day, which has close relationship with charging strategies.

2) E-EPP is able to determine the maximum/minimum available response capacity of EVs from a spatial and temporal perspective under three charging strategies, considering the travelling comfort levels of EV owners.

3) The gradual recharging strategy can ensure smaller fluctuations on VSM, which can solve the second disturbance problem (caused by sudden increase in EV charging power) on power system.

4) EVs and wind farms have great effect on the emission reduction of carbon dioxide, which can promote the low-carbon economy to some extent.

Acknowledgments This work was supported in part by the National Natural Science Foundation of China (collaborating with EPSRC of UK) (Nos. 51361130152 and EP/L001039/1), the National Science and Technology Support Program of China (No. 2013BAA01B03), and Research on Reactive Power Control and Comprehensive Evaluation Technique of Large Scale Integration of Wind/Photovoltaic Power Generation (No. NY71-14-035).

References

[1] Parkinson S, Wang D, Djilali N (2012) Toward low carbon energy systems: the convergence of wind power, demand response, and the electricity grid. In: Proceedings of the 2012 IEEE international conference on innovative smart grid technologies—Asia (ISGT Asia'12), Tianjin, 21–24 May 2012, 8 pp

[2] Vithayasrichareon P, MacGill IF (2011) Generation portfolio analysis for low-carbon future electricity industries with high wind power penetrations. In: Proceedings of the 2011 IEEE Trondheim PowerTech, Trondheim, 19–23 June 2011, 6 pp

[3] Boulanger AG, Chu AC, Maxx S et al (2014) Vehicle electrification: status and issues. Proc IEEE 99(6):1116–1138

[4] Doolan R, Muntean GM (2014) Reducing carbon emissions by introducing electric vehicle enhanced dedicated bus lanes. In: Proceedings of the 2014 IEEE intelligent vehicles symposium, Dearborn, 8–11 Jun 2014, pp 1011–1016

[5] Jiang LP, Chi YN, Qin HY et al (2011) Wind energy in China. IEEE Power Energy Mag 9(6):36–46

[6] Mu YF, Wu JZ, Ekanayake J et al (2013) Primary frequency response from electric vehicles in the Great Britain power system. IEEE Trans Smart Grid 4(2):1142–1150

[7] Oudalov A, Chartouni D, Ohler C (2007) Optimizing a battery energy storage system for primary frequency control. IEEE Trans Power Syst 22(3):1259–1266

[8] US Environmental Protection Agency (2012)

[9] Yamashita D, Niimura T, Yoshimi K et al (2012) Optimal strategy to support the development of charging infrastructure for electric vehicles towards low carbon emissions. In: Proceedings of the 2014 IEEE Power and Energy Society General Meeting, San Diego, 22–26 July 2012, 9 pp

[10] Vittal E, O'Malley M, Keane K (2010) A steady-state voltage stability analysis of power systems with high penetrations of wind. IEEE Trans Power Syst 25(1):433–442

[11] Ben-Kilani K, Elleuch M (2013) Structural analysis of voltage stability in power systems integrating wind power. IEEE Trans Power Syst 28(4):3785–3794

[12] Hossain MJ, Pota HR, Mahmud MA et al (2012) Investigation of the impacts of large-scale wind power penetration on the angle and voltage stability of power systems. IEEE Syst J 6(1):76–84

[13] Mi ZQ, Tian HF, Yu Y et al (2009) Study on voltage stability of power grid with large scale wind farm interconnected. In: Proceedings of the 2009 IEEE international conference on sustainable power generation and supply (SUPERGEN'09), Nanjing, 6–7 April 2009, 6 pp

[14] Kumar VSS, Reddy KK, Thukaram D (2014) Coordination of reactive power in grid-connected wind farms for voltage stability enhancement. IEEE Trans Power Syst 29(5):2381–2390

[15] Le HT, Santoso S, Nguyen TQ (2012) Augmenting wind power penetration and grid voltage stability limits using ESS: application design, sizing, and a case study. IEEE Trans Power Syst 27(1):161–171

[16] Arulampalam A, Barnes M, Jenkins N et al (2006) Power quality and stability improvement of a wind farm using STATCOM supported with hybrid battery energy storage. IEE P-Gener Transm Distrib 153(6):701–710

[17] El Moursi MS, Bak-Jensen B, Abdel-Rahman MH (2011) Coordinated voltage control scheme for SEIG-based wind park utilizing substation STATCOM and ULTC transformer. IEEE Trans Sustain Energy 2(3):246–255

[18] Fu R, Sun WP, Yang J et al (2013) Study on application of STATCOM in voltage stability of wind farm incorporated system. In: Proceedings of the 32nd Chinese control conference (CCC'13), Xi'an, 26–28 July 2013, pp 8868–8873

[19] Ma YC, Houghton T, Cruden A et al (2012) Modeling the benefits of vehicle-to-grid technology to a power system. IEEE Trans Power Syst 27(2):1012–1020

[20] Luo X, Chan KW (2014) Real-time scheduling of electric vehicles charging in low-voltage residential distribution systems to minimise power losses and improve voltage profile. IET Gener Transm Distrib 8(3):516–529

[21] Foster JM, Trevino G, Kuss M et al (2013) Plug-in electric vehicle and voltage support for distributed solar: theory and application. IEEE Syst J 7(4):881–888

[22] Mu YF, Jia HJ (2010) An approach to determining the local boundaries of voltage stability region with wind farms in power injection space. Sci China Tech Sci 53(12):3232–3240

[23] Chiang HD, Jin LC, Varghese M et al (2009) Linear and nonlinear methods for contingency analysis in on-line voltage security assessments. In: Proceedings of the 2009 IEEE Power and Energy Society General Meeting, Calgary, 26–30 July 2009, 6 pp

[24] Chiang HD, Flueck AJ, Shah KS et al (1995) CPFLOW: a practical tool for tracing power system steady-state stationary behavior due to load and generation variations. IEEE Trans Power Syst 10(2):623–634

[25] Affonso CM, da Silva LCP, Lima FGM et al (2004) MW and MVar management on supply and demand side for meeting voltage stability margin criteria. IEEE Trans Power Syst 19(3):1538–1545

[26] Tremblay O, Dessaint LA, Dekkiche AI et al (2007) A generic battery model for the dynamic simulation of hybrid electric vehicles. In: Proceedings of the 2007 IEEE vehicle power and propulsion conference (VPPC'07), Arlington, 9–12 Sept 2007, pp 284–289

[27] Qian KJ, Zhou CK, Allan M et al (2011) Modeling of load demand due to EV battery charging in distribution systems. IEEE Trans Power Syst 26(2):802–810

[28] Ball R, Keers N, Alexander M et al (2010) Modeling electric storage devices for EV. Project MERGE, WP 2 Task 2.1 Deliverable D2.1

[29] Mu YF, Wu JZ, Jenkins N et al (2014) A spatial-temporal model for grid impact analysis of plug-in electric vehicles. Appl Energy 144:456–465

[30] United Kingdom generic distribution system (UKGDS) (2006) Typical load patterns, DG output data. Distributed Generation and Sustainable Electrical Energy Centre, University of Strathclyde, Glasgow, 2007

[31] National Grid (2010) Metered half-hourly electricity demands. National Grid, London

[32] Transport Statistics Bulletin-Vehicle Licensing Statistics (2010) National statistics. Department for Transport, London

[33] Yu DY, Huang HL, Lei M et al (2012) CO_2 reduction benefit by coordinated dispatch of electric vehicle charging and wind power. Autom Electr Power Syst 36(10):14–18 (in Chinese)

Mingshen WANG is a doctor student majoring in Power System and its Automation in Tianjin University. His main research interests include power system stability analysis, renewable energy integration, and smart grids.

Yunfei MU is a lecturer of Tianjin University. His special fields of interests include power system stability analysis and control, renewable energy integration and electric vehicle.

Hongjie JIA is a professor of Tianjin University. His special fields of interests include power system stability analysis and control, distribution network planning, renewable energy integration and smart grids.

Jianzhong WU is currently a Reader in Institute of Energy, Cardiff School of Engineering, Cardiff University, UK. His research activities are focused on energy infrastructure and smart grids.

Xiuping YAO is a senior engineer of Xinjiang Electric Power Dispatching and Communication Center. Her special fields of interests include power system stability analysis.

Xiaodan YU is now a lecture of Tianjin University. Her special fields of interests include power system stability analysis, nonlinear dynamic circuit.

Janaka EKANAYAKE is a professor of Peradeniya University, and a senior lecturer in Institute of Energy, Cardiff School of Engineering, Cardiff University, UK. His main research interests include power electronic applications for power systems, renewable energy generation and its integration. He is also a fellow of IET.

14

A preliminary study of impact of reduced system inertia in a low-carbon power system

Di WU (✉), Milad JAVADI, John Ning JIANG

Abstract One of the important features of low-carbon electric power system is the massive deployment of renewable energy resources in the advent of a new carbon-strained economy. Wind generation is a major technology of generating electric power with zero carbon dioxide emission. In a power system with the high penetration of wind generation, the displacement of conventional synchronous generators with variable speed wind turbines reduces system inertia. This leads to larger system frequency deviation following a loss of large generation. In this paper, the impact of the reduction of system inertia on system frequency is analyzed as the result of the integration of a significant amount of wind generation into power systems. Furthermore, we present a preliminary study of the impact of the distribution of the inertia contributions from those online conventional synchronous generators on the rate of change of frequency (ROCOF) based on the total energy injected into the system due to the fault. The total fault energy is represented using Hamiltonian formulism. With the IEEE 39-bus system, it is shown that for a fault with the given injected total energy, clearing time, and location, the distribution of inertia contributions can significantly affect the magnitude of ROCOF. Moreover, for such a fault with different locations, the average of the magnitudes of ROCOF caused by the fault at different locations is larger when the distribution of the inertia contributions is more dispersed.

D. WU, M. JAVADI, J. N. JIANG, Electricity and Energy Reliability and Risk Laboratory, School of Electrical and Computer Engineering, University of Oklahoma, Norman, OK, USA
(✉) e-mail: dee.d.wu@ou.edu

Keywords Carbon dioxide emission, Wind generation, System inertia, Rate of change of frequency

1 Introduction

The world-wide concern about carbon dioxide (CO_2) emission has led to the increasing interest in the generation technologies of renewable energy sources. Decarbonization of power generation is crucial for reducing CO_2 emission [1–3]. It is reported that 41 % of the energy-related CO_2 emission are currently from the fossil-fuel power plants, and the percentage is expected to be 44 % in 2030 [4]. To reduce CO_2 emission, the power industry is gradually changing the generation technologies from fossil fuel to renewable energy sources. Among these generation technologies of renewable energy sources, wind generation grows rapidly. It is predicted that, by 2020, the total worldwide power generated by wind generation would be more than 1,261 GW, which can supply about 12 % of total worldwide electricity demand [5].

Such a projected high penetration of wind generation in power systems can affect the frequency stability. Nowadays, most of wind turbines (e.g. Doubly Fed Induction Generators) employ variable speed constant frequency generation technology to produce electric power from variable wind [6]. Normally, variable speed wind turbines do not contribute to system inertia since the rotor of the variable speed wind turbines is running at a variable asynchronous speed and thus not sensitive to the change of system frequency [7–9]. The displacement of a large number of conventional synchronous generators with variable speed wind turbines reduces system inertia. When the system is subjected to an unexpected disturbance that causes a significant system-wide power imbalance, the reduced system inertia can cause a larger and faster

ROCOF and frequency nadir, which are two critical variables to set the relays for maintaining frequency stability [10–12].

In the study presented in this paper, the impact of the reduction of system inertia on system frequency is analyzed as the consequence of the integration of wind generation into power systems. Then, we investigate the impact of the distribution of the inertia contributions on system frequency in order to answer questions like, 1) when a number of conventional synchronous generators have to be turned off to accommodate more wind power, how important the distribution of the inertia contributions from the remaining online synchronous generators is? 2) for the same amount of system inertia, do different distributions of inertia contributions have different impacts on ROCOF? The answers to these types of questions are useful to better coordinate a conventional synchronous generator fleet for the variable wind power generation. Also, the answers can contribute to developing methods for planning the power generation expansion towards low-carbon economy [13, 14]. Specifically, the investigation is based on the total fault injected energy, which is the additional energy injected into the power system during the fault period. The total fault injected energy is modeled based on the so-called Hamiltonian formulism. The results of the study are demonstrated and explained with the IEEE 39-bus system.

The rest of the paper is organized as follows: in Sect. 2, we analyze the impact of the reduction of system inertia on system frequency; in Sect. 3, a model of total fault injected energy is developed for our study; in Sect. 4, the impact of the distribution of inertia contributions on system frequency is analyzed; Sect. 5 includes the numerical simulations and results of analyses with the IEEE 39-bus system; and Sect. 6 concludes.

2 Impact of reduction of system inertia on frequency

In this section, the impact of the reduction of system inertia is analyzed as the result of the integration of a significant amount of wind generation into power systems.

2.1 Model of power system dynamics

For a power system with N_g interconnected generators and N_b buses, its dynamics can be described by a set of differential and algebraic equations (DAE). That is,

$$\frac{d\phi_i}{dt} = \omega_i - \omega_R, \quad i = 1, 2, \ldots, N_g \tag{1}$$

$$M_i \frac{d(\omega_i - \omega_R)}{dt} = P_{mi} - P_i, \quad i = 1, 2, \ldots, N_g \tag{2}$$

$$P_i + P_{Li} = 0, \quad i = N_g + 1, N_g + 2, \ldots, N_g + N_b \tag{3}$$

$$Q_i + Q_{Li} = 0, \quad i = N_g + 1, N_g + 2, \ldots, N_g + N_b \tag{4}$$

$$P_i = \sum_{j=1}^{N_g+N_b} B_{ij} V_i V_j \sin(\phi_i - \phi_j), \quad i = 1, 2, \ldots, N_g + N_b$$

$$Q_i = -\sum_{j=1}^{N_g+N_b} B_{ij} V_i V_j \cos(\phi_i - \phi_j), \quad i = 1, 2, \ldots, N_g + N_b$$

where ϕ_i is the ith element of $(N_g + N_b) \times 1$ vector $\boldsymbol{\phi}$ in which the first N_g elements are the rotor angles of the generators in radians and the remaining elements are the bus voltage angles in radians; V_i is the ith element of $(N_g + N_b) \times 1$ vector $\boldsymbol{V}$ in which the first N_g elements are the magnitudes of the generator internal voltages in per unit and the remaining elements are the magnitudes of the bus voltages in per unit; ω_i is the angular velocity of the ith generator in radians per second; ω_R is the rated angular velocity in radians per second; $M_i = 2H_i/\omega_R$ in which H_i is the inertia constant of the ith generator in seconds; P_{mi} is the mechanical power input of the ith generator in per unit; P_i is the real power input at the ith bus in per unit; Q_i is the reactive power input at the ith bus in per unit; P_{Li} and Q_{Li} are the voltage dependent active and reactive power outputs at the ith load bus in per unit, respectively; B_{ij} is the imaginary part of the negative of the branch admittance between buses i and j in per unit; and B_{ii} is the sum of imaginary parts of all branch admittances connected to bus i in per unit.

2.2 Impact of reduction of system inertia on frequency

The impact of the reduction of system inertia on system frequency is analyzed using ROCOF. The ROCOF is the initial slope of the system frequency with respect to time. The ROCOF is an important variable used for relay setting of protection system. Also, the ROCOF is an indicator to measure the seriousness that a disturbance affects system frequency. The ROCOF can be derived from the center of inertia of the system, which has angular velocity below [15],

$$\bar{\omega} = \frac{\sum_{i=1}^{N_g} M_i \omega_i}{\sum_{i=1}^{N_g} M_i} \tag{5}$$

With (5), the ROCOF can be represented as

$$\frac{d\bar{f}}{dt} = \frac{-f_R \Delta P_F(0^+)}{\sum_{i=1}^{N_g} 2H_i} \tag{6}$$

where $d\bar{f}/dt$ is the ROCOF in hertz per second; f_R is rated system frequency and is selected as 60 Hz; and $\Delta P_F(0^+)$ is the change in real power output at fault bus F from $t = 0$ to $t = 0^+$ in per unit. The derivation of (6) is presented in Appendix 1.

Equation (6) indicates that system inertia (i.e., the sum of the inertia constants of all generators) can affect the ROCOF. For a fault with a given size, the magnitude of the ROCOF is inversely proportional to system inertia. That is, the lower system inertia is, the larger the magnitude of the ROCOF is. The larger the magnitude of the ROCOF means the more severe frequency drop; thus, the lower system inertia can cause the more severe frequency drop, which may result in frequency instability.

The analysis above shows that frequency stability is a concern in a power system with a large penetration of variable speed wind turbines. Normally, the variable speed wind turbines do not contribute to system inertia. When a large penetration of variable speed wind turbines into a power system reduces the power supplied by conventional synchronous generators, system inertia correspondingly reduces. According to (6), it is known that the reduced system inertia can cause frequency instability when the system is subjected to a significant power imbalance. On the other hand, (6) does not show how the distribution of the inertia contributions from the remaining online synchronous generators affects the ROCOF. The distribution of the inertia contributions may further increase the magnitude of the ROCOF and thus the seriousness of frequency instability. To understand the impact of the distribution of the inertia contributions on system frequency, the ROCOF defined in (6) will be represented based on the total energy injected into the system during the fault period in Sect. 4. A model of the total fault injected energy will be developed in next section.

3 Modeling of total fault injected energy

In this section, a model of total fault injected energy is developed based on Hamiltonian formulism and a classic model of conservative power system.

3.1 Representations of energy components

Let t_c be the fault clearing time and t_s be the time at which the system following a fault is stabilized. The total kinetic and potential energies of the dynamic power system described in (1)–(4) during the period of interest $t \in [t_c, t_s]$ can be summarized as follows:

Total kinetic energy (W_{KE}) can be expressed as,

$$W_{KE} = \frac{1}{2}\sum_{i=1}^{N_g} M_i[\omega_i(t) - \omega_R]^2 \tag{7}$$

Note that the kinetic energy above is represented with the angular velocity relative to the rated angular velocity. More detailed discussion of the total kinetic energy can be found in many references such as [16].

For total potential energy (W_{PE}), which is the electromagnetic energy stored in various components of the power grid, including generators ($W_{PE,1}$), loads ($W_{PE,2}$), and branches ($W_{PE,3}$) in the transmission network, which can be written as,

$$W_{PE} = W_{PE,1} + W_{PE,2} + W_{PE,3} \tag{8}$$

The detailed derivation of (8) can be found in Appendix 2.

According to electric circuit theory [17], these types of potential energy can be further expressed as follows:

1) Potential energy of generators ($W_{PE,1}$) is related to the change in the rotor angle positions of generators, which can be represented as,

$$W_{PE,1} = -\sum_{i=1}^{N_g} P_{mi}(\phi_i(t) - \phi_i(t_s))$$

2) Potential energy of loads ($W_{PE,2}$) includes the potential energy related to both active power ($W_{PE,21}$) and reactive power ($W_{PE,22}$), which can be further written as,

$$\begin{aligned} W_{PE,2} &= W_{PE,21} + W_{PE,22} \\ &= \sum_{i=N_g+1}^{N_g+N_b} \int_{t_s}^{t} P_{Li}\frac{\mathrm{d}\phi_i(\tau)}{\mathrm{d}\tau}\mathrm{d}\tau + \sum_{i=N_g+1}^{N_g+N_b} \int_{V_i(t_s)}^{V_i(t)} \frac{Q_{Li}}{x_i}\mathrm{d}x_i \end{aligned}$$

3) Potential energy stored in branches ($W_{PE,3}$) can be represented as,

$$\begin{aligned} W_{PE,3} = &-\frac{1}{2}\sum_{i=1}^{N_g+N_b}\sum_{j=1}^{N_g+N_b} B_{ij}V_i(t)V_j(t)\cos(\phi_i(t) - \phi_j(t)) \\ &+\frac{1}{2}\sum_{i=1}^{N_g+N_b}\sum_{j=1}^{N_g+N_b} B_{ij}V_i(t_s)V_j(t_s)\cos(\phi_i(t_s) - \phi_j(t_s)) \end{aligned}$$

When the offset between the potential energy of loads and branches is considered, the total potential energy can further be described by the potential energy of generators and of branches connected to internal generator buses. Thus, we can have,

$$\begin{aligned} W_{PE} = &-\sum_{i=1}^{N_g} P_{mi}(\phi_i(t) - \phi_i(t_s)) \\ &+\sum_{i=1}^{N_g}\int_{t_s}^{t}\left(\sum_{j=1}^{N_g+N_b} B_{ij}V_i(\tau)V_j(\tau)\sin(\phi_i(\tau) - \phi_j(\tau))\right) \\ &\times\frac{\mathrm{d}\phi_i(\tau)}{\mathrm{d}\tau}\mathrm{d}\tau = \sum_{i=1}^{N_g}\int_{t_s}^{t}(P_i - P_{mi})\frac{\mathrm{d}\phi_i(\tau)}{\mathrm{d}\tau}\mathrm{d}\tau \end{aligned} \tag{9}$$

The detailed derivation of (9) can be found in Appendix 3.

In order to better represent the amount of energy injected into the system during the fault period, which is manifested in the form of kinetic energy, we further separate the integral of (9) into two periods, $[t_s, t_c]$ and $[t_c, t]$. That is, (9) can be rewritten as,

$$W_{PE} = \sum_{i=1}^{N_g} \int_{t_s}^{t_c} (P_i - P_{mi}) \frac{\mathrm{d}\phi_i(\tau)}{\mathrm{d}\tau} \mathrm{d}\tau + \sum_{i=1}^{N_g} \int_{t_c}^{t} (P_i - P_{mi}) \frac{\mathrm{d}\phi_i(\tau)}{\mathrm{d}\tau} \mathrm{d}\tau \tag{10}$$

In (10), it can be seen that the first integral over $[t_s, t_c]$ is a time invariant function with respect to time t while the second integral is a time varying function with respect to time t. The former is considered as an energy constant and selected as a reference. Then, the total potential energy (W_{PE}) of interest can be further simplified as,

$$W_{PE} = \sum_{i=1}^{N_g} \int_{t_c}^{t} (P_i - P_{mi}) \frac{\mathrm{d}\phi_i(\tau)}{\mathrm{d}\tau} \mathrm{d}\tau \tag{11}$$

3.2 Total energy of dynamic power system (Hamiltonian)

The total energy of the power system described in (1)–(4) or Hamiltonian, T, can be constructed based on the total kinetic energy and potential energy described in (7) and (11), respectively. That is,

$$\begin{aligned} T &= W_{KE} + W_{PE} \\ &= \frac{1}{2}\sum_{i=1}^{N_g} M_i[\omega_i(t) - \omega_R]^2 + \sum_{i=1}^{N_g} \int_{t_c}^{t} (P_i - P_{mi}) \frac{\mathrm{d}\phi_i(\tau)}{\mathrm{d}\tau} \mathrm{d}\tau \end{aligned} \tag{12}$$

The Hamiltonian described in (12) has some features similar to those of the general Hamiltonian in classical mechanics, which is summarized in [18]. The general Hamiltonian is a function with respect to n generalized position coordinates and n generalized momentum coordinates, and the motion characteristics of a dynamic system without dissipation can be described by a set of Hamilton's equations,

$$\frac{\mathrm{d}q_i}{\mathrm{d}t} = \frac{\partial T}{\partial p_i}(q_1, q_2 \ldots, q_n, p_1, p_2 \ldots . p_n), \quad i = 1, 2, \ldots, n \tag{13}$$

$$\frac{\mathrm{d}p_i}{\mathrm{d}t} = -\frac{\partial T}{\partial q_i}(q_1, q_2 \ldots, q_n, p_1, p_2 \ldots, p_n), \quad i = 1, 2, \ldots, n \tag{14}$$

where q_i is the ith generalized position coordinate and p_i is the ith generalized momentum coordinate. For a dynamic system consisting of n rigid bodies with mass m_i, $p_i = m_i\dot{q}_i$, $i = 1,2,\ldots,n$.

Let $q_i = \delta_i$ and $p_i = M_i(\omega_i - \omega_R)$. The Hamiltonian in (12) can be viewed as a function with respect to position coordinates and momentum coordinates; with (13)–(14), the Hamilton's equations of the power system described by DAEs in (1)–(4) can be expressed as,

$$\frac{\mathrm{d}\phi_i}{\mathrm{d}t} = \frac{1}{M_i}\frac{\partial(W_{KE} + W_{PE})}{\partial(\omega_i - \omega_R)} = \omega_i - \omega_R, \quad i = 1, 2, \ldots, N_g \tag{15}$$

$$M_i \frac{\mathrm{d}(\omega_i - \omega_R)}{\mathrm{d}t} = -\frac{\partial(W_{KE} + W_{PE})}{\partial\phi_i} = P_{mi} - P_i, \quad i = 1, 2, \ldots, N_g \tag{16}$$

Thus, the derivative of Hamiltonian in (12) with respect to time t can be written as,

$$\begin{aligned} \frac{\mathrm{d}T}{\mathrm{d}t} &= \sum_{i=1}^{N_g}\left(\frac{\partial T}{\partial q_i}\frac{\mathrm{d}q_i}{\mathrm{d}t} + \frac{\partial T}{\partial p_i}\frac{\mathrm{d}p_i}{\mathrm{d}t}\right) \\ &= \sum_{i=1}^{N_g}\left(\frac{\partial(W_{KE}+W_{PE})}{\partial\phi_i}\frac{\mathrm{d}\phi_i}{\mathrm{d}t} + \frac{\partial(W_{KE}+W_{PE})}{\partial(\omega_i-\omega_R)}\frac{\mathrm{d}(\omega_i-\omega_R)}{\mathrm{d}t}\right) \\ &= 0 \end{aligned} \tag{17}$$

Equation (17) confirms that the total system energy during the period $t \in [t_c, t_s]$ is conserved in the power system described by (1)–(4) since the Hamiltonian defined in (12) is a constant during the period. Such confirmation indicates that (12) is a complete expression of the total system energy.

The Hamiltonian during the period $t \in [t_c, t_s]$ is equal to the amount of fault-related energy that is injected into the system at $t = t_c$. Thus, the Hamiltonian during the period $t \in [t_c, t_s]$ can be represented with the total kinetic and potential energies at $t = t_c$ which are described in (7) and (11), respectively,

$$T(t) = T(t_c) = W_{KE}(t_c) + W_{PE}(t_c) = \frac{1}{2}\sum_{i=1}^{N_g} M_i[\omega_i(t_c) - \omega_R]^2 + \sum_{i=1}^{N_g}\int_{t_c}^{t_c}(P_i - P_{mi})\frac{\mathrm{d}\phi_i(\tau)}{\mathrm{d}\tau}\mathrm{d}\tau = W_{KE}(t_c) \tag{18}$$

Equation (18) is a general representation of the total energy that is injected into the power system during the fault period.

3.3 Numerical estimation of Hamiltonian

The Hamiltonian at $t = t_c$ can be calculated with (18) if the angular velocity of each generator at $t = t_c$ is obtained. Let us consider a small fault with short duration during which the acceleration power for a generator can be considered a constant. Then, (19) can be obtained from the previous dynamical Eqs. (1)–(4),

$$\omega_i(t_c) - \omega_R \approx -\frac{E_i V_f B_{iF} \cos(\delta_i(0) - \theta_F(0)) \Delta P_F(0^+)}{\sum_{i=1}^{N_g} E_i V_F B_{iF} \cos(\theta_F(0) - \delta_i(0))} \frac{1}{M_i} t_c \tag{19}$$

where E_i is the internal voltage magnitude of the ith generator in per unit; V_F is the pre-fault voltage magnitude at fault bus F in per unit; B_{iF} is the imaginary part of the negative of the transfer admittance between generator internal bus i and fault bus F in per unit; $\delta_i(0)$ is the pre-fault rotor angle of the ith generator at $t = 0$ in radians; and $\theta_F(0)$ is the pre-fault voltage angle of fault bus F in radians. The derivation of (19) is presented in Appendix 4.

Plugging (19) into (18), we have

$$\bar{T}_F(t_c) = \frac{t_c^2}{2} \cdot [\Delta P_F(0^+)]^2 \cdot \sum_{i=1}^{N_g} \frac{1}{M_i} \left[\frac{E_i V_F B_{if} \cos(\delta_i(0) - \theta_F(0))}{\sum_{i=1}^{N_g} E_i V_f B_{iF} \cos(\theta_F(0) - \delta_i(0))} \right]^2$$
$$F = 1, 2, \ldots, N_b \tag{20}$$

where $\bar{T}_F(t_c)$ is the estimated Hamiltonian of the system at $t = t_c$ in per unit when a fault occurs at bus F.

Equation (20) indicates that the Hamiltonian is related to three fault factors—fault size, clearing time, and location. The relationship between the Hamiltonian and these three fault factors can be clearly shown by simplifying (20) with assuming $E_i \approx V_F \approx 1$ p.u. and $\cos(\delta_i(0) - \theta_f(0)) \approx 1$,

$$\bar{T}_F(t_c) = \frac{1}{2} \cdot t_c^2 \cdot [\Delta P_F(0^+)]^2 \cdot \sum_{i=1}^{N_g} \frac{B_{iF}^2}{M_i \left[\sum_{i=1}^{N_g} B_{iF} \right]^2},$$
$$F = 1, 2, \ldots, N_b \tag{21}$$

Replacing M_i with $H_i/\pi f_R$, (21) becomes

$$\bar{T}_F(t_c) = \frac{\pi f_R}{2} \cdot t_c^2 \cdot [\Delta P_F(0^+)]^2 \cdot \sum_{i=1}^{N_g} \frac{B_{iF}^2}{H_i \left[\sum_{i=1}^{N_g} B_{iF} \right]^2},$$
$$F = 1, 2, \ldots, N_b \tag{22}$$

4 Impact of distribution of inertia contributions on frequency

As the result of the integration of a significant amount of wind generation into power systems, the impact of the distribution of inertia contributions from online conventional synchronous generators on system frequency can be analyzed based on the total fault injected energy defined in (22) and the ROCOF defined in (6).

By rewriting (6) based on (22), the impact of the distribution of inertia contributions on the ROCOF can be shown by the following equation:

$$\frac{d\bar{f}}{dt} = -\frac{1}{\sum_{i=1}^{N_g} H_i} \left[\frac{\bar{T}_F(t_c) f_R}{2\pi t_c^2 D_F} \right]^{\frac{1}{2}} \tag{23}$$

where

$$D_F = \sum_{i=1}^{N_g} \frac{B_{iF}^2}{H_i \left[\sum_{i=1}^{N_g} B_{iF} \right]^2}, \quad F = 1, 2, \ldots, N_b \tag{24}$$

Equations (23) and (24) indicate that the distribution of inertia contributions can affect the ROCOF by factor D_F. Equation (23) shows that, for a fault with a given injected energy and clearing time, when system inertia is constant, the magnitude of the ROCOF is inversely proportional to factor D_F. Equation (24) further shows that, for a given fault location, factor D_F depends on the reciprocal of the inertia of each generator, which is related to the distribution of inertia contributions. Thus, for a constant system inertia and a fault with a given injected energy, clearing time and location, when the characteristic of the distribution of inertia contributions increases factor D_f, the magnitude of the ROCOF reduces; when the characteristic of the distribution of inertia contributions reduces factor D_f, the magnitude of the ROCOF increases.

5 Case studies

In this Section, the impact of the distribution of inertia contributions on system frequency is further discussed with example of IEEE 39-bus system shown in Fig. 1. In the system, there are 34 lines, 12 transformers, 39 buses, 10 generators, and 17 loads [19]. Table 1 presents the inertia constant of each generator. In the system, the following two cases are investigated: in Case A, the impact is analyzed in the system under a fault at a given location; in Case B, the impact is analyzed in the system under the fault at different fault locations. In the investigation, the system frequency response is evaluated based on dynamic simulations.

5.1 Impact of distribution of inertia contributions on frequency

At a given fault location, the impact of the distribution of the inertia contributions on frequency response is

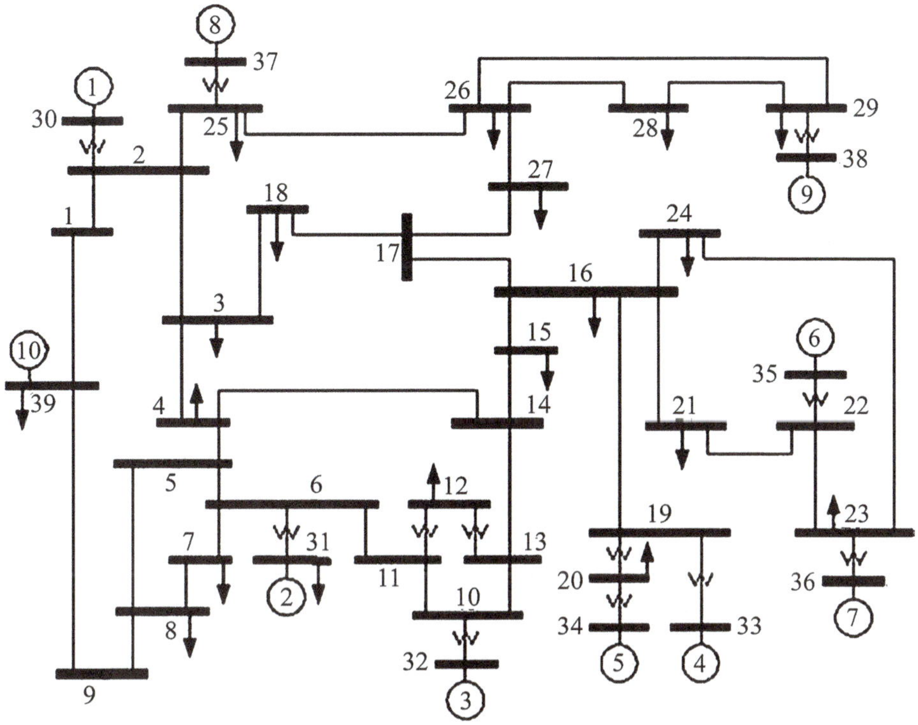

Fig. 1 One-line diagram of the IEEE 39-bus system

Table 1 Generator inertia constants in the IEEE 39-bus system in seconds

H_1	H_2	H_3	H_4	H_5	H_6	H_7	H_8	H_9	H_{10}
42.00	30.50	35.80	28.60	26.00	34.80	26.40	24.30	34.50	500.00

investigated using different scenarios of the inertia constants, which are shown in Table 2. In each scenario, system inertia is the same, but the dispersion of the distribution of the inertia constants is different. The dispersion is quantified by calculating the standard deviation of all inertia constants. For a given scenario, the three-phase fault with a given injected energy is applied in bus 39 at $t = 0$ s and then is cleared at $t_c = 0.15$ s in the system. Figure 2 shows the change of the ROCOF with the dispersion of the distribution of the inertia contributions. Figure 3 shows the change of factor D_F in (23) with the dispersion of the distribution of the inertia contributions.

It is observed from Fig. 2 that the distribution of the inertia contributions significantly affects the magnitude of the ROCOF. In Fig. 2, the magnitude of the ROCOF increases with the dispersion of the distribution of the inertia contributions when the system is subjected to a fault with a given injected energy, location, and clearing time. This indicates that the seriousness of the impact of the fault at bus 39 on the system frequency response increases with the dispersion of the distribution of the inertia contributions.

Figure 3 shows that the observation from Fig. 2 is consistent with the analysis results of (23) and (24). According to (23), it is known that the magnitude of the ROCOF increases with the decrease in factor D_F when the system with a constant system inertia is subjected to a fault with a given injected energy, location, and clearing time. Equation (24) shows that the distribution of inertia contributions can change factor D_F. In Fig. 3, factor D_F reduces when the dispersion of the distribution of the inertia contributions increases. According to (23), the magnitude of the ROCOF increases with the dispersion of the distribution of the inertia contributions. This is consistent with the observation from Fig. 2.

5.2 Impact of distribution of inertia contributions and fault locations on frequency

At different fault locations, the impact of the distribution of inertia contributions on frequency response is investigated using the same scenarios of the inertia constants shown in Table 2. For a given scenario, the three-phase fault that is used in Sect. 5.1 is separately applied to buses 30, 33, 35, and 36. Figures 4, 5 show the changes of the ROCOF and factor D_F with the dispersion of the distribution of the inertia contributions in the system under the fault at different locations, respectively. Figure 6 shows the

Table 2 Different scenarios of generator inertia constants in the IEEE 39-bus system in seconds

	S_1	S_2	S_3	S_4	S_5	S_6	S_7	S_8	S_9	S_{10}	S_{11}
H_1	88.67	84.00	79.33	74.67	70.00	65.33	60.67	56.00	51.33	46.67	42.00
H_2	76.97	72.30	67.63	62.97	58.30	53.63	48.97	44.30	39.63	34.97	30.30
H_3	82.47	77.80	73.13	68.47	63.80	59.13	54.47	49.80	45.13	40.47	35.80
H_4	75.27	70.60	65.93	61.27	56.60	51.93	47.27	42.60	37.93	33.27	28.60
H_5	72.67	68.00	63.33	58.67	54.00	49.33	44.67	40.00	35.33	30.67	26.00
H_6	81.47	76.80	72.13	67.47	62.80	58.13	53.47	48.80	44.13	39.47	34.80
H_7	73.07	68.40	63.73	59.07	54.40	49.73	45.07	40.40	35.73	31.07	26.40
H_8	70.97	66.30	61.63	56.97	52.30	47.63	42.97	38.30	33.63	28.97	24.30
H_9	81.17	76.50	71.83	67.17	62.50	57.83	53.17	48.50	43.83	39.17	34.50
H_{10}	80.00	122.00	164.00	206.00	248.00	290.00	332.00	374.00	416.00	458.00	500.00
Sum	782.70	782.70	782.70	782.70	782.70	782.70	782.70	782.70	782.70	782.70	782.70
Standard deviation	5.19	15.46	29.04	42.89	56.81	70.77	84.73	98.71	112.69	126.68	140.67

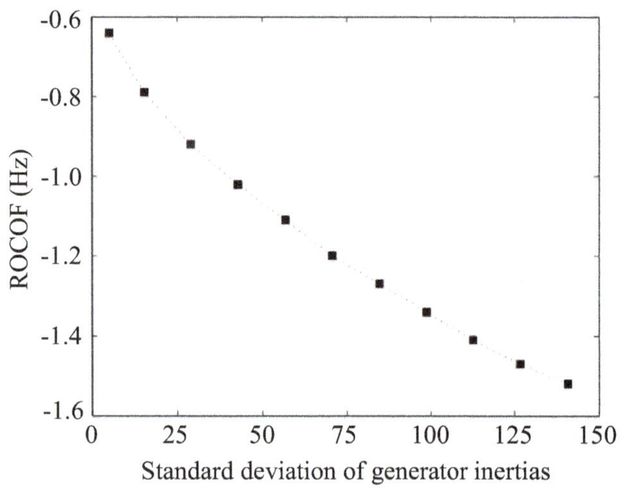

Fig. 2 Change of the ROCOF with the dispersion of the distribution of the inertia contributions in the IEEE 39-bus system under the fault at bus 39

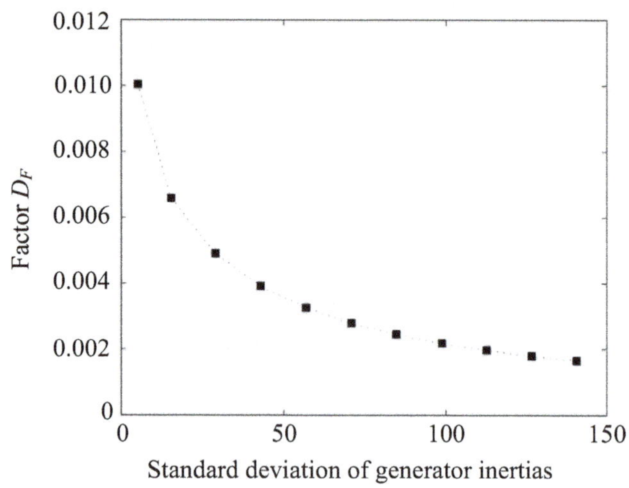

Fig. 3 Change of factor D_F with the dispersion of the distribution of the inertia contributions in the IEEE 39-bus system under the fault at bus 39

comparison of the ROCOFs at different fault locations in different scenarios of the inertia constants.

Figures 4, 5 show that the changes of the ROCOF with the dispersion of the distribution of the inertia contributions in the system under the fault at different locations are consistent with the analysis results of (23) and (24). According to (23) and (24), it is known that the magnitude of the ROCOF reduces with the increase in factor D_F when the system with a constant system inertia is subjected to a fault with a given injected energy, location, and clearing time. In Fig. 5, when the system is subjected to the fault at buses 30, 33, 35, or 36, factor D_F increases with the dispersion of the distribution of the inertia contributions. Thus, when the system is subjected to the fault at these buses, the magnitude of the ROCOF reduces with the increase in the dispersion of the distribution of the inertia contributions. This is consistent with the results shown in Fig. 4.

More importantly, Fig. 6 and Table 2 show that the average of the magnitudes of the ROCOF caused by the fault at different fault locations increases with the dispersion of the distribution of the inertia contributions. It can be seen from Fig. 6 and Table 2 that when the fault occurs at bus 30, 33, 35, or 36, the magnitude of the ROCOF reduces with the increase in the dispersion of the distribution of the inertia contributions; however, when the fault occurs at bus 39, the magnitude of the ROCOF increases with the dispersion of the distribution of the inertia contributions. Also, Fig. 6 shows that, in most of scenarios of inertia constants, the magnitude of the ROCOF caused by the fault at bus 39 is much larger than that caused by the fault at bus 30, 33, 35, or 36. Thus, the average of the magnitudes of the ROCOF caused by the fault at these buses increases with the dispersion of the distribution of the inertia contributions.

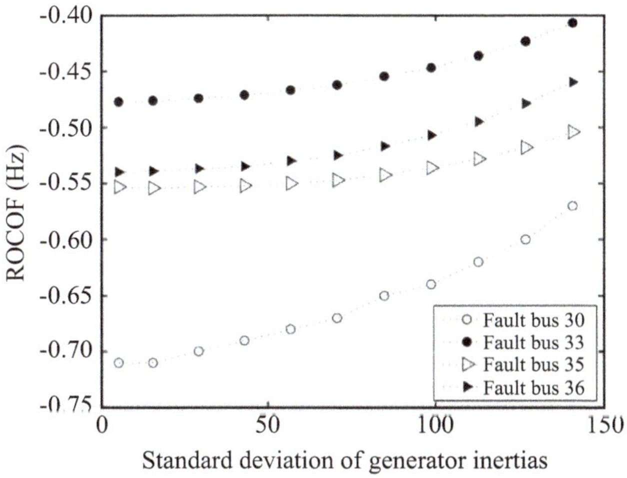

Fig. 4 Changes of the ROCOF with the dispersion of the distribution of the inertia contributions in the IEEE 39-bus system under the fault at different locations

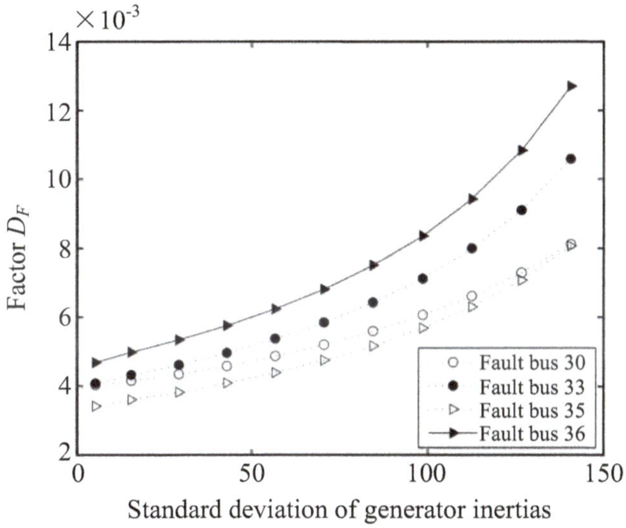

Fig. 5 Changes of factor D_F with the dispersion of the distribution of the inertia contributions in the IEEE 39-bus system under the fault at different fault locations

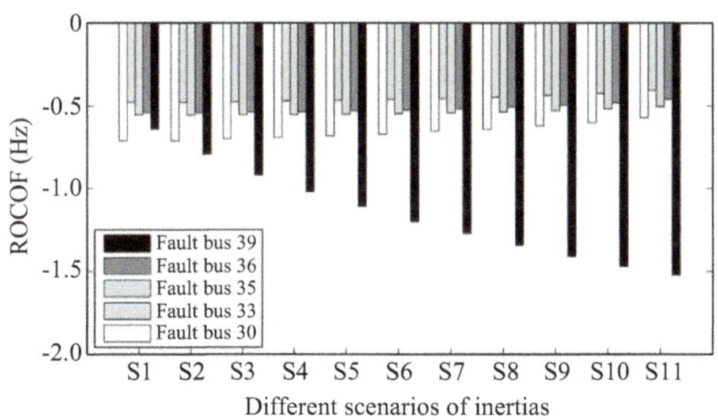

Fig. 6 Comparison of the ROCOFs at different fault locations in different scenarios of the inertia constants

6 Conclusions

This paper discussed the impact of the reduction of system inertia on system frequency as the result of the high penetration of wind generation into power systems. Moreover, a preliminary study of the impact of the distribution of inertia contributions on system frequency was included. In the study, the impact of the distribution of inertia contributions on the ROCOF was investigated based on the total energy injected to the power system during the fault period. The total fault energy was represented using Hamiltonian formulism. With the IEEE 39-bus system, it was shown that, for a fault with a given injected energy, clearing time, and location, the distribution of the inertia contributions can significantly affect the magnitude of the ROCOF; furthermore, for a given (sized) fault at different locations, the average of the magnitudes of the ROCOF caused by the fault is larger when the distribution of the inertia contributions is more dispersed.

The implications of the study presented in this paper to the development of the low-carbon electric power system are: 1) if the concern about frequency stability resulting from the reduction of system inertia and the distribution of inertia contributions are indeed a serious concern in the development of the low-carbon power system, the online synchronous generator fleet needs a better coordination to accommodate the variable wind power; 2) frequency stability should be seriously studied for the development of the low-carbon electric power system with the high penetration of renewable energy sources, especially for those associated with the initial frequency response in the system following a power imbalance.

Appendix 1

Derivation of (6)

Based on (5), the mean acceleration of all generators in the system at $t = 0^+$ can be represented as,

$$\frac{\mathrm{d}\bar{\omega}(0^+)}{\mathrm{d}t} = \frac{\sum_{i=1}^{N_g} M_i \frac{\mathrm{d}\omega_i(0^+)}{\mathrm{d}t}}{\sum_{i=1}^{N_g} M_i} \tag{25}$$

In (25), the acceleration of the ith generator at $t = 0^+$ can be expressed as,

$$\frac{\mathrm{d}\omega_i(0^+)}{\mathrm{d}t} = \frac{-\Delta P_{ei}(0^+)}{M_i} \tag{26}$$

where $\Delta P_{ei}(0^+)$ is the change in the real power output of the ith generator from $t = 0$ to $t = 0^+$ in per unit and can be represented with (27) below by using a typical treatment of a small fault.

$$\Delta P_{ei}(0^+) = -E_i V_F B_{iF} \cos(\delta_i(0) - \theta_F(0))\Delta\theta_F \tag{27}$$

where $\theta_F(0)$ can be further represented as,

$$\Delta\theta_F = \frac{-\Delta P_F(0^+)}{\sum_{i=1}^{N_g} V_i V_f B_{iF} \cos(\theta_F(0) - \delta_i(0))} \tag{28}$$

With (26)–(28), (25) can be rewritten as,

$$\frac{\mathrm{d}\bar{\omega}(0^+)}{\mathrm{d}t} = -\frac{\Delta P_F(0^+)}{\sum_{i=1}^{N_g} M_i} \tag{29}$$

Replacing $\bar{\omega}(0^+)$ and M_i with $2\pi\bar{f}$ and $H_i/\pi f_R$, respectively, (29) becomes

$$\frac{\mathrm{d}\bar{f}}{\mathrm{d}t} = -\frac{f_R \Delta P_F(0^+)}{\sum_{i=1}^{N_g} 2H_i} \tag{30}$$

Appendix 2

Derivation of (8)

In (8), the components of potential energy are defined according to (31) below, which is derived from (1)–(4).

$$\begin{aligned}
\sum_{i=1}^{N_g} \frac{1}{2} M_i[\omega_i(t) - \omega_R]^2 &= \sum_{i=1}^{N_g} P_{mi}[\phi_i(t) - \phi_i(t_s)] \\
&- \int_{t_s}^{t} \sum_{i=N_g+1}^{N_g+N_b} P_{Li}(V_i)\frac{\mathrm{d}\phi_i(\tau)}{\mathrm{d}\tau}\mathrm{d}\tau - \int_{t_s}^{t} \sum_{i=N_g+1}^{N_g+N_b} \frac{Q_{Li}}{V_i}\frac{\mathrm{d}V_i}{\mathrm{d}\tau}\mathrm{d}\tau \\
&+ \frac{1}{2}\sum_{i=1}^{N_g+N_b}\sum_{j=1}^{N_g+N_b} [B_{ij}V_i(t)V_j(t)\cos(\phi_i(t) - \phi_j(t))] \\
&- \frac{1}{2}\sum_{i=1}^{N_g+N_b}\sum_{j=1}^{N_g+N_b} [B_{ij}V_i(t_s)V_j(t_s)\cos(\phi_i(t_s) - \phi_j(t_s))] \\
&= -W_{PE,1} - (W_{PE,21} + W_{PE,22}) - W_{PE,3} \\
&= -(W_{PE,1} + W_{PE,2} + W_{PE,3})
\end{aligned} \tag{31}$$

Equation (31) is derived as follows:

1) Both sides of (2) are multiplied by $(\mathrm{d}\phi_i/\mathrm{d}t)$ and then the summation of N_g generator equations yields

$$\sum_{i=1}^{N_g} M_i(\omega_i - \omega_R)\frac{\mathrm{d}(\omega_i - \omega_R)}{\mathrm{d}t} = \sum_{i=1}^{N_g} (P_{mi} - P_i)\frac{\mathrm{d}\phi_i}{\mathrm{d}t} \tag{32}$$

When $\omega_i(t_s) = \omega_R$, the integral of both sides of (32) from t to t_s can be represented as,

$$\begin{aligned}
&\sum_{i=1}^{N_g} \frac{1}{2} M_i[\omega_i(t) - \omega_R]^2 = -W_{PE,1} \\
&- \int_{t_s}^{t} \sum_{i=1}^{N_g}\sum_{j=1}^{N_g+N_b} B_{ij}V_iV_j\sin(\phi_i - \phi_j)\frac{\mathrm{d}\phi_i}{\mathrm{d}\tau}\mathrm{d}\tau = 0
\end{aligned} \tag{33}$$

2) Both sides of (3) are multiplied by $(\mathrm{d}\phi_i/\mathrm{d}t)$, and then the summation of N_b equations yields the following equation,

$$\sum_{i=N_g+1}^{N_g+N_b} P_i \frac{\mathrm{d}\phi_i}{\mathrm{d}t} + \sum_{i=N_g+1}^{N_g+N_b} P_{Li}\frac{\mathrm{d}\phi_i}{\mathrm{d}t} = 0 \tag{34}$$

The integral of both sides of (34) from t to t_s can be represented as follows,

$$\begin{aligned}
&- \int_{t_s}^{t} \sum_{i=N_g+1}^{N_g+N_b}\sum_{j=1}^{N_g+N_b} B_{ij}V_iV_j\sin(\phi_i - \phi_j)\frac{\mathrm{d}\phi_i}{\mathrm{d}\tau}\mathrm{d}\tau \\
&- \int_{t_s}^{t} \sum_{i=N_g+1}^{N_g+N_b} P_{Li}\frac{\mathrm{d}\phi_i}{\mathrm{d}\tau}\mathrm{d}\tau \\
&= - \int_{t_s}^{t} \sum_{i=N_g+1}^{N_g+N_b}\sum_{j=1}^{N_g+N_b} B_{ij}V_iV_j\sin(\phi_i - \phi_j)\frac{\mathrm{d}\phi_i}{\mathrm{d}\tau}\mathrm{d}\tau \\
&- W_{PE,21} = 0
\end{aligned} \tag{35}$$

3) Both sides of (4) are divided by V_i and then multiplied by $(\mathrm{d}V_i/\mathrm{d}t)$; the summation of N_b equations yields the equation below,

$$\sum_{i=N_g+1}^{N_g+N_b} \frac{Q_{Li}}{V_i}\frac{\mathrm{d}V_i}{\mathrm{d}t} + \sum_{i=N_g+1}^{N_g+N_b} \frac{Q_i}{V_i}\frac{\mathrm{d}V_i}{\mathrm{d}t} = 0 \tag{36}$$

Since $\mathrm{d}E_i/\mathrm{d}t = 0$ at the internal generator bus, the left-hand-side second term of (36) can be represented with the summation of $N_g + N_b$ equations; then (36) can be changed as,

$$\sum_{i=N_g+1}^{N_g+N_b} \frac{Q_{Li}}{V_i}\frac{\mathrm{d}V_i}{\mathrm{d}t} + \sum_{i=1}^{N_g+N_b} \frac{Q_i}{V_i}\frac{\mathrm{d}V_i}{\mathrm{d}t} = 0 \tag{37}$$

The integral of both sides of (37) from t to t_s can be represented with the equation below

$$\begin{aligned}
&- \int_{t_s}^{t} \sum_{i=N_g+1}^{N_g+N_b} \frac{Q_{Li}}{V_i(\tau)}\frac{\mathrm{d}V_i(\tau)}{\mathrm{d}\tau}\mathrm{d}\tau + \int_{t_s}^{t} \sum_{i=1}^{N_g+N_b}\sum_{j=1}^{N_g+N_b} \\
&\times B_{ij}V_j(\tau)\cos(\phi_i(\tau) - \phi_j(\tau))\frac{\mathrm{d}V_i(\tau)}{\mathrm{d}\tau}\mathrm{d}\tau = -W_{PE,22} \\
&+ \int_{t_s}^{t} \sum_{i=1}^{N_g+N_b}\sum_{j=1}^{N_g+N_b} B_{ij}V_j(\tau)\cos(\phi_i(\tau) - \phi_j(\tau))\frac{\mathrm{d}V_i(\tau)}{\mathrm{d}\tau}\mathrm{d}\tau = 0
\end{aligned} \tag{38}$$

4) Eq. (31) can be obtained by summating (33), (35), and (38). Note that in addition to the total kinetic energy and negative potential energy components $W_{PE,1}$, $W_{PE,21}$, $W_{PE,22}$, the remanding terms in these equations can be integrated and then represented with negative potential energy component $W_{PE,3}$. The details can be found in Appendix 3.

Appendix 3

Derivation of (9)

The potential energy of branches ($W_{PE,3}$) can be reorganized as follows:

$$
\begin{aligned}
W_{PE,3} = & -\sum_{i=1}^{N_g+N_b} \frac{1}{2} B_{ii} V_i^2(t) + \sum_{i=1}^{N_g+N_b} \frac{1}{2} B_{ii} V_i^2(t_s) \\
& - \sum_{i=1}^{N_g+N_b-1} \sum_{j=i+1}^{N_g+N_b} B_{ij} V_i(t) V_j(t) \cos\left(\phi_i(t) - \phi_j(t)\right) \\
& + \sum_{i=1}^{N_g+N_b-1} \sum_{j=i+1}^{N_g+N_b} B_{ij} V_i(t_s) V_j(t_s) \cos\left(\phi_i(t_s) - \phi_j(t_s)\right)
\end{aligned} \tag{39}
$$

Equation (39) can be represented with the integral form below,

$$
\begin{aligned}
W_{PE,3} = & -\int_{t_s}^{t} \sum_{i=1}^{N_g+N_b} B_{ii} V_i(\tau) \frac{dV_i(\tau)}{d\tau} d\tau \\
& - \int_{t_s}^{t} \sum_{i=1}^{N_g+N_b-1} \sum_{j=i+1}^{N_g+N_b} B_{ij} \cos\left(\phi_i(\tau) - \phi_j(\tau)\right) \\
& \times \left(V_j(\tau) \frac{dV_i(\tau)}{d\tau} + V_i(\tau) \frac{dV_j(\tau)}{d\tau} \right) d\tau \\
& + \int_{t_s}^{t} \sum_{i=1}^{N_g+N_b-1} \sum_{j=i+1}^{N_g+N_b} B_{ij} V_i(\tau) V_j(\tau) \sin\left(\phi_i(\tau) - \phi_j(\tau)\right) \\
& \frac{d\left(\phi_i(\tau) - \phi_j(\tau)\right)}{d\tau} d\tau \\
= & -\int_{t_s}^{t} \sum_{i=1}^{N_g+N_b} B_{ii} V_i(\tau) \frac{dV_i(\tau)}{d\tau} d\tau \\
& - \int_{t_s}^{t} \sum_{i=1}^{N_g+N_b} \sum_{j=1, j\neq i}^{N_g+N_b} B_{ij} V_j(\tau) \cos\left(\phi_i(\tau) - \phi_j(\tau)\right) \frac{dV_i(\tau)}{d\tau} d\tau \\
& + \int_{t_s}^{t} \sum_{i=1}^{N_g+N_b} \sum_{j=1}^{N_g+N_b} B_{ij} V_i(\tau) V_j(\tau) \sin\left(\phi_i(\tau) - \phi_j(\tau)\right) \frac{d\phi_i(\tau)}{d\tau} d\tau
\end{aligned}
$$

With (3) and (4), the equation above can be simplified as,

$$
\begin{aligned}
W_{PE,3} = & -\int_{t_s}^{t} \sum_{i=N_g+1}^{N_g+N_b} \frac{Q_{Li}}{V_i} \frac{dV_i}{d\tau} d\tau - \int_{t_s}^{t} \sum_{i=N_g+1}^{N_g+N_b} P_{Li} \frac{d\phi_i(\tau)}{d\tau} d\tau \\
& + \int_{t_s}^{t} \sum_{i=1}^{N_g} \sum_{j=1}^{N_g+N_b} B_{ij} V_i(\tau) V_j(\tau) \sin(\phi_i(\tau) \\
& - \phi_j(\tau)) \frac{d\phi_i(\tau)}{d\tau} d\tau = -W_{PE,2} + \int_{t_s}^{t} \sum_{i=1}^{N_g} \sum_{j=1}^{N_g+N_b} \\
& \times B_{ij} V_i(\tau) V_j(\tau) \sin\left(\phi_i(\tau) - \phi_j(\tau)\right) \frac{d\phi_i(\tau)}{d\tau} d\tau
\end{aligned} \tag{40}
$$

Plugging (40) into (8), we have (9). That is,

$$
\begin{aligned}
W_{PE} = & -\sum_{i=1}^{N_g} P_{mi}(\phi_i(t) - \phi_i(t_s)) \\
& + \sum_{i=1}^{N_g} \int_{t_s}^{t} \left(\sum_{j=1}^{N_g+N_b} B_{ij} V_i(\tau) V_j(\tau) \sin(\phi_i(\tau) - \phi_j(\tau)) \right) \\
& \times \frac{d\phi_i(\tau)}{d\tau} d\tau = \sum_{i=1}^{N_g} \int_{t_s}^{t} (P_i - P_{mi}) \frac{d\phi_i(\tau)}{d\tau} d\tau
\end{aligned}
$$

Appendix 4

Derivation of (19)

The angular velocity of a generator relative to the rated angular velocity at $t = t_c$ can be represented with the change in angular velocity of the generator from $t = 0$ to $t = t_c$,

$$
\omega_i(t_c) - \omega_R = \Delta\omega_i(t_c) = (\omega_i(t_c) - \omega_R) - (\omega_i(0) - \omega_R), \quad i = 1, 2, \ldots, N_g \tag{1}
$$

where the angular velocity of a generator is assumed to be equal to the rated angular velocity (i.e., $\omega_i(0) = \omega_R$) before a fault occurs.

The change in the angular velocity of the generator in (41) can be represented with the integral of the acceleration of the generator during the period $t \in (0, t_c)$,

$$
\Delta\omega_i(t_c) = \int_{0^+}^{t_c} \frac{d\omega_i(\tau)}{d\tau} d\tau, \quad i = 1, 2, \ldots, N_g \tag{42}
$$

For a short duration of fault, the acceleration power of a generator can be considered as a constant that is equal to the change in the real power output of the generator from $t = 0$ to $t = 0^+$. Thus, the acceleration of the generator can be considered as a constant during the duration of fault. Then, (42) can be represented as,

$$\Delta\omega_i(t_c) = \int_{0^+}^{t_c^-} \frac{\mathrm{d}\omega_i(\tau)}{\mathrm{d}\tau}\mathrm{d}t = \frac{\mathrm{d}\omega_i(0^+)}{\mathrm{d}t}t_c = -\Delta P_{ei}(0^+)\frac{1}{M_i}t_c,$$
$$i = 1, 2, \ldots, N_g \tag{43}$$

With (27)–(28) and (43), the angular velocity of a generator relative to the rated angular velocity at $t = t_c$ can be approximately represented as,

$$\omega_i(t_c^-) - \omega_R = -\frac{E_i V_F B_{iF}\cos(\delta_i(0) - \theta_F(0))\Delta P_F(0^+)}{\sum_{i=1}^{N_g} V_i V_F B_{iF}\cos(\theta_F(0) - \delta_i(0))}\frac{1}{M_i}t_c$$

References

[1] Metz B (2005) IPCC special report on carbon dioxide capture and storage. Cambridge University Press, Cambridge

[2] Intergovernmental Panel for Climate Change (2005) Fourth Assessment Report. Cambridge University Press, Cambridge

[3] Committee on Climate Change (2008) Building a low-carbon economy—the UK's contribution to tackling climate change. The Stationary Office, London

[4] Internal Energy Agency (2006) World energy outlook 2006. Renouf Pub Co Ltd, Rochford

[5] European Wind Energy Association (EWEA) (2002) Wind Force 12. EWEA, Brussels

[6] Muller S, Deicke M, De Doncker RW (2002) Doubly fed induction generator systems for wind turbines. IEEE Ind Appl Mag 8(3):26–33

[7] Lalor G, Mullane A, O'Malley M (2005) Frequency control and wind turbine technologies. IEEE Trans Power Syst 20(4):1905–1913

[8] Reza M, Morren J, Schavemaker PH et al (2005) Power electronic interfaced DG units: impact of control strategy on power system transient stability. In: Proceedings of the 3rd IEEE international conference on reliability in transmission and distribution network (RTDN'05), 15–17 Feb 2005, London, pp 179–182

[9] Morren J, Pierik J, De Hann SWH (2006) Inertial response of variable speed wind turbines. Electr Power Syst Res 76(11):980–987

[10] Anderson PM, Mirheydar M (1992) An adaptive method for setting underfrequency load shedding relays. IEEE Trans Power Syst 7(2):647–655

[11] Drukin CJ, Eberle ER, Zarakas P (1969) An underfrequency relay with frequency decay compensation. IEEE Trans Power Appar Syst 88(6):812–820

[12] Farrokhseresht N, Orostica HC, Hesamzadeh MR (2014) Determination of acceptable inertia limit for ensuring adequacy under high levels of wind integration. In: Proceedings of the 11th international conference on the European energy market (EEM'14), 28–30 May 2014, Krakow, 5 pp

[13] Careri F, Genesi C, Marannino P et al (2011) Generation expansion planning in the age of green economy. IEEE Trans Power Syst 26(4):2214–2223

[14] Chen QX, Kang CQ, Xia Q et al (2010) Power generation expansion planning model towards low-carbon economy and its application in China. IEEE Trans Power Syst 25(2):1117–1125

[15] Anderson PM, Fouad AA (2003) Power system control and stability. IEEE Press, Piscataway

[16] Sauer PW, Pai MA (1998) Power system dynamics and stability. Prentice-Hall, Englewood Cliffs

[17] Alexander CK, Sadiku MNO (2007) Fundamentals of electric circuits, 3rd edn. McGraw-Hill, Boston

[18] Greiner W (2010) Classical mechanics: Systems of particles and Hamiltonian dynamics. Springer, Berlin

[19] Pai MA (1989) Energy function analysis for power system stability. Kluwer Academic Publishers, Boston

Di WU is currently the research scientist and instructor at the Electricity and Energy Reliability and Risk Laboratory in the School of Electrical and Computer Engineering at the University of Oklahoma. He received his first Ph.D degree in electrical engineering from Zhejiang University in China, and another Ph.D degree in applied complex networks from Politecnico Di Torino in Italy. His research interests include power grid cascading failure analysis and application of advanced complex networks methods in power systems.

Milad JAVADI is currently working toward his Ph.D. degree in the School of Electrical and Computer Engineering at the University of Oklahoma. His research interests include power system operation, statistical and data analysis in power system, and power economics.

John Ning JIANG is OG&E chair professor in electrical engineering and director of the Electricity and Energy Reliability and Risk Laboratory in the School of Electrical and Computer Engineering at the University of Oklahoma. His current research interest is to understand various types of risks in power system operation and the power market.

A quasi-automated generation control strategy for multiple energy storage systems to optimize low-carbon benefits

Gangui YAN (✉), Xudong ZHENG,
Gang MU, Junhui LI, Yang CUI, Yue WANG

Abstract Integrating a battery energy storage system (ESS) with a large wind farm can smooth the intermittent power obtained from the wind farm, but the smoothing function will not be achieved if multiple ESSs are used to smooth the fluctuations in individual wind power plants in a distributed pattern. Therefore, this study focuses on the development of a control strategy to optimize the use of multiple ESSs to accelerate the adoption of wind energy resources. This paper proposes a quasi-automated generation control (QAGC) strategy to coordinate multiple ESSs, which responds to the grid dispatch demand rather than smoothing out the intermittent power from individual wind farms. The aims of QAGC are to ensure that multiple ESSs provide a service that is as balanced as possible, so more wind power systems at various scales can be accepted by the grid, as well maximizing the low-carbon benefits of ESSs. The effectiveness of QAGC is demonstrated by using data from an actual gigawatt scale cluster of wind plants.

Keywords Control strategy, Low-carbon benefits, Multiple battery energy storage systems, Quasi-automated generation control

1 Introduction

Increasing the ratio of renewable energy in terminal consumption is an effective way of achieving a low-carbon energy supply. In China, the ratio of renewable energy in end use will reach 20% by 2020 [1] according to the National Plan for Medium and Long Term Renewable Energy Development. Therefore, the energy structure, low-carbon benefits, energy conservation, and emission reduction problems are receiving increasing attention [2, 3].

Several incentive policies for promoting wind energy have been issued in China. In 2007, the National Development and Reform Commission, the Environmental Protection Administration, and the Energy Office jointly formulated the "Energy Generation Scheduling Approach (Trial Implementation)" [4], in which the wind power was highlighted as having the main scheduling priority.

However, the wind power is intermittent and the uncertainty of the wind power can have many adverse effects on the economical operation of power systems, possibly endanger the safe operation of power systems [5, 6]. Thus, the scales of wind power systems integrated into the grid are restricted at present, and addressing the uncertainty of the wind power and its integration remains a problem.

The development of wind power prediction systems is a valid approach for dealing with the wind power uncertainty. The National Energy Bureau has asked operators to improve the wind power prediction accuracy and to enhance the management of wind power sources. However, the wind power forecasting accuracy is affected by many factors at present and it is still difficult to satisfy the secure and economic operation requirements of power systems.

Energy storage systems (ESSs) can transform energy usage in terms of time and space, and they are regarded as an important way of improving the characteristics of wind power sources [7, 8]. Demonstration projects such as the Hebei Zhangbei wind storage demonstration project and Liaoning Woniushi wind storage demonstration project have improved the wind power output characteristics, as

G. YAN, X. ZHENG, G. MU, J. LI, Y. CUI, Y. WANG,
School of Electrical Engineering, Northeast Dianli University,
Jilin 132012, China
(✉) e-mail: Yangg@mail.nedu.edu.cn
X. ZHENG
e-mail: minnyzxd@163.com

well as providing valuable experience in terms of running large-scale ESS applications [9, 10].

In fact, the integration of wind power can have adverse effects on power systems, which are caused by volatility in the overall wind power network, rather than the power output volatility of local or single wind farms. This is because that a power system is a large inertial system with a strong anti-interference capacity. However, if the wind power fluctuations exceed the adjustment capacity of the system, the wind power must be stabilized and regulated, but the ESS control strategies used to smooth wind power fluctuations on single wind farms might lead to the excessive regulation.

Given recent breakthroughs and the lower prices of large-scale energy storage technology, it has broader application prospects. Indeed, a power system auxiliary service market based on large-scale ESSs is developing. Multiple ESSs must address the task of managing control functions and increasing the different scales of wind power systems that can be integrated into the grid.

Previously, researchers have focused on the operational control technology for single type ESSs [11–13] and the control performance optimization for hybrid ESSs [14–16], improving the storage system optimization configuration to enhance the performance of wind power access systems [17–20]. Thus, many studies have investigated large-scale ESSs, but the problem of coordinating control for multiple ESSs has not been addressed previously.

The integration of large-scale wind power systems requires support from the flexible power system auxiliary service market. In the present study, a control strategy based on a power system auxiliary service for large-scale ESSs is described, in which the offset effect that occurs with multiple ESSs is demonstrated when smoothing wind power fluctuations in a distributed pattern. A quasi-automated generation control (QAGC) strategy is proposed for multiple ESSs, which satisfies the power system control requirements for electric power and relaxes the bottleneck to accommodate the wind power in the grid. This approach regulates the performance of ESSs, broadens the scale of wind power systems accepted by the grid, and optimizes the low-carbon benefits of ESSs.

2 Offset effect in wind power storage systems with distributed control

Due to space–time differences in the energy distribution, the wind power outputs vary among wind farms. In contrast to smoothing the total deviation in the power of a cluster of wind farms, smoothing the wind power fluctuations of multiple ESSs based on the distributed control leads to energy offsets due to charging and discharging by the ESSs, which is referred to as the “offset effect”.

The offset effect is shown in Fig. 1 based on smoothing the power output fluctuations of two wind farms using ESSs. For a cluster of two wind farms, the clustering output power is $P_{\Sigma\text{wind}}$, the total reference power is $P_{\Sigma\text{ref}}$, and the deviation is ΔP_{Σ}. With the distributed control, the corresponding individual power outputs are $P_{\text{wind.1}}$ and $P_{\text{wind.2}}$, the individual reference powers are $P_{\text{ref.1}}$ and $P_{\text{ref.2}}$, and the individual deviations are ΔP_1 and ΔP_2.

As shown in Fig. 1, wind storage system No. 1 needs to use its power to charge an ESS in order to store surplus wind power, whereas wind storage system No. 2 needs to discharge the power to compensate for the lack of the wind power, the equations are shown as follows:

$$P_{\text{BESS.1}} = P_{\text{wind.1}} - P_{\text{ref.1}} \tag{1}$$

$$P_{\text{BESS.2}} = -(P_{\text{wind.2}} - P_{\text{ref.2}}) \tag{2}$$

where $P_{\text{BESS.}i}$ is the ESS charging power for part i ($i = 1,2,\ldots,n$).

After smoothing the differences in power between the two wind farms in the cluster, the charge power of the ESS is

$$P_{\Sigma\text{BESS}} = P_{\text{wind.1}} - P_{\text{ref.1}} + P_{\text{wind.2}} - P_{\text{ref.2}} \tag{3}$$

and the following equation is always true.

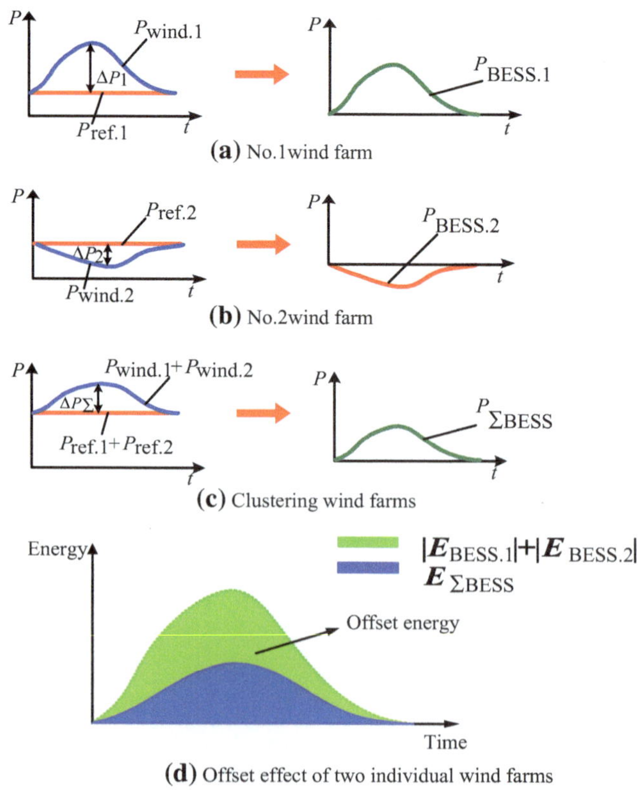

Fig. 1 Schematic diagram of offset effect when smoothing outputs of two individual wind farms

$$|P_{\Sigma\mathrm{BESS}}| \le |P_{\mathrm{wind.1}} - P_{\mathrm{ref.1}}| + |P_{\mathrm{wind.2}} - P_{\mathrm{ref.2}}| \tag{4}$$

In the same manner, when clustering wind storage systems that comprise n wind farms with ESSs, the following is always true.

$$|P_{\Sigma\mathrm{BESS}}| \le \sum_{i=1}^{n} |P_{\mathrm{BESS}.i}| \tag{5}$$

Thus, compared with smoothing the total power deviation, the ESS control cost is greater when smoothing the deviations in the power output from individual wind farms.

We propose the concepts of offset energy and an offset energy factor to measure the degree of ESS energy offset during the distributed control, which are shown as follows:

$$E_{\mathrm{offset}} = \sum_{i=1}^{n}\left[\int |(P_{\mathrm{wind}.i} - P_{\mathrm{ref}.i})|\mathrm{d}t\right] - \int |P_{\Sigma\mathrm{wind}} - P_{\mathrm{ref}}|\mathrm{d}t \tag{6}$$

$$k = \frac{E_{\mathrm{offset}}}{E_{\Sigma\mathrm{BESS}}} = \frac{E_{\mathrm{offset}}}{\sum_{i=1}^{n}\left[\int |(P_{\mathrm{wind}.i} - P_{\mathrm{ref}.i})|\mathrm{d}t\right]} \tag{7}$$

where $P_{\Sigma\mathrm{wind}}$ and P_{ref} are the total wind output power and total wind reference power, respectively. If the offset energy E_{offset} is higher, the stated offset energy when charging and discharging among ESSs is greater with the distributed control. The offset effect factor k reflects the ratio of the offset energy in the total charge and discharge energy with the distributed control.

3 A QAGC strategy for a large-scale ESS

The adverse effects of the wind power integration into a grid are attributable to the wind power uncertainty of the whole network rather than fluctuations in the output power of single wind farms or local wind farms. In order to improve the networking performance of wind farm groups, it is only necessary to regulate the overall wind power of the network to ensure that it does not deviate from the allowable value. Using the fluctuating component of the system payload (the wind power is treated as a negative load) as a regulatory criterion to determine the allowable wind power value for the whole network can fully exploit the wind power potential accepted by the grid, and effectively improve the safety of the wind power integration into the grid. The proposed QAGC strategy allows multiple ESSs to satisfy the power system control requirements and it relaxes the bottleneck to accommodate the wind power produced at various scales in the grid.

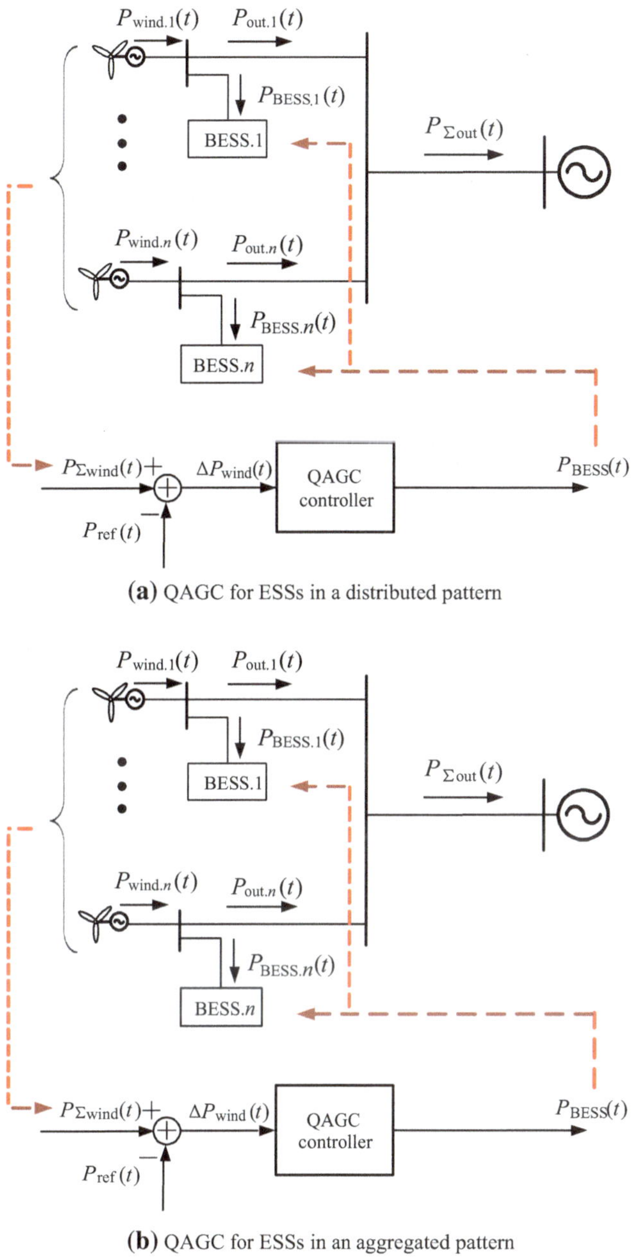

(a) QAGC for ESSs in a distributed pattern

(b) QAGC for ESSs in an aggregated pattern

Fig. 2 QAGC strategies for ESSs in response to grid dispatch commands

Due to recent breakthroughs in the large-scale energy storage technology and lower prices, a power system auxiliary service market based on large-scale ESSs is developed. Irrespective of whether ESSs are installed in the vicinity of wind farms in a distributed pattern at the beginning of the application or installed in an aggregated pattern, both approaches can facilitate regulatory participation in a power system auxiliary service market, where control strategies will assist service regulation for the overall power system. Figures 2 and 3 show examples of ESS installation patterns and control strategies.

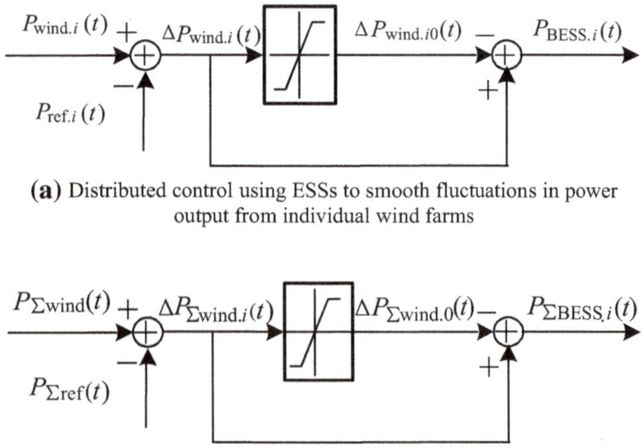

(a) Distributed control using ESSs to smooth fluctuations in power output from individual wind farms

(b) Aggregated control using ESSs to smooth fluctuations in power output from a cluster of wind farms

Fig. 3 Use of ESSs in distributed and aggregated control patterns

For multiple storage systems with a distributed pattern, as shown in Fig. 2a, we assume that the state of charge for each ESS unit is the same, and thus by assessing the total charging power according to the proportion of the overall capacity, the ESS charging power using the QAGC strategy is:

$$\begin{cases} P_{BESS}(t) = \begin{cases} P_{\Sigma wind}(t) - P_{ref}(t) & P_{\Sigma wind}(t) > P_{ref}(t) \\ 0 & P_{\Sigma wind}(t) \leq P_{ref}(t) \end{cases} \\ P_{BESS.i}(t) = \dfrac{P_{BR.i}}{\Sigma P_{BR.i}} P_{BESS}(t) \\ E_{BESS.i}(t) = E_{BESS.i}(t - \Delta t) + P_{BESS.i}(t)\Delta t \eta \end{cases} \tag{8}$$

where the constraint conditions are:

$$\begin{cases} 0 \leq P_{BESS.i}(t) \leq P_{BR.i} \\ 0 \leq E_{BESS.i}(t) \leq E_{ER.i} \end{cases}$$

P_{BESS} is the total ESS charging power; $P_{BR.i}$ and $E_{BR.i}$ are the rated power and the rated energy capacity for part i ($i = 1,2,\ldots,n$), respectively.

When the overall wind power of the network exceeds the allowable value for the whole network, the ESSs are controlled to absorb the surplus wind power.

When the state of charge (SOC) of ESSs reaches to the maximum value (e.g. 1.0), the ESSs are controlled during the higher load period to discharge for possessing the maximum upward reserve capacity.

With multiple ESSs in an aggregated pattern, as shown in Fig. 2b, the control is achieved more directly without any requirements for coordinating communication.

Figure 3 compares the usage of an ESS to smooth the power output from a wind farm in a distributed control pattern and the power output from a wind farm cluster in an aggregated control pattern. In Fig. 3, the reference input $P_{ref.i} = P_{forecast.i}$ corresponds to the distributed control to smooth the wind power according to the predicted power deviation, and the reference input $P_{\Sigma ref} = P_{\Sigma forecast}$ corresponds to the aggregated control to smooth the overall wind power of a cluster of wind farms based on the predicted power deviation.

The corresponding control parameters for the three ESS control patterns are shown in Table 1. In the table, $P_{windspace}$ is the wind power accommodation capacity of the power grid.

As shown in Fig. 3 and Table 1, when $P_{ref.i} = P_{forecast.i}$, and by using ESSs to smooth the predicted wind power error $\Delta P_{wind.i}$ for an individual wind farm in a distributed pattern, the controlled variable is $P_{BESS.i}$ and the wind output power is $P_{out.i}$ after smoothing. This control pattern improves the certainty of the power output from an individual wind farm, but the effective regulation of the ESSs is reduced severely due to the offset effect.

When $P_{\Sigma ref} = P_{\Sigma forecast}$, the control strategy described above can improve the certainty of the power output from the wind farm cluster. Storage systems are used to improve the source characteristics of wind farms. The available wind power capacity of the grid is not considered at this time, thus ESSs may still be over-regulated, which reduces the effectiveness of the regulation. The contribution of ESSs to increasing the system capacity when handling the uncertainty of power regulation is also ignored.

When $P_{ref} = P_{windspace}$, the potential wind power accepted by systems can be fully exploited. Most of the time, the ESSs do not need to be activated and the ESSs are only required to charge to absorb the excess wind energy when the wind power exceeds the accommodation ability of the power grid. Thus, the ESSs are equivalent to "spinning" reserve for the system. If the capacity configuration reaches a sufficient level, it can greatly improve the ability of the system to handle uncertain power fluctuations.

Table 1 Comparison of three control strategies

Control pattern	P_{ref}	Control variable	Control object	Output
Distributed control	$P_{forecast.i}$	$P_{BESS.i}$	$P_{wind.i}$	$P_{out.i}$
Aggregated control	$P_{\Sigma forecast}$	$P_{\Sigma BESS.i}$	$\sum P_{wind.i}$	$\sum P_{out.i}$
Quasi-AGC control	$P_{windspace}$	$P_{\Sigma BESS.i}/P_{BESS}$	$\sum P_{wind.i}$	$\sum P_{out.i}$

4 Simulation example

In this example, a gigawatt (GW)-scale cluster of wind farms with a storage system is considered. The distributed control is compared to smooth fluctuations in the power output of individual plants and aggregated control to smooth the power output of clustering wind farms by using QAGC to adapt to the needs of the power system control. To facilitate the large-scale integration of wind power in the grid, an important reference is provided for configuring the capacity of multiple ESSs and control target selection.

The GW-scale cluster of wind storage systems is shown in Fig. 4, which comprises wind farms and their support ESSs (charge/discharge efficiency is 83%, initial state of charge is 0.5), while the installed capacities of the wind storage systems are shown in Table 2. The capacity of the wind farm cluster storage system is 59.66–1193 MW, or 119.30 MW/h.

The data length is 24 h and the sampling interval is 5 min (24 × 12 = 288) in this example. The curves used to predict the wind generation by eight wind farms and the actual power curves are shown in Appendix 1, and assume that the maximum allowable prediction error ΔP is ±0.05 p.u.

For comparison conveniently, the given grid-connected wind power space is greater than the actual wind power of the wind farm group, as shown in Fig. 5.

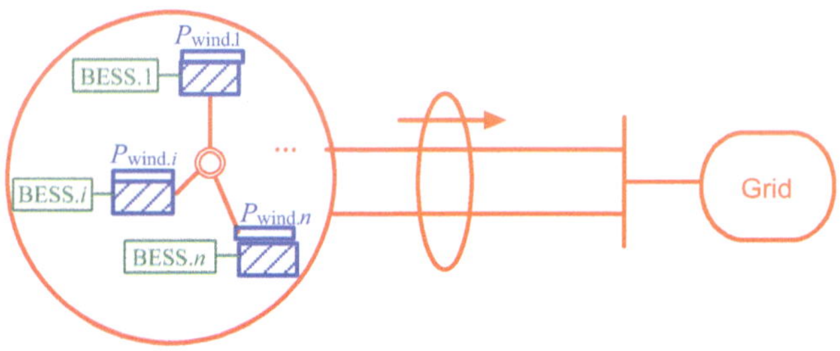

Fig. 4 Multiple wind-storage systems integrated into grid

Table 2 Installed capacity of the wind-storage systems

No.	Installed wind capacity (MW)	Installed of ESS	
		Rated power (MW)	Energy capacity (MWh)
1	349	17.45	34.90
2	349	17.45	34.90
3	99	4.95	9.90
4	99	4.95	9.90
5	99	4.95	9.90
6	99	4.95	9.90
7	49.5	2.48	4.95
8	49.5	2.48	4.95
Total	1193	59.66	119.30

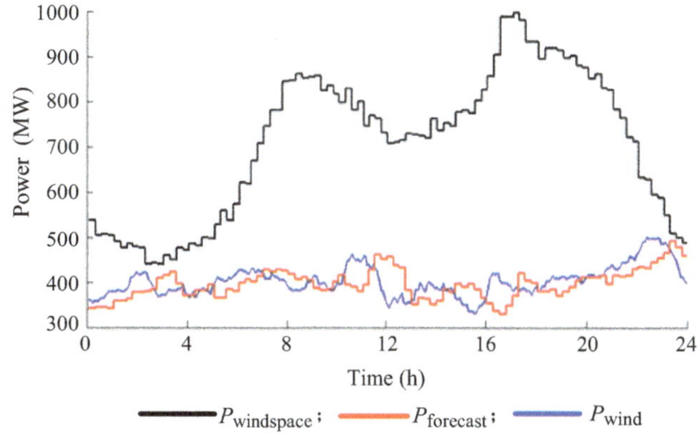

Fig. 5 Curves for grid connected to wind power space, wind power and predicted power for wind farm cluster

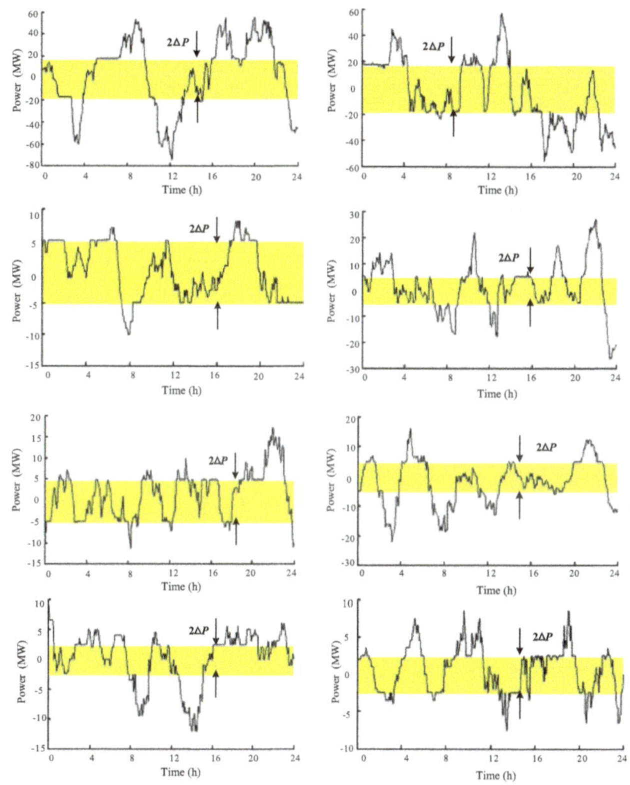

Fig. 6 Curves of distributed smoothed predicted error of wind power for each wind plant

When employing QAGC strategy, ESSs are controlled to discharge during the higher load period, e.g. about 6 a.m.–9 a.m. in this case study.

4.1 ESSs regulation effects of three different control strategies

1) Distributed control using power error of single wind plants as reference input

Figure 6 shows the curves of the distributed smoothed predicted error of wind power for each wind plant. Figure 7

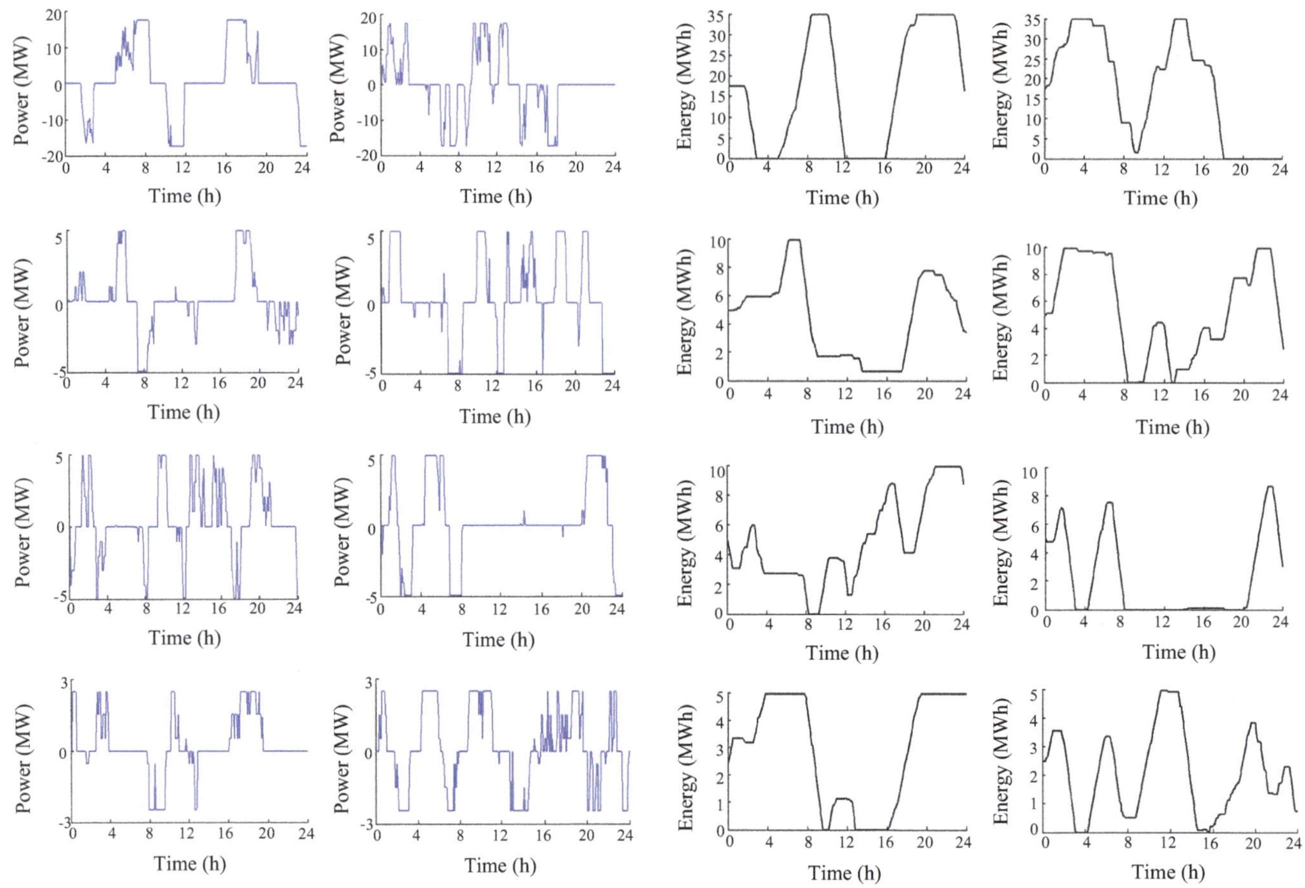

(a) Curves of charge-discharge power for each ESS with distributed control strategy

(b) Curves of energy for each ESS with distributed control strategy

Fig. 7 Curves of charge–discharge power and energy for each ESS with distributed control strategy

shows the curves for the charge–discharge power and energy for each ESS with the distributed control strategy.

As shown in Fig. 6, the generation power characteristics of the wind plants are different. For a given storage capacity and allowable prediction error, the integrated wind power cannot be constrained within the range of the allowable error with the distributed control strategy.

As shown in Fig. 7, each ESS is regulated frequently (action frequency is 127/24 rates/h, accumulative throughput capacity is 425.50 MW/h); thus, the energy "offset effect" between charging and discharging among multiple ESSs occurs during some periods.

2) Integrated control using power error of a wind farm group as reference input

Figure 8 shows the curves of the smoothed predicted error of power output for a wind farm group. Figure 9 shows the curves for the charge–discharge power and energy for the total ESSs with an aggregated control strategy.

As shown in Figs. 8 and 9, the total power integrated from the wind farm group exceeds the allowable error

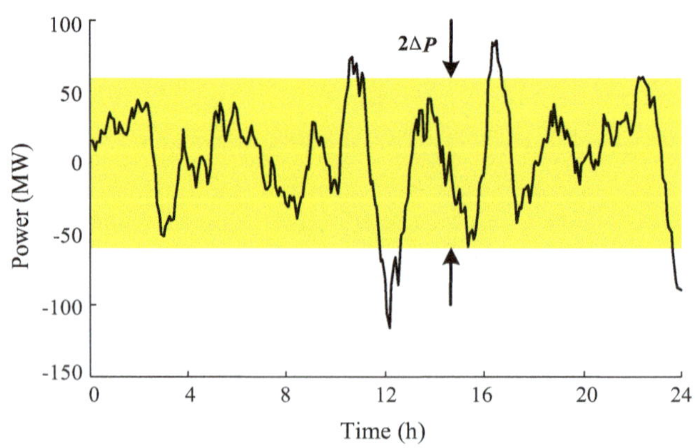

Fig. 8 Curves of the predicted error power for a cluster of wind farms

range only for a small period of time. Thus, the ESS control frequency is decreased greatly and the effect of control is obviously improved (action frequency is 7/24 rates/h and accumulative throughput capacity is 43.17 MW/h).

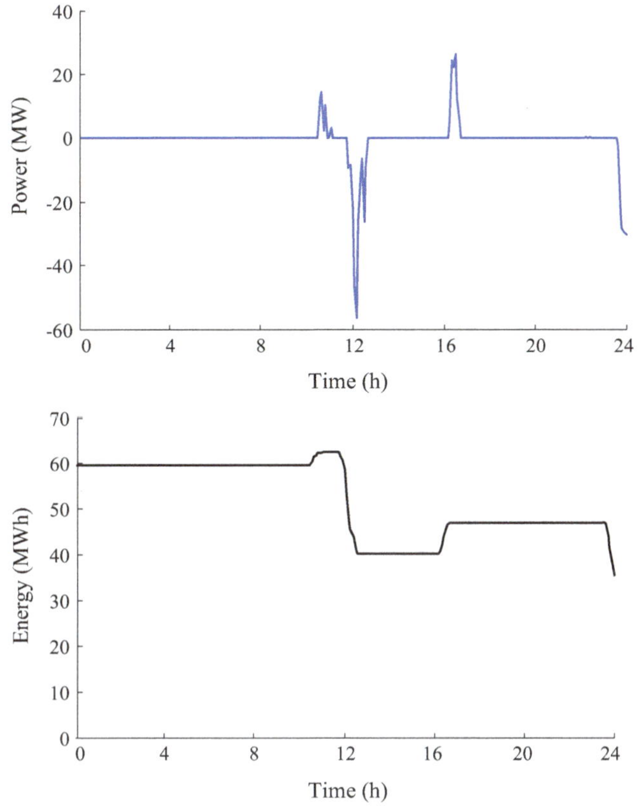

Fig. 9 Curves of charge–discharge power and energy for total ESSs with aggregated control strategy

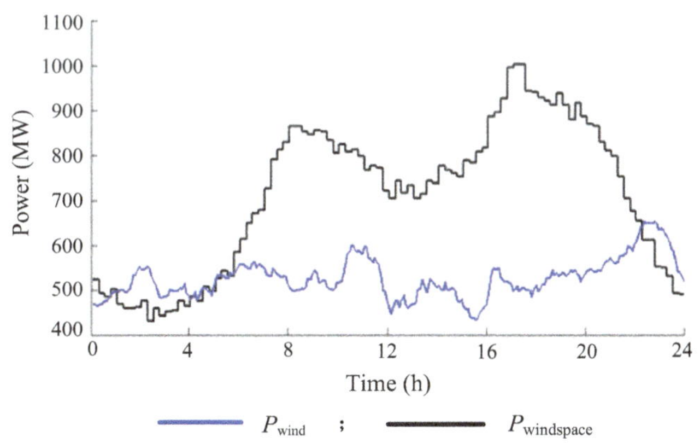

Fig. 10 Curves of consequent wind power and grid-connected wind power space with 1.3 times installed capacity (1551 MW)

3) QAGC using grid-connected wind power space as reference input

As shown in Fig. 5, the ESSs do not need to be regulated since the actual wind power output is within the grid-connected wind power space and the wind power can be integrated into the grid completely.

Giving the wind power output is proportional to the installed capacity and enlarging the installed wind capacity to 1.3 times (1551 MW) of that shown in Table 2, the consequent wind power output shown in Fig. 10 occasionally exceeds the grid-connected wind power space and the ESSs need to be charged. As a result, the scale of the grid-connected wind power can be increased. The corresponding curves for the charge–discharge power and the energy change in the ESSs are shown in Figs. 11 and 12, respectively.

The maximum charging power of the ESSs is 59.55 MW and the control frequency is reduced greatly using the QAGC strategy, as shown in Figs. 11 and 12. The cumulative throughput capacity is 92.46 MW/h with more than 2869.3 (MWh)/day wind energy accepted, and the effect of regulation is obviously improved.

4.2 Comparison of ESS smoothing using three control strategies

If the excess wind power is viewed as curtailed wind power, the control effects of different ESS control strategies can be evaluated using three indices: the curtailed wind energy, the energy throughput, and the reduced equivalent CO_2 emissions. Table 3 shows

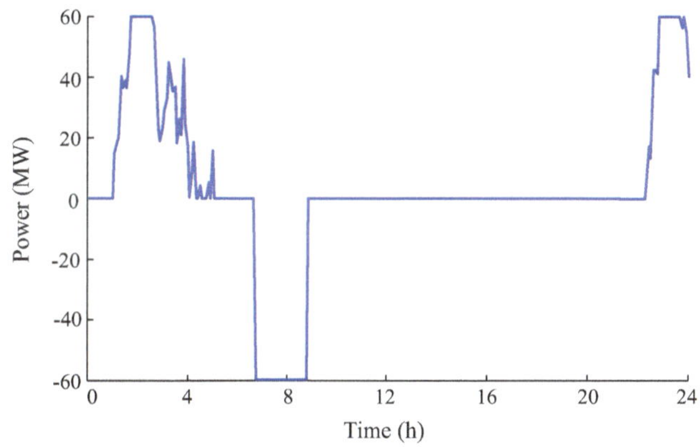

Fig. 11 Charge power curves for ESSs using QAGC strategy

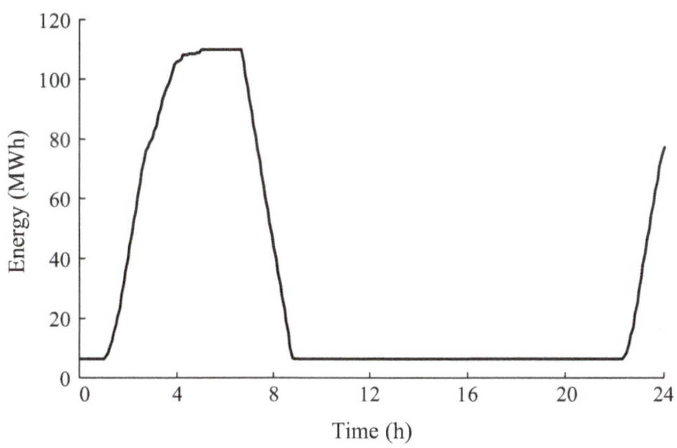

Fig. 12 Change in energy curves of ESSs using QAGC strategy

Table 3 ESS performance comparison using three control patterns

Control pattern	Curtailed wind power (MWh)	Throughput capacity (MWh)	CO_2 emission (kg)
Distributed control	1066.34	425.50	1041814
Aggregated control	113.71	43.17	111095
Quasi-AGC control	−2869.3	92.46	−2803306

Note: negative curtailed wind power values represent more connected wind power, and negative CO_2 values represent reduced emissions

the evaluation results for these different control strategies.

As shown in Table 3, based on the same storage configuration, the curtailed wind energy and energy throughput of the ESSs are reduced by 952.63 and 382.33 MW/h, respectively, using the aggregated control strategy compared with the distributed control strategy. Theoretically, 1 kW/h of electricity is equivalent to 0.977 kg CO_2 emissions, thus the CO_2 emissions would be reduced by 930719 kg. The grid-connected wind energy increment and the corresponding reduction in the CO_2 emissions are 2983.01 MW/h and 2914400 kg, respectively, using the QAGC strategy compared with the aggregated control strategy. Clearly, the low-carbon benefits of ESSs are optimized by the QAGC strategy.

5 Conclusion

Based on an analysis of the wind storage system control problem, "offset effect" is identified among multiple ESSs when smoothing wind power fluctuations using a distributed pattern. Thus, the QAGC strategy is proposed for multiple ESSs to address the need for power system control. A GW-scale wind storage cluster system is used as an example. In contrast to distributed control, the "offset effect" is handled with multiple ESSs and the curtailed wind power caused by wind power fluctuations is avoided by using the QAGC strategy. Thus, the regulatory burden on the storage system is reduced, while the CO_2 emissions are also reduced by 2914400 kg. The proposed method enhances the regulatory efficiency and improves the low-carbon benefits of ESSs.

Acknowledgment This work is supported by National Basic Research Program of China (973 Program) (No. 2013CB228201).

Appendix 1

See Figs. 13 and 14.

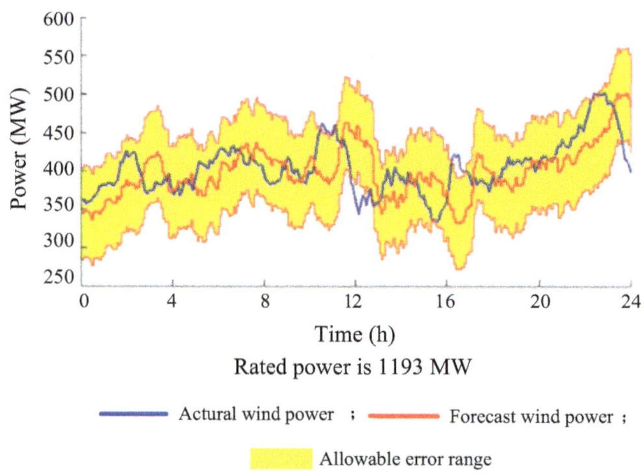

Fig. 13 Cluster of wind plants: actual power, predicted power, and accepted predicted error range

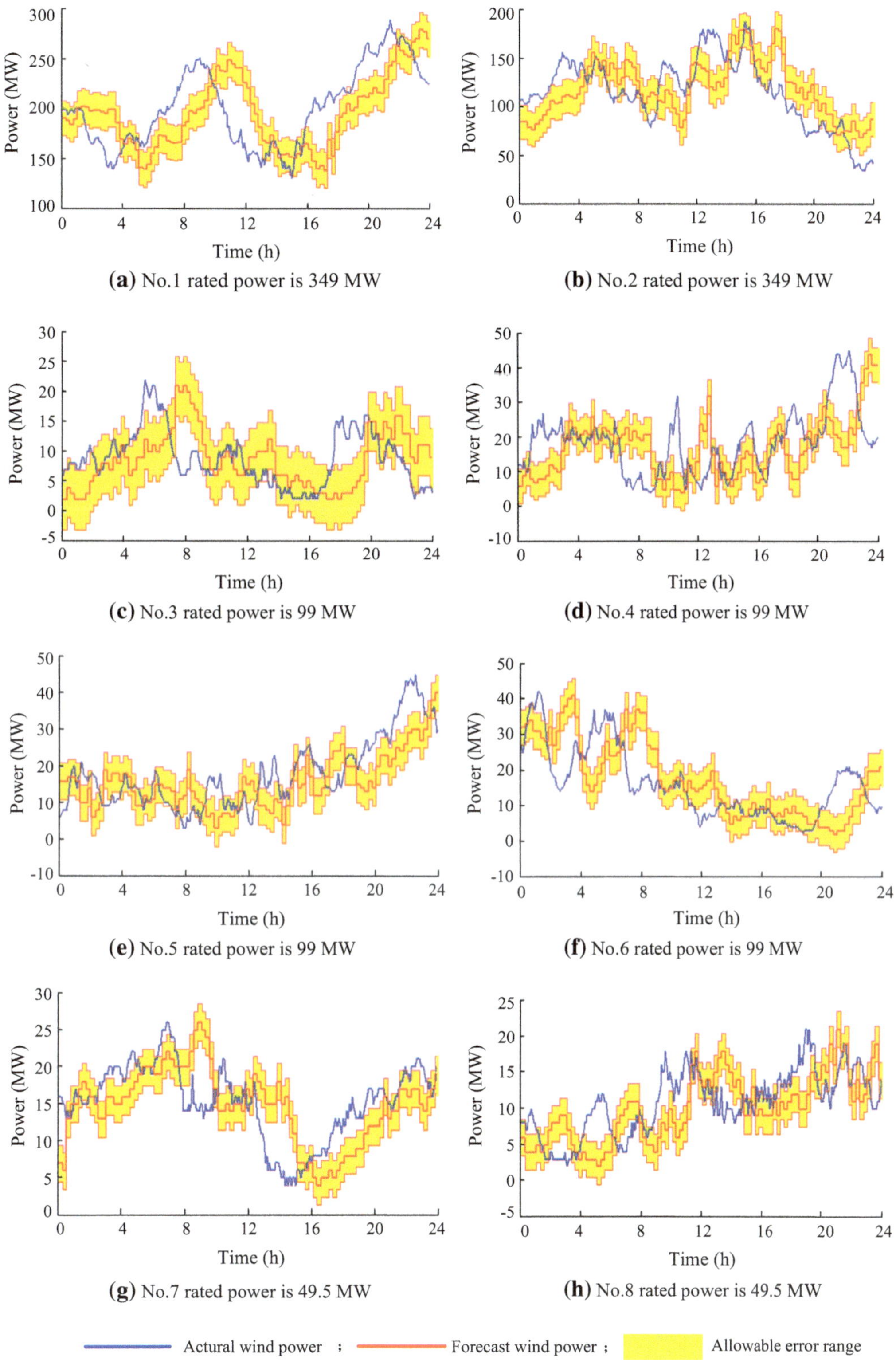

Fig. 14 Eight wind plants: actual power, predicted power and accepted predicted error range

References

[1] China National Energy Administration (2011) The Twelfth Five-Year Plan for national energy technology development (in Chinese)

[2] Kang CQ, Chen QX, Xia Q (2009) Prospects of low-carbon electricity. Power Syst Technol 33(2):1–7 (in Chinese)

[3] Ding R, Kang CQ, Zhou TR et al (2011) Analysis and prospect on technical approaches for low carbon power grid. Power Syst Technol 35(10):1–8 (in Chinese)

[4] National Development and Reform Commission, Ministry of Environmental Protection of the People's Republic of China, State Electricity Regulatory Commission et al (2007) Energy generation scheduling approach (Trial Implementation) (in Chinese)

[5] Eriksen PB, Ackermann T, Abildgaard H et al (2005) System operation with high wind penetration. IEEE Power Energy Mag 3(6):65–74

[6] Sorensen P, Cutululis NA, Vigueras-Rodríguez A et al (2007) Power fluctuations from large wind farms. IEEE Trans Power Syst 22(3):958–965

[7] Kyung SK, McKenzie KJ, Liu YL et al (2006) A study on applications of energy storage for the wind power operation in power systems. In: Proceedings of the IEEE Power Engineering Society general meeting, Montréal, Canada, 18–22 June 2006, 5 pp

[8] Barton JP, Infield DG (2004) Energy storage and its use with intermittent renewable energy. IEEE Trans Energy Convers 19(2):441–448

[9] Wang WL, Qin M, Liu W (2010) Study on the application of large-scale energy storage technology in wind power generation. In: Proceedings of the 12th annual meeting of China Association for Science and Technology, vol 2, Fuzhou, China, 1–3 November 2010 (in Chinese)

[10] Yoshimoto K, Nanahara T, Koshimizu G (2009) Analysis of data obtained in demonstration test about battery energy storage system to mitigate output fluctuation of wind farm. In: Proceedings of the 2009 CIGRE/IEEE PES joint symposium integration of wide-scale renewable resources into the power delivery system, Calgary, Canada, 29–31 July 2009, 5 pp

[11] Jiang QY, Wang HJ (2013) Two-time-scale coordination control for a battery energy storage system to mitigate wind power fluctuations. IEEE Trans Energy Convers 28(1):52–61

[12] Islam F, Al-Durra A, Muyeen SM (2013) Smoothing of wind farm output by prediction and supervisory-control-unit-based FESS. IEEE Trans Sustain Energy 4(4):925–933

[13] Abbey C, Joos G (2007) Supercapacitor energy storage for wind energy application. IEEE Trans Ind Appl 43(3):769–776

[14] Yang GG, Zhu XX, Li JH et al (2013) Control strategy design for hybrid energy storage system with intrinsic operation life measurement and calculation. Autom Electr Power Syst 37(1):110–114 (in Chinese)

[15] Lee H, Shin BY, Han S et al (2012) Compensation for the power fluctuation of the large scale wind farm using hybrid energy storage applications. IEEE Trans Appl Supercond 22(3):5701904/1–4

[16] Esmaili A, Novakovic B, Nasiri A et al (2013) A hybrid system of Li-Ion capacitors and flow battery for dynamic wind energy support. IEEE Trans Ind Appl 49(4):1649–1657

[17] Li H, Fu B, Yang C et al (2013) Power optimization distribution and control strategies of multistage vanadium redox flow battery energy storage systems. Proc CSEE 33(16):70–77 (in Chinese)

[18] Yan GG, Jia Liu, Cui Y et al (2013) Economic evaluation on improving wind power scheduling scale by using energy storage system. Proc CSEE 33(22):45–52 (in Chinese)

[19] Han T, Lu JP, Qiao L et al (2010) Optimized scheme of energy-storage capacity for grid-connected large-scale wind farm. Power Syst Technol 34(1):169–173 (in Chinese)

[20] Yan GG, Feng XD, Li JH et al (2012) Optimization of energy storage system capacity for relaxing peak load regulation bottlenecks. Proc CSEE 32(28):27–35 (in Chinese)

Gangui YAN received his Ph.D. degree in Electrical Engineering from Tsinghua University, China, in 2003. He is a Professor and Ph.D. Supervisor at Northeast Dianli University. His research interests are renewable energy systems, energy storage system and power electronic technique.

Xudong ZHENG is a Master Candidate at Northeast Dianli University. His research interests are the use of energy storage with renewable sources and large-scale wind power generation-grid coordinated operating strategies.

Gang MU received his Ph.D. degree in Electrical Engineering from Tsinghua University. He is a Professor and Ph.D. Supervisor at Northeast Dianli University. His research includes large-scale wind power integration, power system stability analysis, security and economic operation.

Junhui LI received his Ph.D. degree from North China Electric Power University in 2012. He is an Associate Professor in Northeast Dianli University. His research interests are energy storage system and renewable energy generation.

Yang CUI received his Ph.D. degree in Electrical Engineering from North China Electric Power University. He is an Associate Professor in Northeast Dianli University. His research interest is large-scale wind power integration.

Yue WANG is a Master Candidate at Northeast Dianli University. Her research interests are energy storage system and renewable energy generation.

16

Bidding strategy for wind generation considering conventional generation and transmission constraints

Hao HUANG (✉), Fangxing LI

Abstract Under the environmental crisis of global warming, more efforts are put in application of low carbon energy, especially low-carbon electricity. Development of wind generation is one potential solution to provide low-carbon electricity source. This paper researches operation of wind generation in a de-regulated power market. It develops bidding models under two schemes for variable wind generation to analyze the competition among generation companies (GENCOs) considering transmission constraints. The proposed method employs the supply function equilibrium (SFE) for modeling the bidding strategy of GENCOs. The bidding process is solved as a bi-level optimization problem. In the upper level, the profit of an individual GENCO is maximized; while in the lower level, the market clearing process of the independent system operator (ISO) is modeled to minimize the production cost. An intelligent search based on genetic algorithm and Monte Carlo simulation (MCS) is applied to obtain the solution. The PJM five-bus system and the IEEE 118-bus system are used for numerical studies. The results show when wind GENCOs play as strategic bidders to set the price, they can make significant profit uplifts as opposed to playing as a price taker, because the profit gain will outweigh the cost to cover wind uncertainty and reliability issues. However, this may result in an increase in total production cost and the profit of other units, which means consumers need to pay more. Thus, it is necessary to update the existing market architecture and structure considering these pros and cons in order to maintain a healthy competitive market.

Keywords Low carbon, Electricity market, Game theory, Generator bidding, Intermittency, Locational marginal pricing

1 Introduction

For almost half century, global warming is always one of many top challenges to human beings all over the world. many efforts have been made to in order to avoid disasters resulting from global warming, such as polar iceberg melting, sea level increasing, coast area recession, environment deterioration and extremely abnormal climate etc. One important milestone is the conclusion of "Kyoto Protocol" under collaborative efforts of international community in 1997 [1]. Particularly, carbon dioxide is widely believed as one of greenhouse gas and its greenhouse effect was confirmed by lots of experiments. Therefore, it is natural to advocate massive application of low-carbon energy in electricity sector, such as wind, solar, and hydro generation etc., in practice to alleviate carbon emission. Especially wind generation is quickly developing in quantity and still has huge potential to increase its total capacity. Recently, there are several studies about future prospect and possible challenge of developing low-carbon electricity [2–6].

Over the past decades, the old vertically structured power industry throughout the world has been gradually re-designed for market de-regulation [7]. The generation and transmission systems have been split into different entities to introduce competition into the power market. The purpose is to increase investment efficiency and to reduce the cost of power supply [8].

H. HUANG, Customized Energy Solutions, Philadelphia, PA 19102, USA
(✉) e-mail: hhuang@ces-ltd.com
F. LI, University of Tennessee, Knoxville, TN 37996, USA

A variety of market operation models have been proposed and practiced in various countries. Among all different models, the power pool market structure is the most popular one [8–14]. This power pool is managed by a market operator or an independent system operator (ISO) to collect the bids of energy suppliers from generation companies (GENCOs) and the offers of load consumers from load serving entities (LSEs). Then, a market clearing price (MCP) is calculated as the bid price of the most expensive supplier that is needed to completely meet the demand [10]. This market structure is built to encourage suppliers to bid their energy price close to their marginal cost to ensure economic efficiency. Further, to address the transmission constraints, an economic dispatch model is applied to minimize total generation supply cost while satisfying the system reliability and security requirements. Then, the locational marginal pricing (LMP) method [15] is usually employed to calculate the generation profit and load payment and to manage the transmission constraints.

Therefore, to model the dispatch function of the ISO and the individual behavior of the GENCO, it is naturally to split the bidding process into two parts [10–12]. The first part is the ISO market clearing process: the ISO collects all necessary information such as bids and offers from GENCOs and LSEs, and then performs security constrained economic dispatch (SCED) to set the market price. The second step is the self-scheduling of the GENCO for their own payoff optimization such that they can present the best bidding strategy in the forthcoming market.

Modeling and solving bidding strategies problem has been a hot research topic for a long time. In [9], a probability based Monte Carlo (MC) method is proposed to solve competitive generator game with imperfect information, but without transmission constraints. In [10], a mathematical analysis based on a Lagrangian Relaxation is proposed. In [16], a cooperative game is analyzed with potential coalitions and collusions of participants in electricity markets. A prime-dual interior point iteration based on sensitivity was developed to update bidding strategies for GENCOs in [11]. Bidding with transmission constraint was solved in [11–13]. Also in [12], it is shown that the feasibility of Bender Decomposition to solve bidding strategy problems in two parts. In [11], an incomplete information case combined with transmission constraint was carried out. A bidding strategy problem was solved by Monte Carlo simulation and genetic algorithm (GA) in [17]. Intelligent heuristic search such as GA is also a good way to deal with bidding strategy problems in [17, 18]. Further, for a multi-Nash equilibrium of multiplayer games in electricity markets, all Nash equilibria, if exist, could be calculated based on solving polynomial equations in [19]. An analytical approach of transmission-constrained residual demand derivative is used for a power market bidding problem solution in [20].

Strategies for wind power trading were studied in [21]. Two types of bid scenarios are proposed as linear bid and block bid trading for wind power generation, but the model did not consider transmission constraints and competition with other types of generators. In [22, 23], a trading strategy is given for wind power producers to minimize their imbalance cost in short-term, but the transmission constraints as well as competition with other types of generators are not considered. In [24], the uncertainty of wind power generation was modeled in constraints of an optimization problem instead of in the objective function. However, it did not consider the wind power generation as a variable in the objective function of the optimization problem.

The goal of this paper is to develop a bidding strategy model for wind generation participating in the competition with conventional generators. Here, the difference between two types of generators is the high uncertainty of wind generation. Thus, probabilistic approach is taken for the bidding strategy model. Also, the transmission constraints are considered. To solve the overall problem, a bi-level optimization model is formulated where the upper-level subproblem maximizes the payoffs of the GENCOs and the lower-level subproblem solves the market clearing problem of the ISO including economic dispatch and pricing. The Monte Carlo simulation (MCS) method is used to describe the wind generation statistical characteristic, linear programming (LP) is used to solve the lower-level subproblem, and GA is used to solve the upper-level subproblem.

The paper is organized as follows. Section 2 presents the problem formulation including the proposed wind generation bidding strategy model. Section 3 discusses the GA, a simplified Monte Carlo method and their applications to the solution of the proposed problem formulation. Section 4 shows the numerical examples with the PJM 5-bus system and the IEEE 118-bus system. Section 5 concludes the discussion.

2 Problem formulation

In a complete information game, all players know the bidding strategy of other players and their payoff functions. Equilibrium is reached when no player can increase its payoff by unilaterally changing its strategy.

Some assumptions commonly employed in bidding strategy study are listed as follows:

1) Each GENCO has only one generator unit and bids a constant price for a single block for simplicity, while in practice a monotonically increasing multi-block bid model is commonly used.
2) GENCO uses supply function equilibrium (SFE) model.

3) Load is always inelastic and constant for simplicity because load's bids can be essentially modeled as negative generation if needed.
4) Power losses on transmission lines are neglected and the transmission limit is considered in this paper.
5) The LMPs in day-ahead market are the same as those in the real-time market.

2.1 GENCO's bidding strategy model

GENCOs cannot decide the price just by themselves. It is the ISO to clear the market and determine the price. However, GENCOs can affect the price via their bidding strategies. Hence, the whole bidding process is a bi-level optimization problem. The first level is that each GENCO maximizes its own profit, and the second level is a transmission constrained economic dispatch by ISO to minimize total production cost under all security constraints.

Suppose all conventional GENCOs have a convex quadratic production cost function as follows

$$C_i = C(G_i) = a_i G_i^2 + c_i G_i + d_i \tag{1}$$

The marginal cost is calculated as

$$\frac{dC_i}{dG_i} = 2a_i G_i + c_i \tag{2}$$

where a_i is the generation cost coefficients of conventional GENCO i (\$/MWh^2), c_i is the marginal cost of conventional GENCO i (\$/MWh), $C_i = C(G_i)$ is the generation production cost function of conventional GENCO i (\$), d_i is the generation cost coefficients of conventional GENCO i (\$), G_i is the scheduled generation of conventional GENCO i (MWh).

It is a linear function of its scheduled generation G_i. Obviously, GENCOs can make their strategic bids by changing a_i and c_i. For simplicity, in this paper, only c_i will be changed and also let a_i equals to zero based on assumption at the beginning of this section. Therefore, each GENCO will submit generator bids to the ISO according to the following linear supply function for Generator i.

$$f_i = b_i \cdot \frac{\partial C_i}{\partial G_i} = b_i \cdot c_i \tag{3}$$

where b_i is the unknown bidding strategic coefficient variable of conventional GENCO i (it equals to 1 for non-strategic bidders), f_i is the bidding price of conventional GENCO i (\$/MWh).

All of the wind GENCO shares the same description as in (1), (2) and (3). The only difference is that we use the subscript j for wind GENCOs, while other GENCOs use the subscript i.

2.2 Market clearance model

Suppose the ISO uses a transmission constrained economic dispatch to clear the market after collecting all bids and to calculate the market price based on the locational marginal pricing (LMP) model. If the wind power generation output is taken as a deterministic variable, the classic general DCOPF dispatch model is given as follows:

$$\min \sum_{i=1}^{H} b_i \cdot c_i \cdot G_i + \sum_{j=H+1}^{T} b_j \cdot c_j \cdot G_j \tag{4}$$

subject to

$$\sum_{l=1}^{n} G_l = \sum_{l=1}^{n} D_l \tag{5}$$

$$G_{i\min} \le G_i \le G_{i\max}, G_{j\min} \le G_j \le G_{j\max} \tag{6}$$

$$\sum_{l=1}^{n} F_{k\_l} \cdot (G_l - D_l) \le L_k \quad \text{for } k = 1, 2, \cdots, m \tag{7}$$

where D_l is the load demand at bus l (MWh), $G_{i\min}$ and $G_{i\max}$ are the minimum and maximum generation output of conventional GENCO i (MWh) respectively, G_l is the generation at bus l (MWh), $F_{k\text{-}l}$ the generation shift factor to line k from bus, L_k is the transmission limit of line k.

The control variables are b_i, b_j, G_i and G_j. The GENCO production cost is minimized in (4). Constraint (5) ensures the balance of supply and demand. Constraint (6) represents the generation capacity limit. Constraint (7) represents the transmission line constraints.

After the economic dispatch is solved, LMP at each bus l (i.e. M_l) can be calculated as follows [15]:

$$M_l = \lambda + \left(\sum_{k=1}^{m} \mu_k \cdot F_{k\_l} \right) \tag{8}$$

where -λ is the Lagrange multiplier of (5), and -μ_k the Lagrangian multiplier of (7). Note we take -λ and -μ_k as the Lagrangian multipliers such that we have positive signs when calculating LMP as shown in (8).

Once the energy market is cleared by ISO, each GENCO i will be paid according to its LMP and its dispatched generation. The payoff function for the conventional GENCO i and the wind GENCO j is given by

$$Y_i = M_i \cdot G_i - c_i \cdot G_i \tag{9}$$

$$Y_j = M_j \cdot G_j - c_j \cdot G_j \tag{10}$$

where Y_i, Y_j is the profit (\$) of conventional GENCO i and wind GENCO j respectively.

2.3 Probabilistic model of wind generation output

Wind generation output at a specific time spot is usually uncertain and cannot be described as a deterministic variable, so it is broadly accepted to use a random variable, subject to a statistical distribution, to represent it [25]. However, it is difficult to determine the distribution type due to insufficient historical data. Since wind speed forecast error is usually considered normally distributed and the wind speed and wind generation output can be considered linearly correlation in a small region, the wind generation output is assumed to roughly follow normal distribution from the viewpoint of the day-ahead operation [26]. Thus, the wind generation output distribution is given by

$$G_j(t) \sim N\left(\mu_j(t), \sigma_j(t)^2\right) \tag{11}$$

$$\varphi(x) = \frac{1}{\sigma_j\sqrt{2\pi}}\mathrm{e}^{-\frac{(x-\mu_j)^2}{2\sigma_j^2}} \tag{12}$$

$$\Phi(x) = \int_{-\infty}^{x} \varphi(u)\mathrm{d}u \tag{13}$$

where $\mu_j(t)$ is the mean value of $G_j(t)$, $\sigma_j(t)$ is the variance of $G_j(t)$, $\varphi(x)$ is the probability density function (PDF) of $G_j(t)$, $\Phi(x)$ is the cumulative density function (CDF) of $G_j(t)$.

2.4 Wind generation bidding schemes

From the perspective of the ISO, the increasing penetration of renewable energy such as wind and solar generation presents great challenges because of its intermittency and uncertainty. This makes it harder than conventional generation to be controlled in practice. For example, in [20], the wind generation is considered undispatchable and sampled in different scenarios. For each scenario, it is taken as a deterministic negative load in the power balance constraint rather than in the objective function of the economic dispatch program. The ISO runs security constrained economic dispatch to find the output of conventional GENCOs. Finally, the expectation will be taken to combine the results of all scenarios together. As a matter of fact, it suffices to consider wind generation as a zero production cost source. It means that the wind generation will always be dispatched first because it often has the lowest production cost in reality. And this also matches with current practical dispatch policy, i.e., to dispatch renewable energy in priority to meet the percentage of wind penetration in the renewable portfolio standards (RPS). This is probably a legitimate model when the wind generation penetration level in the grid is low and insignificant.

However, with an increasing penetration of wind generation integrated into the grid, the above simple treatment of wind generation as a negative load is likely neither feasible nor reasonable. Also, this treatment tends to discourage wind power suppliers producing more wind power or making more profits. Although the advantage of this dispatch scheme is its easy implementation in practice, this scheme also excludes wind generation as a bidder in electric power market. Therefore, in this paper, two schemes are modeled to consider wind GENCOs as constraints (always dispatched first and being price takers) and as strategic bidders, respectively.

2.4.1 Scheme I: wind generation as constraint in dispatch

Suppose the mean and variance values can be assessed beforehand. The ISO may use its mean value and its bidding price to carry out economic dispatch. When wind generation is considered into this bidding model, it should have some important adjustment. First, since wind source is intermittent, it is hard to use only one deterministic scenario to represent its performance. We have to consider its probabilistic characteristics, i.e., its expectation and variance.

In this paper, a Monte Carlo simulation model, elaborated in the next section, is employed to model the randomness. Suppose we take S samples for wind generation output $G_j(t)$, and each sampled scenario s has a corresponding probability P_s and a corresponding wind generation output G_{j_s} for wind GENCO j. Note that $\sum_s P_s = 1$ for probability and $\mu_j(t) = E\left(G_j(t)\right) = G_j$ based on the proposed assumption.

Based on the previous discussion, for each Monte Carlo scenario s, the economic dispatch scheme from ISO's perspective can be described mathematically as follows:

$$\min \sum_{i=1}^{H} b_{i\_s} \cdot c_i \cdot G_{i\_s} \tag{14}$$

subject to

$$\sum_{l=1}^{n} G_l = \sum_{l=1}^{n} D_l \tag{15}$$

$$G_{i\min} \leq G_{i\_s} \leq G_{i\max} \tag{16}$$

$$\sum_{l=1}^{n} F_{k\_l} \cdot (G_l - D_l) \leq L_k \quad \text{for } k = 1, 2, \cdots, m \text{ and all } s \tag{17}$$

where b_{i_s} is the unknown bidding strategic coefficient variable of conventional GENCO i in scenario s (it equals to 1 for non-strategic bidders), G_{i_s} the scheduled generation of conventional GENCO i in scenario s (MWh).

The control variables are b_{i_s} and G_{i_s}. The difference between (4) and (14) is that wind generation variables are removed in (14). In fact, wind power generation cost could be viewed as zero cost in this case. In addition, the wind generation capacity constraint is removed in (16), while (15) and (17) remains the same as (5) and (7). Also, $\sum_{l=1}^{n} D_l$ in (15) and (17) is the total load subtracts the total wind generation offset. After this transmission constrained economic dispatch process, the LMP calculation still follows (8) and the payoff function for conventional GENCO i and wind GENCO j is the same as (9) and (10), respectively. For each scenario s, the profit of conventional GENCO i is as follows:

$$Y_{i\_s} = M_{i\_s} \cdot G_{i\_s} - c_i \cdot G_{i\_s} \tag{18}$$

Since the wind GENCO j is a price-taker in this case, its profit function at scenario s is calculated as follows:

$$Y_{j\_s} = M_{j\_s} \cdot G_{j\_s} - c_j \cdot G_{j\_s} \tag{19}$$

where Y_{i_s}, Y_{j_s} is the profit ($) of conventional GENCO i or wind GENCO j in scenario s respectively.

Therefore, the whole bidding process can be rewritten as a bi-level optimization problem as follows:

$$\max_{i,s} Y_{i\_s} = \max_{i,s}(M_{i\_s} \cdot G_{i\_s} - c_i \cdot G_{i\_s}) \tag{20}$$

subject to

$$b_{i\min} \leq b_{i\_s} \leq b_{i\max} \tag{21}$$

$$\min \sum_{i=1}^{H} b_{i\_s} \cdot c_i \cdot G_{i\_s} \tag{22}$$

subject to

$$\sum_{l=1}^{n} G_l = \sum_{l=1}^{n} D_l \tag{23}$$

$$G_{i\min} \leq G_{i\_s} \leq G_{i\max} \tag{24}$$

$$\sum_{l=1}^{n} F_{k-l} \cdot (G_l - D_l) \leq L_k \quad \text{for } k = 1, 2, \cdots, m \text{ and all } s \tag{25}$$

The control variables are b_{i_s} and G_{i_s}. The objective function for a strategic bidder i at scenario s is given by (20). The first constraint (21) is to set a limitation with b_{i_s} selection to be realistic; otherwise, the bidder can have infinite market power in theory.

Thus, the total profit expectation of conventional GENCO i for all scenarios is calculated as follows:

$$Y_i = E[Y_{i\_s}] = \sum_{s} P_s \cdot Y_{i\_s} \tag{26}$$

And the total profit expectation of wind GENCO j for all scenarios is calculated as follows:

$$Y_j = E[Y_{j\_s}] = \sum_{s} P_s \cdot Y_{j\_s} \tag{27}$$

2.4.2 Scheme II: wind generation as strategic bidder

In this scheme, the randomness of wind power is also modeled via Monte Carlo simulation. This is the same as in Scheme I.

The difference is that wind GENCOs are taken as strategic bidders in this scheme. Since wind generation is not a constant power source, its payoff function needs to be modified for each sampled scenario s as follows:

a) $G_j > G_{j\_s}$

$$Y_{j\_s} = M_{j\_s} \cdot G_{j\_s} - c_j \cdot G_{j\_s} + M_{j\_s}(G_{j\_s} - G_j) \tag{28}$$

b) $G_j < G_{j\_s}$

$$Y_{j\_s} = M_{j\_s} \cdot G_j - c_j \cdot G_{j\_s} + M_{j\_s}(G_{j\_s} - G_j) = M_{j\_s} \cdot G_{j\_s} - c_j \cdot G_{j\_s} \tag{29}$$

Next, the objective function in (28) and (29) is explained. After sampling, it is a deterministic process for each scenario. At the end of the market clearing process of the ISO, all LMPs and generation dispatches will be settled. The wind GENCO will get its revenue as shown by the first item on the left hand side of (28) and (29). The second item in (28) and (29) is its production cost.

The third item in (28) is the obligation penalty cost if it cannot meet the dispatch requirement in day-ahead market subject to its output uncertainty, because it has to purchase the gap amount of power from the real-time spot market. If it has more generation than required in day-ahead market as in (29), it is assumed to earn extra profits from selling it to the real-time spot market with the day-ahead price. This approach represents the penalty or extra profit due to insufficient or extra output in real-time. Since the goal of this paper is to compare the two schemes, as long as they are based on the same assumption (no price difference between day-ahead and real-time), the comparison is fair.

Therefore, the whole bidding process can be rewritten as a bi-level optimization problem shown below:

if it is a conventional GENCO

$$\max_{i} Y_i = \max_{i,s}(M_{i\_s} \cdot G_{i\_s} - c_i \cdot G_{i\_s}) \tag{30}$$

or if it is a wind GENCO with $G_j > G_{j\_s}$

$$\max_{j} Y_j = \max_{j,s}\begin{pmatrix} M_{j\_s} \cdot G_{j\_s} - c_j \cdot G_{j\_s} \\ + M_{j\_s}(G_{j\_s} - G_j) \end{pmatrix} \tag{31}$$

or if it is a wind GENCO with $G_j < G_{j\_s}$

$$\max_j Y_j = \max_{j,s}\begin{pmatrix} M_{j\_s}\cdot G_{j\_s} \\ -c_j\cdot G_{j\_s}\end{pmatrix} \tag{32}$$

subject to

$$b_{i\min}\le b_{i\_s}\le b_{i\max},\, b_{j\min}\le b_{j\_s}\le b_{j\max} \tag{33}$$

$$\min\sum_{i=1}^{H} b_{i\_s}\cdot c_i\cdot G_{i\_s}+\sum_{j=H+1}^{T} b_{j\_s}\cdot c_j\cdot G_{j\_s} \tag{34}$$

subject to

$$\sum_{l=1}^{n} G_l=\sum_{l=1}^{n} D_l \tag{35}$$

$$G_{i\min}\le G_{i\_s}\le G_{i\max},\, G_{j\min}\le G_{j\_s}\le G_{j\max} \tag{36}$$

$$\sum_{l=1}^{n} F_{k\_l}\cdot(G_l-D_l)\le L_k \quad \text{for } k=1,2,\cdots,m \text{ and all } s \tag{37}$$

The control variables are b_{i_s}, b_{j_s}, G_{i_s} and G_{j_s}. If it is a conventional GENCO, the upper level objective function is (30), while if it is a wind GENCO, the upper level objective function should be replaced as (31) and (32) instead. Again, (33) is to set a limitation with the b_{i_s} and b_{j_s} selections to avoid the bidder to have infinite market power in theory. In the lower level optimization, LMP calculation still follows (8) and the expected payoff function for conventional GENCO i and wind GENCO j follows (26) and (27) respectively.

Therefore, its final profit expectation of wind GENCO j for all wind generation output scenarios is considered as

$$\begin{aligned} &a)\; G_j > G_{j\_s}\\ &Y_j = E\left[Y_{j\_s}\right]\\ &\quad=\sum_s P_s\cdot\begin{bmatrix} M_{j\_s}\cdot G_{j\_s}-c_j\cdot G_{j\_s}\\ +M_{j\_s}(G_{j\_s}-G_j)\end{bmatrix}\end{aligned} \tag{38}$$

$$\begin{aligned} &b)\; G_j < G_{j\_s}\\ &Y_j = E\left[Y_{j\_s}\right]\\ &\quad=\sum_s P_s\cdot\left[M_{j\_s}\cdot G_{j\_s}-c_j\cdot G_{j\_s}\right]\end{aligned} \tag{39}$$

3 Monte Carlo simulation and genetic algorithm

3.1 Monte Carlo simulation

The Monte Carlo simulation, applicable to both Scheme I and II, is implemented as follows:

1) The PDFs of wind GENCOs are obtained as the inputs of the Monte Carlo simulation.
2) Take s repeated random samplings for the PDF of each wind GENCO j to obtain G_{j_s} and P_s for each Monte Carlo sampled scenario s.
3) For each scenario s, perform a deterministic optimization based on GA and then LP to calculate all bidding strategies, all GENCO costs, dispatches, all LMPs, and the profits for all GENCOs (conventional and wind).
4) Aggregate the results to get the profit expectation of all GENCOs.

The Monte Carlo simulation will stop when a predefined convergence threshold ε has been reached. The stopping criterion is shown mathematically as follows [27]:

$$\frac{\sigma[E(X)]}{E(X)}=\frac{\sigma(X)}{\sqrt{s}\cdot E(X)}\le\varepsilon \tag{40}$$

where X is the random variable representing the wind generation profit, $E(X)$ the mean value of X, and $\sigma(X)$ the standard deviation of X.

3.2 GA

Essentially, this proposed optimization problem is to find the global Nash equilibrium to each bidder. The formation of the proposed solution is a bi-level optimization, which is depicted in Fig 1. There are various approaches to solve this non-linear, non-convex, bi-level optimization. Here the GA is used in this paper. Note that the term 'biological generation' instead of the commonly used term 'generation' in the GA algorithm discussion, is used to avoid possible confusion with the electrical generation.

Generally, the algorithm terminates when either a maximum number of generations is reached, or a satisfactory fitness level has been reached for the population. If the algorithm is terminated due to a maximum number of

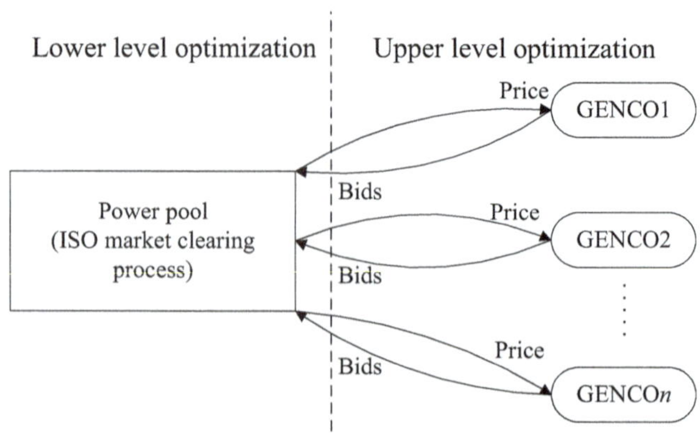

Fig. 1 Framework of the proposed bi-level GA optimization process

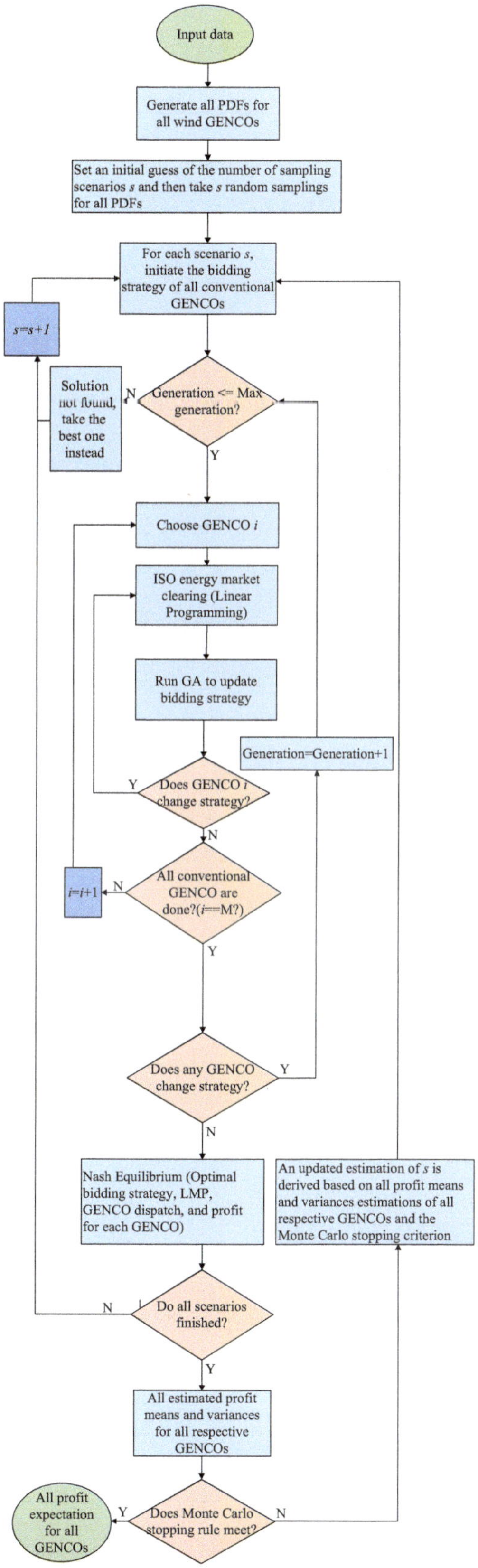

Fig. 2 Flowchart of the proposed GA and Monte Carlo simulation for wind generation bidding scheme I

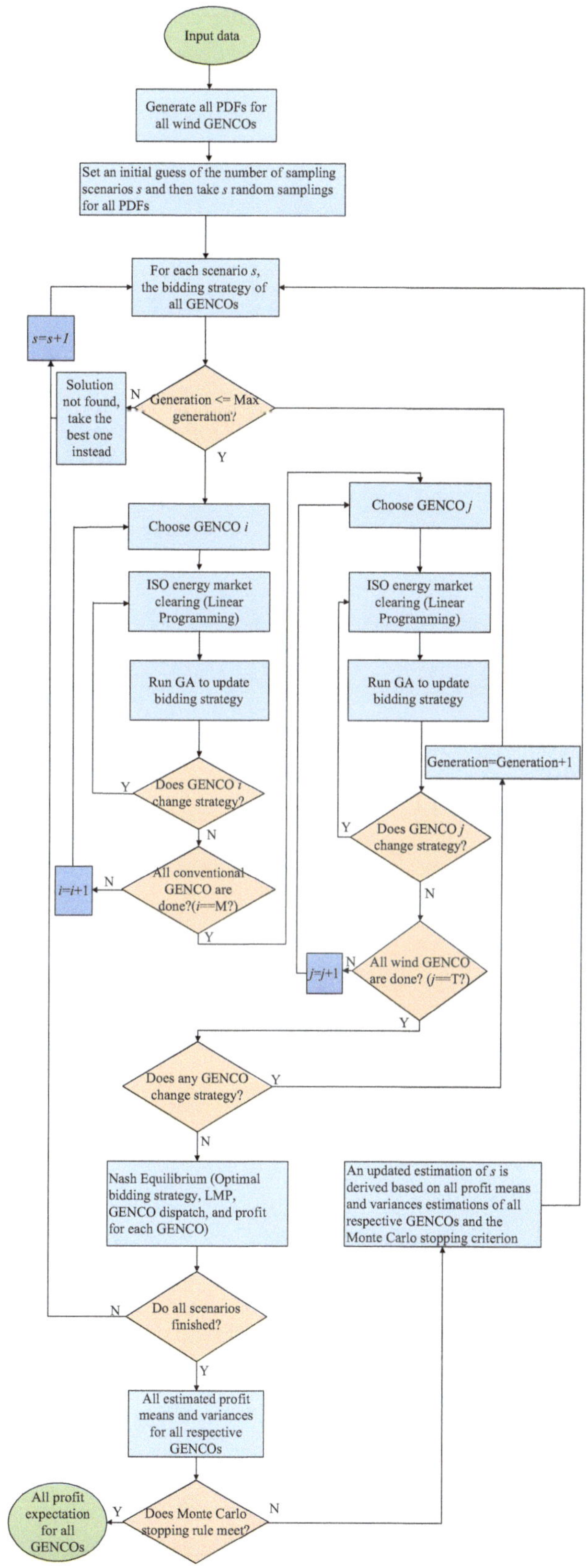

Fig. 3 Flowchart of the proposed GA and Monte Carlo simulation for wind generation bidding scheme II

generations, a satisfactory solution may or may not have been reached.

By combining the results of all scenarios based on Monte Carlo simulation and GA, the expected results can be derived. The flowcharts of Scheme I and II are shown in Figs. 2–3.

4 Numerical example analysis

The stopping criterion ε of Monte Carlo simulation is set to 0.01 for all cases below [27]. In addition, for the parameters associated with GA applied to the cases below, we set the total biological generation to 100, the population size to 50, crossover rate to 0.5, mutation rate to 0.01, and eight bits for bidding strategy coefficients (b_i and b_j). Also, the GA stopping criterion is that the relative difference between the previous profit and the current profit of each GENCO is less than 1% of the current profit for all scenarios in each case.

4.1 PJM five bus system

This modified PJM five-bus system is shown in Fig 4 [15]. The transmission line profiles are shown in Table 1. In the modified system, its total load level is 900 MW. Alta is a wind generator which observes normal distribution

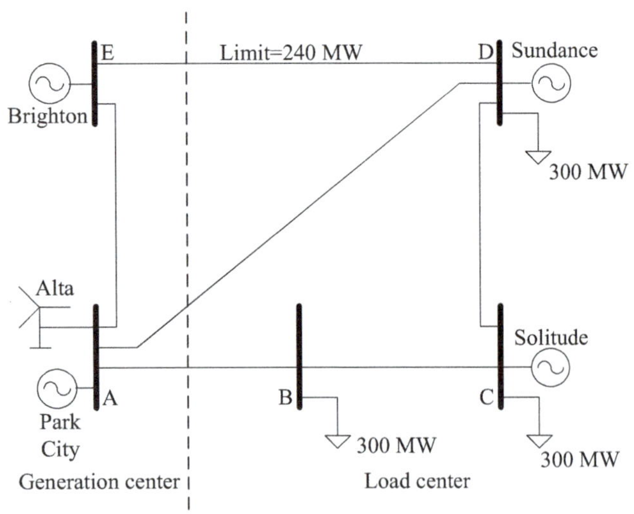

Fig. 4 Modified PJM five-bus system

Table 1 Line impedance and flow limit

Line	AB	AD	AE	BC	CD	DE
X (%)	2.81	3.04	0.64	1.08	2.97	2.97
Limit (MW)	999	999	999	999	999	240

Table 2 Generator data

Generator	Alta	Park City	Solitude	Sundance	Brighton
Type	Wind	Hydro	Gas	Gas	Steam
P_{min} (MW)	50	0	0	0	10
P_{max} (MW)	150	100	520	300	600
Marginal cost ($/MWh)	7	15	30	35	10

with a mean output 100 MW and standard variance 16.67 MW. Considering the practical limit of wind output is three times of standard variance, the wind generation output will be within the interval [50 MW, 150 MW] with 99.7% confidence based on normal distribution. All other generators are conventional units, with unit minimum and maximum generation and cost listed in Table 2 according to the type of each generator.

4.1.1 Scheme I: wind generation as a constraint

Let Park City and Sundance be the two bidders involved within this case. The two bidding strategy coefficients for Park City and Sundance are constrained to be in interval [1, 3] and [1, 1.5], respectively. They may bid up to 45 $/MWh and 52.5 $/MWh, respectively. Therefore, they have a wider range to set the price.

Wind generation is considered to be a negative load in this scheme. 1000 sampling scenarios are taken in this case. The profit, generation and price expectation for each generator are shown in Table 3.

The GA convergence rate (i.e., ratio between number of scenarios converged to a Nash equilibrium and total scenarios) is 1. It means that the probability of reaching a Nash equilibrium solution under current system conditions is 100%. The average total generation cost of this scheme is $12959.

4.1.2 Scheme II: wind generation as a bidder

All the assumptions and parameters are the same as in the previous case in Section 4.1.1. The only difference is

Table 3 Profit generation and price expectation for each generator

Generator	Alta	Park City	Solitude	Sundance	Brighton
Expected profit ($)	1129	0	0	0	1664
Expected output (MW)	100	0	212.82	0	587.19
Expected price ($/MW)	18.78	18.78	30	38.56	12.77

Table 4 Profit and price expectation for each generator

Generator	Alta	Park City	Solitude	Sundance	Brighton
Expected profit ($)	1893.8	0	0	0	9593.3
Expected output (MW)	69.84	0	238.28	0	591.77
Expected price ($/MW)	27.39	27.39	30	31.99	25.99

that Alta is also a price bidder of the whole bidding process at this time with bidding strategic coefficient constrained in interval [1, 5] such that the wind unit's bid can be up to 35 $/MWh, which is in a comparable range of the bids of the other strategic units. Thus, there are three bidders in this case. The profit expectation is shown in Table 4.

The GA convergence rate in this case is 0.99, i.e. the probability of reaching a Nash equilibrium solution is 99% under the current system conditions. The average total generation cost of this system is $13773, which is about 6.3% higher than the previous case in Section 4.1.1.

4.1.3 *Analysis of results with sensitivity study*

If we compare the results of the above two cases in Tables 3 and 4, we can conclude that the profit of wind generator Alta in the Scheme II is more than in the Scheme I case even with less expected generation because the profit gains weight much more than the possible losses when under-production occurs. This implies that allowing wind unit to bid may financially help them cover their own uncertainty and reliability issues. The profit of generator Solitude is 0, which means the LMP on this bus is always the same as its marginal cost in both cases. In addition, generator Park City and Sundance also earn no profit due to zero production.

However, the generator Brighton is the biggest winner in Scheme II, because the LMP at its bus doubles and its expected output stays the same. Also, the total generation cost goes up by 6.3% which is also significant. Thus, there are pros and cons for allowing wind generation to participate in bidding. This implies the need of an update in the power market architecture and structure for adapting high penetration of wind generation.

Based on different total system load levels, the sensitivity study is performed. The comparison of the profit of wind generator Alta and the total cost of the system are shown in the following figures.

As shown in Figs. 5–6, the load level increases in a 50 MW step. At different load levels, the wind generation bidder consistently earns in Scheme II more than in Scheme I, while the total system cost is in Scheme II higher than in Scheme I as well. At the studied four load levels, the GA always has a high convergence rate (i.e., ratio between the number of scenarios converged to a Nash equilibrium and the total number of scenarios) more than 99.5%, which guarantees the validity of the results.

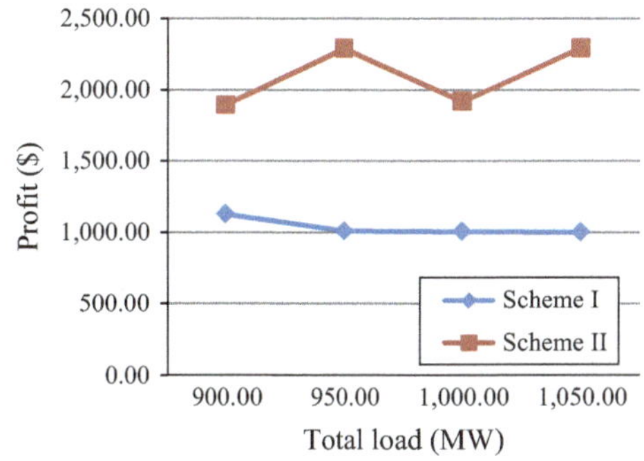

Fig. 5 Comparison of the profit of wind generator Alta

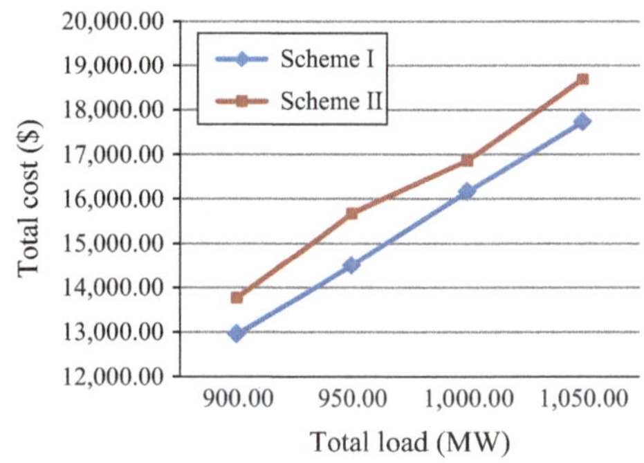

Fig. 6 Comparison of the total cost of the system

When the total system load level goes beyond 1100 MW or even more, the convergence rate of GA is lower than 50%. This means the total generation capacity is not sufficient. Thus, the sensitivity analysis is stopped at 1050 MW load level.

4.2 IEEE 118-bus system

There are 186 branches, 91 loads, and 54 generators in the IEEE 118-bus test system. All the detailed information can be found in [28]. The original IEEE 118-bus system data does not contain the information of generator marginal costs and branch thermal limits. Therefore, generator marginal costs are constructed in this paper as follows: 2 wind generators with marginal cost $8, 20 cheap generators with marginal cost from $10 to $19.5 with $0.5 increment; 18 expensive generators with marginal cost from $20 to $23 and $26 to $39 with $1 increment; and 14 extremely expensive generators with marginal cost from $40 to $53 with $1 increment. In addition, five thermal limits are introduced into the transmission system: 345 MW for lines

69–77, 630 MW for lines 68–81, 106 MW for lines 83–85 and 94–100, 230 MW for lines 80–98 [26]. Also, the maximum total generation is more than twice of the total load. In order to show the efficiency of wind generation, each load is scaled up at 1.8 times of its original value in the whole system.

Suppose two wind GENCOs are located at buses 59 and 61, respectively. Their generation output mean values are 155 MW and 160 MW, respectively. Also, they have the same standard variance at 33.33 MW to make possible wind generation output in the interval [55 MW, 255 MW] and [60 MW, 260 MW], respectively, with 99.7% confidence based on the normal distribution property. In addition, the two wind generations are independent random variables each other. All other generators are conventional.

4.2.1 Scheme I: wind generation as a constraint

Let generators at buses 65, 66 and 69 be the strategic players involved in this case. The three bidding strategy coefficients are constrained to be in interval [1, 2] such that the involved units may bid up to $56/MWh which gives a sufficiently wide range for the simulation.

Wind generation is considered to be a negative load in this scheme with 1000 sampling scenarios. The profit expectation is shown in Table 5. (Here only the results of the strategic bidders and wind generator owners are listed).

The GA convergence rate (i.e., ratio between the number of scenarios converged to a Nash equilibrium and the total number of scenarios) is 1, which means that the probability of finding a Nash equilibrium solution is 100% under the current system conditions. The average generation cost is $176430.

Table 5 Profit expectation for each generator

Generator at bus	59	61	65	66	69
Expected profit ($)	4577.4	4727.9	5354.8	4839.9	6981.7
Expected output (MW)	153.74	158.66	490.9	491	794.75
Expected price ($/MW)	37.85	37.87	37.9	37.86	37.84

Table 6 Profit expectation for each generator

Generator at bus	59	61	65	66	69
Expected profit ($)	4838.5	4952.5	5826.4	5328.9	7876.7
Expected output (MW)	108.04	115.36	491	491.51	801.91
Expected price ($/MW)	38.84	38.85	38.87	38.84	38.83

4.2.2 Scheme II: wind generation as a bidder

All the assumptions and parameters are the same as in the previous case in 4.2.1. The only difference is that the two wind units at buses 59 and 61 are market players in the entire bidding process with the bidding factor in the range of interval [1, 6]. Thus, we will have five bidders in this case. Similar to the previous case in 4.2.1, 1000 sampling scenarios are taken. The profit expectations of five strategic bidders are shown in Table 6.

The GA convergence rate is 99.5%, which means the probability of having a Nash equilibrium solution is 99.5%. The average total generation cost is $178080, which is about 1% increase from Scheme I. Since the five strategic bidders represent a small portion of the total units, this 1% increase is considerable.

4.2.3 Analysis of results

If we compare the results of these two cases in Table 5 and Table 6, the observation shall be very similar to the one from the previous PJM 5-bus system study. Both wind generation bidders may have tremendous profit uplift as a marginal unit even though the probabilistic uncertainty is considered, because the gains from wind strategic bidding outweigh the cost of purchasing power due to insufficient wind production. From this perspective, the renewable generation will be encouraged to aggressively play in the power market to gain more profits.

However, also similar to the PJM 5-bus case study, the above benefit is at the cost of increased total production cost and more profit of other conventional generation bidders. Thus, the consumers will pay more.

Therefore, this implies the power market architecture and structure should be updated to better accommodate high penetration of wind power. Also, the high GA convergence rate guarantees the validity of the results.

5 Conclusions

The contribution of this paper can be summarized as follows:

1) Two bidding strategy schemes are modeled in this paper to consider wind GENCOs, conventional GENCOs, and transmission constraints, while few literatures has studied the impact of wind GENCOs to bidding strategy. The first scheme considers wind power as negative loads, which is aligned with the ongoing practice that wind generation must be dispatched with higher priority. The second scheme regards wind GENCOs as possible strategic bidders,

which is aligned with the common expectation that wind power owners may participate in market competition in the future.

2) In each scheme, a comprehensive bidding strategy model is proposed in a probabilistic approach using Monte Carlo simulation. In each Monte Carlo sample, a bi-level optimization model is employed with different objective functions for wind GENCOs. The GA is employed as the solution method.
3) Simulation results show that, when wind GENCOs play as strategic bidders to set the price, they can make significant profit gains as opposed to playing as a price taker. Note this result considers the probabilistic variability of wind generation output. However, this is at the cost of increased production cost and the profit of other generators, which means the consumers will pay more. Thus, we can draw an important conclusion that when there is a high-penetration of wind generation, it is very necessary to update the existing market architecture in terms of competitive electricity market for high-penetration renewables.

The future work may include the bidding strategy model with ancillary service models, detailed penalty model and the effect of GENCO's collusion and coalition in the power electricity market.

Acknowledgments The authors would like to acknowledge the support from the Shared Facilities and the Industry Partnership Program by CURENT, an Engineering Research Center (ERC) Program of the US NSF and US DOE under the NSF Grant (No. EEC-1041877).

References

[1] United Nations (1997) Kyoto protocol to the United Nations framework convention on climate change. New York

[2] Kang CQ, Chen QX, Xia Q (2009) Prospects of low-carbon electricity. Power Syst Technol 33(2):1–7 (in Chinese)

[3] Yuan J, Hu ZG (2011) Low carbon electricity development in China–An IRSP perspective based on super smart grid. Renew Sustain Energy Rev 15(6):2707–2713

[4] Kahrl F, Williams J, Ding JH et al (2011) Challenges to China's transition to a low carbon electricity system. Energy Policy 39(7):4032–4041

[5] Yao L (2013) Guest editorial: special issue on wind power-technologies. J Mod Power Syst Clean Energy 1(1):1

[6] Yuan X (2013) Overview of problems in large scale wind integrations. J Mod Power Syst Clean Energy 1(1):22–25

[7] Yang J, Li FX, Freeman L (2003) A market simulation program for the standard market design and generation/transmission planning. In: Proceedings of the IEEE PES general meeting, vol 1, Toronto, 13–17 Jul 2003, pp 442–446

[8] Wen FS, David AK (2002) Optimally co-ordinated bidding strategies in energy and ancillary service markets. IEEE Proc Gener Transm Distrib 149(3):331–338

[9] Wen FS, David AK (2001) Optimal bidding strategies and modeling of imperfect information among competitive generators. IEEE Trans Power Syst 16(1):15–21

[10] Zhang DY, Wang YJ, Luh PB (2000) Optimization based bidding strategies in the deregulated market. IEEE Trans Power Syst 15(3):981–986

[11] Li T, Shahidehpour M, Li Z (2005) Bidding strategy for FTR obligations in transmission markets. IEE Proc Gener Transm. Distrib 152(3):422–428

[12] Yamin H, Al-Agtash S, Shahidehpour M (2004) Security-constrained optimal generation scheduling for GENCOs. IEEE Trans Power Syst 19(3):1365–1372

[13] Peng TS, Tomsovic K (2003) Congestion influence on bidding strategies in an electricity market. IEEE Trans Power Syst 18(3):1054–1061

[14] Bompard E, Ma YC, Napoli R et al (2010) Comparative analysis of game theory models for assessing the performances of network constrained electricity markets. IET Gener Transm Distrib 4(3):386–399

[15] Li FX, Bo R (2007) DCOPF-based LMP simulation: Algorithm, comparison with ACOPF, and sensitivity. IEEE Trans Power Syst 22(4):1475–1485

[16] He Y, Song YH (2003) The study of the impacts of potential coalitions on bidding strategies of GENCOs. IEEE Trans Power Syst 18(3):1086–1093

[17] Gountis VP, Bakirtzis AG (2004) Bidding strategies for electricity producers in a competitive electricity marketplace. IEEE Trans Power Syst 19(1):356–365

[18] Richter CW Jr, Sheble GB (2000) A profit-based unit commitment GA for the competitive environment. IEEE Trans Power Syst 15(2):715–721

[19] Yang Y, Zhang Y, Li FX et al (2012) Computing all Nash equilibria of multiplayer games in electricity markets by solving polynomial equations. IEEE Trans Power Syst 27(1): 81–91

[20] Xu L, Baldick R (2007) Transmission-constrained residual demand derivative in electricity markets. IEEE Trans Power Syst 22(4):1563–1573

[21] Singh SN, Erlich I (2008) Strategies for wind power trading in competitive electricity markets. IEEE Trans Energ Conver 23(1):249–256

[22] Matevosyan J, Soder L (2006) Minimization of imbalance cost trading wind power on the short-term power market. IEEE Trans Power Syst 21(3):1396–1404

[23] Morales JM, Conejo AJ, Perez-Ruiz J (2010) Short-term trading for a wind power producer. IEEE Trans Power Syst 25(1): 554–564

[24] Liu X, Xu W (2010) Economic load dispatch constrained by wind power availability: a here-and-now approach. IEEE Trans Sustain Energy 1(1):2–9

[25] Chao H, Li F, Trinh LH, et al (2004) Market based transmission planning considering reliability and economic performance. In: Proceeding of the 8th IEEE international conference on probabilistic methods applied to power systems (PMAPS'04), Ames, 12–16 Sept 2004, pp 557–562

[26] Bo R, Li FX (2009) Probabilistic LMP forecasting considering load uncertainty. IEEE Trans Power Syst 24(3):1279–1289

[27] Billinton R, Gan L (1993) Monte Carlo simulation model for multiarea generation system reliability studies. IEEE Proc Gener Transm Distrib 140(6):532–538

[28] Electrical Engineering Department, University of Washington, Power system test case archive (2007) Seattle.

Hao HUANG is now a power system engineer at Customized Energy Solutions in Philadelphia, PA. He was previously with EnerNex Corporation, Knoxville, TN. He received the BSEE and MSEE degrees from Zhejiang University in 2003 and 2006, and then the PhD degree from the University of Tennessee in 2014. His major interests include power market, controllable load, and generation control.

Fangxing LI (M'01, SM'05), also known as Fran Li, is presently an Associate Professor at The University of Tennessee at Knoxville. He received his BSEE and MSEE degrees from Southeast University in 1994 and 1997, respectively, and then his PhD degree from Virginia Tech in 2001. From 2001 to 2005 he worked at ABB Electric Systems Consulting (ESC) prior to joining UTK. His current interests include renewable energy integration, demand response, power markets, distributed energy resources, and smart grid. He is a registered Professional Engineer in North Carolina, an Editor of IEEE Transactions on Renewable Energy, and a Fellow of IET.

17

Steady-state security in distribution networks with large wind farms

Dilan JAYAWEERA (✉), Syed ISLAM

Abstract Aging network assets, forced and unforced outages, and the way the networks are operated in a deregulated market are of significant concerns to integrate large wind farms in a distribution network. In many cases, the constrained network capacity is a potential barrier to the large-scale integration of wind power. This paper probabilistically assesses the steady-state security in a distribution network in the presence of large wind farms. The approach incorporates active distribution network operating conditions, including intermittent power outputs, random outages, demand fluctuations, and dynamic interactions and exchanges, and then assesses the steady state security using Monte Carlo simulation. A case study is performed by integrating large wind farms into a distribution network. The results suggest that intermittent outputs of large wind farms in a distribution network can impact the steady-state security considerably. However, the level of impact of wind farms does not necessarily correlate with the installed capacity of them.

Keywords Monte Carlo simulation, Network integration capacity, Random outages, Security of supply, Wind power generation

1 Introduction

Conventional distribution networks are less stressed compared to modern distribution networks because of centrally generated power is aimed at transmitting through network assets to end-use customers. However, the usability of central generation also depends on the availability of in-service assets of a distribution network. Modern distribution networks are moving towards the active operation where the voltage and power flows are also controlled by distributed generation (DG). Among the DG, the wind and photovoltaic (PV) generated electricity dominates the load share in most cases. Outputs of wind and PV are naturally intermittent and they can potentially impact the security of supply to customers.

Integration of wind and PV requires additional generation supports to mitigate impacts of intermittent effects. This additional generation can be supplied in the form of standby generation. On the other hand, energy storage solutions can also be used to mitigate impacts of intermittent effects. However, the economic competitiveness of large capacities of storage technologies against high penetration of wind power is not necessarily attractive. Intermittent effects can also produce intermittent stresses onto a distribution network, resulting in early retirement of assets.

With the paradigm shifts, the research problems of integrating wind generation into active distribution networks are explored in a wider spectrum. The DG penetration based on the harmonic limit, increase in DG penetration with the energy storage, protection coordination with the DG penetration, optimal allocation of DG with mathematical programming, capacity sufficiency with DG integrations, and reliability levels of active distribution networks are explored in [1–9]. The spinning reserve requirement for a significant penetration of wind power generation is explored in [10].

Power restoration practices of active distribution networks reached a new paradigm, and they are investigated in [11]. Some researchers developed new techniques to model wind power generation characteristics [12]. Security with the intermittent wind power is explored in [13].

D. JAYAWEERA, S. ISLAM, Curtin University, Perth, WA 6102, Australia
(✉) e-mail: dilan.jayaweera@ieee.org

Economic analysis of a DG mixture, long term simulation models of intermittent DG, and the dispersed generation and their correlation factors are investigated in [14–16]. A method is proposed in [17] to assess reliability of utility and industrial power generation systems. In [18], results from a customer survey are presented. Its findings suggest that customers with a backup system pay more per event than those without.

In recent years, the attention of the Plug-in Hybrid electric vehicles (PHEV) is increased as a mobile source of DG. They can also be used as a strategic source to mitigate stresses. In [19], adaptability of electric vehicle aggregation to electricity market is explored in the context of enhancing the security of power supply. In [20], the impact of PHEV behavior on the electric grid is analyzed by taking into account daily driving schedules. In [21], impacts of PHEV charging patterns on the power system operation are explored using stochastic unit commitment models. In [22], an approach is proposed to model PHEV home charging patterns, taking into account the stochastic nature of individual loads. In [23], a distributed framework for demand response and the user adaptation is explored in the context of smart grids, taking into account the pricing of the congestion. In [24], a three-state weather model is proposed for the reliability assessment. In [25], a technique is presented to reduce the errors in the short-term load forecasting. Reference [26] argues that number of transmission line outages is not necessarily proportional to the physical lengths of lines for many transmission systems. In [27], the value of security is quantified with inter-regional transmission lines. Security impacts with the large-scale integration of wind power are explored in [28]. In [29], a multi objective probabilistic risk index is proposed to capture the likelihood and consequences of events. Fuzzy and Monte Carlo simulation based hybrid technique is proposed in [30] for the assessment of power system risk. Multi-objective optimization based algorithm is presented in [31] for the active distribution network planning, taking into account uncertainties in DG and the demand response. Reference [32] argues that the statistical fitting techniques are not necessarily applicable to characterize the observed learning behavior.

The steady state security can be assessed deterministically as well as probabilistically. In a deterministic security assessment, the number of contingencies can be limited to N-1 or N-2 criterion and each component of the distribution network is assumed to have an equal probability of failure or in other words no probabilistic terms is taken into account. In contrast, in a probabilistic assessment, the number of outages is determined by state random sampling.

This paper proposes a probabilistic approach to assess the steady-state security in an active distribution network in the presence of large wind farms. The approach quantifies the impact of intermittent outputs of large wind farms as a function of severity and the duration of impacts. The Monte Carlo simulation is the main engine of the probabilistic approach that incorporates the minimum load-shedding routines at various stages of remedial actions. The active distribution network operating conditions are modeled by incorporating random outages, intermittent power outputs, dynamic interactions and exchanges, and sector customer load demand fluctuations. The paper also extends the approach in the context of capacity credits with large wind farms.

The remaining sections of the paper are organized as follows. Section 2 presents the approach. It also describes the Monte Carlo simulation steps and the minimum load shedding formulations. Section 3 presents a case study and critically analyses the results. Section 4 presents conclusions.

2 The approach

There are many uncertainties in an active distribution network and most of them are stochastic in nature. Due to the level of uncertainties exists in an active distribution network, a probabilistic approach is necessary to capture the rational characteristics of large wind farms that can impact the steady-state security. The approach proposed in this paper incorporates non-sequential Monte Carlo simulation to capture uncertainties of operating conditions of an active distribution network [27].

In a power distribution network, the network assets can experience stresses as a consequence of disturbances. The approach classifies the stresses into two types. The first type is the stress that arises from outages and other events excluding intermittent effects of DG and dynamic interactions. The second type is the stress that can occur due to intermittent effects of DG outputs and the dynamic interactions of stochastic power generating units with conventional units such as diesels.

The severity of disturbances in a network that leads to the first and the second types of stresses can be quantified using different metrics. One of the metric that can be used to quantify the stress is the level of shed-load. This is because if a network experiences a stress that cannot be eliminated without shedding loads, the ultimate decision of a network operator is to initiate the load shedding. The constraint violation in a network can also be considered as an early indication of a system that heads towards a stressed operating condition, however, it is a tentative indication of stress because not all constraint violations lead to load shedding.

The shed load does not completely reflect the system stress because the system stress should also capture the electricity customers' stress, which is the product of the magnitude of the shed-load and the time to restore the shed-load. Thus, the approach considering the restoration time of the shed-load is also as a part of the system stress quantification metric.

In addition, in a power transmission network, the shed-load can be restored using a step-by-step restoration scheme that uses different restoration rates (MW/minute) or a continuous restoration scheme that uses a single restoration rate. Typically, most power transmission network loads are aggregated loads that represent distribution networks. The security standards of power transmission systems recommend the rate of restoration of a shed-load and the time threshold to restore the curtailed load. However, in distribution networks, the shed-load can be restored using a single rate of restoration scheme because of the loads that are to be restored are considerably lower in magnitude compared with power transmission networks. Thus, in a trial in Monte Carlo simulation, a single rate of restoration can be applied to calculate the restoration time of shed loads of a network. However, multiple restoration rates can also be applied as per the requirement of a network.

A network can experience stresses when the intermittency of wind power generation coincides with critical operating conditions. Therefore, the shed-loads resulting through all disturbances are referred in this paper as the total shed-load. The total shed-load can be categorized into four elements of shed-loads. They are the shed-load due to inadequacy of wind power generation (excluding dynamic interactions and exchanges resulting shed-load), dynamic interactions and exchanges, contingencies, and unavailability of standby power generation. A detailed description of the dynamic interactions and exchanges is given in section 2.2. The first two shed-load components are influenced by intermittency of wind power generation. Therefore, the term 'shed-load due to intermittency of wind power' refers to the sum of the first two elements of shed-loads.

The second shed-load type arises as a consequence of interaction of intermittency of wind power with other disturbances. The sign of the second element can be either positive or negative as it depends on the network operating condition and the criticality of the post-disturbance state. Some of the post disturbance operating conditions are less stressed and can also lead to reduce the level of shed-loads.

2.1 Modeling load demand and wind power output

The load demand and intermittency of wind power output can be characterized by using time series of their outputs. In order to reduce the processing time of Monte Carlo simulation, the approach incorporates characteristic

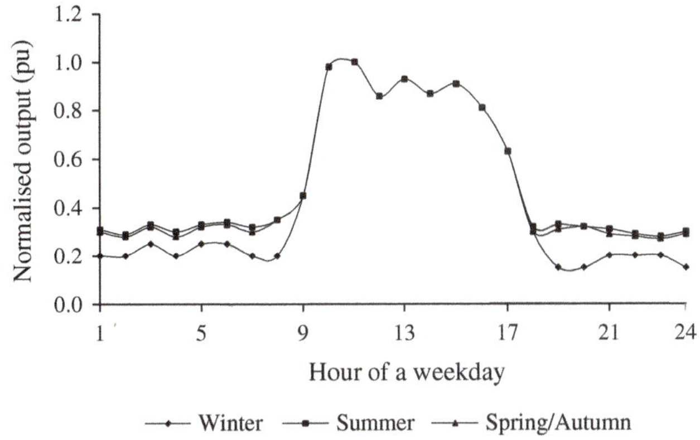

Fig. 1 An example of weekday's load demand characteristic profile of an industrial customer

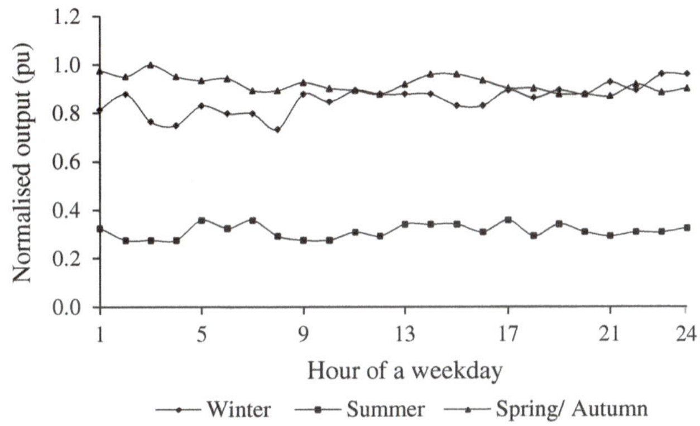

Fig. 2 An example of weekday's generation characteristic profile of an onshore wind farm

profiles of the load demand and the wind-power outputs. Incorporation of characteristic profiles does not compromise the accuracy of the estimation because the zero wind power generation states are captured through random state sampling of wind units (as it does in conventional unit outage modeling) and other generation states are modeled through characteristic profiles. This limits the averaging effects of characteristic profiles. The conventional unit outage modeling process is described in section 2.2.

Characteristic profiles are calculated by truncating yearly profiles into characteristic days. Days in each week are categorized into characteristic Weekday, Saturday, and Sunday. Seasons in each year are categorized into characteristic winter, summer, and spring/autumn. Thus, each season can produce characteristic Weekday, Saturday, and Sunday and in total the process produces nine types of days in a year. In other words, the process produces 216 sample-hours to represent a characteristic profile. Figures 1 and 2 show weekday's characteristic profiles of an industrial type customer and onshore wind farm respectively. Both profiles are derived from yearly profiles.

On the other hand, the entire time series of wind and the sector customer load demand can also be incorporated into Monte Carlo simulation without truncating but reducing the variance of the estimation by applying a variance reduction technique.

2.2 Monte Carlo simulation

Monte Carlo simulation incorporates a healthy network model to begin the simulation. The healthy network model is used as the base case and the model should be free from voltage limit and thermal limit violations. On top of the base-case, random outages are modeled. Random outages are modeled by using random sampling of states where uniformly distributed random numbers are generated between 0 and 1.0 for each network component and then they are compared with the probability of the outage of each component. If the generated random number for the equipment is less than the probability of the outage of the equipment then the equipment status is set as out of service. The probabilities of outage of equipment are calculated using statistical data of component failure rates. Following the convergence of Monte Carlo simulation, the failure rates generated by random numbers are almost equal to the statistical failure rates of equipment. Occurrences of random outages can change the network configuration which leads to re-assess the power balance using A/C power flow solution. Loads are shedded if the operating condition leads to the divergence of the power flow solution [27, 33].

Some of the bus and branch outages may trip neighboring branches. Similar to the components outage modeling, the protective relay actions are modeled by using random sampling of relay operations. This requires the statistics of tripping of neighboring branches. Clearing a fault can initiate dynamic interactions and exchanges at some operating conditions. These interactions mostly occur due to the existence of large wind turbine generators because the rapid responsive generators respond faster than slow units (e.g. wind). The plants that interact dynamically share the network load in an oscillatory pattern due to the constant low frequency power angle oscillations interacting in the system, which can also continue for an extended period of time. Depending on time constants of units and their controls, the resulting operating condition may even destabilize weak units and disconnect them from the network. The instability tripping of units due to dynamic interactions are also modeled by using random sampling of stability of units.

On the other hand, performing the transient stability analysis in parallel with Monte Carlo trials can also extract the units that are to be disconnected due to the dynamic interactions and exchanges. Similar philosophy can be applied for the protection relay operation to determine the tripping of neighboring branches. These options reduce the necessity of random sampling and the outcome can directly be applied to the sample trial of Monte Carlo simulation. However, the parallel processing of the entire transient processes can demand a significantly large processing time. Such issues can be limited by applying a probabilistic approach.

Next, the resulting operating condition is re-assessed to establish the power balance. Any divergence condition is rectified by applying the load shedding routine. The mathematical framework to shed the minimum level of loads is given through (1) to (6).

Minimize

$$\sum_{k\in ng-flex} C_k P_k^{gen-flex} + \sum_{k\in ng-nonflex} C_k P_k^{gen-nonflex} + \sum_{k\in nld} W_k C_k^{shed} P_k^{shed} \tag{1}$$

Subject to:
Power flow through a branch,

$$T_j = \sum_{k=1}^{nb} A_k^j (P_k^{gen-flex} + P_k^{gen-nonflex} - P_k^{load} + P_k^{shed}); \quad (j = 1, \ldots, nbr) \tag{2}$$

Power balance at a bus,

$$\sum_{k\in ng-flex} P_k^{gen-flex} + \sum_{k\in ng-nonflex} P_k^{gen-nonflex} + \sum_{k\in nld} P_k^{shed} = \sum_{k\in nld} P_k^{load} \tag{3}$$

Generation limit constraint,

$$P_k^{gen-flex_{\min}} \le P_k^{gen-flex} \le P_k^{gen-flex_{\max}}; \quad (k \in ng) \tag{4}$$

Load shedding constraint,

$$0 \le P_k^{shed} \le P_k^{load}; \quad (k \in nld) \tag{5}$$

Branch flow constraint,

$$|T_j| \le T_j^{\max}; \quad (j = 1, \ldots, nbr) \tag{6}$$

where $P_k^{gen-flex}$, $P_k^{gen-nonflex}$, P_k^{load} and P_k^{shed} are the flexible generation (dispatchable), non-flexible (including intermittent DG), load, and load shedding at the kth bus respectively. A_k^j is the elements of the connectivity matrix between branch flows and power injections for the trial. T_j is the power flow through the jth branch. $P_k^{gen-flex_{\min}}$ and $P_k^{gen-flex_{\max}}$ are the lower and upper capacity limits of generation at the kth bus respectively. $T_j^{\max}$, W_k, C_k^{shed} and C_k are the thermal limit of the jth branch, weighting factor that reflects the importance of the kth bus for load shedding, cost of the shed load factor at bus k, and unit generation cost for the kth bus generation respectively. *nld*, *ng–flex*, *ng–nonflex*, *nb* and *nbr* are the total number of load buses, flexible generator buses, non-flexible generator, buses, and branches in the network respectively.

Wind plants have two dispatch options: use the entire power output of plants of the trial or curtail the output if that is feasible. Curtailment of output of wind plants is possible if there are a number of wind turbine generators that share a network demand. If there is a single wind turbine generator, then it is disconnected from the network because of the practical limitations of curtailing outputs. However, wind plants can also be considered as limited-flexible plants if the alternative platforms are available to control the outputs of wind plants entering the power grid. These alternative platforms include energy storage solutions, micro grids, and flexible controls that can adjust the active power output of a wind farm.

The solution of the linear programming problem is accepted as a feasible solution if the operating condition of the network relating to the solution is within voltage and frequency statutory limits. The compliance is checked before leaving the load shedding routine of a trial. If the network achieved a stable operating condition, then the calculated results should indicate the existence of any thermal or voltage limit violations. The thermal limit violations are corrected by reconfiguring the network and re-dispatching generation by taking into account the merits and the flexible order of units. Then, on load tap changing and shunt compensation actions are applied if they are required to be deployed. At last, the loads can be shedded if the operating condition of the network is critical.

A set of other actions are also modeled to control over and under voltage conditions of the network. The over-voltage conditions are mitigated by switching-in shunt reactors, taking off switchable capacitor banks, reducing taps of connected transformers, and at last opening the lightly loaded lines if they are relevant. Under voltage conditions are controlled by switching shunt reactors off, switching capacitor banks in, switching line reactors off, tap operation of transformers, and at last loads are shedded only if the condition is critical.

Shed-load due to unavailability of standby power is calculated by following the steps described earlier but excluding the outages and dynamic interactions and exchanges of standby power units. The entire process described above occurs in a sample trial in Monte Carlo simulation. At the end of each trial, ENS (energy not supplied) for the trial is calculated and EENS (expected ENS) for the completed trials and impact factor on steady-state security due to large wind farms are estimated. Calculation of ENS, EENS, and impact factor due to large wind farms are described in 2.3 to 2.5.

Monte Carlo simulation performs a number of trials to estimate the shed-load components and to satisfy the convergence criteria. It stops if the expected degree of confidence of the shed-load is within the confidence limit and the number of trials being processed is within the range of the minimum and maximum trials. It can also be stopped if the simulation reaches the maximum number of trials without satisfying the degree of confidence and confidence interval. In that case, the estimated value does not have an acceptable degree of confidence and is inconclusive. Figure 3 shows the basic steps of calculating the impact of wind power on the steady-state security in a sample trial in Monte Carlo simulation, where 'corrective action' referred to applying load-shedding routine described above and the 'limit violation correction' referred to correcting the constraint violation using the actions described above.

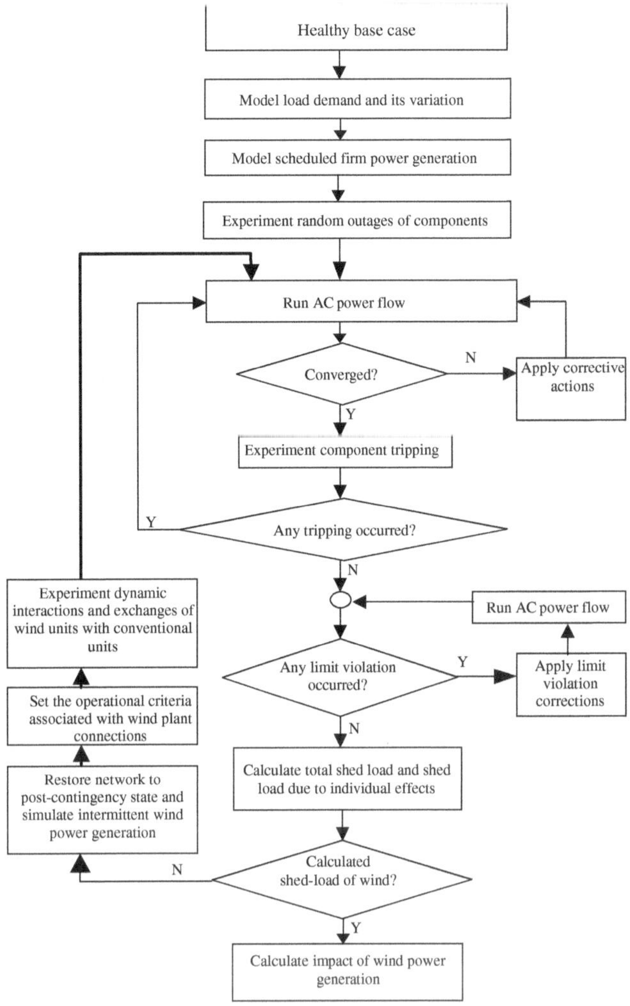

Fig. 3 A sample trial calculation of impact of wind power on steady-state security

2.3 Calculation of ENS

The energy not supplied is defined as the product of the magnitude of the shed-load and the time to restore the shed-load. The total energy not supplied in a trial of Monte Carlo Simulation is mainly due to the intermittency of generation output and outages. These two types of events are not linearly associated with the total ENS due to the fact that the restoration schemes operate based on the total magnitude of the shed-load. However, the total shed-load is linearly associated with the shed-load due to intermittent effects and outages, and they can be weighted in accordance with their contributions. Therefore, the weighting factors are calculated with the shed-load, and then they are incorporated into the calculation of ENS. Thus, the ENS due to intermittency of wind power generation at a trial is calculated by using

$$ENS^i_{\text{int}} = \frac{\left(P^i_{shed_{\text{int}}} \times ENS^i_{tot}\right)}{P^i_{shed_{tot}}} \tag{7}$$

where ENS^i_{tot} is the total ENS at a trial i and calculated by using

$$ENS^i_{tot} = \sum_{k=1}^{m} P^k_{shed_{tot}} \times T^k_{rest} \tag{8}$$

where $P^k_{shed_{tot}}$ is the total shed-load that is restored during the time T^k_{rest}, and m is the number of shed-load components that form the total shed-load. Shed-load due to intermittency ($P^i_{shed_{\text{int}}}$) is the difference between the total shed-load ($P^i_{shed_{tot}}$) and the shed-load due to contingencies ($P^i_{shed_{cont}}$). $P^i_{shed_{\text{int}}}$ includes the shed-load due to interactions of wind plant dynamics with other disturbances. $P^i_{shed_{cont}}$ excludes interactions of the shed-load of wind plant dynamics with other disturbances. $P^i_{shed_{tot}}$ is due to all disturbances during the trial i.

2.4 Estimation of ENS (EENS)

The energy not supplied due to intermittency of wind power generation and dynamic interactions and exchanges is estimated by using

$$EENS_{\text{int}} = \frac{1}{n}\sum_{i=1}^{n} (P^i_{shed_{\text{int}}} \times ENS^i_{tot})/P^i_{shed_{tot}} \tag{9}$$

where $P^i_{shed_{\text{int}}}$ is the shed due to the intermittency of wind-power output and dynamic interactions and exchanges at the trial i, and n is the number of Monte Carlo trials being processed. The EENS due to the unavailability of the standby power generation to eliminate intermittent effects is estimated by using (10)

$$EENS_{stdb} = \frac{1}{n}\sum_{i=1}^{n} (P^i_{shed_{stdb}} \times ENS^i_{tot})/P^i_{shed_{tot}} \tag{10}$$

where $P^i_{shed_{stdb}}$ is the shed-load that is purely due to the inadequacy of the wind-power generation to supply the demand at the trial i. On the other hand, it can also be defined as the required standby power generation to preserve the firm power generation capacity with wind plants.

2.5 Impacts of large wind farms

The impact of large wind farms on steady-state security is calculated by increasing the installed capacities of wind farms gradually (from zero to feasible limits of installed capacities) and then estimating the ENS at each of the increments. The impact factor of large wind farms on steady-state security (μ) can be calculated by using (11)

$$\mu = \frac{EENS_{\text{int}}}{EENS_{tot}}; \quad 0 \leq \mu \leq 1.0 \tag{11}$$

where $EENS_{tot}$ is estimated by using

$$EENS_{tot} = \frac{1}{n}\sum_{i=1}^{n} ENS^i_{tot} \tag{12}$$

The factor (μ) gives the impact of wind generated electricity on the steady state security with the presence of large wind farms in a scale of zero to one. For the central connection of wind farms, the lowest impact factor that provides the largest installed capacity is the maximum capacity that can be integrated into the network safely because it gives the least impact on the steady-state security yet absorbing a relatively larger installed capacity of wind. On the other hand, a medium level impact factor with a medium level installed capacity of wind can also be benefitted for a network. For the decentralized connection of wind farms, the same rule can be applied, however, the impact factor due to large wind farms can be lower compared to central integration because of the ability of decentralized connections of wind farms to provide firm power outputs with auxiliary resources.

2.6 Generation capacity credits

The wind capacity credits can be seen through the proposed impact factor of large wind farms. In order to receive capacity credits with the integration of large wind farms, there should be more than one wind plant site that shares the demand of the network. This is because the combination of all wind farm outputs makes the minimum capacity, which enables to maintain the same or a better security of supply compared with the case without wind plants. Therefore, the total installed capacity that provides the largest firm capacity with a smallest impact factor can be considered as the capacity credit that does not compromise the security of supply to customers.

3 Case study

The aim of the case study is to demonstrate the performance of the proposed approach and to probabilistically assess the wind integration capacity that provides the lowest impact on the steady-state security in an active distribution network. Figure 4 shows a single-line diagram

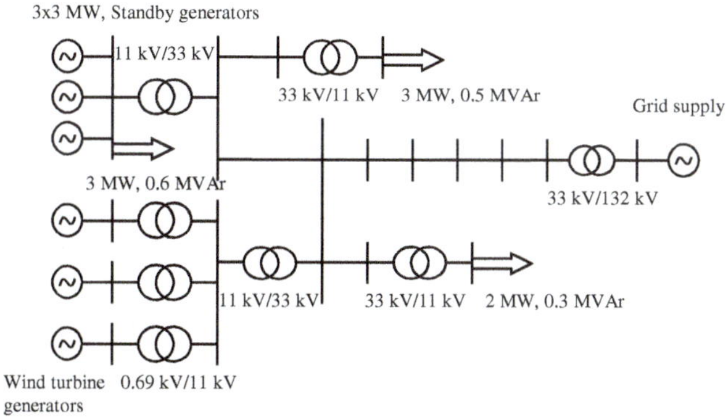

Fig. 4 Single line diagram of a rural distribution network

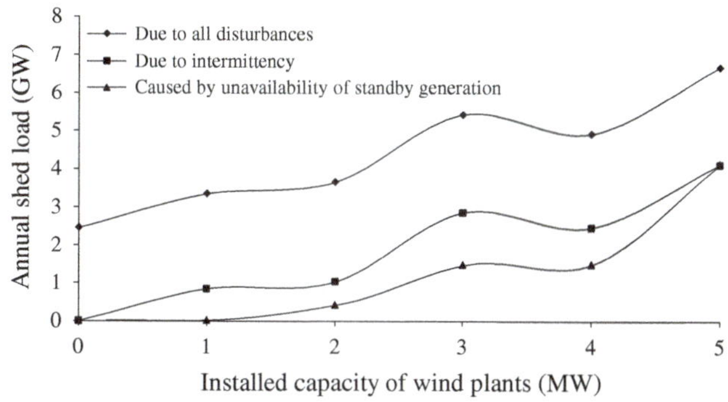

Fig. 5 Annual shed-load due to disturbances

of a rural distribution network that facilitates the connection of wind plants. The network has 17 buses and eight transformers of which the 15 MVA 33/132 kV transformer is a voltage regulating transformer with on-load tap changing facility.

The generating capacity of the standby generating plant is 9 MW. It is connected to the 33 kV grid via a 10 MVA 11 kV/33 kV transformer. There are three wind plants, which have similar operating characteristics. Wind plants operate at the unity power factor. The load demand of the network is shared by wind/standby or wind/grid generating unit combinations.

Four types of customers are connected at each load bus. They are residential restricted, residential unrestricted, industrial, and commercial. The peak active and reactive power loads on the network are 8 MW and 1.4 MVAr respectively. The loads are susceptible to fluctuations.

The network load and the wind farm capacity factors are 0.48 and 0.28 respectively. The minimum and the maximum number of trials of Monte Carlo simulation were set as 8,000 and 100,000 respectively. Monte Carlo simulation stops if it meets the stopping criteria detailed in section 2. Truncated characteristic profiles of loads and wind power generation are applied during the simulation study.

3.1 Scenarios

The scenarios are built upon a case where the grid shown in Fig. 4 is planning to supply an increase in load demand that requires upgrading the grid. The need for upgrading the grid can be differed by increasing the wind generation capacity at the site shown in Fig. 4. However, the impact of large wind farms on steady state security is to be estimated in order to justify the case. The proposal is accepted if the impact on steady state security does not compromise the existing level of security. It is also required to assess the installed capacity of wind units that meets the said requirement.

Six scenarios were designed to assess the viability of the proposal. The first scenario represents the base case which has no wind plants connected and is used to compare merits of other scenarios. The second through six scenarios have installed capacities of wind units from 1 MW to 5 MW in 1 MW steps respectively.

The methodology proposed in section 2 is applied to each scenario. A trial of the Monte Carlo simulation represents an hour-long duration. The expected certainty and confidence limit are set as 95 % and 5 % respectively. Thus, the simulation results should have a 95 % confidence of the estimated mean (shed-load) within 5 % of the true mean for conclusive arguments.

3.2 Results

Each scenario lasted a time between 1 to 2 hours to converge. Figures 5, 6 and 7 show the annual shed-load, annual EENS, and impact factor (μ) for the increased installed capacities of wind units. Annual shed-load and annual EENS almost follow a similar trend because of the level of loads to restore in a distribution network is relatively lower compared to power transmission systems and most of the restorations can be done with a single step of restoration. There are exceptional cases, which are not considered for the studies.

3.3 Analysis

The key message of the results is that the steady-state security impact of wind power in an active distribution network is non-linear and a quantitative assessment is needed for assessing true impacts. The impact factor proposed in the paper has the ability to signal about the accommodation capacity that provides the lowest stress to a network. Increase in installed capacity of a wind power generating unit does not necessarily increase the system stress. A network can have a threshold capacity of wind power beyond which the impact factor of wind can be increased significantly.

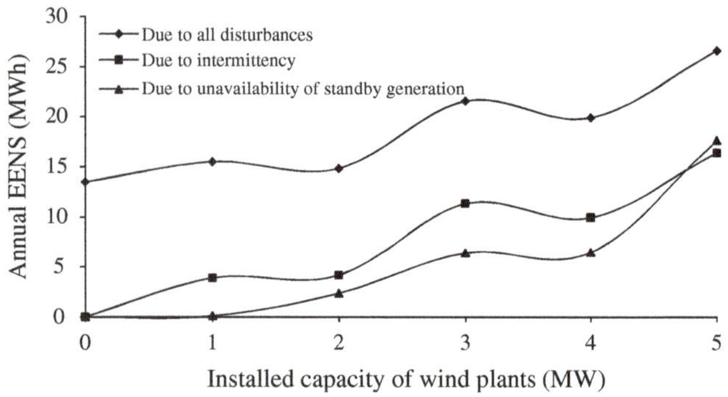

Fig. 6 Annual EENS due to disturbances

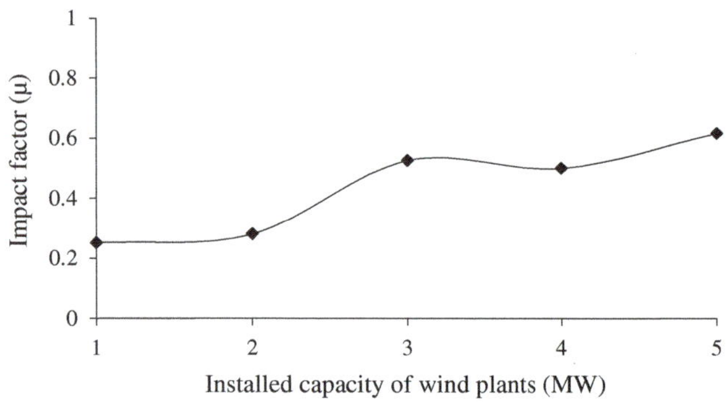

Fig. 7 Impact factor of large wind farms

Results in Figs. 5 and 6 argue that worth of using standby power generation increases with the increase in installation capacity of a wind unit. In this particular network, the contingency contribution to shed-load is approximately consistent with the increased installed capacity. Results further suggest that the increase in wind installed capacity does not necessarily increase the impacts of the intermittency of wind power. For example, installation of 2 MW wind plant provides more benefit than the installation of 1 MW. However, if the standby generation is not available in the system then, the system can absorb a 1 MW capacity than 2 MW. The fifth scenario (4 MW installed capacity) releases a better network capacity to accommodate wind compared to the fourth scenario (3 MW installed capacity) because of its reduced impact on the steady-state security. The reason for this occurrence is that the generation constraint at the fourth scenario limits supplying the full energy demand at some sample trials compared to the fifth scenario, and some of the post-fault operating conditions in the fifth scenario are also less stressed.

The impact factor in Fig. 7 suggests that integrating 2 MW of wind capacity at the site delivers most benefits compared to any other case because it can limit the stress of intermittent wind power output as in 1 MW case and helps reducing the need for fossil-fuelled power generation. Similarly, installation of 3 MW and 4 MW of wind provides similar stresses. However, they carry a higher level of impact compared to the impact of the scenario of 2 MW installed capacity.

The steady-state impact of wind power output remains consistent for 1 MW to 2 MW of installed capacities of wind. However, the impacts considerably increase when the penetration level exceeds said level. Then, the impact marginally reduces at the 4 MW of installed capacity of wind and thereafter it continues to increase. Such a variation occurs due to the severity of intermittency with the increased penetration of wind power and associated constraints of assets in the network.

The study results also affirm on the potential ability of the approach to quantify the steady-state security with the integration of large wind farms in an active distribution network. As an auxiliary outcome, the approach also produces the latent capacity of a network that delivers the lowest impact.

With the increased trend in modernizing traditional distribution networks towards smart distribution networks, the shares of wind and other intermittent generation technologies to supply the load demands can be increased considerably. In that context, the quantification of steady-state stresses due to wind power outputs can potentially provide vital information for network operators to take actions against uncontrollable conditions of a network. Such actions can also limit the interruption of power supply to electricity consumers.

4 Conclusions

The paper presents a Monte Carlo simulation based probabilistic approach to assess the steady-state security in an active distribution network. The approach quantifies the stresses caused by integration of large wind farms against other events.

Case study suggests that impacts of large wind farms on steady state security are inconsistent and a quantitative assessment can suggest the level of inconsistency.

Availability of standby power can potentially mitigate the impacts of intermittency. However, increase in wind power penetration does not necessarily increase the need for standby power. The effectiveness of standby-power to mitigate security impacts can be significant in high wind-power penetrated distribution networks.

The approach can be applied to benchmark active distribution networks based on impacts on steady-state security with large wind farms.

References

[1] Bhowmik A, Maitra A, Halpin SM et al (2003) Determination of allowable penetration levels of distributed generation resources based on harmonic limit considerations. IEEE Trans Power Deliv 18(2):619–624

[2] Brahm SM, Girgis AA (2004) Development of adaptive protection scheme for distribution system with high penetration of distributed generation. IEEE Trans Power Deliv 19(1):56–63

[3] Chilvers I, Jenkins N, Crossley P (2005) Distance relaying of 11 kV circuits to increase the installed capacity of distributed generation. IEE P-Gener Trans Dis 152(1):40–46

[4] Keane A, O'Malley M (2005) Optimal allocation of embedded generation on distribution networks. IEEE Trans Power Syst 20(3):1640–1646

[5] Billinton R, Bai G (2004) Generating capacity adequacy associated with wind energy. IEEE Trans Energy Convers 19(3): 641–646

[6] Chiradeja P, Ramakumar R (2004) An approach to quantify the technical benefits of distributed generation. IEEE Trans Energy Convers 19(4):764–773

[7] Barton JP, Infield DG (2004) Energy storage and its use with intermittent renewable energy. IEEE Trans Energ Conver 19(2): 441–448

[8] Karki R, Billinton R (2004) Cost-effective wind energy utilization for reliable power supply. IEEE Trans Energy Convers 19(2):435–440

[9] Quezada VHM, Abbad JR, Roman TGS (2006) Assessment of energy distribution losses for increasing penetration of distributed generation. IEEE Trans Power Syst 21(2):533–540

[10] Ortega-Vazquez MA, Kirschen DS (2009) Estimating the spinning reserve requirements in systems with significant wind power generation penetration. IEEE Trans Power Syst 24(1):114–124

[11] Pham TTH, Besanger Y, Hadjsaid N (2009) New challenges in power system restoration with large scale of dispersed generation insertion. IEEE Trans Power Syst 24(1):398–406

[12] Vallee F, Lobry J, Deblecker O (2008) System reliability assessment method for wind power integration. IEEE Trans Power Syst 23(3):1288–1297

[13] Wang J, Shahidehpour M, Li Z (2008) Security-constrained unit commitment with volatile wind power generation. IEEE Trans Power Syst 23(3):1319–1327

[14] Brown PD, Peas Lopes JA, Matos MA (2008) Optimization of pumped storage capacity in an isolated power system with large renewable penetration. IEEE Trans Power Syst 23(2):523–531

[15] Banakar H, Luo C, Ooi BT (2008) Impacts of wind power minute-to-minute variations on power system operation. IEEE Trans Power Syst 23(1):150–160

[16] Vallee F, Lobry J, Deblecker O (2007) Impact of the wind geographical correlation level for reliability studies. IEEE Trans Power Syst 22(4):2232–2239

[17] Chowdhury AA, Koval DO (2005) Impact of photovoltaic power sources on a power system's capacity reliability Levels. In: Proceedings of the 2005 IEEE industrial and commercial power systems technical conference (I&CPS'05), Saratoga Springs, 8–12 May 2005, pp 32–37

[18] Chowdhury AA, Mielnik TC, Lawton LE et al (2009) System reliability worth assessment using the customer survey approach. IEEE Trans Ind Appl 45(1):317–322

[19] Ortega-Vazquez MA, Bouffard F, Silva V (2013) Electric vehicle aggregator/system operator coordination for charging scheduling and services procurement. IEEE Trans Power Syst 28(2):1806–1815

[20] Lee T, Adornato B, Filipi ZS (2011) Synthesis of real-world driving cycles and their use for estimating PHEV energy consumption and charging opportunities: case study for midwest/US. IEEE Trans Veh Technol 60(9):4153–4163

[21] Liu C, Wang J, Botterud A et al (2012) Assessment of impacts of PHEV charging patterns on wind-thermal scheduling by stochastic unit commitment. IEEE Trans Smart Grid 3(2):675–683

[22] Grahn P, Munkhammar J, Widen J et al (2013) PHEV home-charging model based on residential activity patterns. IEEE Trans Power Syst 28(3):2507–2515

[23] Fan Z (2012) A distributed demand response algorithm and its application to PHEV charging in smart grids. IEEE Trans Smart Grid 3(3):1280–1290

[24] Billinton R, Singh G (2006) Application of adverse and extreme adverse weather: modelling in Transmission and distribution system reliability evaluation. IEE P-Gener Trans Distrib 153(1): 115–120

[25] Fay D, Ringwood JV (2010) On the influence of weather forecast errors in short-term load forecasting models. IEEE Trans Power Syst 25(3):1751–1758

[26] Shen S, Koval D, Shen S (1999) Modelling extreme-weather-related transmission line outages. In: Proceedings of the 1999 IEEE Canadian conference on electrical and computer engineering (CCECE'99), vol 3. Edmonton, 9–12 May 1999, pp 1271–1276

[27] Rios MA, Kirschen DS, Jayaweera D et al (2002) Value of security: modeling time-dependent phenomena and weather conditions. IEEE Trans Power Syst 17(3):543–548

[28] Grijalva S, Dahman SR, Patten KJ et al (2007) Large-scale integration of wind generation including network temporal security analysis. IEEE Trans Energy Convers 22(1):181–188

[29] Xiao F, McCalley JD (2009) Power system risk assessment and control in a multiobjective framework. IEEE Trans Power Syst 24(1):78–85

[30] Li W, Zhou J (2009) Power system risk assessment using a hybrid method of fuzzy set and Monte Carlo simulation. In: Proceedings of the IEEE Power and Energy Society general meeting (PES'09), Calgary, 26–30 July 2009, p 1

[31] Martins VF, Borges CLT (2011) Active distribution network integrated planning incorporating distributed generation and load response uncertainties. IEEE Trans Power Syst 26(4): 2164–2172

[32] Duffey RB, Ha T (2013) The probability and timing of power system restoration. IEEE Trans Power Syst 28(1):3–9

[33] Cochran WG (1977) Sampling techniques, 3rd edn. Wiley, New York

Dilan JAYAWEERA received the M.Sc. and Ph.D. degrees in electrical power engineering from the University of Manchester Institute of Science and Technology (UMIST), Manchester, U.K., in 2000 and 2003, respectively. He was with Imperial College, London and University of Strathclyde, Glasgow. Currently he works at Curtin University, Australia as a Senior Lecturer. He is a senior member of the IEEE and a chartered engineer in the United Kingdom.

Syed ISLAM received the M.Sc. and PhD degree in electrical power engineering from the King Fahd University of Petroleum and Minerals, Saudi Arabia, in 1983, and 1988 respectively. He is a Professor at Curtin University. He is a Fellow of Engineers Australia, Fellow of IET, and a chartered engineer in the United Kingdom.

18

Low-carbon unit commitment with intensive wind power generation and carbon capture power plant

Jiaming LI, Jinyu WEN (✉), Xingning HAN

Abstract The paper proposes a stochastic unit commitment (UC) model to realize the low-carbon operation requirement and cope with wind power prediction errors for power systems with intensive wind power and carbon capture power plant (CCPP). A linear re-dispatch strategy is introduced to compensate the wind power deviation from the spot forecast. The robust optimization technique is employed to obtain a reliable commitment plan against all realizations of wind power within the uncertainty set given by probabilistic forecast. The proposed model is validated with IEEE 39-bus system. The advantages of flexible CCPPs are compared to the normal coal-fueled plants and the impacts of robustness controlling are discussed.

Keywords Low-carbon unit commitment, Carbon capture and storage, Linear re-dispatch strategy, Robust optimization

1 Introduction

The global warming attracts more and more attentions in the recent few decades [1]. Relevant researches advocated that the excessive emission of greenhouse gases is the main cause [2]. As a result, the United Nations framework convention on climate change (UNFCCC) was proposed and Kyoto Protocol has been signed. According to the protocol, low-carbon development aiming to reduce CO_2 emission is essential for the contracting parties including China. In 2005, an inter-governmental panel on climate change (IPCC) special report [3] showed that the electrical power generation accounts for over 34% of the global CO_2 emissions owing to the popularity of fossil-fueled power plants. Hence, power system should be one of the main frontiers in the low-carbon revolution.

It can be noticed that there are two available solutions when it comes to low-carbon development in electric power generations. One is to control the CO_2 emission in the existing fossil-fueled plant, while the other one is to seek for alternative clean power sources. Therefore, the carbon capture and storage (CCS) technique is applied to absorb and seal the CO_2 produced by power plants while the renewable energy sources including wind and solar power are rapidly exploited throughout the world.

Undoubtedly, the integration of carbon capture power plant (CCPP) and renewable energy is bringing out new challenges and ideas to power system operation and dispatch. As for CCS technique, a mix-integer linear model for CCPP dispatch was proposed in [4], only considering the fixed CO_2 capture rate. The potential of flexible control in CCS consumption power was reported to swiftly adjust the net output of power plant in [5]. In [6], typical methods in flexible CCS control were summarized and the corresponding model was proposed for flexible CCPP in steady-state operation analysis. In [7] and [8], the optimal strategies were discussed to maximize the operation profits from flexible CCPPs and the results showed that premium profits may be obtained by properly adjusting the CO_2 capture rate according to the electricity and carbon prices. As for renewable energy, the focus is mainly on coping with the forecast errors. The scenarios, intervals, and risk index for the descriptions of wind power forecast errors were introduced in [9], and it is argued that forecast errors are inevitable and not negligible because of the stochastic nature of wind speed and the parameter uncertainties in the practical

J. LI, J. WEN, X. HAN, State Key Laboratory of Advanced Electromagnetic Engineering and Technology, School of Electrical and Electronic Engineering, Huazhong University of Science and Technology, Wuhan 430074, China
(✉) e-mail: jinyu.wen@hust.edu.cn

wind power curve. Consequently, the spinning reserves were adopted to cover the possible forecast deviations in [10, 11], but this modification in spinning reserves is theoretically proved to be inadequate in large deviation situations. Hence, several stochastic unit commitment (UC) models have been proposed, including scenarios-based optimization [12, 13], robust optimization with interval uncertainty [14, 15] and chance-constrained stochastic UC [16], to inherently utilize the different descriptions in wind power forecast uncertainties. All the literatures above are also suitable to involve the stochastic nature of other renewable resources.

After combining low-carbon technique and renewables integration, it is claimed that fossil-fueled plants equipped with flexible CCS devices possess higher ramping rate and lower minimum output, which is favorable for wind generation accommodation in [17, 18]. A UC model for power system with flexible CCPP and wind generation was proposed in [19], and the frequency response was considered in [20]. However, few literatures have discussed the effects on wind power forecast uncertainties, putting the system in unplanned operation status when the wind power deviation is significantly large.

This paper proposes a stochastic UC model considering intensive wind power generation and coal-fueled plants with flexible CCS devices. The economical operation with CO_2 emission cost is set as the optimization target. A linear re-dispatch strategy is introduced to compensate the wind power deviations to cover the uncertainties. Robust optimization theory is thus applied to provide the reliable countermeasures for power system operators to deal with the wind power deviations.

The paper is organized as follows: Section 2 elaborates relevant issues in low-carbon unit commitment (LCUC) model. Section 3 presents the mathematical formulation of the proposed LCUC model. The transformation of LCUC model into a deterministic MILP problem based on linear robust optimization is presented in Section 4. Case studies in New England 39-bus system are illustrated in section 5 and the conclusions are given in section 6.

2 Relevant issues in LCUC

2.1 Characteristic of carbon capture power plant

In this paper, the coal-fueled power plant equipped with post-combustion solvent/sorbent carbon capture system, which is the most commercially available and efficient type of CCS techniques nowadays, is considered [3]. The CCPP can flexibly control the CO_2 capture rate along with its consumption power by venting channel or sorbent/ solvent storage tanks [6].

According to the previous studies, the output power of CCPPs can be written as follows [8]:

$$P_{i,t}^{\mathrm{CCP}} = P_{i,t}^{\mathrm{Coal}} - P_{i,t}^{\mathrm{CCS}} \tag{1}$$

where $P_{i,t}^{\mathrm{CCP}}$ is the total output of carbon capture plant; $P_{i,t}^{\mathrm{Coal}}$ is the output of plant without CCS system.

$P_{i,t}^{\mathrm{CCS}}$ is the power required by CCS devices of CCPP i in time t, which can be expressed as

$$\begin{cases} P_{i,t}^{\mathrm{CCS}} = P_{i\mathrm{FIX}}^{\mathrm{CCS}} + \lambda_i^{\mathrm{GE}} \beta_{i,t}^{\mathrm{CCS}} e_i^{\mathrm{G}} P_{i,t}^{\mathrm{Coal}} \\ \beta_{i\min}^{\mathrm{CCS}} \le \beta_{i,t}^{\mathrm{CCS}} \le \beta_{i\max}^{\mathrm{CCS}} \end{cases} \tag{2}$$

where $P_{i\mathrm{FIX}}^{\mathrm{CCP}}$ is the fixed power penalty independent from the operation state; λ_i^{GE} is the electric energy consumption to capture 1 ton of CO_2; e_i^{G} is the gross emission of CO_2 when generating 1 MWh of electric energy, and λ_i^{GE} is the adjustable CO_2 capture rate ($\beta_{i,t}^{\mathrm{CCS}}$ and e_i^{G} can be assumed as constants for a specific CCPP).

For simplicity, coal-fueled plant without CCS is not independently modeled in this paper, since it can be seen as a special CCPP when $\beta_{i\min}^{\mathrm{CCS}} = \beta_{i\max}^{\mathrm{CCS}} = 0$ and $P_{i\mathrm{FIX}}^{\mathrm{CCS}} = 0$. In this paper, 'coal-fueled plant' only refers to the part of CCPP excludes CCS device.

2.2 Description of wind power forecast error

The stochastic nature of wind power leads to the poor accuracy in short-term spot forecast. Hence probabilistic approaches providing expected wind power values along with quantitative uncertainty description may be better choices.

As mentioned in Section 1, there are roughly three methods to quantify the forecast uncertainty. The most widely used one is the interval description, which gives the bounds of possible wind power under different confidence levels [9].

The wind power $W_{m,t}$ is expressed as

$$\begin{cases} W_{m,t} = W_{m,t}^{\mathrm{F}} + \Delta W_{m,t} \\ \underline{W}_{m,t} \le W_{m,t} \le \overline{W}_{m,t} \end{cases} \tag{3}$$

where $W_{m,t}^{\mathrm{F}}$ is the expected wind output of plant m at time t; $\overline{W}_{m,t}$ is the upper limit; $\underline{W}_{m,t}$ is the lower limit; $\Delta W_{m,t}$ is the power deviation from $W_{m,t}^{\mathrm{F}}$, which is unknown in day-ahead UC planning. Thus, $W_{m,t}$ should be treated as a random variable.

2.3 Wind power curtailment

The possible wind power curtailment is always considered as an economical choice or a measure to control exceeded range of wind power fluctuations. In this paper, the curtailment power $W_{m,t}^{\mathrm{Cur}}$ in the UC model is an uncertain variable defined by a given ratio $\mu_{m,t}^{\mathrm{W}}$ of the available wind power as follows.

$$\begin{cases} W_{m,t}^{\mathrm{Cur}} = \mu_{m,t}^{\mathrm{W}} W_{m,t} \\ 0 \le \mu_{m,t}^{\mathrm{W}} \le 1 \end{cases} \tag{4}$$

2.4 Linear re-dispatch strategy

The deviation of wind power from spot forecast value must be compensated by other power plants to maintain normal frequency. In this paper, a linear re-dispatch strategy is proposed to fulfill this power regulation, and it can be expressed as

$$\begin{cases} \Delta P_{i,m,t}^{\mathrm{Coal}} = K_{i,m,t}^{\mathrm{Coal}} \Delta W_{m,t} \\ \Delta P_{i,m,t}^{\mathrm{CCS}} = K_{i,m,t}^{\mathrm{CCS}} \Delta W_{m,t} \end{cases}$$
$$\sum_{i \in G} \left(\Delta P_{i,m,t}^{\mathrm{Coal}} + \Delta P_{i,m,t}^{\mathrm{CCS}} \right) + \left(1 - \mu_{m,t}^{\mathrm{W}} \right) \Delta W_{m,t} = 0 \tag{5}$$

where $K_{i,m,t}^{\mathrm{Coal}}$ and $K_{i,m,t}^{\mathrm{CCS}}$ are the coefficients with regard to the output adjustment in CCP i to compensate the output deviation of wind plant m; $\Delta P_{i,m,t}^{\mathrm{Coal}}$ and $\Delta P_{i,m,t}^{\mathrm{CCS}}$ are the power adjustment conducted by both the coal-fueled plant and the CCS device; $K_{i,m,t}^{\mathrm{Coal}}$ and $K_{i,m,t}^{\mathrm{CCS}}$ are variables determined by solving the UC problem; G is the set of CCPPs.

Practically, small but fast changing power unbalance may still exist after re-dispatching and it is hard to predict or manually handle. As a result, AGC and primary frequency response will step in, but it is beyond the concern of UC model in this paper.

2.5 UC robustness

A robust UC means a day-ahead commitment plan that can provide feasible regions for real-time dispatching under all possible scenarios of wind power series. In this paper, it is equivalent to a specific unit commitment with the corresponding re-dispatching plan which is suitable for any wind power values within the uncertainty interval.

Usually the wind power is hardly to realize the extreme value at the same time in all the wind plants. Hence, Γ_t^{W} is introduced to restrict the simultaneous deviations of different random variables from the expected values as follows:

$$\underline{W}_{m,t}^{\Gamma} \le W_{m,t} \le \overline{W}_{m,t}^{\Gamma}$$
$$\begin{cases} \underline{W}_{m,t}^{\Gamma} = W_{m,t}^{\mathrm{F}} - \varphi_{m,t} \left(W_{m,t}^{\mathrm{F}} - \underline{W}_{m,t} \right) \\ \overline{W}_{m,t}^{\Gamma} = W_{m,t}^{\mathrm{F}} + \varphi_{m,t} \left(\overline{W}_{m,t} - W_{m,t}^{\mathrm{F}} \right) \\ \sum_{m \in W} \varphi_{m,t} \le \Gamma_t^{\mathrm{W}}, |W| \ge \Gamma_t^{\mathrm{W}} \ge 0 \end{cases} \tag{6}$$

where W is the set of wind plants; $|W|$ is the number of wind plants in the power system; $\varphi_{m,t}$ is the decisive variable controlling the variation range of wind plant output according to Γ_t^{W}. Under such an assumption, the actual limits of wind power $W_{m,t}$ are decisive variables to be optimized, represented by Γ-limits as $\overline{W}_{m,t}^{\Gamma}$ and $\underline{W}_{m,t}^{\Gamma}$.

3 Low-carbon unit commitment model

3.1 Optimization objective

The optimization objective of LCUC is to minimize the total cost for daily fuel consumption and CO_2 emission under expected wind power output.

$$\min \sum_{t}^{T_{\mathrm{UC}}} \left(C_{\mathrm{CO_2}} \sum_{i \in G} E_{i,t}^{\mathrm{CO_2}} + C_{\mathrm{Coal}} \sum_{i \in G} F_{i,t}^{\mathrm{Coal}} + \sum_{i \in G} C_{i,t}^{\mathrm{SU}} + \sum_{i \in G} C_{i,t}^{\mathrm{SD}} \right) \Bigg|_{W_{m,t} = W_{m,t}^{\mathrm{F}}} \tag{7}$$

where $C_{\mathrm{CO_2}}$ is the CO_2 emission price; C_{Coal} is the coal price; $E_{i,t}^{\mathrm{CO_2}}$ is the CO_2 equivalent emission; $F_{i,t}^{\mathrm{Coal}}$ is the coal consumption; $C_{i,t}^{\mathrm{SU}}$ and $C_{i,t}^{\mathrm{SD}}$ are the startup and shut down prices for CCP i at time t.

3.2 Constraints

The following constraints are included in the LCUC model.

1) Power balance

The power balance should always be maintained between generations and load requirements.

$$\sum_{i \in G} \left(P_{i,t}^{\mathrm{Coal}} - P_{i,t}^{\mathrm{CCS}} \right) + \sum_{m \in W} \left(W_{m,t} - W_{m,t}^{\mathrm{Cur}} \right) = \sum_{k \in L} P_{k,t}^{\mathrm{Load}} \tag{8}$$

where $P_{k,t}^{\mathrm{Load}}$ is the power consumption of load k at time t; L is the set of loads.

2) Spinning reserve

The online generators are available to cover load prediction errors as well as possible power shortage induced by forced-tripping of generators.

$$\begin{cases} \sum_{i \in G} \left(\overline{P}_{i,t}^{\mathrm{Coal}} - \underline{P}_{i,t}^{\mathrm{CCS}} \right) + \sum_{m \in W} \left(W_{m,t} - W_{m,t}^{\mathrm{Cur}} \right) \ge \sum_{k \in L} P_{k,t}^{\mathrm{Load}} + \alpha_t^{\mathrm{up}} \\ \sum_{i \in G} \left(\underline{P}_{i,t}^{\mathrm{Coal}} - \overline{P}_{i,t}^{\mathrm{CCS}} \right) + \sum_{m \in W} \left(W_{m,t} - W_{m,t}^{\mathrm{Cur}} \right) \le \sum_{k \in L} P_{k,t}^{\mathrm{Load}} - \alpha_t^{\mathrm{dn}} \end{cases} \tag{9}$$

where $\overline{P}_{i,t}^{\mathrm{Coal}}$ and $\overline{P}_{i,t}^{\mathrm{CCS}}$ are the currently available upper power limits of coal-fueled plant and CCS system, respectively; $\underline{P}_{i,t}^{\mathrm{Coal}}$ and $\underline{P}_{i,t}^{\mathrm{CCS}}$ are the corresponding lower power limits; α_t^{up} and α_t^{dn} are the upward and downward spinning reserve requirements.

3) Coal-fueled plant output range

The power output range of coal-fueled plant is decided by (10)–(12), which correspond to the minimum and maximum technical power limits, online ramping limits and power limits for startup or shut down operations respectively.

$$\begin{cases} \underline{P}_{i,t}^{\mathrm{Coal}} \le P_{i,t}^{\mathrm{Coal}} \le \overline{P}_{i,t}^{\mathrm{Coal}} \\ \overline{P}_{i,t}^{\mathrm{Coal}} \le u_{i,t} P_{i\max}^{\mathrm{Coal}} \\ u_{i,t} P_{i\min}^{\mathrm{Coal}} \le \underline{P}_{i,t}^{\mathrm{Coal}} \end{cases} \tag{10}$$

where binary decisive variable $u_{i,t}$ is the status of CCPP i at time t, CCPP is online when $u_{i,t} = 1$ while CCPP is offline when $u_{i,t} = 0$; $P_{i\max}^{\mathrm{Coal}}$ and $P_{i\min}^{\mathrm{Coal}}$ are the technical maximum and minimum power limits.

$$\begin{cases} \overline{P}_{i,t}^{\mathrm{Coal}} \le P_{i,t-1}^{\mathrm{Coal}} + U_i^{\mathrm{Coal}} + M_i^{\mathrm{Coal}}\left(1-u_{i,t-1}\right) + M_i^{\mathrm{Coal}}\left(1-u_{i,t}\right) \\ \underline{P}_{i,t}^{\mathrm{Coal}} \ge P_{i,t-1}^{\mathrm{Coal}} - D_i^{\mathrm{Coal}} - M_i^{\mathrm{Coal}}\left(1-u_{i,t-1}\right) - M_i^{\mathrm{Coal}}\left(1-u_{i,t}\right) \end{cases} \tag{11}$$

where U_i^{Coal} and D_i^{Coal} are the maximum upward and downward ramping power of coal-fueled plant, respectively. $M_i^{\mathrm{Coal}} = P_{i\max}^{\mathrm{Coal}} + U_{i,t}^{\mathrm{Coal}} + D_{i,t}^{\mathrm{Coal}}$ is a large number required to disable these constraints when the generator is starting up or shutting down.

$$\begin{cases} \overline{P}_{i,t}^{\mathrm{Coal}} \le P_{i\mathrm{SU}}^{\mathrm{Coal}} + M_i^{\mathrm{Coal}} u_{i,t-1} \\ \overline{P}_{i,t}^{\mathrm{Coal}} \le P_{i\mathrm{SD}}^{\mathrm{Coal}} + M_i^{\mathrm{Coal}} u_{i,t+1} \end{cases} \tag{12}$$

where $P_{i\mathrm{SU}}^{\mathrm{Coal}}$ and $P_{i\mathrm{SD}}^{\mathrm{Coal}}$ are the maximum power right after startup and before shutting down, respectively.

4) CCS consumption power range

The consumption power range for CCS is identical to those of the coal-fueled plants as

$$\begin{cases} \underline{P}_{i,t}^{\mathrm{CCS}} \le P_{i,t}^{\mathrm{CCS}} \le \overline{P}_{i,t}^{\mathrm{CCS}} \\ \overline{P}_{i,t}^{\mathrm{CCS}} \le u_{i,t} P_{i\mathrm{FIX}}^{\mathrm{CCS}} + \lambda_i^{\mathrm{GE}} \beta_{i\max}^{\mathrm{CCS}} e_i^{\mathrm{G}} P_{i,t}^{\mathrm{Coal}} \\ u_{i,t} P_{i\mathrm{FIX}}^{\mathrm{CCS}} + \lambda_i^{\mathrm{GE}} \beta_{i\min}^{\mathrm{CCS}} e_i^{\mathrm{G}} P_{i,t}^{\mathrm{Coal}} \le \underline{P}_{i,t}^{\mathrm{CCS}} \end{cases} \tag{13}$$

$$\begin{cases} \overline{P}_{i,t}^{\mathrm{CCS}} \le P_{i,t-1}^{\mathrm{CCS}} + U_i^{\mathrm{CCS}} + M_i^{\mathrm{CCS}}\left(1-u_{i,t-1}\right) + M_i^{\mathrm{CCS}}\left(1-u_{i,t}\right) \\ \underline{P}_{i,t}^{\mathrm{CCS}} \ge P_{i,t-1}^{\mathrm{CCS}} - D_i^{\mathrm{CCS}} - M_i^{\mathrm{CCS}}\left(1-u_{i,t-1}\right) - M_i^{\mathrm{CCS}}\left(1-u_{i,t}\right) \end{cases} \tag{14}$$

$$\begin{cases} \overline{P}_{i,t}^{\mathrm{CCS}} \le P_{i\mathrm{SU}}^{\mathrm{CCS}} + M_i u_{i,t-1} \\ \overline{P}_{i,t}^{\mathrm{CCS}} \le P_{i\mathrm{SD}}^{\mathrm{CCS}} + M_i u_{i,t+1} \end{cases} \tag{15}$$

where U_i^{CCS} and D_i^{CCS} are the maximum upward and downward power change of CCS, respectively. M_i^{CCS} is a large number just like M_i^{Coal} and $M_i^{\mathrm{CCS}} = P_{i\mathrm{FIX}}^{\mathrm{CCS}} + \lambda_i^{\mathrm{GE}} \beta_{i\max}^{\mathrm{CCS}} e_i^{\mathrm{G}} P_{i\max}^{\mathrm{Coal}} + U_{i,t}^{\mathrm{Coal}} + D_{i,t}^{\mathrm{Coal}}$.

5) Minimum up and down time constraint

For coal-fueled plants, it is impossible to restart the generators immediately after shutting down, or vice versa. The status of generators to consider the minimum up and down time constraints is restricted as

$$\begin{cases} u_{i,t} = 1, \ t \le T_{i\,\mathrm{on}}^{\mathrm{For}} \\ u_{i,t} = 0, \ t \le T_{i\,\mathrm{off}}^{\mathrm{For}} \end{cases} \tag{16}$$

$$\begin{aligned} &1 - \left(u_{i,t} - u_{i,t-1}\right) \ge u_{i,k} \quad \forall t > T_{i\,\mathrm{off}}^{\mathrm{For}}, \\ &\quad k = t - T_{i\mathrm{UP}} + 1, \ldots, t-1 \end{aligned} \tag{17}$$

$$\begin{aligned} &u_{i,k} \ge u_{i,t-1} - u_{i,t} \quad \forall t > T_{i\,\mathrm{on}}^{\mathrm{For}}, \\ &\quad k = t - T_{i\mathrm{DN}} + 1, \ldots, t-1 \end{aligned} \tag{18}$$

where $T_{i\,\mathrm{on}}^{\mathrm{For}}$ and $T_{i\,\mathrm{off}}^{\mathrm{For}}$ are the time period to compel CCPP i to keep online or offline at the beginning of commitment plan; $T_{i\mathrm{UP}}$ and $T_{i\mathrm{DN}}$ are the minimum up and down time, respectively.

6) Wind curtailment percentage

The wind curtailment percentage can be expressed as

$$\sum_t^{T_{\mathrm{UC}}} W_{m,t}^{\mathrm{Cur}} \le \rho_m^{\mathrm{Wcur}} \sum_t^{T_{\mathrm{UC}}} W_{m,t} \tag{19}$$

where ρ_m^{Wcur} is the maximum percentage of daily wind energy curtailment for wind farm m within any possible wind power scenarios.

7) CO_2 emission

The emission of CO_2 constraint is an optional requirement, expressed as

$$\begin{cases} E_{i,t}^{\mathrm{CO_2}} = e_i^{\mathrm{G}} P_{i,t}^{\mathrm{Coal}} - \left(P_{i,t}^{\mathrm{CCS}} - u_{i,t} P_{i\mathrm{FIX}}^{\mathrm{CCS}}\right) \Big/ \lambda_i^{\mathrm{GE}} \\ \sum_t^{T_{\mathrm{UC}}} \sum_{i \in G} E_{i,t}^{\mathrm{CO_2}} \le E_{\mathrm{CO_2}}^{\max} \end{cases} \tag{20}$$

where $E_{\mathrm{CO_2}}^{\max}$ is the maximum CO_2 emission for a whole day.

8) Transmission capacity

DC power flow model is introduced here to represent the limits of power flow in all transmission lines.

$$\begin{cases} f_h = \sum_{r \in N} B_{rh} \left[\sum_{i \in G(r)} \left(P_{i,t}^{\mathrm{Coal}} - P_{i,t}^{\mathrm{CCS}}\right) + \sum_{m \in W(r)} \left(W_{m,t} - W_{m,t}^{\mathrm{Cur}}\right) - \sum_{k \in L(r)} P_{k,t}^{\mathrm{Load}} \right] \\ -\overline{f}_h \le f_h \le \overline{f}_h \end{cases} \tag{21}$$

where $\overline{f}_h$ is the capacity of transmission line h; B_{rh} is the sensitivity between injection power into bus r and power flow over line h; $G(r)$, $W(r)$ and $L(r)$ are the sets of CCPPs, wind plants and loads connected to bus r.

In summary, the proposed LCUC model consists of the optimization objective (7) and regular constraints (8)–(21) along with key constraints (2)–(6) mentioned above.

4 Linearization and robust counterparts

4.1 Linearization

The original LCUC model has an optimization objective with non-linear cost of CCPP fuel consumption. Hence, the approximate linearization of the CCPP fuel cost by piecewise functions [21] is performed to apply linear robust optimization theory and to seek for better efficiency when solving the problem.

4.2 Robust counterpart

The linearized LCUC problem still contains lots of random variables introduced by wind power uncertainty, which is impossible to be solved directly. Hence it must be transformed into an equivalent deterministic form by duality theory, commonly called robust counterpart [22, 23].

The robust counterpart of (22) was presented in [23] with a_v as the random coefficients bearing known bounds a_v^L, a_v^U and expected values $\bar{a}_v$. Decisive variables φ_v and Γ are introduced just as (6) to control the variation range of a_v. It has been proved that (22) with a_v in a restricted variation range as (23) is equivalent to robust counterpart (24), which has no random coefficients any more. z and p_v are dual variables without physical meanings.

$$\forall \sum_v a_v x_v \le b, \quad a_v \in \left[a_v^L, \quad a_v^U\right], \quad E(a_v) = \bar{a}_v \tag{22}$$

$$\begin{cases} \Re(\Gamma) = \bar{a}_v - \varphi_v(\bar{a}_v - a_v^L) \le a_v \le \bar{a}_v + \varphi_v(a_v^U - \bar{a}_v) \\ 0 \le \varphi_v \le 1, \sum_v \varphi_v \le \Gamma \end{cases} \tag{23}$$

Since reference [23] gives that $t_v^F = a_v^U - \bar{a}_v$ and $t_v^F = \bar{a}_v - a_v^L$

$$\begin{cases} \sum_v \bar{a}_v x_v + \Gamma z + \sum_v p_v \le b \\ z + p_v \ge \max\left[(a_v^U - \bar{a}_v)x_v, -(\bar{a}_v - a_v^L)x_v\right] \quad z \ge 0, p_v \ge 0 \end{cases} \tag{24}$$

$$\begin{cases} W_{m,t} = W_{m,t}^F + \Delta W_{m,t} \\ W_{m,t}^{Cur} = \mu_{m,t}^W W_{m,t}^F + \mu_{m,t}^W \Delta W_{m,t} \\ P_{i,t}^{Coal} = P_{i,t}^{F,Coal} + \sum_{m \in W} K_{i,m,t}^{Coal} \Delta W_{m,t} \\ P_{i,t}^{CCS} = P_{i,t}^{F,CCS} + \sum_{m \in W} K_{i,m,t}^{CCS} \Delta W_{m,t} \end{cases} \tag{25}$$

Before applying this robust counterpart transformation in the LCUC model, all the random power output must be rewritten as (25).

Then power balance constraint is converted without further transformation as

$$\sum_{i \in G}\left(P_{i,t}^{F,Coal} - P_{i,t}^{F,CCS}\right) + \sum_{m \in W}\left(1 - \mu_{m,t}^W\right)W_{m,t}^F = \sum_{k \in L} P_{k,t}^{Load} \tag{26}$$

However, (9)–(11), (13)–(14), (19) and (21) are still required to be transformed for solution. The upward spinning reserve constraint in (9) is demonstrated as

$$-\sum_{i \in G}\left(\overline{P}_{i,t}^{Coal} - \underline{P}_{i,t}^{CCS}\right) - \sum_{m \in W} W_{m,t}^F\left(1 - \mu_{m,t}^W\right) - \sum_{m \in W} \Delta W_{m,t}^{Cur}\left(1 - \mu_{m,t}^W\right) \le -\sum_{k \in L} P_{k,t}^{Load} - \alpha_t^{up} \tag{27}$$

Secondly, the controllable variation range of random variables $\Delta W_{m,t}$ are expressed as

$$\begin{cases} -\varphi_{m,t}\left(W_{m,t}^F - \underline{W}_{m,t}\right) \le \Delta W_{m,t} \le \varphi_{m,t}\left(\overline{W}_{m,t} - W_{m,t}^F\right) \\ 0 \le \varphi_{m,t} \le 1, \sum_{m \in W} \varphi_{m,t} \le \Gamma_t^W \end{cases} \tag{28}$$

Finally, robust counterpart according to (22)–(24) can be written as

$$\begin{aligned} &\sum_{i \in G}\left(\overline{P}_{i,t}^{Coal} - \underline{P}_{i,t}^{CCS}\right) - \Gamma_t^W z_t^{USR} - \sum_{m \in W} p_{m,t}^{USR} \ge \sum_{k \in L} P_{k,t}^{Load} + \alpha_t^{up} \\ &\begin{cases} z_t^{USR} + p_{m,t}^{USR} \ge -\left(1 - \mu_{m,t}^W\right)\left(\overline{W}_{m,t} - W_{m,t}^F\right) \\ z_t^{USR} + p_{m,t}^{USR} \ge \left(1 - \mu_{m,t}^W\right)\left(W_{m,t}^F - \underline{W}_{m,t}\right) \\ z_t^{USR} \ge 0, p_{m,t}^{USR} \ge 0 \end{cases} \end{aligned} \tag{29}$$

The transformations of other constraints are similar so as to neglect the details in this paper. The LCUC problem is eventually changed into a MILP representation which is ready to be efficiently solved.

5 Case study

The case studies are conducted based on the modified New England 39-bus system. The generator and CCS data are listed in Table 1 and Table 2. In all the cases, three 800 MW wind farms are connected to the bus 30, 32 and 39, respectively. The forecast error for wind power predictions are assumed to be 30% while the curtailment limit ρ_m^{Wcur} for each wind farm is set to be 1%. The branch and base load data are adopted from 'case39' in Matpower 4.1 toolbox [24] for MATLAB. IBM ILOG CPLEX V12.4 is used as the MILP solver.

The load and wind farm output forecasts are given in per unit value in Fig. 1.

For simplicity, loads at different nodes share the same per unit forecast data. Finally, the price of CO_2 emission is 30 \$/tCO_2; the CO_2 emission limit of a whole day is 100000 tCO_2; the spinning reserves α_t^{up} and α_t^{dn} are set as 10% of the total loads and Γ_t^W is set to be 3 which leads to the most robust solution which means maximum ranges of wind power variations are considered.

5.1 Validation of LCUC model

Fig. 2 indicates that the available generation range is able to cover the net load variation including the wind power forecast errors and spinning reserves.

The maximum load-carrying rates of all branches are obtained with wind power scenarios generated within the forecast intervals by Monte Carlo simulation. In Fig. 3, the maximum load-carrying rates are normalized by the transmission capacities. It can be noticed that all the branch loads are kept in a safe range.

If all the loads are served by the coal-fueled plants without CCS, then 82097 tons of CO_2 will be produced in a whole day. Even with the wind power assisted, the CO_2 emission will reach at least 70737 tons. However, with the help of CCS, the CO_2 produced can be further cut down to 32449 tons, as demonstrated in Fig. 4.

5.2 Advantages of flexible CCS

The CCS devices can swiftly reduce their power output to provide more available margin for upward spinning reserves, as shown in Fig. 5.

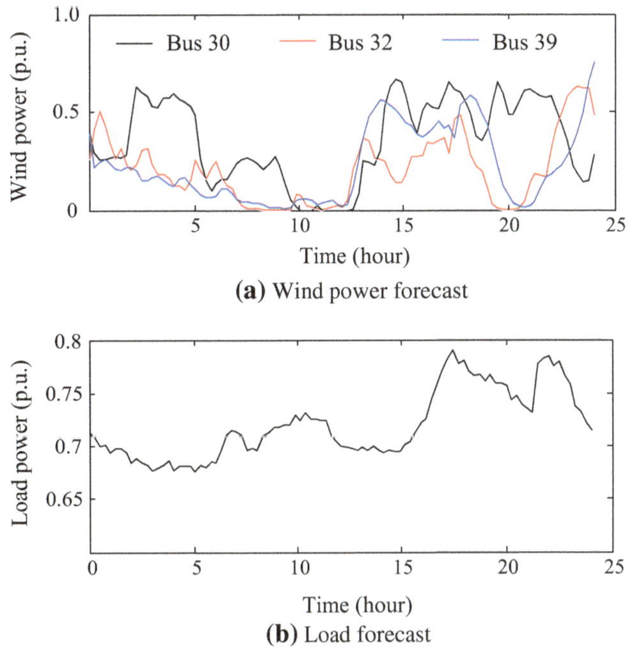

(a) Wind power forecast

(b) Load forecast

Fig. 1 Load and wind farm output forecasts

Table 1 CCS data in the base case

Bus number	$P^{CCS}_{i\,FIX}$ (MW)	CO_2 capture rate		λ^{GE}_i (MWh/tCO_2)	e^G_i (tCO_2/MWh)	15 min ramp range (MW)			
		$\beta^{CCS}_{i\max}$	$\beta^{CCS}_{i\min}$			U^{CCS}_i	D^{CCS}_i	P^{CCS}_{iSU}	P^{CCS}_{iSD}
30	5.2	0.9	0	0.23	0.76	122	122	120	120
31	3.23	0.9	0	0.23	0.76	76	76	75	75
32	3.62	0.9	0	0.23	0.76	85	85	85	85
33	3.26	0.9	0	0.23	0.76	76	76	75	75
39	5.5	0.9	0	0.23	0.76	130	130	130	130

Table 2 Generator data in the base case

Bus number	Technical limits (MW)		15 min ramp range (MW)				Up and down time (hour)		Fuel cost coefficients ($/MWh)			Startup/shut down cost ($)	
	$P^{Coal}_{i\max}$	$P^{Coal}_{i\min}$	U^{Coal}_i	D^{Coal}_i	P^{Coal}_{iSU}	P^{Coal}_{iSD}	T_{iUP}	T_{iDN}	a	b	c	A^{SU}_i	A^{SD}_i
30	1040	416.0	156	156	420	420	10.0	9.0	0.0140	20	500	7000	7000
31	646	258.4	120	120	270	270	7.5	6.0	0.0200	20	380	5500	5500
32	725	290.0	130	130	300	300	8.5	7.5	0.0194	20	42	5500	5500
33	652	260.8	110	110	270	270	7.0	7.0	0.0200	20	380	5000	5000
34	508	203.2	80	80	210	210	6.0	6.0	0.0255	20	295	4000	4000
35	687	274.8	105	105	280	280	7.5	7.5	0.0210	20	400	5500	5500
36	580	232.0	90	90	240	240	6.5	6.0	0.0230	20	350	4500	4500
37	564	225.6	90	90	230	230	6.5	6.0	0.0222	20	330	4500	4500
38	865	364.0	150	150	370	370	9.5	9.0	0.0150	20	490	7000	7000
39	1100	440.0	200	200	450	450	10.0	9.0	0.0140	20	500	7000	7000

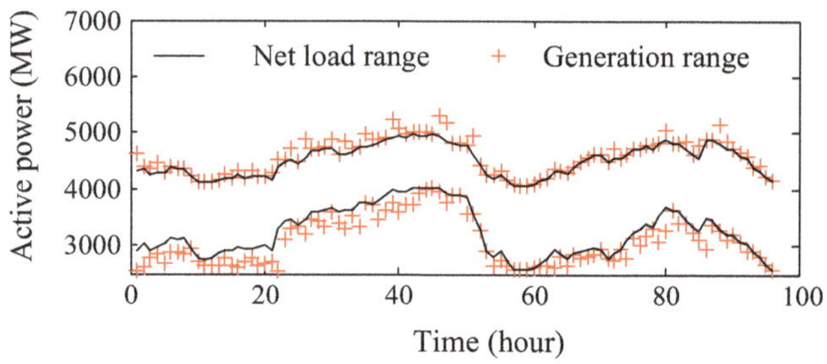

Fig. 2 Power balance between generation and net load

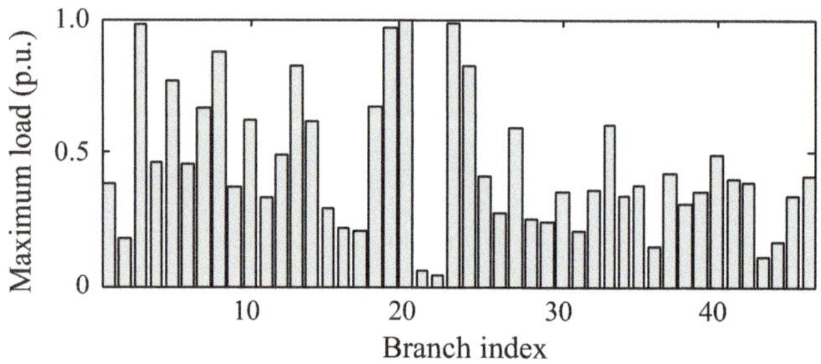

Fig. 3 Maximum load-carrying rate of the branches

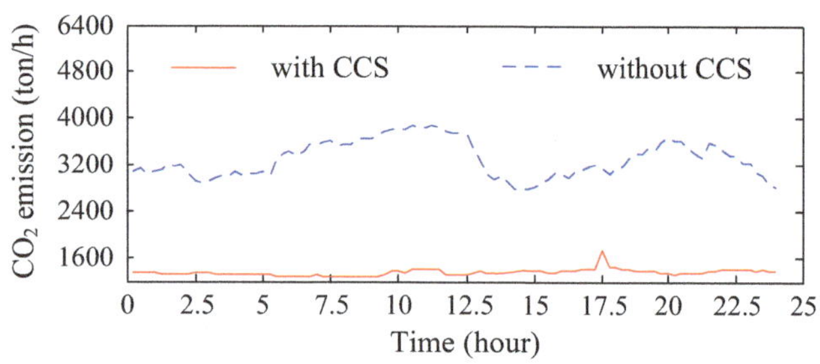

Fig. 4 CO_2 emission reduction

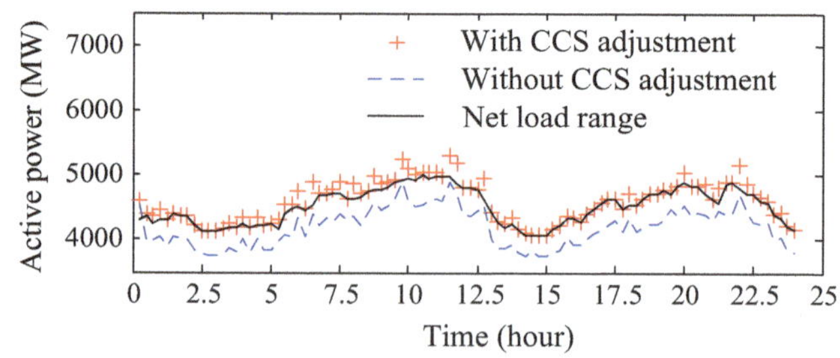

Fig. 5 Upward spinning reserve increased by CCS

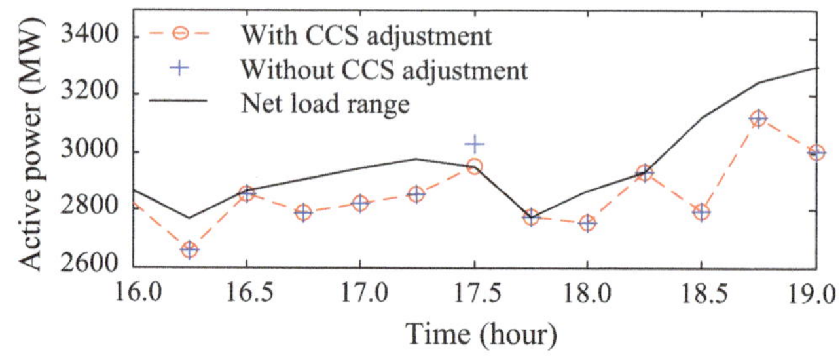

Fig. 6 Downward spinning reserve increased from 16:00 to 19:00

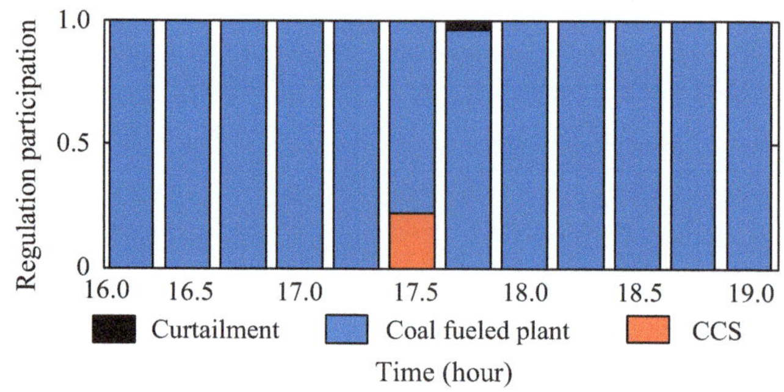

Fig. 7 Wind power regulation participation from 16:00 to 19:00

Furthermore, the CCS devices are allowed to operate at a derating status preparing to increase the power penalty when needed to provide extra downward spinning reserves.

However, the continuous period in derating operation of CCS devices will severely compromise the CO_2 reduction objective. Hence, in the proposed LCUC model, the downward spinning reserve provided from CCS is optimized to be only an emergency precaution, as shown in Fig. 6 and Fig. 7.

5.3 Controlling the robustness

The robustness of the UC plan can be controlled by parameter Γ_t^W aforementioned, which can be chose from 0 (no uncertainty of wind power forecast is considered, namely the least robust) to the wind farm number (the whole wind power uncertainty intervals are considered, namely the most robust). The optimization objectives and constraint violation probabilities with regard to different Γ_t^W are shown in Table 3.

It is shown that Γ_t^W provides a tradeoff between operation security and target optimality. With a smaller range of uncertainty considered, a more optimistic UC plan will be obtained. However the system operation situation may be more tend to risk emergency in the extreme scenarios. The results also indicate that calculation burden will increase when uncertainty of wind power is introduced. When Γ_t^W increases, the time required to solve the LCUC trends to increase.

6 Conclusions

A low-carbon unit commitment model considering flexible CCS and wind power forecast error uncertainties is proposed in this paper. The novel model is finally

Table 3 Tradeoff between operation price and robustness

Γ_t^W	Total cost ($)	CO_2 emission (tons)	Violation probability	Calculation time (s)
3	4081292.5	32449.3	0%	600.29
2.5	4075222.8	32119.8	27.7%	244.42
2.2	4071959.8	31969.4	71.64%	176.91
2.1	4070902.9	31919.3	83.31%	225.55
2	4069851.7	31869.2	92.34%	142.26
1.5	4062824.9	31414.3	100%	149.78
0	4042371.0	29375.8	100%	3.43

converted into a deterministic MILP problem for solution, with a proposed linear re-dispatch strategy and application of linear robust optimization as well. Case studies in New England 39-bus system indicate that the model can efficiently provide reliable and low-carbon UC as well as re-dispatch plan against all realizations of wind power scenarios. Moreover, the robustness of the UC plan can be adjusted to avoid the conservatism which is an attractive feature in practical applications.

Further discussion shows that CCS technique is favorable for intensive wind power accommodation by increasing equivalent ramping rate of generators leading to more adequate spinning reserve. However its adverse effect on constantly changing the CO_2 absorption rate must be carefully considered.

Acknowledgment This work was supported by the National Basic Research Program of China (No. 2012CB215106) and State Grid Corporation of China (No. 52150014006W).

References

[1] Stocker TF, Qin D, Plattner GK et al (2014) Climate change 2013: the physical science basis. IPCC, Geneva

[2] Bernstein L, Bosch P, Canziani O et al (2008) Climate change 2007: synthesis report. IPCC, Geneva

[3] Metz B, Davidson O, de Coninck HC et al (2005) IPCC special report on carbon dioxide capture and storage. Cambridge University Press, Cambridge

[4] Martens P, Delarue E, D'haeseleer W (2012) A mixed integer linear programming model for a pulverized coal plant with post-combustion carbon capture. IEEE Trans Power Syst 27(2):741–751

[5] Ziaii S, Cohen S, Rochelle GT et al (2009) Dynamic operation of amine scrubbing in response to electricity demand and pricing. Energy Proc 1(1):4047–4053

[6] Chen QX, Kang CQ, Xia Q (2010) Modeling flexible operation mechanism of CO_2 capture power plant and its effects on power-system operation. IEEE Trans Energy Convers 25(3):853–861

[7] Chen QX, Kang CQ, Xia Q et al (2012) Optimal flexible operation of a CO_2 capture power plant in a combined energy and carbon emission market. IEEE Trans Power Syst 27(3): 1602–1609

[8] Lu SY, Lou SH, Wu YW et al (2013) Power system economic dispatch under low-carbon economy with carbon capture plants considered. IET Gener Transm Distrib 7(9):991–1001

[9] Zhang Y, Wang JX, Wang XF (2014) Review on probabilistic forecasting of wind power generation. Renew Sustain Energy Rev 32:255–270

[10] Lee TY (2007) Optimal spinning reserve for a wind-thermal power system using EIPSO. IEEE Trans Power Syst 22(4):1612–1621

[11] Chaiyabut N, Damrongkulkumjorn P (2014) Optimal spinning reserve for wind power uncertainty by unit commitment with EENS constraint. In: Proceedings of the 2014 IEEE PES innovative smart grid technologies conference (ISGT'14), Washington, 19–22 Feb 2014, 5 pp

[12] Pappala VS, Erlich I, Rohrig K et al (2009) A stochastic model for the optimal operation of a wind-thermal power system. IEEE Trans Power Syst 24(2):940–950

[13] Wang JH, Shahidehpour M, Li ZY (2008) Security-constrained unit commitment with volatile wind power generation. IEEE Trans Power Syst 23(3):1319–1327

[14] Zhao CY, Wang JH, Watson JP et al (2013) Multi-stage robust unit commitment considering wind and demand response uncertainties. IEEE Trans Power Syst 28(3):2708–2717

[15] Wang Y, Xia Q, Kang CQ (2011) Unit commitment with volatile node injections by using interval optimization. IEEE Trans Power Syst 26(3):1705–1713

[16] Wang QF, Guan YP, Wang JH (2012) A chance-constrained two-stage stochastic program for unit commitment with uncertain wind power output. IEEE Trans Power Syst 27(1):206–215

[17] Chalmers H, Gibbins J (2006) Potential for synergy between renewables and carbon capture and storage. In: Proceedings of the 29th IAEE international conference, Potsdam, 7–10 Jun 2006, 16 pp

[18] van der Wijk PC, Brouwer AS, van den Broek M et al (2014) Benefits of coal-fired power generation with flexible CCS in a future northwest European power system with large scale wind power. Int J Greenhouse Gas Contrib 28:216–233

[19] Ji Z, Kang CQ, Chen QX et al (2013) Low-carbon power system dispatch incorporating carbon capture power plants. IEEE Trans Power Syst 28(4):4615–4623

[20] Yamashita D, Niimura T, Yokoyama R et al (2011) Thermal unit scheduling for CO_2 reduction including significant wind power penetration. In: Proceedings of the 2011 IEEE Power and Energy Society general meeting, San Diego, 24–29 Jul 2011, 5 pp

[21] Carrión M, Arroyo JM (2006) A computationally efficient mixed-integer linear formulation for the thermal unit commitment problem. IEEE Trans Power Syst 21(3):1371–1378

[22] Bertsimas D, Sim M (2004) The price of robustness. Oper Res 52(1):35–53

[23] Kang SC (2008) Robust linear optimization using distributional information. Boston University, Boston

[24] Zimmerman RD, Murillo-Sanchez CE (1997) MATPOWER: a MATLABTM power system simulation package, version 2.0, user's manual. Cornell University, Ithaca

Jiaming LI received the B.Eng. degree in electrical engineering from Huazhong University of Science and Technology (HUST) in 2012. He is now a Ph.D. candidate in HUST. His research interests include power system restoration after cascading failure and power system operation with significant renewable energy.

Jinyu WEN received the B.Eng. and Ph.D. degrees in electrical engineering from Huazhong University of Science and Technology (HUST), Wuhan, China, in 1992 and 1998, respectively. He was a visiting student from 1996 to 1997 and research fellow from 2002 to 2003 at the University of Liverpool, UK, and a senior visiting researcher at the University of Texas at Arlington, USA in 2010. From 1998 to 2002 he was a director engineer in XJ Electric Co. Ltd. in China. In 2003 he joined the HUST and now is a professor at HUST. His current research interests include renewable energy integration, energy storage application, DC grid, and power system operation and control.

Xingning HAN received the B.Eng. degree in electrical engineering from Hohai University, Nanjing, China, in 2011. Currently, she is pursuing the Ph.D. degree in electrical and electronic engineering from Huazhong University of Science and Technology (HUST), Wuhan, China. Her current research interest is the optimization with energy storage system.

19

Low-carbon benefit analysis on DG penetration distribution system

Shaoyun GE, Li XU, Hong LIU (✉),
Jian FANG

Abstract In response to the severe energy and environmental issues, CO_2 emission reduction and low-carbon development are inevitable. China has become the biggest CO_2 emitter in the world since 2006. As a major CO_2 emission source in China, the power industry is facing greater pressure for carbon emission abatement. By applying various low-carbon power technologies and mechanisms, the potential for CO_2 emission reduction in power systems is considerable. This paper proposes a pseudo-sequential Monte Carlo simulation method for the low-carbon benefit evaluation of distribution system including distributed wind turbines, solar array and battery energy storage systems. The hourly sequential models and state variation models for wind turbines, solar array and battery are established. The non-power components are sequentially sampled and the components of the wind turbine, the solar array and the battery are non-sequentially sampled. The failure modes and effect analysis (FMEA) procedure for the system are discussed and the heuristic load shedding approach is used, then the low-carbon benefit evaluation procedure is illustrated. Based on this, the system state transition sampling method for calculating the loss of expected energy index with high DG penetration is proposed. The state transition models for non-active elements and DGs are established. The validity of the proposed method is demonstrated by a study case. The method can be used for the low-carbon benefit analysis on DG penetration distribution networks.

Keywords Low-carbon, Distribution generation penetration, Distribution system, Monte Carlo simulation method

1 Introduction

Climate change and energy crises have become two important challenges for the sustainable development of human society since entering the new millennium. Excessive use of non-renewable fossil fuels is a main cause for concern [1–6]. In China, the proportion of thermal power installed capacity has exceeded 70%, and CO_2 emission of thermal power generation accounts for over 95% of the total emission of power industry. Hence, implementation of carbon emission reduction in China's power industry is critical and fundamental to coping with the challenges of global warming. China has embarked on a low-carbon and resource efficient development pathway, with the focuses on developing clean energy and addressing environmental problems [7–9]. As a significant part in energy supply, the distribution system is vital for the development of clean distribution energies, such as distributed wind turbines and solar array [10–14]. The Chinese Government decided to build smart grids and develop domestic emission trading system to facilitate the deployment of green energies and to reduce CO_2 emissions [15–19].

As an effective complement to centralized power generation, distributed generation (DG) technologies are becoming mature. With flexible operational mode and environment-friendly features, more and more distributed generations are integrated into the distribution network,

S. GE, L. XU, H. LIU, Key Laboratory of Smart Grid,
Tianjin University, Tianjin 300072, China
(✉) e-mail: liuhong@tju.edu.cn
L. XU
e-mail: xuligodfather@gmail.com
J. FANG, Electric Power Research Institute, Guangzhou Power
Supply Company, Guangzhou 510000, China

which brings benefit on the carbon reduction of distribution system [20]. As two basic methods of loss of expected energy (LOEE) for the distribution system with DG, the studies of analytical method [21–29] and Monte Carlo method [30–36] have made some progress. These studies indicate that the adoption of DG is actually equals to increasing the number and capacity of the backup of distribution system, thus improving the carbon reduction of the distribution system.

In Section 2, the boundary conditions of the study are confined. In Section 3, the DGs mathematical modeling are introduced, and the Monte Carlo simulation methods designed for energy without the high penetration case are proposed in detail. In Section 4, a study case is used to exemplify the methods mentioned above. Finally, the conclusion is drawn in Section 5.

2 System configuration

Alternative architectures are possible for the electrical interconnection of DG components, including the possibility for DC interconnection of PV stations to the battery inverters, and all components are connected to the AC distribution network, due to its superior flexibility. In any case, this is not a central issue, from the point of view of energy calculations. To achieve high DG penetration levels, the system must be capable of operating in "DG-only" mode, i.e., with the upper level energy out of supplied. For this purpose, the BESS inverter should provide frequency and voltage regulation, while additional technical requirements may also exist, including the provision of sufficient short-circuit capacity for proper operation of the network overcurrent protection. For this purpose, additional devices, such as synchronous condensers, may be foreseen, while flywheels may also be used for primary frequency response, to reduce pressure on the BESS system. While such technical issues are fundamental for a real-world system, in the analysis presented in this paper emphasis is placed on the energy transactions between system components on relatively long time scales (hours and beyond), and hence, all dynamic phenomena and associated components have been ignored.

3 Modeling and simulation

3.1 Simulation principles

Modeling of the system accounts for the energy transactions between the various components, while all electric transients are ignored, so that long-term simulations (e.g., annual) are possible. A one hour simulation time step is used, and scheduling decisions are made at the beginning of each simulation hour.

In low DG penetration situation, the upper level power supply is considered as sufficient to the load, the DG act as a supplement to the system, the system is operated as a distribution system. When in high DG penetration situation, the DG act as a main power source, and the upper level power supply needs not to be sufficient, when the system components are in normal condition, the system is actually operated as a generation system, the reliability assessment is used as energy sufficiency one.

FMEA method is used for simplify the network, it only matters when the high DG penetration system state is in fault, and the system reliability assessment process is the same as in low DG penetration system at this moment and the FMEA method was used the same as be mentioned in the simulation of the low penetration. When the high DG penetration system is in normal state, the system transfers to the generation state, and the generation reliability assessment process does not need to use FMEA method to be simplified. And the FMEA method is just simply mentioned as [37] in low DG penetration section in this paper because it is a mature tool to simplify the network and it is not the core work of this paper.

The low DG integration system reliability assessment uses the pseudo-sequential Monte Carlo simulation method because it only needs to focus on fault incident of the system, and ignore normal state of the system, thus making the assessment simple and efficient. But for the high DG penetration system, the reliability assessment needs to calculate the energy balance of the system at every hour, for tens of thousands of years simulation, the simulation for every component is meaningless and redundancy. So in high DG penetration system, we adopt the system state transition sampling method, the sample object is the system state instead of the state of each component, thus simplifying the simulation work.

3.2 DG modeling

Real-time wind speeds are generated in each period of a year with autoregressive moving average model in [38], based on which real-time output of wind turbines is calculated by output curve of wind turbine generation (WTG) [39].

Real-time output of photovoltaic (PV) panels is calculated by real-time solar radiation and output curve [40] like output model of wind turbines.

This paper chooses the lead-acid battery as the battery energy storage system (BESS). A variant of the kinetic battery model (KiBaM) [41] is used to simulate the lead-acid BESS.

Using real time outputs of DG was to coordinate the real time load model to make the simulation closer to the reality situation, the time interval was set as 1 hour because for the Monte Carlo simulation method, the simulation time would be reach tens of thousands of years to obtain the convergence result of reliability, longer time interval would make the result inaccurate, and shorter time interval would not be more accurate, but increase calculation work enormously. Considering the repair time of network equipment is in hours, we use 1 hour interval as DG output interval. And for the real-time wind speed model, we chose the widely used autoregressive moving average model in [38] to obtain the real time wind speeds, the outcome result of autoregressive moving average model for the wind speed is in 1 h time interval.

3.3 CO_2 emission issue

According to fuel combustion data of coal, a complete combustion of coal equivalent CO_2 emission factors was 2.62 kg CO_2/kg. The coal consumption of load center and power generation base were both 310 kg/(MWh) [7].

3.4 Simulation of low penetration

3.4.1 Sequential sampling for non-source component

All non-source components used the two-state Markov model, as shown in Fig. 1, and the Monte Carlo Simulation Method for state duration sampling was adopted for its sequential sampling [42, 43].

3.4.2 Sampling for DG

Without considering the limitation of upper system capacity, only when the non-source components in system are in fault, the outage of DG and BESS can exacerbate the power off risk of the user. Compared with normal state, the fault duration of the non-source components is very short, DG and BESS states can be considered to be unchanged during the fault, so their state can be sampled non-sequentially.

Single wind turbine is still using two-state Markov model in Fig. 1. For wind turbine set consisting of m wind turbine units, m random number between [0, 1] is first generated on behalf of its sample state, $R_i > P_d$ represents that the unit is working.

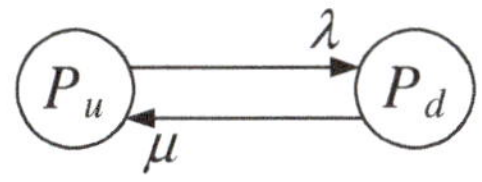

Fig. 1 Two-state Markov model

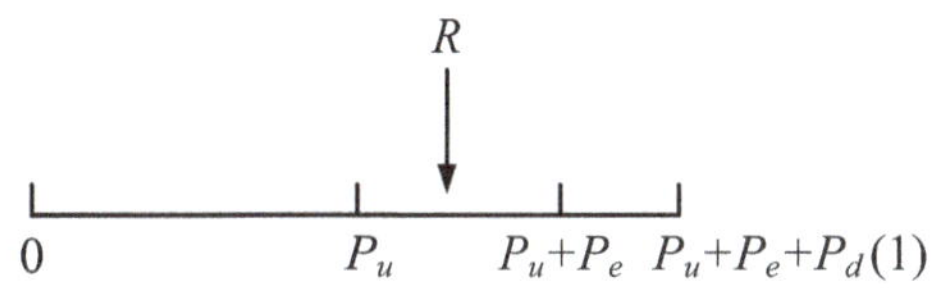

Fig. 2 Sampling process for the three-state element

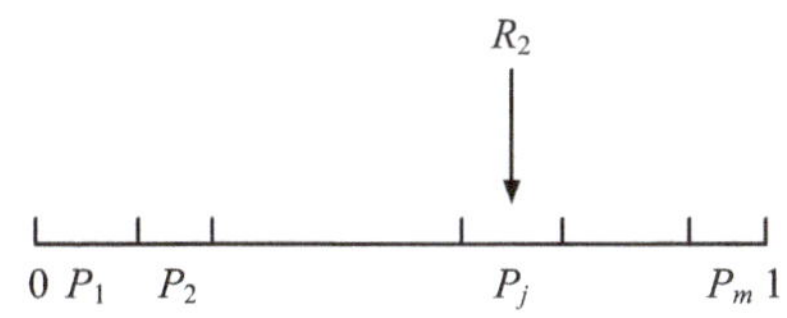

Fig. 3 Explanation of system state transition sampling

PV array consists of a set of photovoltaic panels, to reflect a portion of PV array fault, on the basis of two-state model, the derating state is added, its probability is P_e. Sampling process is shown in Fig. 2. Firstly, the probability of all the states is cumulatively placed on number axes between [0,1], the random number R represents the state, that is, PV array. For example, in Fig. 3, the photovoltaic array is in derating state.

The sample of the BESS state consists of two parts: one is sample of the running state, the other is sample of the state of charge (SOC). Sampling methods for the running state is same with PV array, the derating state represents a part of BESS failure. When sampling the SOC, SOC is converted from a random number in accordance with the SOC changing law on integration mode.

3.4.3 Failure mode and effect analysis (FMEA) process

FMEA analysis is the basis of LOEE assessment. The DG integration will have an impact on the traditional FMEA process. The FMEA process in [37] is used in this paper.

3.4.4 Load shedding strategy

Load shedding determination conditions: at any moment of battery discharge during island, the greatest power of all the battery can be released to be no less than the net exchange power in island:

$$\sum_{i=1}^{n} P^{i}_{out,1}(t)\eta_d \geq P_{ex}(t) \quad P_{ex}(t) \geq 0,\ t \in [t_{st}, t_{end}] \tag{1}$$

where $P^{i}_{out,1}(t)$ is the maximum power which can be released by the battery pack i during time interval t; n is the number of battery pack; $P_{ex}(t) = L(t) + S_l(t) - D(t)$, in which $L(t)$ is the total load; $D(t)$ is the sum of all DG output; S_l(t) is the active power loss within island, to simplify the flow calculation, $S_l(t)$ can be set at about 5% of the current load [44].

In order to improve simulation speed, this paper takes the following heuristic load shedding strategies: Firstly, assuming that all load points are not shed, when the case does not meet the following conditions: (1) the smallest load is shed firstly, repeat this process until it satisfies (1). During load shedding process, the dynamic switching was not considered so the shedding state was unchanged during islanding time. When (1) is satisfied, each battery power is released by the proportional ratio of $P^i_{out,1}(t)$, and the power is to be absorbed by proportional ratio of $\max(P^i_{in,1}(t), -P^i_{in,2}(t), -P^i_{in,3}(t))$. It should be noticed that, for the load points in downstream quarantine segment and downstream active islanding segment, if a load point is shed in both parts of the two island time, the power off frequency is 1, and the power outage time is the sum of two part.

3.4.5 Simulation process

1) Input the network structure, use the method in Section 3.4.3, automatically traversing feeder segments classification of all non-faulty components by computer, establish FMEA table.

2) Set initial analog clock value for 0, at this time all system components are in normal state. Each non-source system component generates a random number, use the state transition model to generate TTF for each component.

3) Find the component with the smallest TTF (TTFmin), and advance the analog clock to TTFmin. If some components are of the same TTF, randomly select one of them.

4) Generate a new random number and convert it into TTR for this component. At the same time, generate power isolation and recovery time (RT).

5) By using the FMEA table generated in Step 1, check the feeder segments under this component outage. The outage time of the fault segment load point is TTR, the upstream quarantine segment load point outage time is RT, and record the time duration of power outages and the insufficient of power supply.

6) For each upstream active island segment, sample DG state and BESS's SOC at TTFmin, calculate real-time output of DG and load during [TTFmin, TTFmin + RT], use the method in Section 3.4.4 to shed load, record the frequency and duration of load point outage.

7) Similar to Step 6, in [TTFmin, TTFmin + RT], simulate the downstream active island; in [TTFmin + RT, TTFmin + TTR], simulate the downstream big island, and record the frequency and duration of the load point outage.

8) For the faulty component in Step 3, re-generate a random number and convert it into TTFN for a new running time. At the moment, the component's TTF is updated to TTFmin + TTR + TTFN.

9) If the analog clock is less than predetermined simulation time, return to Step 3; otherwise, statistically count the outage frequency, duration and insufficient of each load points, and then calculate the LOEE index.

3.5 Simulation of high penetration

3.5.1 Feature of high penetration system

High penetration distribution network with DG will have both power generation and distribution properties, so-called as active distribution network, the key of its LOEE assessment, is how to properly combine power generation and distribution properties together. To solve this problem, this paper uses the sequential Monte Carlo sampling method based on the system state transition sampling to conduct the LOEE assessment for distributed network at high DG penetration level, thus it can simultaneously calculate the LOEE index of power generation and distribution.

3.5.2 System state transition sampling method

The system state transition sampling method is focused on the state of the whole system rather than that of on component in the system. Assuming that the system is composed of m components consisting of n states. The state space of the system is $G = \{S_1, S_2, \ldots, S_n\}$. Supposing that the present system state is S_k and the state transition rate of each component in the system is λ_i $(i = 1, 2, \ldots, m)$. Since the state duration of each component follows an exponential distribution, the duration of the system state S_k also follows the exponential distribution. The duration of the system state S_k has the probability density function as:

$$f(t) = \sum_{i=1}^{m} \lambda_i \exp\left(-\sum_{i=1}^{m} \lambda_i t\right) \tag{2}$$

It can be seen that the state transition rate to the next state from the system state S_k is the sum of state transition rate of each component in the system state S_k.

State transition of any component in the system can lead to a system state. Consequently, the system has m possible states starting from state S_k. The probability that the system reaches one of these possible states is given as:

$$P_j = \lambda_j / \sum_{i=1}^{m} \lambda_i \tag{3}$$

where λ_j is the state transition rate of the jth component.

It is obviously that:

$$\sum_{j=1}^{m} P_j = 1 \tag{4}$$

Two uniformly distributed random numbers R_1 and R_2 between [0, 1] are generated to sample the duration of the system state S_k and the system state which will be reached. The duration of the system state S_k is given by (5):

$$T_k = -\ln R_1 / \sum_{i=1}^{m} \lambda_i \tag{5}$$

The next system state can be determined by the sampling shown in Fig. 3. The probabilities of m possible future states are successively placed in the interval [0,1]. If R_2 falls into the segment corresponding to P_j. The transition of the jth component leads system to the next system state.

3.5.3 Analysis and simulation of generation state

According to different states of the components in the system, the state space of the distribution system with distributed generation G can be divided into two sub sets G_1 and G_2. In the set G_1, all non-source components are in normal state,in the set G_2, a non-source element is in fault state.

In the traditional distribution system, when all non-source components are in normal state, the system load can be supplied without considering the higher level power supply capacity. But the distribution system with DG is a multiple supply network, thus the load in it will be supplied by both DG and the higher level power. Because DG can supply part of the load, the substation capacity will be reduced in order to reduce carbon emission. So it is necessary to consider the impact of the higher level power supply capacity. In this case, whether the total generation of the system including DG can supply all the load needs to be assessed. It is in fact a part of the LOEE evaluation of generation system.

The LOEE of generation system can be evaluated by simple simulation. The LOEE index is determined by superposing the chronological load curve on the generation capacity curve. Assume that the supply from higher level is $P_s(t)$, and the DG provides $P_{DG}(t)$ during the tth period. The sum of the real time load and the loss is $P_L(t)$ (Assuming that the loss is 5% of the load to simplify the power flow calculation [45]). It means that the system is suffering power shortage when $P_s(t) + P_{DG}(t) < P_L(t)$. The LOEE index of generation system is calculated with sequential simulation during all periods. The method will be improved with the consideration of battery charging and discharging strategy. Two dispatch strategies of battery used in the paper are as follows [46].

Load-following strategy: On the load-following strategy, renewable power sources charge the BESS but the generators do not. The BESS is discharged when DG cannot provide enough supply to the load. The generators serve as the backup.

Cycle charge strategy: When DG output is surplus, the BESS is charged by excess power; when DG output is insufficient, if the maximum output of DG and BESS still cannot meet needs of all the loads, the system will firstly supply the remaining load, then to charge the BESS. In addition, to prevent the battery be in a low SOC, once the battery starts charging, it needs to be charged to a specified K_{SOCset}.

Figures 4 and 5 show the simulation process in load following strategy and in cycle charging strategy, respectively. $P_d^i(t)$ is the maximum output of the ith BESS during the tth period. $P_c^i(t)$ is the maximum active power that the battery could absorb. Considering the difference of the capacity and SOC between batteries, the exchanging power of each battery is proportional to its $P_d^i(t)$ or $P_c^i(t)$. For example, when the DG is charging all the batteries, the ith battery can be obtained:

$$P_{ex}^i(t) = (P_{DG}(t) - P_L(t)) \times P_c^i(t) / \sum_{i=1}^{m} P_c^i(t) \tag{6}$$

However, the power that it really absorbs $P_{in}^i(t)$ is $\min\{\eta_c P_{ex}^i(t), P_c^i(t)\}$. η_c is charge efficiency.

3.5.4 Simulation process

The time sequential Monte Carlo simulation using system state transition sampling evaluates the LOEE of distribution system containing distributed wind energy, solar energy and BESS. The step for simulation is 1 hour. The procedure consists of the following steps:

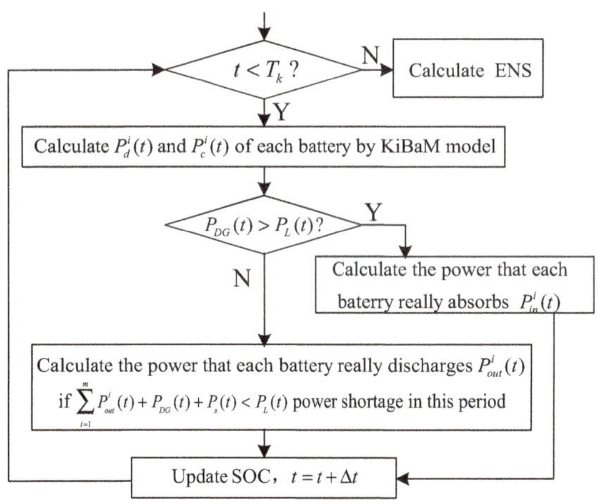

Fig. 4 Simulation process in load-following strategy

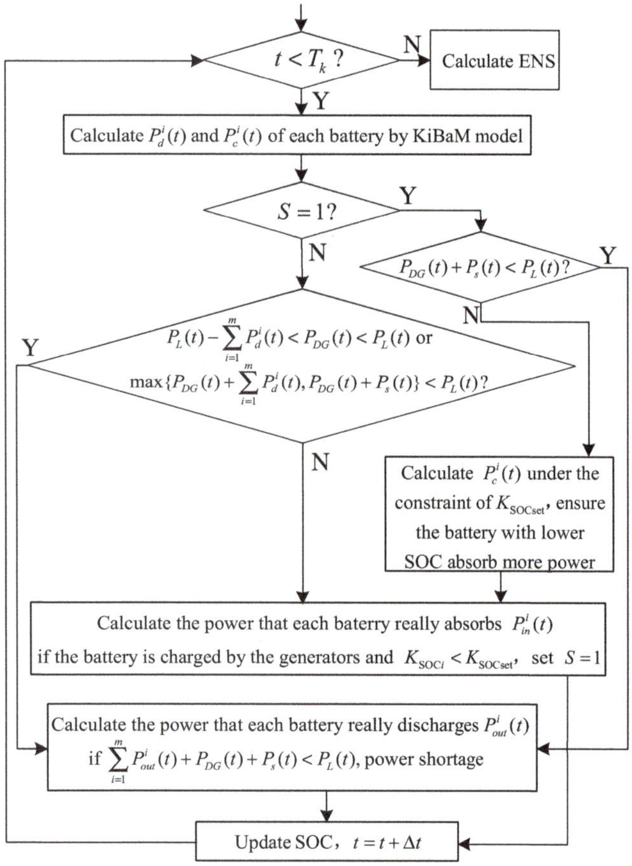

Fig. 5 Simulation process in cycle charging strategy

1) The simulation process starts from the normal system state in which all generating units and transmission components are in the up state at $t_0 = 0$.

2) Calculate the duration of the current system state T_k by (4) with the transition rate λ_i of each component. Get λ_i of a three-state component by adding its two state transition rates. If the present system state is a state in which at least one non-power component is in fault state, set the transition rate of other components zero.

3) If all the non-power components are in the normal state in the current state, call the generation system simulation process, record the energy not supplied ($K_{ENS_k^i}$) during T_k . If there is a non-power component in fault state, call the distribution system simulation process shown in E and record the indices λ_k^i, γ_k^i and $K_{ENS_k^i}$.

4) Set $t_0 = t_0 + T_k$ If the time is shorter than the year limit, sample the next system state by the method in Fig. 1 and go to Step 2. If t_0 is longer than N, stop simulating and calculate the LOEE index (MWh/year):

$$K_{LOEE} = \frac{1}{N}\sum_{k=1}^{P} K_{ENS_k} \quad (7)$$

where P is the simulation times of the generation system state.

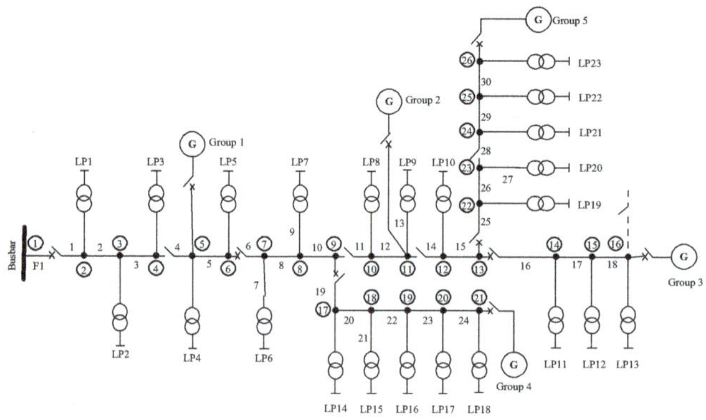

Fig. 6 Modified system under study

4 Study case

This paper uses the multi-branch feeders in the transformed IEEE RBTS Bus6 as examples, as shown in Fig. 6, including a segment bus (substation), 30 feeder segments, 26 nodes, 23 distribution transformers, 23 load points, 5 DGs, and several switches. Each DG includes several fans of the same type, a photovoltaic array and a battery, fan type is Enercon E33, the rated power of each WTG is 335 kW, and the battery type is Hoppecke 24 OPzS 3000.

The length of each line segment and the peak of load point are the same as [47], the total load peak is 11.653 MW, and the time-varying load models are the same as [48]. The feeder failure rate is 0.065 times/year × km, the distribution transformer failure rate is 0.015 times/year, the average repair times are 5 h (repair rate is 0.2 times/h), distributed exponentially. Single wind turbine failure rate is 4.6 times/year, the repair rate is 58.4 times/year. The transfer rate of PV array is the same as the battery, $\lambda = 1.4$, $\mu = 58.4$, $\lambda_d = 3.2$, $\mu_d = 46.72$, $\lambda_f = 11.68$, $\mu_f = 0$, the unit is all times/year, the derating operation capacity is an half of the full state. Bus and switches are completely reliable, fault isolation and recovery time take constantly 1 h, and the number of users for each load point is one.

4.1 Low-carbon benefit of DG

WTG PV and BESS are taken in each DG, the capacity of WTG is 1,005 kW, the capacity of the PV is 1,005 kW and the capacity of the battery is 13,200 kWh, when calculated under different supply substation capacity constraints using this method, the system reliability indices are shown in Table 1. We can see that, when the capacity of the supply substation is less than the total system load peak (11.653 MW), the generating indicators LOEE cease to be 0, indicating that the system will face a shortage risk. With the decrease of the supply substation capacity, power generation and distribution reliability indices are increased,

Table 1 Low-carbon benefit of DG

Charge & Discharge strategy	Substation capacity (MW)	LOEE (MW)	CO_2 emission reduction (t)
Load-following	13	0	0
	11	3.4664	1.6244
	9	156.94	3.2488
	7	1174.6	4.8732
	5	3932.2	6.4976
	3	10653	8.122
	1	24362	9.7464
Cycle charge	13	0	0
	11	0.2931	1.6244
	9	14.24	3.2488
	7	260.84	4.8732
	5	2268.3	6.4976
	3	9145	8.122
	1	24161	9.7464

especially the growth of power generation indices is more sensitive.

From Table 1 we can see that if we appropriately reduce the capacity of the supply substation, the system can still maintain a certain level of LOEE, whereas the CO_2 emission reduction was increased linearly correlation with the reduction of supply capacity. Take cycle charging strategy for example, when supply substation capacity reduces to 9 MW, the system's total lack of power supply LOEE will be 14.24 MWh/year, which is still lower than 49.290 MWh/year which is the number before the DG accessed, and the CO_2 emission reduction was 1.6244 t. This shows that after DG access to the distribution system, when there is no increase in the system load, the capacity of the supply substations may appropriately reduce. In other words, it reduces the CO_2 emission. However, if the supply substation capacity reduce further, or only rely on the renewable DG to supply the load growth, the level of LOEE will increase greatly. For example, when the supply substation capacity is lowered to 7 MW, LOEE in cycle charging strategy will surge to 260.84 MWh/year.

4.2 Impact on substation and DG capacity

Table 2 shows the LOEE calculated with changing supply substation capacity and DG capacity in cycle charging strategy. The DGs access in different locations. Column in Table 2 stands for the supply distribution capacity while row stands for total DG capacity. Each "×1" means that DG consists of WTG for 1,005 kW, PV for 1,005 kW and BESS for 13,200 kWh. The capacity of DG in "×2" is twice of that in "×1". The rest may be deduced by analogy.

From Table 2 we can see that in order to achieve a certain level of system reliability the capacity of supply substation and DG can be comprehensively selected. For example, when the requirement of LOEE is about 23 MWh/year, we can assume that the capacity of the supply substation is 11 MW and the DG capacity is "× 1"; we can also assume that the supply substation capacity is 5 MW and DG capacity is "× 6" . However, due to the volatility of the output of renewable DG, in order to maintain the LOEE level of the system, more DGs are needed to compensate for the reduced superior substation capacity. Compared with the former DGs, DGs in the latter became the main power supply in the system, meanwhile the supply substation capacity can decrease 6 MW, but the total capacity of the wind turbines and photovoltaic in the system will increase approximately 25 MW, while battery capacity will increase approximately 330 MWh. Therefore, when DG is the main power supply of the system, in order to ensure a certain degree of LOEE, total rated capacity of DG should be much higher than the system total load peak, and the cost is higher. According to the actual situation, we should reasonable optimize the capacity of supply substation and DG in order to meet the requirements of carbon emission reduction.

4.3 System reliability assessment with DG integration

Figure 7 shows the system average interruption frequency with the increase of DG capacity. The WTG capacity and the PV capacity units are kW, the Section 4.2 means the BESS, its unit is kWh. Figure 8 shows the

Table 2 Impact on substation and DG capacity

Capacity	CO_2 reduction (t)	×1	×2	×3	×4	×5	×6
11	1.6244	25.097	18.295	17.402	17.03	16.911	16.701
9	3.2488	39.113	23.467	19.442	18.114	17.371	16.876
7	4.8732	286.591	96.966	66.349	42.081	23.958	17.283
5	6.4976	2296.35	852.53	406.22	170.22	58.804	21.871

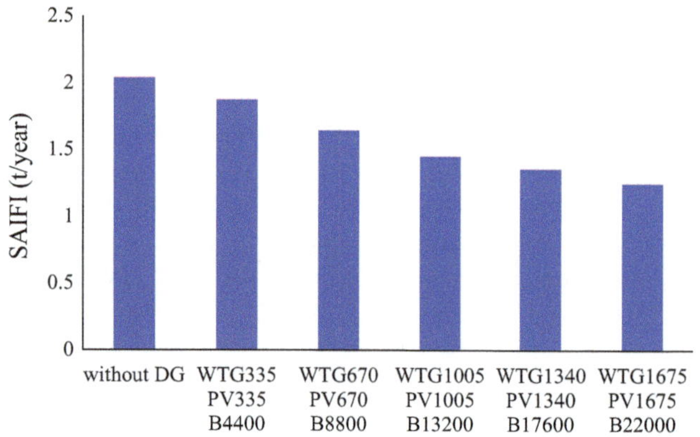

Fig. 7 System average interruption frequency with increase of DG capacity

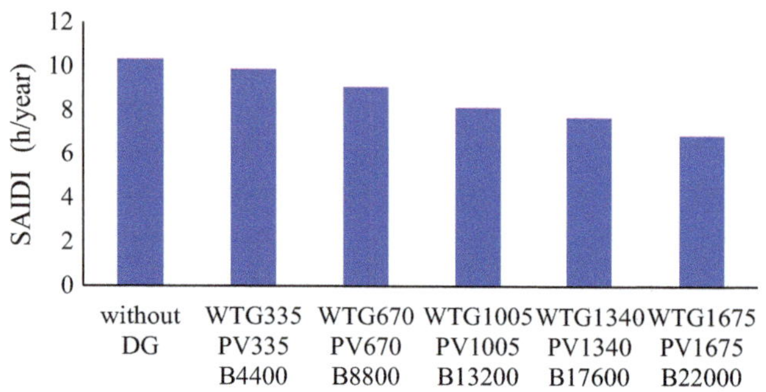

Fig. 8 System average interruption duration with increase of DG capacity

system average interruption duration with the increase of DG capacity.

System Average Interruption Frequency Index (SAIFI) and System Average Interruption Duration Index (SAIDI) are the most representative system reliability index. As can be seen from Fig. 7 and Fig. 8, DG can improve the reliability level of the system.

5 Conclusion

Low-carbon benefit assessment problem in l DG penetration active distribution system is treated by using Monte Carlo simulation method. Results indicate that the integration of DG can reduce the CO_2 emission and increase the system reliability. The method can be applied to the planning and coordination of CO_2 emission requirement of the power system.

Acknowledgements This work is supported by National Natural Science Foundation of China (No. 51477116), National Natural Science Foundation of China (No. 51261130473) and Special science and technology innovation project of Guangzhou Power Supply Company China Southern Power Grid (No. K-GZM2014-020).

References

[1] Zhou TR, Chen QX, Zhu WT et al (2012) A comprehensive low-carbon benefits assessment model for power systems. In: Proceedings of the 2012 IEEE international conference on power system technology (POWERCON'12), Auckland, 30 Oct–2 Nov 2012, 5 pp

[2] Lu SY, Wu YW, Lou SH et al (2014) A model for optimizing spinning reserve requirement of power system under low-carbon economy. IEEE Trans Sustain Energy 5(4):1048–1055

[3] Wu YW, Lou SH, Lu SY (2011) A model for power system interconnection planning under low-carbon economy with CO emission constraints. IEEE Trans Sustain Energy 2(3):205–214

[4] Hu XD, Zhou JH, Tan XD et al (2010) A low-carbon electricity model: Integrated resource strategic planning and its application. In: Proceedings of the 2010 IEEE Power and Energy Society general meeting, Minneapolis, 25–29 Jul 2010, 7 pp

[5] Wang XQ, Ding HW, Qiu MM et al (2011) A low-carbon production scheduling system considering renewable energy. In: Proceedings of the 2011 IEEE international conference on service operations, logistics, and informatics (SOLI'11), Beijing, 10–12 Jul 2011, pp 101–106

[6] Zhou TR, Kang CQ, Chen XY et al (2012) Evaluating low-carbon effects of demand response from smart distribution grid. In: Proceedings of the 3rd IEEE PES international conference and exhibition on innovative smart grid technologies (ISGT Europe'12), Berlin, 14–17 Oct 2012, 6 pp

[7] Zhang QY, Wang J, Li JG et al (2012) Evaluation of the sequence of voltage levels based on low-carbon theory. In: Proceedings of the 2012 China international conference on electricity distribution (CICED'12), Shanghai, 10–14 Sept 2012, 4 pp

[8] Cui MY (2012) Low carbon dispatch of distribution network containing microgrid using chaotic ant swarm. In: Proceedings of the 2012 international conference on control engineering and communication technology (ICCECT'12), Shenyang, 7–9 Dec 2012, pp 818–821

[9] Ochoa LF, Mancarella P (2012) Low-carbon LV networks: challenges for planning and operation. In: Proceedings of the 2012 IEEE Power and Energy Society general meeting, San Diego, 22–26 Jul 2012, 2 pp

[10] Koutitas G (2010) Low carbon network planning. In: Proceedings of the 2010 European wireless conference (EW'10), Lucca, 12–15 Apr 2010, pp 411–417

[11] Ji Z, Kang CQ, Chen QX et al (2013) Low-carbon power system dispatch incorporating carbon capture power plants. IEEE Trans Power Syst 28(4):4615–4623

[12] Fang C, Chen LJ, Zhang Y et al (2014) Operation of low-carbon-emission microgrid considering wind power generation and compressed air energy storage. In: Proceedings of the 33rd Chinese control conference (CCC'14), Nanjing, 28–30 Jul 2014, pp 7472–7477

[13] Dong XW, El-Gorashi T, Elmirghani JMH (2011) Renewable energy for low carbon emission IP over WDM networks. In: Proceedings of the 15th international conference on optical network design and modeling (ONDM'11), Bologna, 8–10 Feb 2011, 6 pp

[14] Huang XH, Zhang W, Ding H (2011) Research on Chinese low-carbon economy mode and system. In: Proceedings of the 2011 international conference on electronics, communications and control (ICECC'11), Ningbo, 9–11 Sept 2011, pp 3881–3884
[15] Zhang XH, Jia RN, Tu GP (2012) Research on low-carbon energy policies selections based on SD modeling. In: Proceedings of the 2012 international symposium on management of technology (ISMOT'12), Hangzhou, 8–9 Nov 2012, pp 688–692
[16] Zhou ZY, Liu Y, Jin CP (2009) Study on construction of low-carbon featured chlor-alkali industry bases energized by non-grid-connected wind power in offshore Jiangsu province. In: Proceedings of the world non-grid-connected wind power and energy conference (WNWEC'09), Nanjing, 24–26 Sept 2009, 4 pp
[17] Cui MY, Ai X, Lei ZL (2010) Study on the low-carbon generation scheduling of distribution network containing microgrid. In: Proceedings of the 5th international conference on critical infrastructure (CRIS'10), Beijing, 20–22 Sept 2010, 4 pp
[18] Parkinson S, Wang D, Djilali N (2012) Toward low carbon energy systems: The convergence of wind power, demand response, and the electricity grid. In: Proceedings of the innovative smart grid technologies-Asia (ISGT Asia'12), Tianjin, 21–24 May 2012, 8 pp
[19] Capros P, Tasios N, Marinakis A (2012) Very high penetration of renewable energy sources to the European electricity system in the context of model-based analysis of an energy roadmap towards a low carbon EU economy by 2050. In: Proceedings of the 9th international conference on the European energy market (EEM'12), Florence, 10–12 May 2012, 8 pp
[20] Bie ZH, Li GF, Wang XF (2011) Review on reliability evaluation of new distribution system with micro-grid. Electr Power Autom Equip 31(1):1–6 (in Chinese)
[21] Liu CQ, Zhang Y (2007) Distribution network reliability considering distribution generation. Autom Electr Power Syst 31(22):46–49 (in Chinese)
[22] Qian KJ, Yuan Y, Zhou CK (2008) Study on impact of distributed generation on distribution system reliability. Power Syst Technol 32(11):74–78 (in Chinese)
[23] Zhu D, Broadwater RP, Tam KS et al (2006) Impact of DG placement on reliability and efficiency with time-varying loads. IEEE Trans Power Syst 21(1):419–427
[24] Bae IS, Kim JO (2007) Reliability evaluation of distributed generation based on operation mode. IEEE Trans Power Syst 22(2):785–790
[25] Billinton R, Li W (1993) A system state transition sampling method for composite system reliability evaluation. IEEE Trans Power Syst 8(3):761–767
[26] Atwa YM, El-Saadany EF (2009) Reliability evaluation for distribution system with renewable distributed generation during Islanded mode of operation. IEEE Trans Power Syst 24(2):572–581
[27] Atwa YM, El-Saadany EF, Guise AC (2010) Supply adequacy assessment of distribution system including wind-based DG during different modes of operation. IEEE Trans Power Syst 25(1):78–86
[28] Wang P, Billinton R (2001) Time-sequential simulation technique for rural distribution system reliability cost/worth evaluation including wind generation as alternative supply. IEE P-Gener Transm Distrib 148(4):355–360
[29] Wang P, Billinton R (2001) Reliability benefit analysis of adding WTG to a distribution system. IEEE Trans Energy Convers 16(2):134–139
[30] Park J, Liang W, Choi J et al (2009) A probabilistic reliability evaluation of a power system including solar/photovoltaic cell generator. In: Proceedings of the 2009 IEEE Power and Energy Society general meeting, Calgary, 26–30 Jul 2009, 6 pp
[31] Maliki G, Wang CS (2005) Reliability analysis on the integration of wind/PV hybrid distributed generation in distribution system. Autom Electr Power Syst 29(23):33–38 (in Chinese)
[32] Wang XD, Lin JK (2010) Reliability evaluation based on network simplification for the distribution system with distributed generation. Autom Electr Power Syst 34(4):38–43 (in Chinese)
[33] Ge SY, Wang HM, Wang YS et al (2012) Reliability evaluation of distribution system including distributed wind energy, solar energy and battery storage. Autom Electr Power Syst 36(5):16–23 (in Chinese)
[34] Savier JS, Das D (2007) Impacts of network reconfiguration on loss allocation of radial distribution systems. IEEE Trans Power Deliv 22(4):2473–2480
[35] Zhu D, Broadwater RP, Tam KS et al (2006) Impact of DG placement on reliability and efficiency with time-varying loads. IEEE Trans Power Syst 21(1):419–427
[36] Mitchell K, Nagrial M, Rizk J (2005) Simulation and optimization of renewable energy systems. Int J Electr Power Energy Syst 27(3):177–188
[37] Savier JS, Das D (2007) Impacts of network reconfiguration on loss allocation of radial distribution systems. IEEE Trans Power Deliv 22(4):2473–2480
[38] Fotuhi-Firuzabad M, Rajabi-Ghahnavie A (2005) An analytical method to consider DG impacts on distribution system reliability. In: Proceedings of the 2005 IEEE/PES transmission and distribution conference and exhibition: Asia and Pacific, Dalian, 14–18 Aug 2005, 6 pp
[39] Manwell JF, McGowan JG (1993) Lead acid battery storage model for hybrid energy systems. Solar Energy 50(5):399–405
[40] Billinton R, Wang P (1999) Teaching distribution system reliability evaluation using Monte Carlo simulation. IEEE Trans Power Syst 14(2):397–403
[41] Broadwater RP, Shaalan HE, Oka A et al (1994) Distribution system reliability and restoration analysis. Electr Power Syst Res 29(3):203–211
[42] Pregelj A, Begovic M, Rohatgi A (2006) Recloser allocation for improved reliability of DG-enhanced distribution network. IEEE Trans Power Syst 21(3):1442–1449
[43] Ge SY, Xu L, Liu H et al (2014) Reliability assessment of active distribution system using Monte Carlo simulation method. J Appl Math, 2014:1–10. Article ID 421347
[44] Lambert T, Gilman P, Lilienthal P (2006) Micropower system modeling with HOMER. In: Farret FA, Simoes MG (eds) Integration of alternative sources of energy. IEEE Press, Hoboken, pp 379–418
[45] Liu H, Han J, Ge SY et al (2014) Improved analytical method of power supply capability on distribution systems. Int J Electr Power Energy Syst 63:97–104
[46] Liu H, Guo YC, Ge SY et al (2014) Impact of DG configuration on maximum use of load supply capability in distribution power systems. J Appl Math, 2014:1–9. Article ID 136726
[47] Billinton R, Jonnavithula S (1996) A test system for teaching overall power system reliability assessment. IEEE Trans Power Syst 11(4):1670–1676
[48] Wang P, Billinton R (1999) Time sequential distribution system reliability worth analysis considering time varying load and cost models. IEEE Trans Power Deliv 14(3):1046–1051

Shaoyun GE received the Ph.D. degree in electrical engineering from Hong Kong Polytechnic University, 1998. Since 1988, he has been a researcher at the Urban Power System Planning, Optimal Operation and Control of Power System, and Smart Grid. He is a Professor at School of Electrical Engineering at Tianjin University, Tianjin, P. R. China.

Li XU received the Undergraduate degree in 2010 and is currently pursuing the Ph.D. degree in the School of Electrical Engineering at Tianjin University, Tianjin, P. R. China. His research area is in Urban Power System Planning and Reliability Assessment of Distribution Power System.

Hong LIU the corresponding author, received the Ph.D. degree in Electrical Engineering from Tianjin University, 2009. Since 2009, he has been a researcher at the Urban Power System Planning, Smart Grid and Electric Vehicle Charging Station. He is an Associate Professor at School of Electrical Engineering at Tianjin University, Tianjin, P. R. China.

Jian FANG has been a researcher at Distribution Power System Technology Supervisor and New Technology Promotion. He is an engineer at electric Power Research Institute, Guangzhou Power Supply Company, Guangzhou, P. R. China.

20

Sequential quadratic programming particle swarm optimization for wind power system operations considering emissions

Yang ZHANG (✉), Fang YAO, Herbert Ho-Ching IU, Tyrone FERNANDO, Kit Po WONG

Abstract In this paper, a computation framework for addressing combined economic and emission dispatch (CEED) problem with valve-point effects as well as stochastic wind power considering unit commitment (UC) using a hybrid approach connecting sequential quadratic programming (SQP) and particle swarm optimization (PSO) is proposed. The CEED problem aims to minimize the scheduling cost and greenhouse gases (GHGs) emission cost. Here the GHGs include carbon dioxide (CO_2), nitrogen dioxide (NO_2), and sulphur oxides (SO_x). A dispatch model including both thermal generators and wind farms is developed. The probability of stochastic wind power based on the Weibull distribution is included in the CEED model. The model is tested on a standard system involving six thermal units and two wind farms. A set of numerical case studies are reported. The performance of the hybrid computational method is validated by comparing with other solvers on the test system.

Keywords Combined economic and emission dispatch, Unit commitment, Particle swarm optimization, Sequential quadratic programming, Weibull distribution, Wind power

1 Introduction

Power system generation scheduling problem can be divided into two sub-problems, unit commitment (UC) and economic dispatch (ED). ED is an important task in the power system operation, which aims to allocate power generation match load demand at minimal possible cost while satisfying all the units and system constraints [1–3]. Suitable improvements in the unit outputs scheduling can contribute to significant cost savings.

Nowadays, with the awareness of environmental pollution contributed by the combustion of fossil fuels, building a low-carbon world has attracted widespread attentions. Many countries are trying to exploit clean energy in order to mitigate the greenhouse effects. The primary source of greenhouse gases (GHGs) is the combustion of fossil fuels. Coal, oil, and gas are the three major types of regular fuels, which produce emissions represented by GHGs, such as CO_2, NO_2, and SOx. In order to reduce the GHGs emissions, the combined economic emission dispatch (CEED) considering with UC was proposed, which can take account of fuel cost and emission tax together. Because the amount of emission from fossil-based thermal generators depends on the amount of generated power, therefore the emission cost increase leads to reduced overall power generated by thermal units, which in turn lowers emissions. Moreover, the natural economic forces will also help to catalyse the move to greater energy efficiency and use of renewable sources. Wind energy is among the major contributors to an overall reduction in GHGs emissions. Dispatch strategies normally can provide quick solutions to improve the current situation of system operation and reduce carbon emissions dramatically. On the other hand, exploiting renewable energy is another effective way to mitigate energy source deficiency, control GHGs emissions, and achieve smart grid vision [4–6]. Wind power being one of the most appealing renewable energy resources has gained widespread concerns during the last decades. Along with the introduction of various emission reduction schemes, increasing number of wind turbines have been installed around the world [7]. However, due to the intermittent and

Y. ZHANG, F. YAO, H. H.-C. IU, T. FERNANDO, K. P. WONG, The University of Western Australia, Perth, WA, Australia
(✉) e-mail: 20738164@student.uwa.edu.au

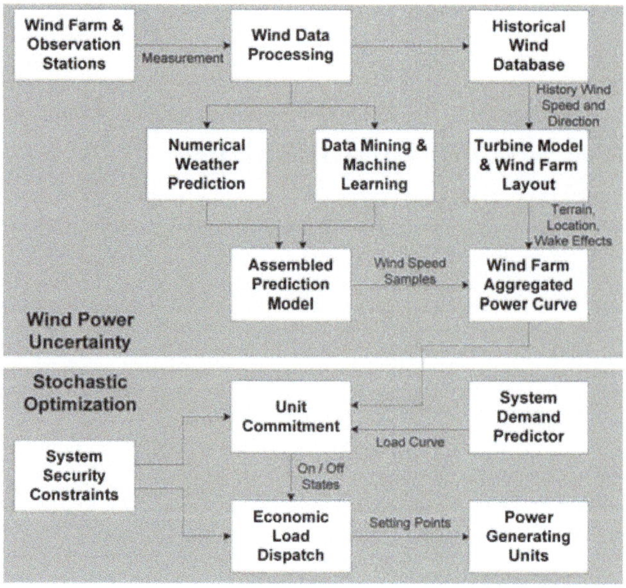

Fig. 1 Computational framework considering wind power uncertainties

stochastic characteristics of wind resource, wind power brings great challenges to power system economic dispatch problems. One of the major challenges is how to effectively accommodate the wind forecasting errors. Because variations of wind speed directly influence the power output of wind farms, which then causes difficulties in estimating suitable system reserve margin to ensure secure and reliable system operations. As a result, high penetration of wind power also causes high potential risks and more difficulties in power system operation. Although wind speed is difficult to forecast by single predictor, composite forecast model can statistically produce an optimal forecast by computing prediction results from a number of different methods. The fundamental concept is that if the errors in the forecasts produced by different methods are unbiased and have a low degree of correlation with each other, the random errors from the individual forecasts will tend to offset each other, with the result that a composite of the forecasts will have lower errors than any individual forecast. Moreover, huge number of publications has indicated that wind speed follows Weibull distribution approximately [8]. In order to assist with management of the uncertainties of wind forecasts, extensive researches have been conducted to develop kinds of probabilistic optimization strategies [9, 10]. In this paper, a schematic representation of computational framework contains wind power forecast and stochastic unit commitment/economic dispatch, which is adopted from [38], is shown in Fig. 1.

In order to accommodate the revised dispatch strategy, more efficient solvers are needed. Different heuristic techniques have been developed to solve the classical ED problems with constraints, to namely simulated annealing (SA) [11], genetic algorithm (GA) [12], evolutionary programming (EP) [13, 14], tabu search (TS) [15], pattern search (PS) [16], particle swarm optimization (PSO) [17, 18], as well as differential evolution (DE) [19, 20]. Based on our experience, when compared with other approaches, the PSO is computationally inexpensive in terms of memory and speed. However, these heuristic methods do not always guarantee discovering globally optimal solutions in finite time, especially when being applied into large-scale optimization problems. Therefore, more sophisticated computational tools are required. Recently, hybrid optimization techniques which combine different approaches receive widespread concerns. In [21], the authors presented a hybrid EP and sequential quadratic programming (SQP) for solving the ED problem with non-smooth fuel cost function. A hybrid self-tuning DE was proposed to solve the ED problem with kinds of constraints in [22]. In [23], a hybrid approach combining DE with biogeography-based optimization (DE/BBO) was developed to address both convex and non-convex ED problem. These hybrid optimization methods were found to be more effective and accurate. SQP is one of best nonlinear-programming method for constrained optimization. SQP–PSO technique is an effective nonlinear-programming method over a large number of test problems in terms of accuracy, efficiency and percentage of successful solutions.

In this paper, a CEED model incorporating wind power to minimize the total cost is proposed. Because of the stochastic nature of wind speed, wind power output is not deterministic. As a sequence, the probability distribution of wind speed must be taken into account in the CEED model.

In our CEED model, wind power is described as the three-parameter Weibull distribution. As ED problem in consideration of emission issue, there are some works reducing the NO_2, as well as SO_2, however, there are few papers on CEED in consideration of the CO_2 emission. In this paper, the reduction of CO_2 emission is one of the main concerns in the CEED model. In terms of the problem solver, we present a hybrid technique which combines SQP and PSO together. In the proposed algorithm, SQP is firstly used to solve the CEED problem without considering the valve-point effects, and then based on the obtained initial solution and boundaries PSO is employed to solve the CEED problem with non-smooth fuel cost function.

The rest of this paper is organized as follows. Section 2 introduces the probability of wind power. Section 3 describes the mathematical formulation of CEED considering UC problem with stochastic wind power. The basic concepts of SQP and PSO are introduced in Section 4, and the hybrid algorithm is also summarized in this section. Section 5 presents the numerical case studies solving the

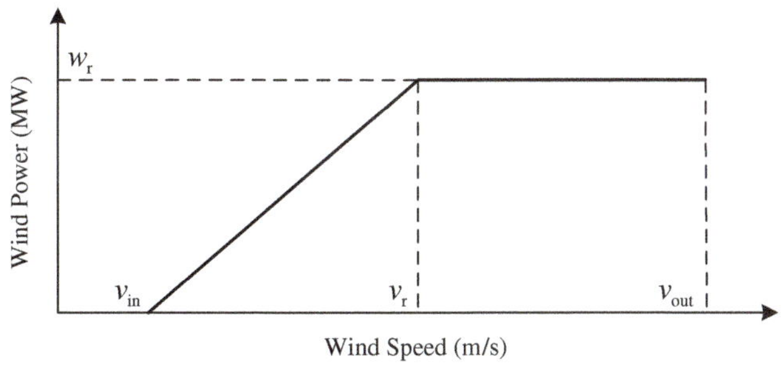

Fig. 2 Simplified wind turbine power curve

ED and CEED problems with and without wind power. Remarks and conclusions are given in Section 6.

2 Probabilistic modeling of wind power for edmodeling

Wind power, one of the most appealing renewable energy sources, has been widely developed in the recent years. Wind power energy has lots of advantages such as no pollution, relatively low capital cost involved, and the short gestation period required. However, the wind resource changes with locations and climates resulting in high uncertainties in the produced energy. The total power available from a wind turbine is equal to the product of the mass flow rate of the wind m_W, and $V^2/2$. Assuming constant area or ducted flow, the continuity equation states that $m_W = \rho AV$, where ρ is the density of the air in kg/m^3, A is the blades area in m^2, and V is the velocity in m/s. Thus, the total wind power becomes $P_W = (m_W V^2)/2 = (\rho A V^3)/2$ (MW). In this equation, the wind speed V is a random variable. Ignoring minor nonlinearities, the function relation between a given wind speed and power output can be described in Fig. 2.

In the above figure, w (MW) is the wind energy conversion systems (WECS) output power; w_r (MW) is the WECS output rated power; v_{in} (m/s), v_r (m/s), v_{out} (m/s) is the WECS cut-in speed, rated speed, and cut-out speed, respectively. Figure 1 shows that there is no power generated at wind speeds below v_{in} or above v_{out}; at wind speeds between v_r and v_{out}, the output is equal to the rated power of the generator; at wind speeds between cut-in wind speed and rated wind speed, the output is a linear function power.

Therefore, the wind power output can be described as,

$$\begin{cases} W = 0, & V < v_{in} \;\; or \;\; V > v_{out} \\ W = aV + b, & v_{in} \le V \le v_r \\ W = w_r, & v_r \le V \le v_{out} \end{cases} \tag{1}$$

where $a = \frac{w_r}{v_r - v_{in}}$, $b = -\frac{v_{in} w_r}{v_r - v_{in}}$.

Weibull distribution is the most popular density function that can be used to describe wind speed frequency curve [24–26]. An extensive review of various probability density functions of wind speed was provided in [26], and comparisons were made. The results indicated that the two-parameter Weibull distribution is the widely accepted model. Using two-parameter Weibull distribution, cumulative distribution function (CDF) and probability density function (pdf) of wind speed are

$$F_V(v) = 1 - \exp\left[-\left(\frac{v}{c}\right)^k\right], \quad (v \ge 0) \tag{2}$$

$$f_V(v) = \frac{k}{c}\left(\frac{v}{c}\right)^{k-1} \exp\left[-\left(\frac{v}{c}\right)^k\right] \tag{3}$$

where $k > 0$ is the shape parameter, $c > 0$ is the scale parameter.

According to (1), three portions of WECS power output can be analysed and the corresponding probabilities (CDF or pdf) can be calculated.

(i) For $V < v_{in}$ or $V > v_{out}$,

$$\begin{aligned} P(W = 0) &= P(V < v_{in}) + P(V > v_{out}) \\ &= F_V(v_{in}) + [1 - F_V(v_{out})] \\ &= 1 - \exp\left[-\left(\frac{v_{in}}{c}\right)^k\right] + \exp\left[-\left(\frac{v_{out}}{c}\right)^k\right] \end{aligned} \tag{4}$$

(ii) For $v_{in} \le V \le v_r$, $W = aV + b = \frac{(V - v_{in}) w_r}{v_r - v_{in}}$, depending on the definition of cumulative distribution function (CDF), the CDF of WECS output power can be described as,

$$\begin{aligned} F_W(w) &= P\{W \le w\} = P\left\{W = \frac{(V - v_{in}) w_r}{v_r - v_{in}} \le w\right\} \\ &= P\left\{V \le \frac{(v_r - v_{in}) w}{w_r} + v_{in}\right\} = F_V\left\{\frac{(v_r - v_{in}) w}{w_r} + v_{in}\right\} \end{aligned} \tag{5}$$

We can obtain the pdf of W by differentiating with respect to w. The chain rule for derivatives can be used, $\frac{dF}{dw} = \frac{dF}{du}\frac{du}{dw}$, where u is the argument of F, $u = \left\{\frac{(v_r - v_{in}) w}{w_r} + v_{in}\right\}$, and then we obtain

$$f_W(w) = \frac{k(v_r - v_{in})}{c w_r} \cdot \left[\frac{\frac{(v_r - v_{in}) w}{w_r} + v_{in}}{c}\right]^{k-1} \cdot \exp\left[-\left(\frac{\frac{(v_r - v_{in}) w}{w_r} + v_{in}}{c}\right)\right]^k \tag{6}$$

(iii) For $v_r \le V \le v_{out}$,

$$\begin{aligned} P(W = w_r) &= P(v_r \le V \le v_{out}) \\ &= F_V(v_{out}) - F_V(v_r) \\ &= \exp\left[-\left(\frac{v_r}{c}\right)^k\right] - \exp\left[-\left(\frac{v_{out}}{c}\right)^k\right] \end{aligned} \tag{7}$$

3 Mathematical formulation of CEED problem with wind power

This section describes the problem formulation of the proposed CEED considering UC model including wind power. The model aims at minimizing the total operation costs (including fuel cost, wind farm cost) and emission cost while satisfying the given constraints. In [27], an economic dispatch (ED) model incorporating wind power is developed. In order to accurately characterize the uncertainty in the availability of wind energy, penalty costs functions for both underestimation and overestimation cases were added. Inspired by the practical application, a similar CEED model are developed with an additional term incorporated to account for government wind farm subsidy. To address the uncertainties in wind power production, wind speed distribution probability functions are applied in formulating the optimization model.

3.1 Objective function

The objective function is formulated to minimize the total system operation costs and greenhouse gases (CO_2 and NO_2) emission costs. A cost function is obtained based on the ripple curve for more accurate modelling which contains higher order nonlinearity and discontinuity due to the valve point effect [16] and should be refined by a sine function [28]. The overall objective function can be expressed as the sum of these two terms,

$$\text{Min.}\ (Cost_1 + Cost_2) \tag{8}$$

1) Total system scheduling costs

$$\begin{aligned} Cost_1 = & \sum_{i=1}^{M} C_i(p_i) + \sum_{j=1}^{N} C_{\mathrm{w},j}(w_{j,\mathrm{av}}) \\ & + \sum_{j=1}^{N} C_{\mathrm{u},j}(W_{j,\mathrm{av}} - w_j) + \sum_{j=1}^{N} C_{\mathrm{o},j}(w_j - W_{j,\mathrm{av}}) \\ & - \sum_{j=1}^{N} C_{\mathrm{s},j}(w_{j,\mathrm{av}}) \end{aligned} \tag{9}$$

$$C_i(p_i) = a_i + b_i p_i + c_i p_i^2 + \left| d_i \sin\left(e_i\left(p_{i,\min} - p_i\right)\right)\right| \tag{10}$$

where M is the number of thermal power generators; N the number of wind turbines; a_i, b_i, c_i the cost coefficients of thermal generator i; $d_{i,}$ e_i are valve-point effects coefficients of thermal generator i; C_i the cost function of thermal generator i; $C_{\mathrm{u},j}$ the cost coefficient for not using all generated wind power due to the underestimation case; $C_{\mathrm{o},j}$ the cost coefficient for purchasing reserve power from other source due to overestimation case; $C_{\mathrm{s},j}$ the government subsidy parameter of turbine j; $C_{\mathrm{w},j}$ the cost coefficient of wind turbine j; p_i the actual power generated by thermal generator i; w_j the predicted wind power generated by turbine j; $W_{j,\mathrm{av}}$ the actual wind power generated by wind turbine j; $C_i(p_i)$ the fuel cost function of thermal generator i; $C_{\mathrm{w},j}(w_{j,\mathrm{av}})$ the wind power cost of the wind farm.

If the wind farm is owned by the system operator, this term may not exist. In this paper, the wind farm is assumed to be owned by the operator, so this cost is equal to zero. The underestimation cost $C_{\mathrm{u},j}(W_{j,\mathrm{av}} - w_j)$ occurs if the actual generated wind power is more than the predicted, thus the system operator should compensate for the surplus wind power cost. On the other hand, if the actual wind power is less the predicted scheduling power, the operator needs to purchase from an alternate source and pay the overestimation cost $C_{\mathrm{o},j}(w_j - W_{j,\mathrm{av}})$. The last term in the Eq. (9) is the wind power subsidy cost $C_{\mathrm{s},j}(w_{j,\mathrm{av}})$. As one of the renewable energy subsidy projects, wind farm in many countries receive a largely covert subsidy. An excellent example is the Renewable Obligation (RO) in UK. The RO is designed to encourage generation of electricity from eligible renewable sources in the UK [29]. In this paper, the wind farm was assumed to receive a fix cost subsidy for generating every MW wind power.

According to [29], the cost of underestimation will be assumed as follow,

$$\begin{aligned} C_{\mathrm{u},j}(W_{j,\mathrm{av}} - w_j) &= C_{\mathrm{u},j} \int_{w_j}^{w_{\mathrm{r},j}} (w - w_j) f_W(w) dw \\ &= C_{\mathrm{u},j} \left[\int_{w_j}^{w_{\mathrm{r},j}} w f_W(w) dw - w_j \int_{w_j}^{w_{\mathrm{r},j}} f_W(w) dw \right] \end{aligned} \tag{11}$$

where $w_{\mathrm{r},j}$ is the rated wind power from wind turbine j.

In terms of overestimation case, the cost equation will be in the similar manner,

$$\begin{aligned} C_{\mathrm{o},j}(w_j - W_{j,\mathrm{av}}) &= C_{\mathrm{o},j} \int_{0}^{w_j} (w_j - w) f_W(w) dw \\ &= C_{\mathrm{o},j} \left[w_j \int_{0}^{w_j} f_W(w) dw - \int_{0}^{w_j} w f_W(w) dw \right] \end{aligned} \tag{12}$$

Equations (11) and (12) can be solved through the wind power probability (4)–(7).

2) Greenhouse gases (GHGs) emission costs

$$Cost_2 = \sum_{i=1}^{M} F_{\mathrm{GHG},i}(p_i) \tag{13}$$

where $F_{\mathrm{GHG},i}(p_i)$ is the emission cost of thermal unit i.

$$F_{\mathrm{GHG},i}(p_i) = h \cdot EM_i(p_i) \tag{14}$$

$$EM_i(p_i) = ef_i\left(f_i + g_i p_i + h_i p_i^2\right) \tag{15}$$

where $EM_i(p_i)$ is the GHGs emissions of thermal generator i; ef_i the fuel emission factor of GHGs for thermal generator i; f_i, g_i, h_i the fuel consumption coefficients of thermal unit; h is the given GHGs emissions price which is determined by regulations and markets. GHGs are CO_2 and NO_2 in this paper

Equation (13) represents the fuel cost function of thermal generators. Equation (14) expresses the GHGs emission cost function.

3.2 System constraints

$$p_{i,\min} \leq p_i \leq p_{i,\max} \tag{16}$$

$$0 \leq w_j \leq w_{r,j} \tag{17}$$

$$\sum_{i=1}^{M} p_i + \sum_{j=1}^{N} w_j = p_d + p_{loss} \tag{18}$$

where p_d is total system loads; and p_{loss} is total transmission losses.

Inequality constraint (16) defines the limitations of thermal units output from the lower to the upper bound. And constraint (17) shows the wind power output limitations. Equation (18) gives the power balance between generations and loads including the transmission losses.

4 Hybrid optimization algorithm

In this section, a hybrid optimization algorithm is presented, which combines SQP and PSO together.

4.1 Sequential quadratic programming (SQP)

Since its popularization in the late 1970s, SQP has arguably become the most successful approaches for solving nonlinearly constrained optimization problems [30]. Backed by a mature and solid theoretical background, SQP has been developed and used to solve a remarkably large number of practical problems. The basic principle of sequential approximations is to replace the given problem by a sequence of quadratic sub-problems that are easier to solve [31, 32]. Consider the application of the SQP methodology to nonlinear optimization problems,

Min. $f(x)$

$$\text{s.t.} \begin{cases} h(x) = 0 \\ g(x) \leq 0 \end{cases} \tag{19}$$

The Lagrangian of this problem can be written as,

$$L(x, \lambda.\mu) = f(x) + \lambda h(x) + \mu^T g(x) \tag{20}$$

where λ and μ are vectors of multipliers. SQP is an iterative procedure which models the problem for a given iterate x^k by a quadratic programming sub-problem, solves that quadratic programming sub-problem, and then uses the solution to construct a new iterate x^{k+1}.

The sub-problem can be constructed by linearizing the constraints of around x^k, and it can be written as,

$$\text{Min } \nabla f(x^k)(x - x^k) + \frac{1}{2}(x - x^k)^T Hf(x^k)(x - x^k)$$

$$\text{s.t.} \begin{cases} h(x^k) + \nabla h(x^k)(x - x^k) = 0 \\ g(x^k) + \nabla g(x^k)(x - x^k) \leq 0 \end{cases} \tag{21}$$

We need to update the estimates of the multipliers, and define the corresponding search directions, and then choose a step size and define the next iterate.

4.2 Particle swarm optimization

PSO is a global search technique originally introduced by Kennedy and Eberhart [33]. It simulates the social evolvement knowledge, probing the optimum by evolving the population which may include candidate solutions. In the classical PSO, each individual is treated as a particle in the space, with position and velocity vectors. The algorithm maintains a swarm of particles, where each particle represents a potential solution to the objective problem. For a given n-dimensional problem, the position and velocity vectors of a particle in the PSO can be represented as

$$\begin{cases} x_j(t) = [x_{j,1}(t), x_{j,2}(t) \ldots, x_{j,n}(t)] \\ v_j(t) = [v_{j,1}(t), v_{j,2}(t) \ldots, v_{j,n}(t)] \end{cases} \tag{22}$$

The core idea of the classical PSO is the exchange of information among the global best, population best, and current particles, which can be done as follows

$$\begin{aligned} v_j(t+1) &= \omega \cdot v_j(t) + \varphi \cdot r_1 \cdot [p_{pb}(t) - x_j(t)] \\ &\quad + \eta \cdot r_2 \cdot [p_{gb}(t) - x_j(t)] \\ x_j(t+1) &= x_j(t) + v_j(t+1) \end{aligned} \tag{23}$$

where v_j is velocity vectors; ω the inertia weight; p_{pb} the local best particle; and p_{gb} is global best particle, $\varphi = 1.65$, $\eta = 1.81$.

4.3 Composite computation approach

The procedures of the proposed hybrid algorithm are summarized as the follows,

Step-1. Load history wind data, generators and wind turbines settings, emission parameters, and forecast wind power output;

Step-2. Solve the ED and CEED problem without considering valve-point effects incorporating wind power using SQP;

Step-3. Calculate the updated constraints using (24) [34], and randomly generate initial population around the solution obtained from SQP for PSO;

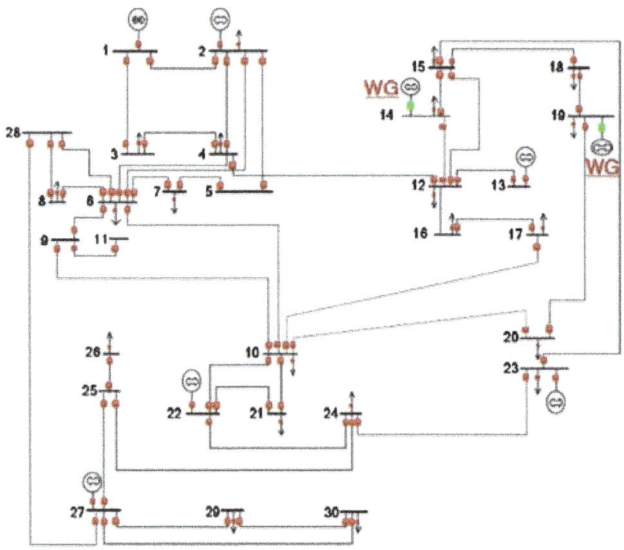

Fig. 3 Modified IEEE 30-bus system

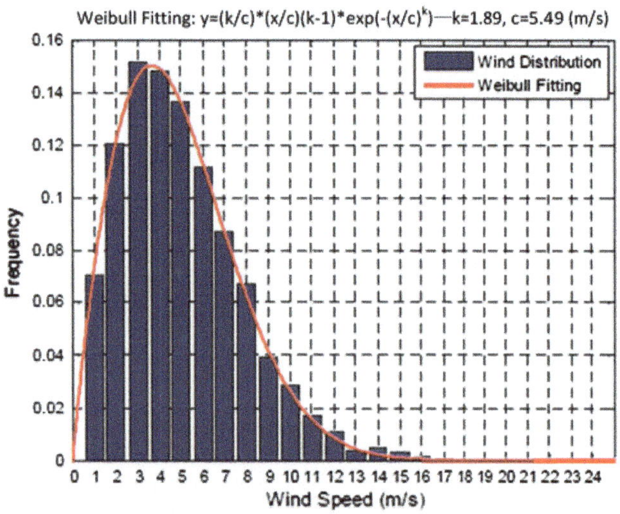

Fig.4 Wind speed distribution and Weibull fitting

$$\begin{cases} p_{i,\min} = \max([p_i - \beta_i], p_{i,\min}) \\ p_{i,\max} = \min([p_i - \beta_i], p_{i,\max}) \\ \beta_i = \pi/(1+\gamma)e_i \end{cases} \tag{24}$$

Step-4. Solve the ED and CEED problem with valve-point effects incorporating wind power using PSO;
Step-5. Save and output final solution. Application of this approach in ED and CEED problem incorporating wind power are presented in the following section.

5 Case studies

The QPSO is implemented on a modified IEEE 30-bus system. The benchmark system consists of 6 thermal generators, 2 wind farms, 41 branches, and 21 loads. These thermal generators include 3 coal-fired units, 2 gas-fired units, and 1 oil-fired unit. The test system is shown in Fig. 3 [35].

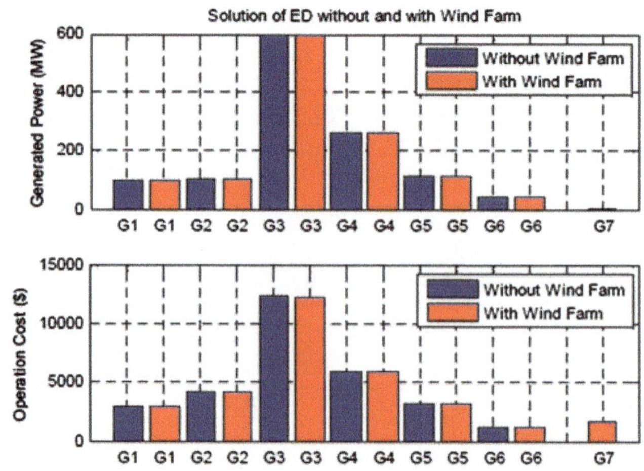

Fig.5 Solutions of ED models without and with wind farm

Table 1 Wind power factors

c	k	θ	v_{in}	v_{out}	v_{r}	w_{r}	$C_{\mathrm{w},j}$	$C_{\mathrm{u},j}$	$C_{\mathrm{o},j}$	$C_{\mathrm{s},j}$
5.5	1.89	0	4	25	16	3	0	60	20	10

In the case study part, the CEED model with wind power was evaluated using the historical wind speed dataset from a wind observation station in Tasmania, Australia. The data was provided by the Australian Bureau of Meteorology [36]. Here we assume that the wind speed data from a large wind farm and use the data to estimate the generated wind power. The wind speed distribution frequency and the corresponding Weibull distribution parameters are presented in Fig. 4.

The Vestas V90 3.0 MW wind turbine is selected for the case studies. It is a pitch regulated upwind wind turbine with active yawing and a three-blade rotor. It has a rotor diameter of 90 m with a generator rated at 3.0 MW. The Vestas V90 3.0 MW is widely used in the wind plants in Australia and has a proven high efficiency. The parameters of the associated Weibull distribution factor and wind farm parameters can be calculated from the wind speed data and are given in Table 1.

The wind farm totally consists of 100 Vestas V90 3.0 MW wind turbines located in a coherent geographic area. The predicted power output for each wind turbine is denoted as w_j and is 15 % of the rated power, which is 0.45 MW. Depended on the actual generated wind power, the extra cost will be determined by overestimation case or underestimation case. According, the maximum capacity of the system under investigation is 2,030 and 2,330 MW incorporating with wind power. The fuel cost coefficients,

Table 2 Fuel cost coefficients

Unit	Fuel cost coefficients				
	a_i	b_i	c_i	d_i	e_i
G1 (Coal)	2,000	10	0.002	200	0.084
G2 (Coal)	2,500	15	0.0025	300	0.035
G3 (Coal)	6,000	9	0.0018	400	0.042
G4 (Gas)	923.4	18	0.00315	150	0.063
G5 (Gas)	950	20	0.0032	100	0.084
G6 (Oil)	124.8	23.4	0.003432	80	0.098

The coefficients of a_i, b_i, c_i and e_i are in \$, \$/MW and \$/MW^2, and \$/MW

Table 3 Fuel consumption coefficients and generator limits

Unit	Fuel consumption coefficients			P_{min}	P_{max}
	f_i	g_i	h_i		
G1 (Coal)	40	0.2	0.00004	20	110
G2 (Coal)	50	0.3	0.00005	20	100
G3 (Coal)	80	0.12	0.000024	120	600
G4 (Gas)	2,462.4	48	0.0084	110	520
G5 (Gas)	2,500	50	0.009	110	500
G6 (Oil)	1.248	0.234	3.43e-05	40	200
G7 (Wind)	0	0	0	0	300

The coefficients of f_i, g_i, and h_i are in t, t/MW and t/MW^2 for coal/oil units. The coefficients of f_i, g_i, and h_i are in m^3, m^3/MW and m^3/MW^2 for gas unit

Table 4 Emission factors of units

Emission factor	Coal (kg/kg)	Gas (kg/m^3)	Oil (kg/kg)
efCO_2	3.1604	1.84	2.8523
efNO_2	1.29e-03	3.4e-04	3.3e-04

generator limits, and fuel consumption coefficients are shown in Tables 2 and 3 [35]. The proposed algorithm is implemented on a test system including 6 thermal generators and 1 large wind farm. There are 3 coal-fired units, 2 gas-fired units, and 1 oil-fired unit in this test system. The wind farm totally consists of 100 Vestas V90 3.0 MW wind turbines located in a coherent geographic area. The predicted power output for each wind turbine is denoted as wj and is 15 % of the rated power, which is 0.45 MW. Depended on the actual generated wind power, the extra cost will be determined by overestimation case or underestimation case. The maximum capacity of the system under investigation is 2,030 and 2,330 MW incorporating with wind power. The fuel cost coefficients, generator limits, and fuel consumption coefficients are shown in Tables 2 and 3 [35].

Table 5 Emission prices

Fuel	CO_2 (\$/t)	NO_2 (\$/kg)
Price	1.5	5.0

Table 6 Forecast system demand and wind farm output

Case index	Case 1	Case 2
Demand (MW)	1,200	1,600
G7	45	45

Table 7 Solution of ELD without wind farm

Unit	Power (MW)	Operation cost (\$)
G1 (Coal)	96.9286	29,69,047
G2 (Coal)	99.4079	4,122.16
G3 (Coal)	593.5730	12,359.45
G4 (Gas)	259.1281	5,808.68
G5 (Gas)	110.6357	3,207.22
G6 (Oil)	40.3266	1,076.58
Total	1,200.0000	29,538.56
Overall cost (\$)	29,538.56	

Table 8 Solution of ELD with wind farm

Unit	Power (MW)	Operation cost (\$)
G1 (Coal)	94.9286	2,967.33
G2 (Coal)	99.9710	4,125.33
G3 (Coal)	592.0273	12,290.35
G4 (Gas)	258.9938	5,802.31
G5 (Gas)	110.0097	3,189.00
G6 (Oil)	40.5473	1,083.54
G7 (Wind)	3.8621	1,657.76
Total	1,200.0000	31,115.63
Overall cost (\$)	31,115.63	

In this paper, two of most concerned GHGs emissions, CO_2 and NO_2 are considered in the model. The emission characteristics of the units and emission allowance price are shown in the Tables 4 and 5. The wind farm forecast system demand and output are shown in Table 6.

5.1 Case-I. ELD model without and with wind farm

In this case study, the system load is 1,200 MW and the system loss power is assumed to be zero. The basic ELD model with and without wind farm are tested on the system and the simulation results are shown in Tables 7, 8 and Fig. 5.

It can be shown that the solution of ED with wind farm succeeds in reducing generated power and operation costs

Table 9 Solution of CEED without wind farm

Unit	Power (MW)	Operation cost ($)	Emission cost ($)
G1 (Coal)	95.5408	2,986.10	3,202.99
G2 (Coal)	20.7747	2,820.83	3,029.61
G3 (Coal)	598.7496	12,414.63	8,641.41
G4 (Gas)	509.7226	10,924.30	852.97
G5 (Gas)	333.1363	7,978.45	590.56
G6 (Oil)	42.0759	1,131.62	495.65
Total	1,600.0000	38,255.93	16,813.19
Overall cost ($)		55,069.12	

Table 10 Solution of CEED with wind farm

Unit	Power (MW)	Operation cost ($)	Emission cost ($)
G1 (Coal)	95.3455	2,980.80	3,200.81
G2 (Coal)	21.3548	2,835.68	3,039.05
G3 (Coal)	569.0520	11,708.60	8,404.66
G4 (Gas)	507.6528	10,885.02	849.54
G5 (Gas)	296.0316	7,159.18	530.05
G6 (Oil)	40.0636	1,068.30	474.47
G7 (Wind)	70.4998	3,230.97	0.00
Total	1,600.0000	39,868.55	16,498.57
Overall cost ($)		56,367.12	

of some fuel units (G1, G3, G4, G5). However, the outputs and scheduling costs of generators (G2, G6) were increased slightly. The reason is that although wind power generators have lots of advantages, the operation cost caused by wind prediction errors is really expensive. With the wind power generator, part of the load of high cost units (G1, G3, G4, G5) is shifted to comparative low cost units (G2, G6). The operation cost of solution of ED with wind farm is highly increased in comparison with solution of ED without wind farm. In addition, the wind power government subsidy is just a little bit due to the low wind power output.

The generated wind power in this case is 3.8621 MW which is far less than the predicted wind power (45 MW), and the cost incurred by overestimation will be applied. The operator needs to purchase more power from another source. Furthermore, the common ED model does not take in account the emission issue. The incorporation of wind power in simple ED problem is not an economic solution due to the really high operation cost of wind power.

5.2 Case-II. CEED model without and with wind farm

In this case study, the system load is 1,600 MW and the system loss power is assumed to be zero. The CEED model with and without wind farm are performed on the test system and the simulation results are shown in Tables 9, 10 and Fig. 6.

The system load is increased to 1,600 MW in this case. But the load is still less than the maximum capacity for

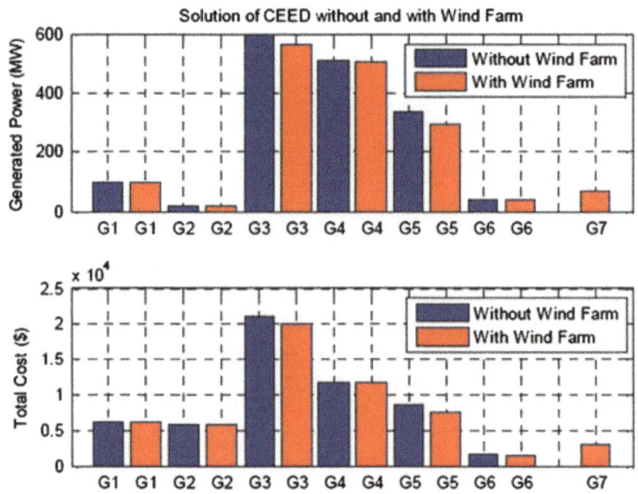

Fig. 6 Solutions of CEED models without and with wind farm

both thermal units and system with wind power. The objective of CEED is to minimize the total system operation costs and greenhouse gases (CO_2 and NO_2) emission costs. It is clear that part of the load of highly polluted fuel fired units (G1$\sim$G6) is shifted to no emission polluted wind power generator (G7). Although the wind power cost is expensive, emission cost were decreased in the solution of CEED with wind farm. The reason is that the government wind power subsidy is directly proportional to the output wind power. In this case, the real generated wind power is 70.4998 MW which is larger than the predicted wind power (45 MW). The underestimation situation will be considerate and the cost for not using all wind power available from wind turbine should be applied. From Tables 9 and 10, we can find that the CEED model with wind farm reduces the emission cost dramatically in comparison with CEED solution without wind power because of the no-emission character of wind energy. In Eq. (9), the government wind power subsidy is directly proportional to the output of wind power. Thus, the overall cost is acceptable from a standpoint of wind farm operator. Therefore, the results have shown that the proposed CEED with wind energy gives a better emission solution efficiently and economically.

5.3 Case-III. Comparisons with other approaches

In order to evaluate the performance of the proposed method, GA, Immune Algorithm (IA) [37], and PSO are employed in the case studies. For comparison purposes, these algorithms are used directly to solve the CEED problem with wind power. For the proposed SQP–PSO algorithm, the population size is 100 and maximum iteration is 3 for PSO. Meanwhile, in order to make a fair comparison of the other approaches, we fixed the same

Table 11 Comparison of different approaches

Algorithm	Best solution ($)	Average solution ($)	Average time (s)
GA	57,369.97	57,916.20	13.28
IA	57,180.98	57,669.57	12.57
PSO	56,714.06	57,417.04	8.01
SQP + PSO	56,367.12	56,538.19	1.29

population size as 100 and tested them to reach maximum iteration 100. The initial crossover and mutation rates for GA and IA were all set as 80 % and 5 %. All the programs were run on a 2.66 GHz, Intel Core 2, with 4G RAM desktop. Table 11 shows the results out of 50 runs with each method

A comparison with other approaches is made to evaluate the proposed algorithm which is shown in Table 11. As shown, we can conclude that, the proposed hybrid approach can greatly enhances the searching ability and ensures quality of average solutions, saves computation time, and also efficiently manages the system constraints.

6 Conclusion

UC and ED problem of wind power will start to affect market price of smart grid system, because it became a factor affecting the operation of smart grid and hence the cost. Wind power has impact on the smart grid, since most wind energy flows on to the transmission grid. This paper developed a hybrid method combining the SQP and PSO to achieve faster and better optimization performance. The method was successfully applied to solve the power system ED problem considering GHGs emissions and wind power in an integrated CEED model, where the valve-point effect is also taken into account. In the present work, the wind speed distribution probability functions are applied in formulating the optimization model to address the uncertainties involved. The proposed hybrid method was applied to solve the CEED problem of a test system involving 6 thermal units and 1 wind farm. The comparisons were made between the classical ED and the proposed CEED model with and without wind farm. The proposed CEED model with wind farm shows a better performance in terms of less emission cost. In addition, the resultant overall dispatching cost is also optimized considering the government subsidy. Furthermore, the proposed hybrid optimization method was compared with other optimization approaches for the studied cases. The simulation results show that the hybrid method is better in terms of the speed and accuracy. Compared to the classical PSO and other methods, it can be concluded that the hybrid method greatly enhances the searching ability and efficiently manages the system constraints, therefore providing a new and efficient tool for the CEED problem.

References

[1] Zhu JZ (2009) Optimization of power system operation. Wiley-IEEE, Piscataway

[2] Wong KP, Wong YW (1995) Thermal generator scheduling using hybrid genetic/simulated-annealing approach. IEE Proc Gener Transm Distrib 142(4):372–380

[3] Wong KP, Wong YW (1996) Combined genetic algorithm/simulated annealing/fuzzy set approach to short-term generation scheduling with take-or-pay fuel contract. IEEE Trans Power Syst 11(1):112–118

[4] Zhao JH, Dong ZY, Li X et al (2007) A framework for electricity price spike analysis with advanced data mining methods. IEEE Trans Power Syst 22(1):376–385

[5] Zhao JH, Wen FS, Dong ZY et al (2012) Optimal dispatch of electric vehicles and wind power using enhanced particle swarm optimization. IEEE Trans Ind Inform 8(4):889–899

[6] Gungor VC, Sahin D, Kocak T (2011) Smart grid technologies: communication technologies and standards. IEEE Trans Ind Inform 7(4):529–539

[7] Dong ZY, Wong KP, Meng K et al (2006) Wind power impact on system operations and planning. In: Proceedings of the 2010 power and energy society general meeting (PES'10), Minneapolis, 25–29 Jul 2010, p 5

[8] Patel MR (2006) Wind and solar power systems. CRC Press, Boca Raton

[9] Hetzer J, Yu DC, Bhattrarai K (2008) An economic dispatch model incorporating wind power. IEEE Trans Energy Convers 23(2):603–611

[10] Liu X, Xu W (2010) Minimum emission dispatch constrained by stochastic wind power availability and cost. IEEE Trans Power Syst 25(3):1705–1713

[11] Wong KP, Fung CC (1993) Simulated annealing based economic dispatch algorithm. IEE Proc Gener Transm Distrib 140(6):509–515

[12] Wong KP, Wong YW (1994) Genetic and genetic/simulated-annealing approaches to economic dispatch. IEE Proc Gener Transm Distrib 141(5):507–513

[13] Vlachogiannis JG, Lee KY (2008) Quantum-inspired evolutionary algorithm for real and reactive power dispatch. IEEE Trans Power Syst 23(4):1627–1636

[14] Tsai MS, Hsu FY (2010) Application of grey correlation analysis in evolutionary programming for distribution system feeder reconfiguration. IEEE Trans Power Syst 25(2):1126–1133

[15] Lin XN, Ke SH, Li ZT (2010) A fault diagnosis method of power systems based on improved objective function and genetic algorithm-tabu search. IEEE Trans Power Deliv 25(3):1268–1274

[16] Al-Sumait JS, Al-Othman AK, Sykulski JK (2007) Application of pattern search method to power system valve-point economic load dispatch. Int J Electr Power Energy Syst 29(10):720–730

[17] Ling SH, Iu HHC, Leung FHF (2008) Improved hybrid particle swarm optimized wavelet neural network for modeling the

development of fluid dispensing for electronic packaging. IEEE Trans Ind Electron 55(9):34–42
[18] Meng K, Wang HG, Dong ZY (2010) Quantum-inspired particle swarm optimization for valve-point economic load dispatch. IEEE Trans Power Syst 25(1):215–222
[19] Yao F, Meng K, Zhao X et al (2011) Differential evolution algorithm for multi-objective economic load dispatch considering minimum emission costs. In: Proceedings of the 2011 IEEE power and energy society general meeting, Detroit, 24–28 Jul 2011, p 5
[20] Duvvuru N, Swarup KS (2011) A hybrid interior point assisted differential evolution algorithm for economic dispatch. IEEE Trans Power Syst 26(1):541–549
[21] Attaviriyanupap P, Kita H, Tanaka E et al (2002) A hybrid EP and SQP for dynamic economic dispatch with non-smooth fuel cost function. IEEE Trans Power Syst 17(2):411–416
[22] Wang SK, Chiou JP, Liu CW (2007) Non-smooth/non-convex economic dispatch by a novel hybrid differential evolution algorithm. IET Gener Transm Distrib 1(5):793–803
[23] Bhattacharya A, Cattopadhyay PK (2010) Hybrid differential evolution with biogeography based optimization for solution of economic load dispatch. IEEE Trans Power Syst 25(4):1955–1964
[24] Masters GM (2005) Renewable and efficient electric power systems. Wiley, Hoboken
[25] Carta JA, Ramirez P, Velazquez S (2009) A review of wind speed probability distributions used in wind energy analysis: case studies in the Canary Islands. Renew Sustain Energy Rev 13(5):933–955
[26] Liu X (2010) Economic load dispatch constrained by wind power availability: a wait-and-see approach. IEEE Trans Smart Grid 1(4):347–355
[27] Meng K, Wang HG, Dong ZY et al (2010) Quantum-inspired particle swarm optimization for valve-point economic load dispatch. IEEE Trans Power Syst 25(1):215–222
[28] Olsson AE (2011) Particle swarm optimization: theory, techniques and applications. Nova Science, New York
[29] Office of Gas and Electricity Markets (Ofgem) (2012) Annual report and accounts 2011–12 (For the year ended 31 March 2012), The Stationery Office, London
[30] Boggs PT, Tolle JW (1996) Sequential quadratic programming. Acta Numer 4:1–52
[31] Hanan SP (1975) A globally convergent method for nonlinear programming. J Optim Theory Appl 22(3):297–309
[32] Powell MJD (1978) A fast algorithm for nonlinearly constrained optimization calculations. In: Watson GA (ed) Numerical analysis, Lecture notes in Mathematics, vol 630. Springer, Berlin, pp 144–157
[33] Kennedy J, Eberhart R (1995) Particle swarm optimization. In: Proceedings of the IEEE international conference on neural networks (ICNN'95), vol 4, Perth, 27 Nov–1 Dec 1995, pp 1942–1948
[34] Duvvuru N, Swarup KS (2011) A hybrid interior point assisted differential evolution algorithm for economic dispatch. IEEE Trans Power Syst 26(2):541–549
[35] Venkatesh P, Lee KY (2008) Multi-objective evolutionary programming for economic emission dispatch problem. In: Proceedings of the 2008 IEEE power and energy society general meeting—conversion and delivery of electrical energy in the 21st century, Pittsburgh, 20–24 Jul 2008, p 8
[36] Australian Bureau of Meteorology.
[37] Farmer JD, Packard N, Perelson A (1986) The immune system, adaptation and machine learning. Phys D Arch 2(1–3):187–204
[38] Yao F, Dong ZY, Meng K et al (2012) Quantum-inspired particle swarm optimization for power system operations considering wind power uncertainty and carbon tax in Australia. IEEE Trans Ind Inform 8(4):880–888

Author Biographies

Yang ZHANG is working towards the Ph.D. degree at the School of Electrical, Electronics and Computer Engineering, University of Western Australia, Perth, Australia. His research interests are power generation and operation optimization of sustainable energy system and smart grid, energy storage.

Fang YAO (S'09) received the Ph.D. degree in Electrical Engineering from the University of Western Australia, Perth, Australia, in 2012. He is with Schneider Electric, Beijing, China. His research interests are wind power generation and optimization of sustainable energy system.

Herbert HO-CHING IU received the B.Eng. (Hons.) degree in Electrical and Electronic Engineering from the University of Hong Kong, Hong Kong, in 1997, and the Ph.D. degree from the Hong Kong Polytechnic University, Hong Kong, in 2000. Since 2002, he has been with the School of Electrical, Electronic and Computer Engineering, University of Western Australia, Perth, Australia, where he was initially a Lecturer and is currently a Professor. He was a Visiting Lecturer with the University of Reims Champagne-Ardenne, Reims, France, in 2004, and a Visiting Assistant Professor with the Hong Kong Polytechnic University in 2006. He is the author of more than 100 published papers. He currently serves as an Associate Editor for *International Journal of Circuit Theory and Applications*, *IEEE Circuits and Systems Society Newsletter*, and *International Journal of Bifurcation and Chaos*, an Editorial Board Member for *Australian Journal of Electrical and Electronics Engineering*. He is a Co-editor of Control of Chaos in Nonlinear Circuits and Systems (World Scientific, 2009). His research interests include power electronics, renewable energy, smart grid, nonlinear dynamics, current sensing techniques, and memristive systems.

Tyrone FERNANDO received the B.Eng. (Hons.) and Ph.D. degrees in Electrical and Electronic Engineering, all from The University of Melbourne, Australia, in 1990 and 1997, respectively. He joined the University of Western Australia in 1996 where he is currently a Professor. He was the Associate Head and Deputy Head in the School of Electrical, Electronic, and Computer Engineering at the University of Western Australia in 2008–2010, respectively. His current research activities are in the areas of systems and control theory and application of control theory to biomedical systems.

Kit PO WONG received the M.Sc., Ph.D., and higher doctorate D.Eng. degrees from the University of Manchester, Institute of Science and Technology, Manchester, UK, in 1972, 1974, and 2001, respectively. He is a Fellow of IEEE. Since 1974, he has been with the School of Electrical, Electronic and Computer Engineering, The University of Western Australia, where he is currently a Winthrop Professor. His current research interests include power system analysis, planning and operations, and smart grids. Prof. Wong received three Sir John Madsen Medals (1981, 1982, and 1988) from the Institution of Engineers Australia, the 1999 Outstanding Engineer Award from IEEE Power Chapter Western Australia, and the 2000 IEEE Third Millennium Award. He was General Chairman of IEEE/CSEE PowerCon2000 conference. He was an Editor-in-Chief of IEE Proceedings in Generation, Transmission and Distribution. Currently he is serving as Editor-in-Chief for IEEE Power Engineering Letters of the IEEE PES Transactions.

Optimal reactive power dispatch with wind power integrated using group search optimizer with intraspecific competition and lévy walk

Yuanzheng LI, Mengshi LI, Qinghua WU (✉)

Abstract This paper presents the mean–variance (MV) model to solve power system reactive power dispatch problems with wind power integrated. The MV model considers the profit and risk simultaneously under the uncertain wind power (speed) environment. To describe this uncertain environment, the Latin hypercube sampling with Cholesky decomposition simulation method is used to sample uncertain wind speeds. An improved optimization algorithm, group search optimizer with intraspecific competition and lévy walk, is then used to optimize the MV model by introducing the risk tolerance parameter. The simulation is conducted based on the IEEE 30-bus power system, and the results demonstrate the effectiveness and validity of the proposed model and the optimization algorithm.

Keywords Mean–variance model, Reactive power dispatch, Wind power, Group search optimizer with intraspecific competition and lévy walk (GSOICLW)

1 Introduction

The optimal reactive power dispatch (ORPD) problem has played significant influence on the economical and secure operation of power system [1, 2]. It aims to seek optimal settings of control variables to minimize a specific objective function, while meeting a set of operational constraints. The control variables consist of both continuous and discrete variables, including generator voltage magnitudes, discrete tap settings of transformers and outputs of reactive compensation devices. The operational constraints are composed of equality and inequality constraints, presented in the power flow equations and the limits of bus voltages, transformer tap-settings and reactive power outputs. Therefore, the ORPD problem is a complex non-linear constrained optimization problem with a mixture of discrete and continuous variables [2].

Many conventional techniques based on mathematics have been applied in dealing with the ORPD problem, such as gradient-based searching method [3], interior point method [4] and quadratic programming [5]. However, these conventional methods cannot guarantee to find the global optimum because of the non-convex, non-differential and non-linear nature of the ORPD problem [6]. In the last decades, many evolutionary algorithms (EAs) have been used for the ORPD problem, such as genetic algorithm (GA) [6], evolutionary programming (EP) [1].

Recently, a novel optimization algorithm, called group search optimizer (GSO), was proposed in [7]. It is a swarm-based algorithm and inspired by animal searching behavior. GSO consists of three kinds of members, i.e., the producer, scroungers and rangers. The producer with the best fitness value acts as the leader and it adopts scanning mechanism to find the optimal resource in the searching space. Scroungers are always following the producer to enjoy the discovered resource. On the other hand, rangers employ random walk (RW) to perform ranging behavior to seek other distributed resources. Therefore, it is the rangers that maintain the group's diversity to increase GSO's possibility to escape from local optima. In [7], it has been proved that GSO outperforms other EAs, such as GA, PSO and EP, especially in terms of optimizing multi-modal benchmark functions.

Y. LI, M. LI, Q. WU, Department of Electric Power Engineering, South China University of Technology (SCUT), Guangzhou 510641, Guangdong, China
Q. WU, Department of Electrical Engineering and Electronics, University of Liverpool, Liverpool L69 3GJ, UK
(✉) e-mail: wuqh@scut.edu.cn

However, GSO's local searching ability is not satisfactory, as shown in its modest performance on optimizing unimodal functions [7]. In order to improve GSO's local searching ability while maintaining its merit in global searching, we have proposed an improved GSO, group search optimizer with intraspecific competition and lévy walk (GSOICLW) [8], incorporating intraspecific competition (IC) [9] and lévy walk (LW) [10] into the GSO model. IC makes the scroungers strengthen local searching while stimulates more rangers to do global searching. Moreover, LW stimulates rangers to perform more efficient global searching than random walk (RW). In this paper, GSOICLW is adopted for the first time to solve the ORPD problem.

Recently, wind energy has been greatly utilized all over the world. It is definite that it is a good alternative to the traditional thermal power generation [11, 12]. However, its inherent nature of uncertainty determines it is extremely difficult to predict wind power [13–15]. Therefore, it is difficult to solve power system dispatch problems (DPs), including the ORPD problem, if large amount of wind power is integrated into power systems [12, 16, 17]. To the best of our knowledge, there have been mainly two methodologies on DPs with wind power integrated, i.e., the fuzzy and the probabilistic methods. In the fuzzy method, the wind power is deemed as the fuzzy variable, and the fuzzy set theory is used to model the corresponding DP issue [17, 18]. However, this method may be subjected to strong subjectiveness, although it can reflect dispatchers' attitude to some extent [17, 18].

On the other hand, the probabilistic method usually assumes that wind speed, wind power or wind forecast error follows a specific probabilistic distribution. The wind speed is often assumed to follow the Weibull distribution [19, 20]. However, the period of DP and ORPD is very short, thus it is not suitable to use the Weibull distribution [19]. Ref. [19] indicates that the wind speed forecast error follows the Gaussian distribution in a short time, which demonstrates that this kind of distribution can be applied in DPs. Therefore, in this paper, the Gaussian distribution is applied to describe the uncertain wind speed in dealing with ORPD.

It is noted that the Monte Carlo (MC) simulation is often applied to generate wind power samples to conduct stochastic optimization in the probabilistic method [21–23]. However, the MC method is notorious for its heavy computation burden. Thanks to the Latin hypercube sampling with Cholesky decomposition (LHS-CD) simulation method, it obtains reliable results with a much smaller sample size, compared with the MC method [24]. Moreover, in terms of the stochastic optimization based on the probabilistic method, the objective function is usually the expected value. In this way, the final dispatch solution can obtain the best average value. However, the risk brought about by the solution is not considered under the uncertain environment. Thanks to the mean-variance (MV) model, which was proposed by Markowitz in the portfolio optimization problem, it can well measure both the profit and the risk brought a portfolio [25] in the uncertain environment. Therefore, we can use this model and LHS-CD to deal with the ORPD problem under the wind power environment.

Therefore, in this paper, the MV model and the sampling method LHS-CD are used to solve the ORPD problem with uncertain wind power integrated, and GSOICLW is applied to optimized the MV model to determine the dispatch solution. The modified IEEE 30-bus power system is employed for carrying out the simulation study to demonstrate the effectiveness of the MV model and the proposed optimization algorithm, GSOICLW.

The rest of the paper is organized as follows. Section 2 introduces the ORPD problem and the MV model. Section 3 presents the algorithm of GSO and GSOICLW. Section 4 carries out the experiments and discusses the simulation results. In the end, Sect. 5 draws conclusions.

2 Mean and variance model for optimal reactive power dispatch

2.1 Optimal reactive power dispatch (ORPD)

The ORPD problem aims to minimize an objective function, usually the transmission loss, by adjusting control variables while satisfying a set of equality and inequality constraints. As a result, the ORPD problem can be presented in the following formulation:

$$\begin{aligned} \min_{u} \quad & J = f(\boldsymbol{u},\boldsymbol{x}) \\ \text{s.t} \quad & \boldsymbol{g}(\boldsymbol{u},\boldsymbol{x}) = 0 \\ & \boldsymbol{h}(\boldsymbol{u},\boldsymbol{x}) \le 0 \end{aligned} \tag{1}$$

where J represents the objective function; $\boldsymbol{x}$ stands for the vector of state variables, including load-bus voltage $\boldsymbol{V}_{\mathrm{L}}$, generator reactive power $\boldsymbol{Q}_{\mathrm{G}}$ and apparent power flow $\boldsymbol{S}_k$; the vector of control variables $\boldsymbol{u}$ comprises generator voltage $\boldsymbol{V}_{\mathrm{G}}$, transformer tap setting $\boldsymbol{T}$, reactive power output of compensative reactive power sources $\boldsymbol{Q}_{\mathrm{C}}$.

The objective function of the ORPD problem is transmission loss, which can be expressed as the following.

$$P_L = \sum_{k=1}^{N_E} g_k(V_i^2 + V_j^2 - V_iV_j\cos(\theta_{ij})) \tag{2}$$

where g_k represents the conductance of branch k; $\boldsymbol{V}_i$ and $\boldsymbol{V}_j$ are the voltage magnitude at bus i and j; θ_{ij} is the voltage angle difference between bus i and j; and N_E is the total number of power system network branches.

As for constraints, ***g(u,x)*** represents the must of power balance in power systems, i.e., the generated active and reactive power must be balanced with their corresponding parts in consuming load.

$$\begin{cases} P_{Gi} = P_{Di} + V_i \sum_{j \in N_i} V_j (G_{ij}\cos\theta_{ij} + B_{ij}\sin\theta_{ij}) \\ Q_{Gi} = Q_{Di} + V_i \sum_{j \in N_i} V_j (G_{ij}\sin\theta_{ij} - B_{ij}\cos\theta_{ij}) \end{cases} \quad (3)$$

where P_{Gi} and P_{Di} are the injected and demanded active power at bus i; Q_{Gi} and Q_{Di} are the injected and demanded reactive power; B_{ij} and G_{ij} are the transfer susceptance and conductance between bus i and j; N_i stands for the total number of buses adjacent to bus i, including bus i. ***h(u,x)*** represents working limits of generators, power transformers and reactive power sources, and emphasizes power system security constraints, such as limits on bus voltages and branch apparent power flow. The details can be referred to [8].

The most common method used to deal with constrained optimization problem is adopting penalty function [2, 6]. Therefore, by adopting this method, the objective function in (1) can be expressed as

$$F = J + \sum_{i \in N_V^{\lim}} \lambda_{V_i}(V_i - V_i^{\lim})^2 + \sum_{i \in N_Q^{\lim}} \lambda_{G_i}(Q_i - Q_{G_i}^{\lim})^2 + \sum_{i \in N_E^{\lim}} \lambda_{S_i}(|S_i| - S_i^{\lim})^2 \quad (4)$$

where λ_{V_i}, λ_{G_i} and λ_{S_i} are penalty factors, and the limited value $x_i^{\lim}(x \in (P_G, V, Q_G, S_k))$ can be defined as the following:

$$x_i^{\lim} = \begin{cases} x_i^{\max} & (x_i > x_i^{\max}) \\ x_i^{\min} & (x_i < x_i^{\min}) \end{cases} \quad (5)$$

2.2 Mean and variance model for optimal reactive power dispatch

The mean-variance (MV) model was firstly proposed by Markowitz in the portfolio optimization problem, which aimed to both maximize the profit and minimize the profit [25]. The MV model has also been used in power systems, Refs. [26, 27] have applied this model to solve the self-scheduling problem under uncertain electricity markets. The effect of risk has been explicitly modelled taken the variance of the market-clearing prices into consideration to address the trade-off of maximum profit versus minimum risk [26, 27]. In [28], the MV model has also been used to deal with the stochastic optimal power flow (OPF) problem with uncertain power demands. As the uncertain wind power is increasingly integrated into power system, it is of great difficulty to solve the DPs [19]. Therefore, in this paper, we use the MV model to analyze the ORPD problem.

Inspired by the MV model, we can use this model to consider both the profit and risk brought by a dispatch solution under the uncertain wind power environment, and determine the optimal dispatch solution by analyzing the trade-off relationship between the profit and risk. However, how to introduce the concept of profit and risk into the ORPD problem? The profit means that a dispatch solution in the uncertain wind power environment brought about a less averaged transmission loss; in other words, we should minimize the profit as small as possible. On the other hand, the risk means that a robust dispatch solution should be obtained, which can adjust to different uncertain wind power samples. Therefore, the MV model applied in ORPD can be expressed as

$$F^{\exp} = \mathrm{E}_{W_1, W_2, \ldots, W_M}\left\{\sum_{i=1}^{N_S} F_i \cdot \mathrm{P}(F_i)\right\} \quad (6)$$

where $\mathrm{E}(\cdot)$ is the expectation operator, computing the average value of the profit function; M is the number of wind farms integrated to a power system, and $(W_1, W_2, \ldots, W_M)$represent power outputs of these wind farms; N_S is the number of wind power samples corresponding to each W_j ($j = 1,2,\cdots,M$) obtained by the LHS-CD method [24]; F_i is the transmission loss when the ith wind power sample of all wind farms has integrated to the power system; and $\mathrm{P}(F_i)$ describes the probability of the transmission loss F_i.

Then, the variance function is formulated as

$$V = \mathrm{Var}_{W_1, W_2, \ldots, W_M}\left\{\sum_{i=1}^{N_S} F_i \cdot \mathrm{P}(F_i)\right\} \quad (7)$$

where $\mathrm{Var}(\cdot)$ is the variance operator.

Once a dispatch solution is determined, then the mean and variance of the profit function can be calculated. Moreover, if the value of variance is high, the transmission loss corresponding to different wind power samples deviates far from the expected value, which means the dispatch solution cannot well adjust the uncertain wind power environment. For example, the transmission loss as for some wind power samples maybe much more than its mean value. Therefore, it is quite necessary for power system dispatchers to take both profit and risk into consideration. Refs. [26, 27] optimize these two indexes by adopting the risk tolerance parameter, in the perspective of the single-objective optimization problem. However, the risk tolerance parameter is difficult to set. Therefore, Ref. [26] uses different risk tolerance parameters to obtain the expected profit and the profit standard deviation, and Ref. [27] analyses the relationship of the risk tolerance parameter and the expected profit, computing different expected profits regarding to different risk tolerance parameters. In this paper, we also use the risk tolerance parameter, λ, to study the MV model. Then the objective function becomes

$$\begin{aligned} \min_{u} \quad & F^{\exp} + \lambda V \\ \text{s.t.} \quad & \boldsymbol{g}(\boldsymbol{u},\boldsymbol{x},W) = 0 \\ & \boldsymbol{h}(\boldsymbol{u},\boldsymbol{x}) \leq 0 \end{aligned} \tag{8}$$

where W consists of P_W and Q_W, which stand for the vector of wind farms' uncertain active and reactive power outputs. It is noted that W mainly affects ORPD on power flow equations, which will be presented in the next subsection.

2.3 Wind power

The Weibull distribution is often used to describe the uncertainty of wind speeds within a long period of time, [29–31]. The ORPD is usually conducted within a short term; therefore, it is not suitable to use the Weibull distribution when solving ORPD with wind power integrated [19]. The actual wind speed can be regarded as the sum of the forecast speed and its error, and the wind speed forecast error follows the Gaussian distribution [19]. Moreover, the Gaussian distribution has been used to depict the wind speed forecast error to solve DP problems [19, 20]. Therefore, in this paper, we adopt the Gaussian distribution:

$$\Delta v \sim N(0, \sigma_v^2) \tag{9}$$

where Δv represents the wind speed forecast error, $N(0, \sigma_v^2)$ stands for the Gaussian distribution with the mean value 0 and the standard deviation σ_v.

Then the actual wind speed is expressed as:

$$v = v_f + \Delta v \tag{10}$$

where v_f indicates the forecast value of the wind speed.

The active power generated by the wind turbine, P_{wt}, is determined by the wind speed [32], which is formulated as:

$$P_{wt} = \begin{cases} 0 & 0 \leq v < v_{ci} \\ a + bv^3 & v_{ci} \leq v < v_{ra} \\ P_{ra} & v_{ra} \leq v < v_{co} \\ 0 & v > v_{co} \end{cases} \tag{11}$$

$$\text{where } a = \frac{P_{ra} v_{ci}^3}{v_{ra}^3 - v_{ci}^3} \quad b = \frac{P_{ra}}{v_{ra}^3 - v_{ci}^3}$$

where v, v_{ci}, v_{ra}, v_{co} are the wind speed, the cut-in wind speed, the rated wind speed, the cut-out wind speed, respectively, and P_{ra} is the rated active power of the wind turbine.

If a wind farm consists of N_{wt} wind turbines, evidently, the total amount of active wind power is $P_{wt} \times N_{wt}$. However, the reactive power generated by wind farms is determined by the control strategy of the farm (turbine) [32]. In this paper, we choose the doubly-fed induction wind power generator with a constant power factor, so the wind farm is a PQ bus [32]. Then the active and reactive power of the wind farm can be expressed as

$$\begin{cases} P_{\text{farm}} = P_{wt} \times N_{wt} \\ Q_{\text{farm}} = \frac{P_{farm}}{\cos\varphi} \sqrt{1 - \cos^2\varphi} \end{cases} \tag{12}$$

The wind power mainly affects the dispatch solution of ORPD by power flow equations at the power system bus (e.g., bus i) connected to a wind farm, as shown in the equation constraints:

$$\begin{cases} P_{Gi} = P_{Di} - P_{farm} + V_i \sum_{j \in N_i} V_j (G_{ij}\cos\theta_{ij} + H_{ij}\sin\theta_{ij}) \\ Q_{Gi} = Q_{Di} - Q_{farm} + V_i \sum_{j \in N_i} V_j (G_{ij}\sin\theta_{ij} - H_{ij}\cos\theta_{ij}) \end{cases} \tag{13}$$

3 Group search optimizer with intraspecific competition and lévy walk

3.1 Group search optimizer

GSO is a swarm-based optimization algorithm, consisting of three kinds of members, i.e., the producer, scroungers and rangers. In each generation, the member with the best fitness value is chosen as the producer, and a number of members except the producer are randomly selected as scroungers, then the rest of members are rangers. The producer is always located in the most promising area and adopts animal scanning to seek the optimal resource. Scroungers perform area copying to join the resource found by the producer, and perform local searching around it. Meanwhile, rangers employ ranging behavior by random walk (RW) in the searching space to increase the GSO's chance to escape from local optima. Therefore, GSO performs much better on multi-modal optimization problems than other EAs, such as GA, PSO, EP [7].

In GSO, each member has its current position $X_i^k \in R^n$ and a scanning angle $\varphi_i^k = \left(\varphi_{i1}^k, \varphi_{i2}^k, \cdots, \varphi_{i(n-1)}^k\right) \in \mathbb{R}^{n-1}$ in the kth bout, and the corresponding unit vector $D_i^k(\varphi_i^k) = \left(d_{i1}^k, d_{i2}^k, \cdots, d_{in}^k\right) \in \mathbb{R}^n$can be calculated [7]. The searching mathematical models regarding to producer, scroungers and rangers can be shown as follows, respectively.

3.1.1 Producer

The member with the best fitness value is chosen as the producer, whose position is X_p^k. Afterwards, it will scan randomly by sampling three points: one is at zero degree, the other two points are at the right and left degrees, respectively. As a result, the producer's searching formulations can be expressed as

$$\begin{cases} X_z = X_p^k + r_1 l_{\max} D_p^k(\varphi^k) \\ X_r = X_p^k + r_1 l_{\max} D_p^k(\varphi^k + r_2\theta_{\max}/2) \\ X_l = X_p^k + r_1 l_{\max} D_p^k(\varphi^k - r_2\theta_{\max}/2) \end{cases} \tag{14}$$

where φ_p^k is scanning angle of the producer, $\theta_{\max} \in \mathbb{R}^{n-1}$and $l_{\max} \in \mathbb{R}^1$ are maximum pursuit angle and distance. $r_1 \in \mathbb{R}^1$is a normally distributed random number with mean 0 and standard deviation 1 and $r_2 \in \mathbb{R}^{n-1}$ is a uniformly distributed random sequence in the range of (0,1).

If the producer finds a better position, it will fly to there. Otherwise, it will stay in the same position and take a new random scanning angle

$$\varphi^{k+1} = \varphi^k + r_2\alpha_{\max} \tag{15}$$

where $\alpha_{\max} \in \mathbb{R}^1$ is the maximum turning angle.

Moreover, if the producer cannot find a better position after a constant number of iterations, its scanning angle will go back to zero degree

$$\varphi^{k+a} = \varphi^k \tag{16}$$

where $a \in \mathbb{R}^1$ is a constant.

3.1.2 *Scroungers and rangers*

Except the producer, a number of group members are randomly selected as scroungers. Scroungers employ the producer-scrounger (PS) model [7] to perform area copying to join the resource found by the producer; in other words, each scrounger keeps moving towards and searches the area around the producer. At the kth iteration, the area copying behavior of the ith scrounger can be modeled by:

$$X_i^{k+1} = X_i^k + r_3 \circ (X_p^k - X_i^k) \tag{17}$$

where $r_3 \in \mathbb{R}^n$ is a normally distributed random number with mean 0 and standard deviation 1 and "$\circ$" is the Hadamard product.

Aside from the producer and scroungers, the rest of GSO members are rangers. They perform RW in the searching space to resort to other resources, which increases the GSO's chance to escape from local optima. At the kth iteration, each ranger generates a random head angle φ_i^k and a random distance $l_{\max}$, then it moves to a new position:

$$X_i^{k+1} = X_i^k + a \cdot r_1 D_i^k(\varphi^{k+1}) \tag{18}$$

It should be noted that the members' roles can be switched between one another in GSO group. For instance, if a better resource is found by a scrounger or a ranger in the next bout compared with the current producer, the scrounger or the ranger will be switched to be the producer; on the other hand, the producer in the previous searching bout will perform scrounging strategies as a scrounger or RW as a ranger.

3.2 Group search optimizer with intraspecific competition and lévy walk

3.2.1 *Intraspecific competition*

Intraspecific competition (IC) is a particular form of competition in which members of the same species struggle for the same resource in an ecosystem [9]. IC is divided into two components, contest competition and scramble competition [9]. In the former competition, each successful competitor obtains all resources it needs for survival while the rest of competitors are deprived of such resources. However, in terms of scramble competition, it happens when competitors are crowded around limited resources, which are not monopolized by successful competitors. This sort of IC stimulates animals to compete for resources seriously [33].

As for GSO, group members are always hunting for the optimal resource which is definitely scarce. Therefore, IC exists inevitably in the form of scramble competition in GSO's searching process and it happens if GSO members are crowded around the producer. In [34], an index is proposed that can well describe the population's crowdedness in some evolutionary algorithms (EAs) dominated by the leading member with the best fitness value, e.g., PSO and GSO. The detailed calculation steps are expressed as follows.

1) Compute the mean distance d_i of each particle i with its position x_i to all the other particles

$$d_i = \frac{1}{N-1}\sum_{j=1,j\neq i}^{N}\sqrt{\sum_{k=1}^{D}(x_i^k - x_j^k)^2} \tag{19}$$

where N is the population size and D is the number of particles' dimensions.

2) Denote d_i of the leading particle as d_g, determine the maximum distance $d_{\max}$ and minimum distance $d_{\min}$ by comparing all d_i in (9). Therefore, f can be determined as

$$f = \frac{d_g - d_{\min}}{d_{\max} - d_{\min}} \in [0,1] \tag{20}$$

It can be easily seen that the leading member is surrounded closely by other members if f is small [34]. Therefore, the index f can be used to describe the population's crowdedness. If f is small, it means that GSO members are crowded around the producer. Then IC happens, and $r_3 = (r_{3_1}, r_{3_2}, \ldots, r_{3_D})$, presented in (21), is suggested to be a random vector with higher value ranged in (0.8,1) for scroungers to manifest this serious competition. Here r_3 is called scrounging coefficient. In GSOICLW, we suggest if f is less than 0.2, IC happens, then the value of r_3 can be determined by the index f as follows.

$$r_{3_i} = \begin{cases} \text{random}(0.8, 1) & f < 0.2 \\ \text{random}(0, 0.8) & f \geq 0.2 \end{cases} \quad i = 1, 2, \ldots, D \tag{21}$$

3.2.2 *Diversifying effect of intraspecific competition*

As mentioned above, IC occurs when animals from one species in crowdedness directly compete for a limited resource. Moreover, Richard has proposed that IC will drive the diversifying effect within a population, i.e., the increasing population density leads to reduced prey availability, in this way, some individuals will resort to alternative prey types [35]. Furthermore, he has pointed out that more intense competition caused by increasing population density compel more members to choose to seek another food resource [35]. Similar phenomena caused by IC have also been observed by other biologists [36, 37].

Therefore, when the group members in GSO are in IC, the diversifying effect will happen, i.e., some members are going to seek another food, acting as the role of rangers to escape this serious competition. In addition, it is noted that the IC become more intense, the more rangers will emerge [35]. In the above subsection, we have introduced the index f describing the crowdedness of population in GSO group, and it can be easily seen that when f becomes smaller, it means GSO members are more crowded, then IC become more intense, leading more individuals to become rangers.

According to the basic GSO algorithm, the ratio of rangers is set as a constant (20%) during the whole searching process. But in practical process of seeking food resources, as mentioned above, the proportion of rangers should vary in accord with the intensity. More specifically, the smaller index f, the more intense of IC; thus the bigger ratio of rangers. Therefore, we propose the ratio of rangers c_f in the proposed algorithm is the function of the index f when the searching group is in IC, otherwise a constant ratio, i.e., 20%, as employed in GSO. Therefore, the formulation of c_f is expressed as follows.

$$c_f = \begin{cases} \frac{1}{2.8571 + 2.5357 \times \sin(f)} & f < 0.2 \\ 0.2 & f \geq 0.2 \end{cases} \tag{22}$$

3.3 Lévy walk

In GSO, ranging animals perform searching strategies by means of RW, which are thought to be one of the most efficient searching method for randomly distributed resources [7]. It can be seen that rangers choose random searching distance in (18), where r_1 is a normally distributed random number with mean 0 and standard deviation 1. As a result, it can be seen rangers in GSO are performing the classical random walk because its variance of step length distribution is finite [38].

However, Viswanathan has claimed that LW is more efficient than classical random walk, because LW owns an inverse square power-law distribution of fight lengths. He has proved his claim by analyzing experimental foraging data on selected insect, mammal and bird species, and found that they were consistent with the predicted inverse square power-law distributions [10]. What is more, Andrew has conducted experiments on honey bees, and proved that the LW constitutes an optimal searching strategy for food resources [39].

Therefore, the LW is chosen as the strategy of ranging behavior in GSOICLW, and its walking length, r, can be drawn from a probability distribution function with a power-law tail as follows [10]

$$P(r) \sim r^{-\mu} \quad (1 < \mu < 3) \tag{23}$$

It is noteworthy to mention that when $\mu \leq 1$, the distribution function cannot correspond to probability distributions that can be normalized; on the other hand, the function will become Gaussian distribution function if $\mu \geq 3$. Therefore, here we choose $\mu = 2$.

In this way, (14) should be changed by replacing the normal random number r_1 with the random number r from lévy distribution function. Then the ith member acting as a ranger moves to a new position

$$X_i^{k+1} = X_i^k + a \cdot rD_i^k(\varphi^{k+1}) \tag{24}$$

In conclusion, GSOICLW incorporates IC and LW into GSO, as illustrated in Fig. 1. In GSO, most members, acting as scroungers, who scrounges the producer, and rangers perform RW to seek other resources. However, when other members are crowding around the producer, IC happens. It leads to more intense scrabble for the resource lying in the position of the producer and more rangers who will disperse with longer step by LW in the searching space. In [8], it has been proved that IC makes the scrounging coefficient and rangers' ratio vary adaptively to balance local and global searching, and LW stimulates rangers to perform more efficient global searching than RW. In other words, GSO achieves better performance than GSO in that it improves GSO's local searching ability while maintaining its global searching performance.

4 Latin hypercube sampling with Cholesky decomposition

To deal with the uncertainty of wind speed or wind power, the MC simulation combined with simple random sampling (SRS) has been widely used in power system dispatch problems [11, 15]. However, the SRS sampling method is inefficient for its heavy computational burden due to the need of a large number of repeated calculations

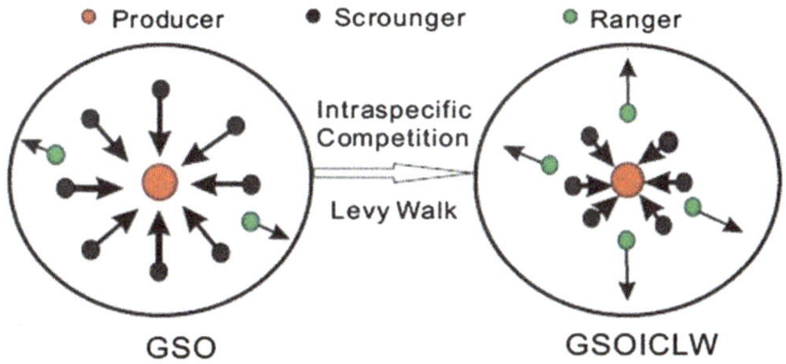

Fig. 1 Illustration of IC and LW in GSOICLW

to obtain a specified accuracy. It has been proved in Ref. [24] that the sampling method of Latin hypercube sampling with Cholesky decomposition (LHS-CD) can obtain more reliable results with relatively small sample size, compared with SRS. Therefore, in this paper, LHS-CD is used to sample wind speed forecast errors, which folllow the Gaussian distribution.

Then the wind speed samples can be obtained to calculate the corresponding power flows to compute the mean and variance of the profit function mentioned above. The detailed LHS-CD calculation steps for power flow can be found in [24], and the calculation flowchart is shown in Fig. 2.

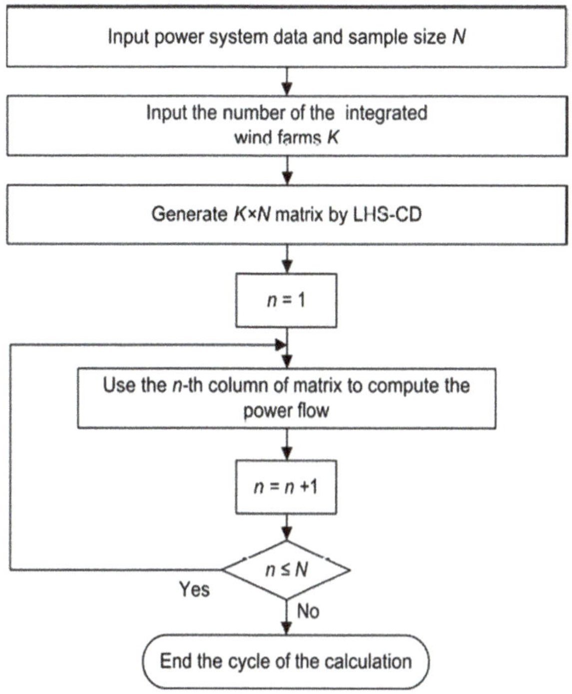

Fig. 2 Calculation flowchart of LHS-CD for power flow

5 Experimental studies

The MV model and the algorithm of GSOICLW have been tested on the IEEE 30-bus test system. This test system consists of 48 branches, 4 transformers, 6 generator-buses, and 22 load-buses. The total system demand is 283.4 MW. The locations for the wind farms are on buses 7, 10, 16, 24 and 30, and predicted wind speeds and number of wind turbines operated in the wind farms are given in Table 1. The standard deviation of the wind speed forecast error is set to be 8% of its corresponding forecast wind speed.

Suppose the doubly-fed induction generator (DFIG) with a constant power factor is used in wind farms. The rated active power is set to be 2 MW, and the rated, cut-in and cut-out wind speed are set to be 12.5 m/s, 4 m/s and 20 m/s, respectively. As for the sampling method of LHS-CD, we set the number of wind speed samples to be 400.

Moreover, in order to demonstrate the effectiveness of GSOICLW, it is tested and compared with GSO, PSO and the algorithm of primal-dual interior point (PDIP) based on the modified IEEE 30-bus test system, considering the predicted wind speeds. In addition, 50 independent runs have been used for GSOICLW, GSO, PSO and PDIP to test their average performances on the ORPD problem. The population sizes of GSOICLW and GSO are set as 47, and the total iterations during each run are both 300, respectively. Therefore, the total number of fitness evaluation is 15,000. The parameters of IC and LW in GSOICLW are listed in Sect. 3. Moreover, we also set the total number of fitness evaluation in terms of PSO and PDIP are the same as that of GSOICLW. Please note that PDIP is sensitive to the initial searching point, which is difficult to select, and we randomly choose it when conducting the 50 independent runs.

Table 1 The forecast wind speeds and number of wind turbines in wind farms

Node	7	10	16	24	30
Predicted wind speed (m/s)	9.3	15	7.6	8.7	6.5
Number of turbines	20	20	20	20	20

In order to further assess the performance of the proposed algorithm in a stochastic search process, the Mann-Whitney U-test is adopted. It is a non-parametric test for comparing two populations, assessing whether the two groups of results are statistically different from each other [40].

Therefore, we conduct two simulation cases: one aims to test the performance of GSOICLW and the other is designed to prove the effectiveness of the MV model.

Case 1: Minimization of the transmission loss with predicted wind speeds.

Table 2 Results from GSOICLW, GSO and PSO in Case 1

Algorithms	Best result	Worst result	Average result	Standard deviation	*p*-value (*h*-value)
GSOICLW	1.3297	1.3701	1.3404	0.0081	N/A
GSO	1.3590	1.4020	1.3780	0.0121	1.99E-12 (1)
PSO	1.3592	1.7134	1.5142	0.0581	1.25E-14 (1)
PDIP	1.3602	3.0459	2.4269	0.5963	2.13E-19 (1)

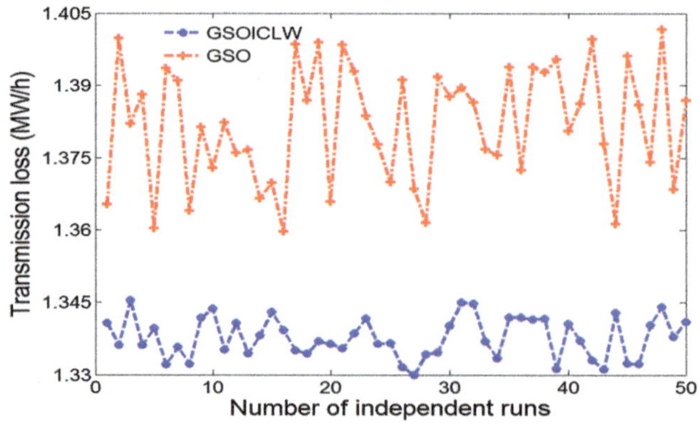

Fig. 3 Convergence results of Case 1 obtained by GSOICLW and GSO in 50 runs

Case 2: Minimization of the transmission loss with uncertain wind speeds.

5.1 Minimization of the transmission loss with predicted wind speeds

In this case, it aims to minimize transmission loss only based on IEEE 30-bus system, which focuses on optimizing the objective function J_1 in (2). The best results, worst results, average results and standard deviations of transmission loss gained by GSOICLW, GSO and PSO from 50 runs are presented in Table 2. It is shown that GSOICLW can find more accurate and robust solution than GSO and PSO because the average result and standard deviation obtained by GSO are 1.3404 and 0.0081, much better than that of GSO, PSO and PDIP. In particular, the average transmission loss obtained by GSOICLW is superior to the best results gained by GSO, PSO and PDIP, which are 1.3590 MW/h, 1.3592 MW/h and 1.3602 MW/h, respectively. As for PSO and PDIP, the worst results (1.7134 MW/h and 3.0459 MW/h) and standard deviations (0.0581 and 0.5963) of transmission loss, much worse than that of GSOICLW, indicate PSO and PDIP easily entrap in local optima during the searching process. Moreover, the gained *p*-values and *h*-values using the Mann-Whitney U-test are shown in Table 2, which prove that the results obtained by GSOICLW are significantly different from that of GSO, PSO and PDIP.

Fig. 3 shows the convergence results in 50 trials in this case for GSO and GSOICLW, respectively. It can be easily seen that GSOICLW can find better solution than GSO. Although GSO can find good results in some trials (i.e., the 16th trial), but in the rest trials the optimized costs are even close to 1.40 MW/h. Consequently, the standard deviation corresponding to GSO is 0.0121, lager than that of GSOICLW.

5.2 Minimization of the transmission loss with uncertain wind speeds

The MV model shown in (7) is solved for different values of the risk tolerance parameter, which allows assigning different weights to the fuel cost term $F^{\exp}$ (representing the profit term) versus the risk term V in the objective function. It can be easily seen that the risk tolerance parameter will put a great impact on the final optimized results. The smaller risk tolerance parameter is, the more emphasis on the profit. Suppose, λ varies from 0.0 to 0.5, and this range is considered wide enough to reflect the variety of risk that power system dispatchers are willing to assume.

Table 3 illustrates dispatching differences for the cases of the maximum risk ($\lambda = 0$) and a low level of risk ($\lambda = 0.5$). Moreover, Fig. 4 depicts transmission loss' expected value and standard deviation versus the risk tolerance parameter. It can be observed that the expected transmission loss increases as variance decreases. The expected transmission loss achieved by a conservative dispatch solution ($\lambda = 0.5$) is 2.2045 MW/h whereas the expected fuel cost with the maximum risk is equal to 1.4431 MW/h ($\lambda = 0$), corresponding to the aggressive dispatch solution.

It can be easily seen that different levels of risk imply different dispatching solutions, and ultimately, different transmission loss under the uncertain wind power environment. It is noted that different risk tolerance parameters correspond to different dispatch solutions. To demonstrate the different effectiveness in terms of the dispatch solutions obtained in Table 3, Fig. 5 shows transmission loss corresponding to different wind speed samples among the LHS-CD simulation, in terms of the dispatch solutions A ($\lambda = 0.5$), B ($\lambda = 0.2$) and C ($\lambda = 0$).

It is clear that if solution C is adopted by power system dispatchers, the expected transmission loss among the 400 wind samples is 1.4431 MW/h, much lower than that of B and C. However, the deviation regarding to solution C is much higher, which demonstrates that this solution cannot adjust all the wind samples well. For instance, the transmission loss regarding to many wind samples are even

Table 3 Mean and variance values with different risk tolerance parameters

λ	0.0	0.1	0.2	0.3	0.4	0.5
Mean	1.4431	1.4786	1.5645	1.7236	1.9421	2.2045
Variance	0.6013	0.4356	0.2693	0.2031	0.1539	0.0989

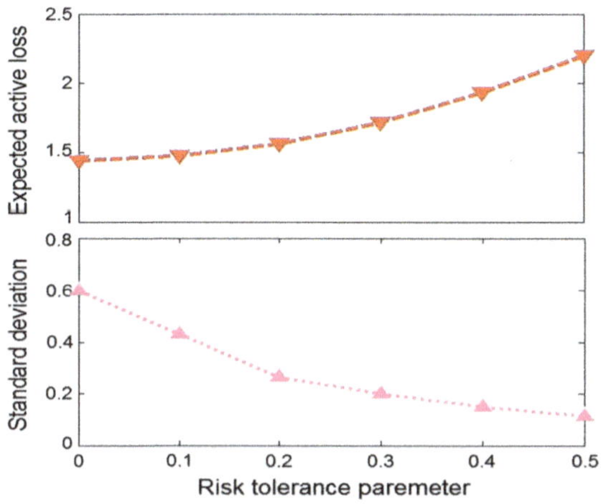

Fig. 4 Mean and variance values with different risk tolerance parameters

more than that of the solution *B*, and the transmission loss corresponding to the 323th and 361th wind power samples are even more than that of the solution *A*, although the expected transmission loss of solution *C* is smaller than *B* and much smaller than *A*.

On the other hand, although the deviation of solution *A* is much smaller, which proves it can adjust all the uncertain wind samples well, the average transmission loss is too high, which is 2.2045 MW/h. It may not be advisable for power system dispatchers to choose this solution for the consideration of economic aspects. In essence, the final dispatch solution is determined mainly by the dispatchers' attitudes toward the trade-off between profit and risk. How to make decision based on dispatchers' attitudes is one of our further research directions.

6 Conclusion

This paper presents the mean-variance (MV) model to solve power system reactive power dispatch (RPD) problems with wind power integrated. The MV model considers the profit and risk simultaneously under the uncertain wind power (speed) environment. To describe this uncertain environment, the Latin hypercube sampling with Cholesky decomposition (LHS-CD) simulation method is used to sample uncertain wind speeds. An improved optimization algorithm, group search optimizer with intraspecific competition and lévy walk (GSOICLW), is then used to optimize the MV model by introducing the risk tolerance parameter. To test the performance of GSOICLW and the effectiveness of the MV model, simulation studies have been carried out on the IEEE-30 bus system in two cases: one is minimization of the transmission loss with predicted wind speeds, and the other intends to minimize the transmission loss with uncertain wind speeds. In the first case, the obtained results have proved that GSOICLW can obtain more accurate and robust optimal solution, in comparison with GSO and PSO. In the second case, simulation results have proved that the expected transmission loss and the corresponding standard deviation vary in different direction with the change of different risk tolerance parameters, which demonstrates the effectiveness and validity of the proposed model.

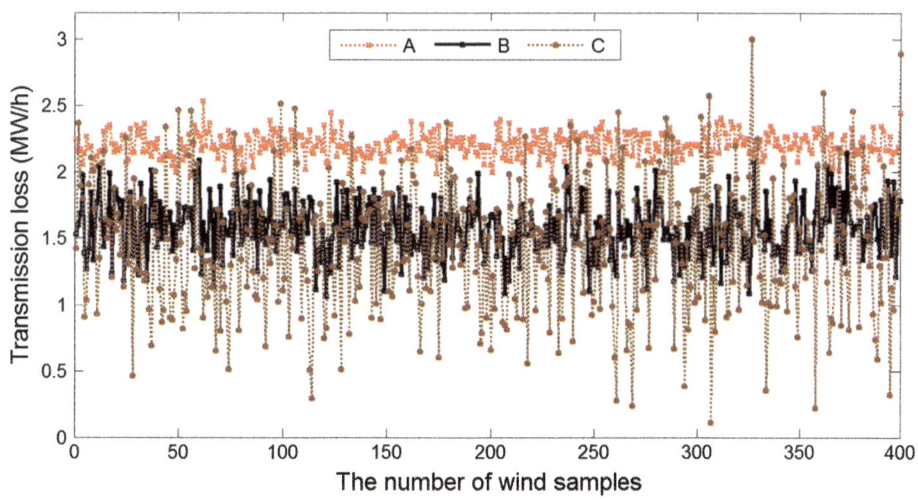

Fig. 5 Transmission loss corresponding to 400 wind speed samples in terms of different dispatch solutions

Acknowledgments The work is funded by Guangdong Innovative Research Team Program (No. 201001N0104744201) and National Key Basic Research and Development Program (973 Program, No. 2012CB215100), China. The first author thanks for the financial support from China Scholarship Council Program (No. 201306150070).

References

[1] Wu QH, Ma JT (1995) Power system optimal reactive dispatch using evolutionary programming. IEEE Trans Power Syst 10(3):1243–1249
[2] Dai CH, Chen WR, Zhu YF et al (2009) Seeker optimization algorithm for optimal reactive power dispatch. IEEE Trans Power Syst 24(3):1218–1231
[3] Deeb N, Shahedipour SM (1990) Linear reactive power optimization in a large power network using the Decomposition approach. IEEE Trans Power Syst 5(2):428–435
[4] Granville S (1994) Optimal reactive dispatch through interior point methods. IEEE Trans Power Syst 9(1):136–146
[5] Momeh JA, Guo SX, Oghuobiri EC et al (1994) The quadratic interior point method solving power system optimization problems. IEEE Trans Power Syst 9(3):1327–1336
[6] Wu QH, Cao YJ, Wen JY (1998) Optimal reactive power dispatch using an adaptive genetic algorithm. Int J Electr Power Energy Syst 20(8):563–569
[7] He S, Wu QH, Saunders JR (2009) Group search optimizer: an optimization algorithm inspired by animal searching behavior. IEEE Trans Evol Comput 13(5):973–990
[8] Li YZ, Li MS, Ji Z et al (2013) Optimal power flow using group search optimizer with intraspecific competition and lévy walk. In: Proceedings of the IEEE symposium on swarm intelligence(SIS'13), Singapore, 16–19 Apr 2013, 256–262 pp
[9] Nicholson AJ (1954) An outline of the dynamics of animal populations. Aust J Zool 2(1):9–65
[10] Viswanathan GM (1999) Optimizing the success of random searches. Nature 401:911–914
[11] Shi LB, Wang C, Yao LZ et al (2012) Optimal power flow solution incorporating wind power. IEEE Syst J 6(2):233–241
[12] Zhou W, Peng Y, Sun H (2011) Optimal wind-thermal coordination dispatch based on risk reserve constraints. Eur Trans Electr Power 21(1):740–756
[13] Hong YY, Chang HL, Chiu CS (2010) Hour-ahead wind power and speed forecasting using simultaneous perturbation stochastic approximation (SPSA) algorithm and neural network with fuzzy inputs. Energy 35(9):3870–3876
[14] De Giorgi MG, Ficarella A, Tarantino M (2011) Assessment of the benefits of numerical weather predictions in wind power forecasting based on statistical methods. Energy 36(7):3968–3978
[15] Ummels BC, Gibescu M, Pelgrum E et al (2007) Impacts of wind power on thermal generation unit commitment and dispatch. IEEE Trans Energy Convers 22(1):44–51
[16] Wang JH, Shahidehpour M, Li ZY (2010) Security-constrained unit commitment with volatile wind power generation. IEEE Trans Power Syst 23(3):1319–1327
[17] Miranda V, Hang PS (2005) Economic dispatch model with fuzzy wind constraints and attitudes of dispatchers. IEEE Trans Power Syst 20(4):2143–2145
[18] Wang LF, Singh C (2008) Balancing risk and cost in fuzzy economic dispatch including wind power penetration based on particle swarm optimization. Electr Power Syst Res 78(8):1361–1368
[19] Lee TY (2007) Optimal spinning reserve for a wind-thermal power system using EIPSO. IEEE Trans Power Syst 22(4):1612–1621
[20] Piperaqkas GS, Anastasiadis AG, Hatziarqyriou ND (2011) Stochastic PSO-based heat and power dispatch under environmental constraints in corporating chp and wind power units. Electr Power Syst Res 81(1):209–218
[21] Boufard F, Galiana FD (2008) Stochastic security for operations planning with significant wind power generation. IEEE Trans Power Syst 23(2):306–316
[22] Sun YZ, Wu J, Li GJ et al (2009) Dynamic economic dispatch considering wind power penetration based on wind speed forecasting and stochastic programming. Proc CSEE 29(4):41–47 (in Chinese)
[23] Siahkali H, Vakilian M (2010) Stochastic unit commitment of wind farms integrated in power system. Electr Power Syst Res 80(9):1006–1017
[24] Yu H, Chung CY, Wong KP et al (2009) Probabilistic load flow evaluation with hybrid latin hypercube sampling and cholesky decomposition. IEEE Trans Power Syst 24(2):661–667
[25] Markowitz HM (1952) Portfolio selection. J Finance 7(1):77–91
[26] Conejo AJ, Nogales FJ, Arroyo JM et al (2004) Risk-constrained self-scheduling of a thermal power producer. IEEE Trans Power Syst 19(3):1569–1574
[27] Yamin HY, Shahidehpour SM (2004) Risk and profit in self-scheduling for gencos. IEEE Trans Power Syst 19(4):2104–2106
[28] Li MS, Ji TY, Wu QH et al(2010) Stochastic optimal power flow using a paired-bacteria optimizer. In: Proceedings of the IEEE Power Energy Society general meeting, Minneapolis, 25–29 July 2010, 7 pp
[29] Justus CG, Hargraves WR, Mikhail A et al (1978) Methods for estimating wind speed frequency distributions. J Appl Meteorol 17(3):350–353
[30] Conradsen K, Nielsen LB, Prahm LP (1984) Review of Weibull statistics for estimation of wind speed distributions. J Climate Appl Meteorol 23(8):1173–1183
[31] Tuller SE, Brett AC (1984) The characteristics of wind velocity that favor the fitting of a weibull distribution in wind speed analysis. J Climate Appl Meteorol 23(1):124–134
[32] Zhao M, Chen Z, Blaabjerg F (2006) Probabilistic capacity of a grid connected wind farm based on optimization method. Renew Energy 31(13):2171–2187
[33] Berryman AA (1999) Principles of population dynamics and their application. Garland Science, New York
[34] Zhan ZH, Zhang J, Li Y et al (2009) Adaptive particle swarm optimization. IEEE Trans Syst Man Cybern B 39(6):1362–1381
[35] Svanbäck R, Bolnick DI (2007) Intraspecific competition drives increased resource use diversity within a natural Population. Proc Roy Soc B 274(1611):839–844
[36] Bolnick DI (2001) Intraspecific competition favours niche width expansion in drosophila melanogaster. Nature 410:463–466
[37] Craig MacLean R (2005) Adaptive radiation in microbial microcosms. J Evol Biol 18(6):1376–1386
[38] Benhamou S (2007) How many animals really do the levy walk? Ecology 88(8):1962–1969
[39] Reynolds AM, Smith AD, Reynolds DR et al (2007) Honeybees perform optimal scale-free searching ights when attempting to locate a food source. J Exp Biol 210(21):3763–3770
[40] Hollander M, Wolfe DA (1973) Nonparametric statistical methods. Wiley, New York

Yuanzheng LI received the M.S. degree in Electrical Engineering from Huazhong University of Science and Technology (HUST), Wuhan, China, in 2011. He is currently pursuing the Ph.D. degree at the School of Electrical Power Engineering, South China University of Technology (SCUT), Guangzhou, China. Now, he is a Visiting Research Assistant in the University of Liverpool, Liverpool, U.K. His research interests include power system decision making, optimal dispatch, stochastic optimization considering large-scale integration of renewable energy into the power system and multi-objective optimization.

Mengshi LI received the M.Sc. (Eng) (Hons.) degree in information and intelligence engineering from the University of Liverpool, Liverpool, U.K. and the Ph.D. degree in electrical engineering from the University of Liverpool, Liverpool, U.K., in 2005 and 2010, respectively. Currently, he is a Lecturer in the School of Electric Power Engineering, South China University of Technology, Guangzhou, China. His research interests include computational intelligence and their applications in power systems.

Qinghua WU obtained a Ph.D. degree in Electrical Engineering from The Queen's University of Belfast (QUB), Belfast, U.K. in 1987. He worked as a Research Fellow and subsequently a Senior Research Fellow in QUB from 1987 to 1991, and as a Lecturer, subsequently he was appointed Senior Lecturer in the Department of Mathematical Sciences, Loughborough University, Loughborough, U.K. in 1991 to 1995. In September, 1995, he joined The University of Liverpool, Liverpool, U.K. to take up his appointment to the Chair of Electrical Engineering in the Department of Electrical Engineering and Electronics. Since then, he has been the Head of Intelligence Engineering and Automation Research Group working in the areas of systems control, computational intelligence and electric power and energy. He is also a Distinguished Professor at the South China University of Technology, China. He has authored and coauthored more than 420 technical publications, including 210 journal papers, 20 book chapters and 3 research monographs. Professor Wu is a Chartered Engineer, Fellow of IET, Fellow of InstMC and Fellow of IEEE. His research interests include nonlinear adaptive control, mathematical morphology, evolutionary computation, machine learning and power system control and operation.

22

Estimating life-cycle energy payback ratio of overhead transmission line toward low carbon development

Xiaoming DONG, Chongqing KANG (✉),
Ning ZHANG, Huaguang YAN, Junxia MENG,
Xinsheng NIU, Xin TIAN

Abstract The energy conservation plays an important role for low carbon development. In order to evaluate the energy conservation in the full life-cycle, a scheme to estimate the energy consumption, or alternatively the energy pay, in constructing an overhead transmission line is proposed in this paper. The analysis of a typical projection is given for demonstration. With new additional overhead transmission lines, the energy consumption, known as the power loss in power network, is expected to be decline, which is defined in this paper as the energy payback. In order to estimate this kind of contribution, the scheme that consisted of load forecast, production simulation for generating systems, load flow simulation and power loss calculation has been proposed. Case studies, based on the IEEE 24-bus test system, are given to demonstrate the efficacy of the schemes. Moreover, several presumptive scenarios are deployed and analysed with the presented schemes for comparison.

Keywords Power system operation, Power system planning, Power system simulation, Power transmission, Life cycle analysis

X. DONG, C. KANG, N. ZHANG, The State Key Lab of Power Systems, Department of Electrical Engineering, Tsinghua University, Beijing 100084, China
(✉) e-mail: cqkang@tsinghua.edu.cn
H. YAN, J. MENG, The China Electric Power Research Institute, Beijing 100192, China
X. NIU, X. TIAN, The State Grid Shandong Power Economic Research Institute, Jinan 250001, China

1 Introduction

For power system reliability and security, the power grid must be intensified as the increase in power demand and generation. Various important analysis works, known as power network planning [1], should be carried out in advance of establishing a new transmission project. Furthermore, power network planning is such an essential work which aims to consolidate the power network. Constructing new overhead transmission lines (OTLs) is considered as one basic part of power network planning.

With the gradual exhaustion of fossil fuels, all kinds of efforts are made to promote the energy conservation [2] for the purpose of low carbon development [3, 4]. Therefore, the energy consumption, which is the major concern of this paper, is an important aspect for the performance of the power network. Allowing for life-cycle [5], an OTL project will go through several steps of manufacturing, installing, operating and dismantling. Each step has the demand of energy consumption. The quantity of the energy consumption at the stages of manufacturing and installing, which is largely determined by the design scheme, is defined here as the energy consumption of construction (ECC). The ECC will not fluctuate too much with an established design scheme, which specify where, when and how to execute the project. Section 2 gives a case study to demonstrate the analysis of ECC by a typical OTL project.

While an OTL integrates into the power network and participates in carrying the power, there is inevitably some electric power loss [6–10]. The total network power loss, which is defined here as the energy consumption of operation (ECO), is largely dependent on the system operation mode rendering the load demand and power source allocation. However, while new OTLs have been constructed and put into operation, the ECO is expected to be decline as the supplemented lines having positive effect on network connectivity. The potential decline in ECO by constructing new OTL is considered as a kind of energy payback. In order to estimate the OTL's contribution to the energy payback in the future, load power should be obtained by load forecast [11–15] and generator outputs should be assigned by production simulation [16–19]. Then, the ECO can be produced by load flow [20] simulation. Section 3 details the steps comprising the above procedures.

This paper gives approximate approaches for estimating the ECC and the ECO of OTL projects. The comparison of those two kinds of energy consumptions is deployed in Section 4. Furthermore, case studies for different scenarios corresponding to optional OTL projects highlight the significant of the presented approaches. In Section 5, the main results are summarized and key conclusions are drawn.

Note that, the energy consumption is not the sole factor to evaluate the performance of a power network. Therefore, a transmission project may have more benefits to the system than just reducing losses. For example, a project could have reliability or stability benefits. A complex multi-objective decision should be made for network planning. However, the primary focus of this study is on the analysis with respect to energy consumption.

2 Energy consumption of construction

2.1 Descriptions of the typical project

Amount of equipment and industrial materials are used to construct an OTL projects. Plenty of energy should be consumed in producing those manufactures. To analyze the energy consumption at the stage of construction, a typical OTL project is elaborated in this section. As shown in Fig. 1, **A** denotes a high-voltage (230 kV) and double-circuit OTL. The new power plant (**B**) is going to be established. To transfer power from **B** to load center, an optimal scheme is to break up **A** and rebuild double π OTLs (**C**) to connect **A** and **B**. The total length of **C** is about 33.8 km.

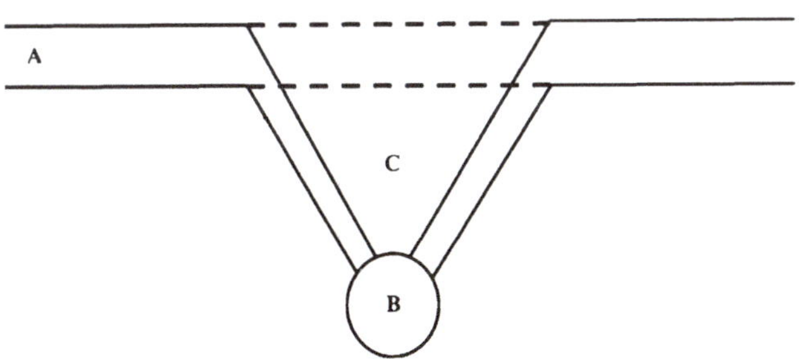

Fig. 1 The schematic diagram of the typical project

2.2 Data of equipment

The equipment and industrial materials that forming the OTL project are various. In advance of carrying out a project, the task of top priority is estimating the plan. In accordance with related demands, such as system security and stability, the construction plan should detail the design concept and equipment selection of the project. The plan of a typical project is given in Table 1 with the type, the model, the quantity and the weight of those equipment and industrial materials elaborated.

2.3 Estimate for ECC

The typical values of energy consumptions (EC) are provided in Table 2. The unit of "kgce" denotes kilograms of standard coal. In accordance with the average coal consumption of thermal power units, the parity of 0.404 kgce is 1 kW h. In order to capture the consumed diesel oil in the process of excavation and transport, the diesel oil consumption is converted into equivalent EC according to the average efficacy of diesel generator.

Then, the ECC can be estimated in accordance with the information shown in Table 1 and related unit energy consumption given in Table 2. Figure 2 shows the proportion of the ECC. Apparently, the equipment with maximal energy consumption is the conductor, which is mainly composed of aluminum.

Table 1 The equipment list of the typical project

Type	Model	Quantity	Weight
Conductor	JL/LHA1-650/295-900	33.8 km	545 t
Ground wire	JLB40-150	33.8 km	24.5 t
OPGW	OPGW-150	33.8 km	26 t
Insulator	U160B/155	11,136 pieces	160.7 t
Insulator	U300B/195	7,680 pieces	
Insulator	U70B/146	1,920 pieces	
Armour clamp	1 × 16WXP-16	456 sets	13 t
Armour clamp	2 × 16WXP-16D	120 sets	
Armour clamp	2 × 16WXP-30	240 sets	
Armour clamp	1 × 16WXP-07	120 sets	
Transmission tower			2,170 t
Foundation (steel)			446.6 t
Foundation (concrete)	$f_c = 14.3$ N/mm^2	8,286 m^3	19,886 t
Cushion (concrete)	$f_c = 7.2$ N/mm^2	1,597.8 m^3	3,834.7t
Grounding (steel)			26 t
Earthworks		17,752 m^3	29,587 t

Table 2 The unit energy consumption

Materials	EC (kgce/t)	EC (kW h/t)
Steel (high quality)	860	2,129
Steel (low quality)	750	1,856
Aluminum	5,818	14,400
Ceramic	1,212	3,000
Cement	44	110
Diesel oil	1,429	3,537

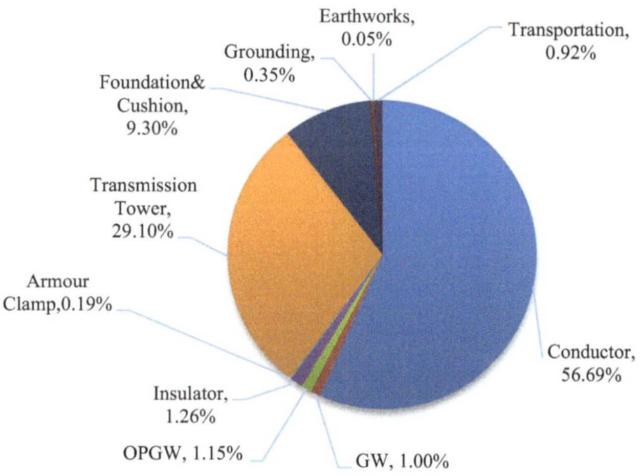

Fig. 2 The proportion of ECC

3 Energy consumption of operation

While new OTLs are going to be constructed and put into operation, the total power loss is expected to be decrease. To capture the contribution of establishing a new OTL, the power system simulation is needed. The process of ECO calculation consists of four steps is shown in Fig. 3.

3.1 Load forecasting

In this step, both the load active power and the load reactive power should be forecasted over a long period of time. Various optional methods, such as time series method and trend extrapolation method, can be qualified for this work.

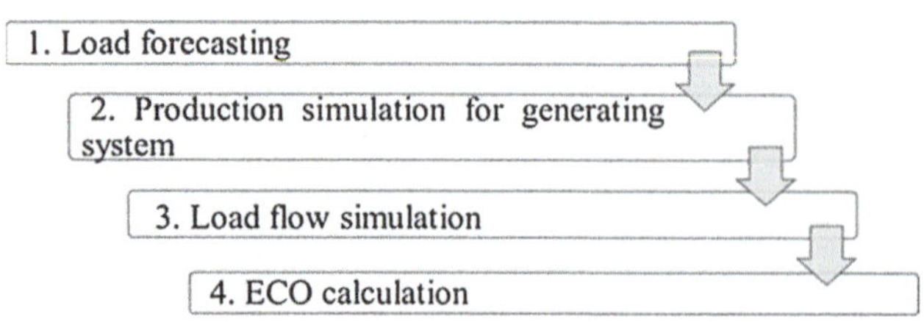

Fig. 3 The steps diagram of the ECO calculation

3.2 Production simulation for generating system

While power demands at receiving buses vary continuously, power balance must be maintained by adjusting outputs of sending generators. Once the load power is determined, the optimal output of generators can be obtained by production simulation allowing for minimizing the operating costs. A security constrained UC model [21] as described by (1) to (12) is adapted.

$$\min C_{sys} = \sum_{t\in N}\Big(\sum C_f P_f^t + \sum C_c P_c^t + \theta \sum C_d D_d^t + \gamma \sum S_f^t\Big) \tag{1}$$

$$\sum P_f^t + \sum P_c^t + \sum D_d^t = \sum D^t \quad \forall t \in N \tag{2}$$

$$\sum P_{f\min} I_f^t \le \sum P_f^t \le \sum P_{f\max} I_f^t \quad \forall t \in N \tag{3}$$

$$\sum P_{c\min} I_c \le \sum P_c^t \le \sum P_{c\max} I_c \quad \forall t \in N \tag{4}$$

$$\sum S_f^t \ge \sum V_f (I_f^t - I_f^{t-1}),\ S_f^t \ge 0 \quad \forall t \in N \tag{5}$$

$$\sum D^t + r_u^t \le \sum P_{f\max} I_f^t + \sum P_{c\max} I_c + \sum D_d^t \quad \forall t \in N \tag{6}$$

$$\sum P_{f\min} I_f^t + \sum P_{c\min} I_c + \sum D_d^t \le \sum D^t - r_d^t \quad \forall t \in N \tag{7}$$

$$-F_{\max}^t \le W(A_{ngf} P_f^t + A_{ngc} P_c^t + D_d^t - D^t) \le F_{\max}^t \quad \forall t \in N \tag{8}$$

$$-F_{sr\max}^t \le A_{sl} W(A_{ngf} P_f^t + A_{ngc} P_c^t + D_d^t - D^t) \le F_{s\max}^t \quad \forall t \in N \tag{9}$$

$$\begin{cases} -\Delta P_{f\,\mathrm{down}} \le P_f^t - P_f^{t-1} \le \Delta P_{f\,\mathrm{up}} \\ -\Delta P_{c\,\mathrm{down}} \le P_c^t - P_c^{t-1} \le \Delta P_{c\,\mathrm{up}} \end{cases} \quad \forall t \in N \tag{10}$$

$$I_f^t \in \{0,1\} \quad \forall t \in N,\ I_c \in \{0,1\} \tag{11}$$

$$D_d^t \ge 0 \quad \forall t \in N \tag{12}$$

where Σ represents the summation of all the corresponding elements; N represents the number of time periods in the daily dispatching model; f and c (subscripts) denote respectively generating units that can start and stop daily and that cannot start and stop daily; C_f and C_c represent the average cost of corresponding type of generation; C_d represents the load shedding cost; V_f is the start-stop cost of generating units that can start and stop daily; S_f^t is the start-stop cost of generating unit that can start and stop daily at t; θ is the amplification coefficient of load shedding penalty; γ is the amplification coefficient of start-stop cost; P_f^t and

P_c^t represent the output of different types of generation at t; $P_{f\,\max}$, $P_{f\,\min}$, $P_{c\,\max}$, $P_{c\,\min}$, represent the maximum and minimum power of the different types of generation; $\Delta P_{f\,\text{down}}$, $\Delta P_{f\,\text{up}}$, $\Delta P_{c\,\text{down}}$, $\Delta P_{c\,\text{up}}$ represent the maximum and minimum ramping rates of units that can start and stop daily and units that cannot start and stop daily, respectively; I_f^t is the state at t of generating unit that can start and stop daily; I_c is the state of unit that cannot start and stop over the dispatching day; D^t is the nodal loads at t; D_d^t is the nodal load shedding at t; r_d^t and r_u^t are system up and down reserve requirements at t; A_{ngf} and A_{ngc} are units and nodes incidence coefficients of the different types of generation; A_{sl} represents the incidence coefficient between transmission lines and transmission sections; W is generation distribution shift factors; $F_{\max}^t$ represents the transmission line capacity at t; $F_{sr\,\max}^t$ and $F_{s\,\max}^t$ represents the forward and backward direction maximum capacity for transmission sections

The objective of the above optimization model is minimizing the total system operating cost, which consists of the costs of the fuel, start-stop and load shedding. The decision variables consist of P_f^t, P_c^t, D_d^t, S_f^t, I_f^t and I_c. (2) represents the power balance between load and generation. (3) and (4) represents the output limits of each generator unit. (5) denotes the start-stop cost of generating units when their states change from stop to work. (6) and (7) respectively represent the system positive and negative reserve constraints. (8) and (9) represent the power constraints for transmission lines and sections capacity. (10) enforces the thermal generation ramp rate limits. (11) defines the states of the generators as discrete variables. (12) enforces the amount of load shedding to be zero or positive. By linearizing the cost functions of the thermal generating units, the mixed integer linear programming algorithm could be used to solve this optimal problem.

3.3 Load flow simulation

With the load power and generator output obtained using load forecast and production simulation (corresponding to step 1 and step 2, respectively), the load flow equations can be initialized and solved using the Newton iteration method. The shunt compensators are taken into account by operating in accordance with the bus voltage magnitude. As is shown in (13), When $V_i > V_{\max}$ (reactive power at bus i is redundant), the shunt reactors at bus i are switched off prior to switching on valid shunt capacitor, and vise versa.

$$V_{\max} \geq V_i \geq V_{\min} \quad (i = 1, 2, \cdots, n) \tag{13}$$

Following steps simulate the operation of the compensators:

Step 1: Solving load flow equations using Newton iterations.

Step 2: Check voltage magnitude of each bus. If all the buses match (1), finish; else go to Step 3.

Step 3: Adjust the shunt compensator. If any status of compensator is changed, go back to Step 1; else, finish.

3.4 Load flow simulation

With the steps shown in Sections 3.1, 3.2 and 3.3, the power loss at various points in time can be obtained. In order to estimate the ECO for a period of time, following formulation can be applied.

$$E_{\text{CO}} = \sum_{i=1}^{n-1} \left(\frac{P_{loss}^{(i)} + P_{loss}^{(i+1)}}{2} \right) \tag{14}$$

where E_{CO} represents the value of ECO corresponding to the hypothetical scenario in which the specified new OTL have been integrated into the network; n denotes the total of time points within a specific time window; P_{loss} denotes the network power loss.

4 Energy payback ratio

With the method presented in Section 3, the ECO can be estimated over a period of time, which is defined here as the time window. It is assumed that all the OTLs and the transformers participate in operation without failure during the time window. Hence, the ECO of a target system for a specific time window is invariable unless the network is changed by establishing new OTLs. Since a new OTL is constructed and put into operation, the ECO is expected to be decline due to the improvement of the power transmission. To evaluate the contribution to this aspect of system operation, the decrease in ECO corresponding to a specific time window, or alternatively the energy payback, is defined by,

$$E_{\text{PB}} = E_{\text{CO}} - E_{\text{CO}}^{(0)} \tag{15}$$

where $E_{\text{CO}}^{(0)}$ represents the base value of ECO for the original case; E_{PB} represents the value of EPB.

To estimate the EPB in life-cycle, the approach presented in Section 2 is available to obtain the ECC in establishing a new OTL. As a result, the energy payback ratio is defined by,

$$R_{\text{PB}} = \frac{E_{\text{PB}}}{E_{\text{CC}}} \times 100\,\% \tag{16}$$

where E_{CC} represents the value of ECC. This ratio is considered as a more comprehensive index to evaluate the contribution for energy conservation because of the consideration in life-cycle.

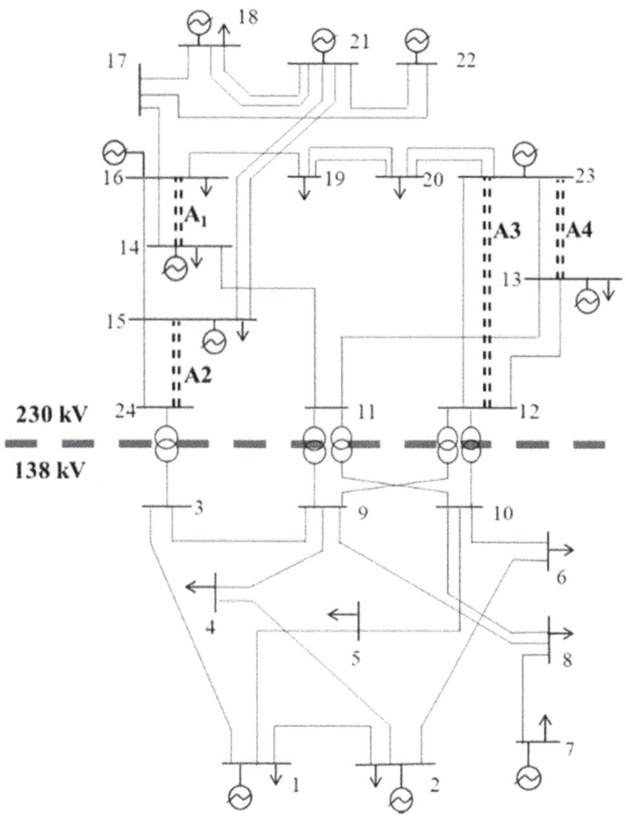

Fig. 4 The IEEE 24-bus test system

5 Case studies

5.1 Description of the test system

In this case, the IEEE 24-bus reliability test system [22] (see Fig. 4) is analyzed using the presented scheme. The S_{ref} is set to 100 MW.

The values of load active power for 8,760 hours (one year) that derived from [22] are used as "the results of load forecast". With the assumption that the power factor of each bus is fixed, the values of load reactive power for 8,760 hours are obtained.

The shunt compensators which are shown in Table 3 are supplemented to adjust the reactive power support. With the supplemented shunt compensators, the voltage magnitude at each bus could maintain in a range (0.9 p.u.$\sim$1.1 p.u.).

5.2 Calculation of ECO

Figure 5 gives the power loss of one OTL (from bus 14 to bus 16) for 24 hours. Obviously, the energy consumption that calculated by (14) is equivalent of the area of the shaded part in Fig. 5. Using (14), the power loss of each line for 8,760 hours could be obtained.

Table 3 The supplemented shunt compensators

Type	Bus	Shunt resistance	Shunt reactance
Capacitor	7	0	−0.4
Capacitor	7	0	−0.4
Capacitor	7	0	−0.4
Capacitor	8	0	−0.4
Capacitor	3	0	−0.4
Capacitor	3	0	−0.4
Capacitor	3	0	−0.4
Capacitor	3	0	−0.4
Capacitor	3	0	−0.4
Reactor	6	0	1
Reactor	5	0	0.3
Reactor	2	0	0.3

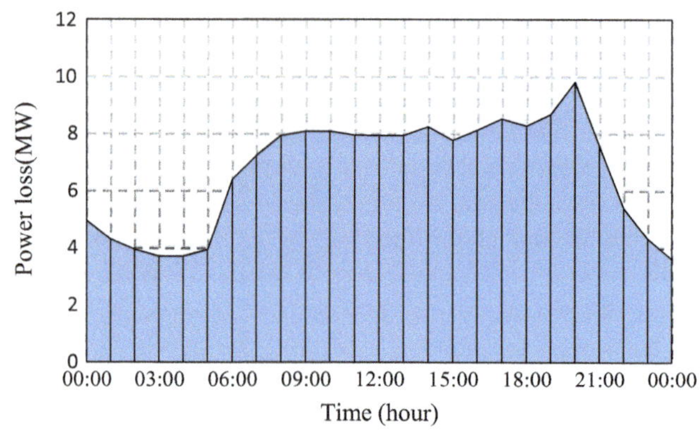

Fig. 5 Power loss calculation

While the total load power for one year is approximately 15,297,075 MW h, the ECO accounts for about 3.5% of the total generation power. The ECO for each transmission line could be much different from each other. As can be seen from Fig. 6, the ECO (line power loss) mainly allocated at the specific transmission lines that performing poorly.

The details of the ECO for the test system are elaborated in Table 4 for comparison. The transmission line list ordered by ECO is beneficial for network planning due to the illustration of the weak links of the grid. In a way, the power loss is closely related to the power network connectivity. Thus, the power loss that shown in Table 4 deserves consideration for power network planning to consolidate the network.

5.3 Comparison for multiple scenarios

It is assumed that one double-circuit OTL is planning to be constructed. There exist four optional schemes which are marked as $\mathbf{A}_1$, $\mathbf{A}_2$, $\mathbf{A}_3$, $\mathbf{A}_4$ in Fig. 4. The type and model of the equipment are specified as those given in Table 1. In

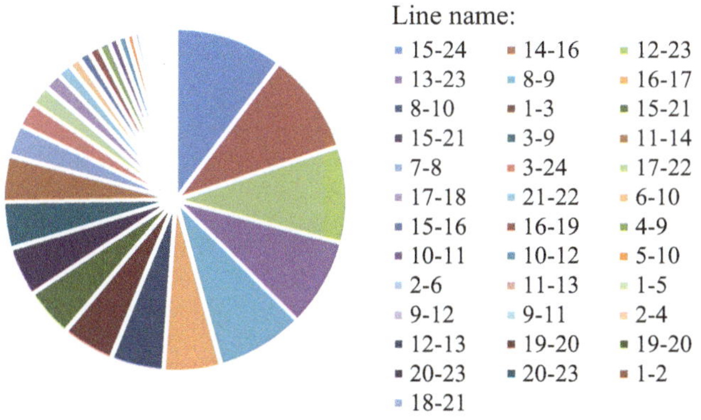

Fig. 6 Allocation of the ECO

this case, the ECCs of those alternative projects are shown in Table 5.

In Section 3, the steps for obtain the ECO could be considered as an approach to foreseen the performance of the network in a specific imaginary scenario. Since one out of the four optional schemes ($\mathbf{A}_1$, $\mathbf{A}_2$, $\mathbf{A}_3$, $\mathbf{A}_4$) should be chosen, it is expected to learn how the network performs if any one of those optional lines has been constructed. Then, different scenarios are defined in Table 6. Scenario S_i indicate the condition while line $\mathbf{A}_i$ has been constructed. The index "i" refers to the ith optional line. Particularly, S_0 indicates the original scenario. The results for the hypothetical scenarios are obtained and compared. Note that the time window for each scenario is specified as 8,760 hours.

As can be seen from Table 6, in each what-if scenario, the energy payback ratio is bigger than 100%, which implied that the ECC of the line is expected to be paid back in less than one year.

6 Discussion

As is known, the service life of an OTL is generally more than 30 years. One concern is that the energy payback for the entire life-cycle of OTL is more valuable. However, it is too difficult to take into account all the uncertainties in such a long period.

7 Conclusions

It is indicated that the energy consumption in constructing an OTL can be estimated by analyzing the components of the project. Then, the analysis of a typical OTL project is given for demonstration. To estimate the prospective energy consumption in system operation, or alternatively the power loss, a hybrid scheme is proposed

Table 4 The ECO for 8,760 hours (1 year)

No.	Bus i	Bus j	Bus name	Base voltage (kV)	Line power loss (MW h)
1	15	24	15–24	230	57,012.5
2	14	16	14–16	230	53,747.6
3	12	23	12–23	230	50,713.4
4	13	23	13–23	230	46,420.0
5	8	9	8–9	138	46,005.8
6	16	17	16–17	230	29,932.0
7	8	10	8–10	138	27,718.6
8	1	3	1–3	138	27,343.4
9	15	21	15–21	230	25,643.3
10	15	21	15–21	230	25,643.3
11	3	9	3–9	138	24,921.5
12	11	14	11–14	230	24,316.2
13	7	8	7–8	138	16,401.7
14	3	24	3–24	138	13,224.6
15	17	22	17–22	230	10,668.7
16	17	18	17–18	230	9,056.8
17	21	22	21–22	230	7,564.0
18	6	10	6–10	138	6,006.5
19	15	16	15–16	230	5,852.1
20	16	19	16–19	230	5,746.0
21	4	9	4–9	138	5,732.5
22	10	11	10–11	138	4,855.2
23	10	12	10–12	138	4,753.8
24	5	10	5–10	138	4,032.0
25	2	6	2–6	138	3,257.2
26	11	13	11–13	230	2,659.5
27	1	5	1–5	138	2,322.9
28	9	12	9–12	138	2,311.3
29	9	11	9–11	138	2,265.6
30	2	4	2–4	138	1,917.5
31	12	13	12–13	230	1,709.4
32	19	20	19–20	230	1,190.7
33	19	20	19–20	230	1,190.7
34	20	23	20–23	230	1,140.8
35	20	23	20–23	230	1,140.8
36	1	2	1–2	138	399.8
37	18	21	18–21	230	256.0
Total ECO					555,329.8

Table 5 The information of alternative lines

Lines name	From bus	To bus	Base voltage (kV)	Length (km)	ECC (MW h)
$\mathbf{A}_1$	14	16	230	43.5	18,189.7
$\mathbf{A}_2$	15	24	230	57.9	24,211.1
$\mathbf{A}_3$	12	23	230	107.8	45,077.0
$\mathbf{A}_4$	13	23	230	96.6	40,393.7

Table 6 The comparison of different scenarios

i	0	1	2	3	4
Scenario (S_i)	S_0	S_1	S_2	S_3	S_4
$ECO^{(i)}$	555,330	512,173	510,826	494,359	501,792
$EPB^{(i)}$		43,156	44,503	60,970	53,537
$Ratio^{(i)}$		237%	184%	135%	133%

for power system simulation. Then, the decrease in network power loss, which caused by integrating new transmission lines into network, is defined as the energy payback. With the case study, which uses the IEEE 24-bus test system, the energy payback is estimated and compared with consumption of construction energy. Results of the case study indicate that an OTL is expected to have a considerable energy payback ratio. Then, allowing for lifecycle, more attention should be paid to the energy payback while planning an OTL.

Acknowledgments This work was supported by the National Science Fund for Distinguished Young Scholars (No. 51325702) and the China Postdoctoral Science Foundation (No. 2014M560968). This study was also supported in part by scientific and technical project of State Grid Corporation of China. The authors gratefully acknowledge the contributions of Hongtao YIN, Ning LIU, Siyuan WANG and Qi HAN from the State Grid Corporation of China.

References

[1] Shu J, Wu J, Shahidehpour M et al (2012) A new method for spatial power network planning in complicated environments. IEEE Trans Power Syst 27(1):381–389
[2] Khan AZ (1996) Electrical energy conservation and its application to a sheet glass industry. IEEE Trans Energy Convers 11(3):666–671
[3] Ji Z, Kang CQ, Chen QX et al (2013) Low-carbon power system dispatch incorporating carbon capture power plants. IEEE Trans Power Syst 28(4):4615–4623
[4] Chen QX, Kang CQ, Xia Q et al (2010) Power generation expansion planning model towards low-carbon economy and its application in China. IEEE Trans Power Syst 25(2):1117–1125
[5] Karlsson D, Wallin L, Olovsson HE et al (1997) Reliability and life cycle cost estimates of 400 kV substation layouts. IEEE Trans Power Delivery 12(4):1486–1492
[6] Xu Y, Dong ZY, Wong KP et al (2013) Optimal capacitor placement to distribution transformers for power loss reduction in radial distribution systems. IEEE Trans Power Syst 28(4): 4072–4079
[7] Ochoa LF, Harrison GP (2011) Minimizing energy losses: optimal accommodation and smart operation of renewable distributed generation. IEEE Trans Power Syst 26(1):198–205
[8] Al-Turki YA, Lo KL (2008) Power loss allocation in pool-based electricity markets. Australasian. In: Universities Power Engineering Conference (AUPEC2008), Sydney, 1–6 Dec 2008, 6pp
[9] Quezeda VHM, Abbad JR, Gomez TGS (2006) Assessment of energy distribution losses for increasing penetration of distributed generation. IEEE Trans Power Syst 21(2):533–540
[10] Federico J, Gonzalez V, Lyra C (2005) Learning classifiers shape reactive power to decrease losses in power distribution networks. In: Proceedings of the 2005 IEEE Power and Energy Society general meeting, San Francisco, 557–562 June 2005, 6 pp
[11] Fidalgo JN, Lopes JAP (2003) Forecasting active and reactive power at substations' transformers. In: IEEE Power Tech Conference Proceedings, Bologna, 23–26 June 2003, 4 pp
[12] Mandal P, Senjyu T, Uezato K, Funabashi T (2004) Forecasting several-hours-ahead electricity demand using neural network. In: Electric utility deregulation, restructuring and power technologies, Proceedings of the 2004 IEEE International Conference, Hong Kong, 515–521 April 2004, 7 pp
[13] Silva DD, Yu X, Alahakoon D et al. (2011) Incremental pattern characterization learning and forecasting for electricity consumption using smart meters. In: Industrial Electronics (ISIE), 2011 IEEE International symposium on, Gdansk, 807–812 June 2011, 6 pp
[14] Wang Y, Niu D, Lee V (2011) Optimizing of BP neural network based on genetic algorithms in power load forecasting. In: IECON 2011—37th Annual Conference on IEEE Industrial Electronics Society, Melboume, 4322–4327 November 2011, 6 pp
[15] Jin YX, Su J (2007) Similarity clustering and combination load forecasting techniques considering the meteorological factors. In: Proceedings of the 6th international conference on instrumentation, measurement, circuits & systems (WSEAS), Hangzhou, 115–119 April 2007, 5 pp
[16] Yamayee ZA, Hakimmashhadi H (1985) Production simulation for power system studies. IEEE Trans Power Appar Syst PAS-104(12):3376–3381
[17] Sondergren C, Ravn HF (1996) A method to perform probabilistic production simulation involving combined heat and power units. IEEE Trans Power Syst 11(2):1031–1036
[18] Khan SU, Lee ST (1981) Estimation of power-pooling benefits from production simulations. IEEE Trans Power Appar Syst PAS-100(4):1759–1765
[19] Li Y, Singh C (1999) A direct method for multi-area production simulation. IEEE Trans Power Syst 14(3):906–912
[20] Stott B (1974) Review of load-flow calculation methods. Proc IEEE 62(7):916–929
[21] Zhang N, Kang C, Kirschen DS et al (2013) Planning pumped storage capacity for wind power integration. IEEE Trans Sustain Energy 4(2):393–401
[22] Subcommittee PM (1979) IEEE reliability test system. IEEE Trans Power Appar Syst PAS-98(6):2047–2054

Xiaoming DONG received his Ph.D. degree from the School of Electrical Engineering at Shandong University, Jinan, China, in 2013. He is currently a research associate at Tsinghua University. His research interests include power system stability, power system control and power system operation.

Chongqing KANG received his Ph.D. degree from Electrical Engineering Department of Tsinghua University, China, in 1997. He is now a professor at the same University. His research interests include renewable energy, low-carbon electricity, load forecasting, power system planning and generation scheduling optimization.

Ning ZHANG is currently a lecturer at Tsinghua University. His research interests include power system planning with renewable energy, renewable energy statistic characteristic analysis and operation simulation, and power system dispatching with renewable energy.

Huaguang YAN is currently a senior engineer at the China Electric Power Research Institute. His research interests include power network planning and energy conservation.

Junxia MENG is currently a senior engineer at the China Electric Power Research Institute. Her research interests include energy conservation and low carbon.

Xinsheng NIU is currently a senior engineer at the State Grid Shandong Power Economic Research Institute. His research interests include power network planning and estimating.

Xin TIAN is currently an engineer at the State Grid Shandong Power Economic Research Institute. His research interests include power network planning, energy conservation and low carbon.

23

Optimal proportion of wind and PV capacity in provincial power systems based on bilevel optimization algorithm under low-carbon economy

**Yue YUAN, Yang CAO (✉), Xinsong ZHANG,
Chun LIU, Yuehui HUANG, Peng LI,
Kejun QIAN**

Abstract In order to boost contributions of power systems to a low-carbon economy, the installed capacity of renewable power generation, such as wind and photovoltaic (PV) power generation should be well planned. A bilevel formulation is presented to optimize the proportion of wind and PV capacity in provincial power systems, in which, carbon emissions of generator units and features of renewable resources are taken into account. In the lower-level formulation, a time-sequence production simulation (TSPS) model that is suitable for actual power system has been adopted. In order to maximize benefits of energy conservation and emissions reduction resulting from renewable power generation, the commercial software called General Algebraic Modeling System (GAMS) is employed to optimize the annual operation of the power system. In the upper-level formulation, the optimal pattern search (OPS) algorithm is utilized to optimize the proportion of wind and PV capacity. The objective of the upper-level formulation is to maximize benefits of energy conservation and carbon emissions reductions optimized in the lower-level problem. Simulation results in practical provincial power systems validate the proposed model and corresponding solving algorithms. The optimization results can provide support to policy makers to make the polices related to renewable energy.

Keywords Time-sequence production simulation, Optimal proportion of renewable energy, Pattern search method, General algebraic modeling system

1 Introduction

The increasing CO_2 emissions have been reported as one main cause of global warming that is widely considered as a great threat to our society in the foreseeable future. Developing low-carbon electricity technologies is the fundamental way to meet the challenge of global warming mainly caused by power generation [1–3]. Fortunately, there are rich wind and solar resources in many provinces of China. Due to the features of sustainability and zero carbon emissions, wind and solar power have developed rapidly in recent years with the stimulus from the government. In 2013, newly added wind power generation capacity is as high as 16.09 GW and the cumulative capacity integrated with power grid has reached 91.4 GW. Both of them have topped other countries in the world [4]. The photovoltaic power generation (PV) integrated with power grid in a variety of ways has claimed a growing proportion of renewable energy capacity. Newly added PV power capacity in 2013 in China is as high as 12.06 GW. China has surpassed Germany to become the top one in the world for the first time [5].

At present, many provinces in China plan their wind power and PV power industry separately and normally ignore inherent complementary features between wind and solar power. The high penetration of wind power and PV power may greatly threaten the way power systems have been operated. If conventional power generation units fail

Y. YUAN, Y. CAO, College of Energy and Electrical Engineering, Hohai University, Nanjing 211100, China
(✉) e-mail: caoyang1988828@126.com
X. ZHANG, College of Electrical Engineering, Nantong University, Nantong 226019, China
C. LIU, Y. HUANG, P. LI, China Electrical Power Research Institute, Beijing 100192, China
K. QIAN, Suzhou Power Supply Company, State Grid Corporation of China, Suzhou 215004, China

to compensate fluctuating wind and PV power, a certain portion of wind and PV power output will have to be curtailed. Reports show that curtailments of wind and PV power in 'Three North' areas (Northeast, North China and Northwest) of China are very serious. Reasonable capacity planning of wind farms and PV stations is therefore crucial to reduce wind and PV power curtailments. In order to achieve the goal, the time-sequence production simulation (TSPS) technology which is based on space-time complementary advantages between wind and solar energy has been utilized in [6–8]. One major advantage of TSPS is the capability of incorporating some annual plan of power systems such as thermal unit maintenance schedule and actual wind and PV power plan [9].

There are efforts published in studying the optimal proportion of wind and PV capacity in power systems. The optimal proportion of wind and PV capacity in a microgrid has been studied in [10–12], in which the proposed methodologies cannot extend to practical provincial power systems comprising several tens of generators because of their limited capacities in solving large-scale problems. Sizing optimization for a hybrid photovoltaic-wind energy system was conducted only based on a typical day without considering annual output characteristics of wind and PV power [13]. In addition, the results reported in [13] tend to be conservative because the difference between the peak and valley load of the typical day is the maximum and accommodation condition for wind and PV power is consequently the worst. Moreover, the methodology presented in [13] is that energy conservation and emission reduction of wind and PV power can not be considered sufficiently. In [14–16], optimal proportion of hybrid wind and PV power generation systems have been studied without considering the features of heating units. As a result, the conclusions in [14–16] may mislead the planning of power grid and the operation of real systems in China, because heating units exist widely in 'Three North' areas of China and always operate in winter. Meanwhile, all references described above leave out of consideration of CO_2 emissions of power systems.

A novel formulation based on bilevel optimization under low-carbon economy is proposed to optimize the proportion of wind and PV power capacity in provincial systems of China [17, 18]. A time-sequence output model of wind and PV power that considers regional resources in the provincial power systems is proposed in lower-level problem. Besides, the objective function of the lower-level model is to maximize energy conservation and emission reduction benefits of the provincial power grid. Load characteristics, peak-load regulation capability, thermo-electric coupling characteristics of different thermal units, transmission capacity and some other factors are also considered in this model. The lower-level problem is an annual TSPS model requiring a single set of binary variables therefore the model is a large-scale mixed integer program problem (MIP). CPLEX solver in General Algebraic Modeling System (GAMS) [19] is applied to optimize the annual operation of power systems. TSPS technology utilized in present paper has following advantages:

1) Both the rationality and credibility of planning results could be improved in this method because it is actually more in line with the power systems.
2) Temporal variation characteristics of renewable output can be reflected.
3) Some annual plan of power system such as thermal unit maintenance schedule or annual wind and PV power plan can also be made by this method.
4) A comprehensive evaluation for annual renewable power generation, energy-saving capacity of power grid as well as thermal power index could be calculated more accurately. A simple model of optimal proportion of wind and PV capacity is developed in the upper-level problem, of which the objective function is maximizing benefits of energy conservation and emissions reductions after solving the lower-level problem. This is because the upper-level problem does not require the gradient of the problem to be optimized. The efficiency and accuracy of optimization can be improved by the optimal pattern search (OPS) method [20].

The remaining sections of this paper are organized as follows. Section 2 describes the lower-level and upper-level formulations. Section 3 presents the bilevel optimization algorithm framework and how it is used in the optimal proportion of wind and PV capacity model. Numerical examples are provided in Section 4. Section 5 concludes the paper with discussions.

2 Model description

2.1 Model of time sequence output

Three kinds of time-sequence outputs are constructed in the lower-level model: annual load time sequence output, annual wind power time sequence output and annual PV power time sequence output. In this paper, the normalized values are adopted for all of these three time series. The definition methods in the planning progress are as follows: First of all, the history data from all substations including wind power output, PV power output and electrical load in one province should be collected and organized. In the next place, all the historical data above are used to divide the wind power installed capacity, the PV power installed capacity and the maximum load in the corresponding year so that the annual load time sequence output, annual wind power time sequence output and annual PV power time sequence output can be obtained respectively. Finally, the value of wind power output, PV power output and load in the planning year can be calculated by using these normalized time series output multiplied by the predicted wind installed capacity, the predicted PV installed capacity and the predicted maximum load.

2.2 Model of annual time sequence production simulation under low-carbon economy in the lower level

2.2.1 Objective function in the lower-level model

Developing the low-carbon electricity technologies for the power industry is a fundamental way to meet challenges of global warming. Minimizing CO_2 emissions from power systems conforms to current trend of national policy and social development in China. The impacts of large scale wind power and PV integration on low-carbon benefits are divided into two aspects, i.e., positive contributions and negative contributions. On one hand, wind and PV power are sustainable and have zero carbon emissions. On the other hand, wind and PV power are intermittent and can be hardly predicted. The fluctuation in wind and PV power output requires more frequent start-stop of thermal units which leads to more CO_2 emissions. Since a large number of cases must be calculated in the planning processes, the coal consumption of thermal units should be modeled as a linear function so that efficiency of model calculation could be greatly improved.

Above all, the objective function in low-carbon economy of the lower-level model is shown below:

$$F_{\text{Lower}} = \min \sum_{t=1}^{T} \sum_{g=1}^{G} [f_{\text{Neg}}(g,t) + f_{\text{Pos}}(t)]\gamma \tag{1}$$

$$\begin{cases} f_{\text{Neg}}(g,t) = A_g U_g^t + B_g V_g^t \\ f_{\text{Pos}}(g,t) = a_g P_g^t + b_g \end{cases} \tag{2}$$

where g denotes the index of the generating units; G denotes total number of thermal units; t is the hourly scheduling intervals; T is the number of periods of the time span in the lower-level problem; γ is the coefficient of CO_2 emissions; U_g^t is a binary decision variable that indicates whether unit g will be started up or not during the period t (1/0 are startup/not startup respectively); V_g^t is a binary decision variable that indicates unit g will be shut down or not during period t (1/0 are shutdown/not shutdown respectively); P_g^t is the output of unit g during the period t; A_g, B_g, a_g and b_g are all constants that denote startup coal consumption of unit g, shutdown coal consumption of unit g, production coal consumption of unit g and coal consumption of unit g respectively.

2.2.2 Thermal unit constraints

$$P_g^{t+1} - P_g^t \le \Delta P_g^{\text{up}} \tag{3}$$

$$P_g^t - P_g^{t-1} \le \Delta P_g^{\text{dow}} \tag{4}$$

$$P_g^{\min} O_g^t \le P_g^t \le P_g^{\max} O_g^t \tag{5}$$

$$P_{\text{BP},g}^t = \alpha_g H_g^t \tag{6}$$

$$\alpha_g H_g^t \le P_{\text{TG},g}^t \le P_{\text{TG},g}^{\max} - \beta_g H_g^t \tag{7}$$

$$U_g^t + \sum_{i=0}^{K} V_g^{t+i} \le 1 \tag{8}$$

$$V_g^t + \sum_{i=0}^{K} U_g^{t+i} \le 1 \tag{9}$$

$$\begin{cases} O_g^t - O_g^{t-1} - U_g^t + V_g^t = 0 \\ -O_g^t - O_g^{t-1} + U_g^t \le 0 \\ O_g^t + O_g^{t-1} + U_g^t \le 2 \\ -O_g^t - O_g^{t-1} + V_g^t \le 0 \\ O_g^t + O_g^{t-1} + V_g^t \le 2 \end{cases} \tag{10}$$

where ΔP_g^{up} and ΔP_g^{dow} are respectively ramp-up rate limit and ramp-down limit of generator g; $P_g^{\max}$ and $P_g^{\min}$ are respectively output power upper and lower bound of generator g; O_g^t is a binary decision variable that indicates on/off status of generator g during the period t; $P_{\text{BP},g}^t$ and $P_{\text{CG},g}^t$ are respectively the output of back pressure generator of generator g and extraction turbo-generator during their heating period; α_g and β_g are ratios between heat output and electricity output; H_g^t is heat output during the period t.

The hourly TSPS model constraints listed above include the unit ramping-up constraints shown in (3), unit shut-down constraints in (4), unit generation capacity constraints in (5), output of heating units during heating period constraints in (6) and (7), unit minimum-up time constraints in (8) (e.g., if a generator is turned on, it should be kept on for at least K periods), unit minimum-down time constraints in (9) (e.g., if a generator is turned off, it should be kept off at least K periods), the logical state of unit start-up, unit shut-down and unit operations constraints in (10).

2.2.3 System constraints

$$P_l^t = P_w^t + P_v^t + \sum_{g=1}^{G} P_g^t \tag{11}$$

$$\sum_{g=1}^{G} P_g^{\max} O_g^t + R_N^t \geq P_l^t + R_{\text{Pos}} \tag{12}$$

$$\sum_{g=1}^{G} P_g^{\min} O_g^t - R_N^t \geq P_l^t - R_{\text{Neg}} \tag{13}$$

$$0 \leq P_w^t \leq C_w^k P_w^{t^*} \tag{14}$$

$$0 \leq P_v^t \leq C_v^k P_v^{t^*} \tag{15}$$

where P_l^t is the electrical load of during the period t; P_w^t and P_v^t are respectively output of wind and PV power during the period t; R_N^t is renewable power credible capacity during the period t. The value of R_N^t is normally decided by renewable power prediction accuracy and is usually set to 60%–80% of the total renewable output [21]. The start-up capacity of thermal unit would be greatly reduced when the renewable power credible capacity is considered so that the concept of "green dispatch" under low-carbon economy could be reflected quite well. R_{Pos} and R_{Neg} are the system positive and negative reserve demand respectively during period t. $P_w^{t^*}$ and $P_v^{t^*}$ are the normalized time sequence output, C_w^k and C_v^k are respectively installed capacity of wind farms and PV stations. It is noted that these two variables are the kth optimization variables in the upper-level optimization formulation that will be described in Section 2.3.

Hourly system constraints listed above include system power balance constraints in (11), system spinning reserve requirement in (12) and (13), wind power and PV power output constraints in (14) and (15).

2.3 Model of optimal proportion of wind and PV capacity in the upper level

The upper-level model of optimal proportion of wind and PV capacity does not need to calculate the derivative so that it could be effectively solved by the OPS method. The objective function and constrains of the upper-level model are as follows.

2.3.1 Objective function in the upper level

$$F_{\text{Upper}} = \min F_{\text{Lower}}(C_w^k, C_v^k) \tag{16}$$

where F_{Upper} is the objective function of the upper-level optimization formulation. It is in fact the energy conservation and emissions reductions of renewable power generation are optimized in the lower-level formulation. Due to the objective function F_{Lower} is determined by the optimization of the TSPS model with different C_w and C_v in every group, the OPS algorithm is employed to update the particle optimization direction according to the value of F_{Lower} $(C^k{}_w, C^k{}_v)$.

2.3.2 Variable solution space constraints

$$\theta_v^{\min} \leq \theta_v \leq \theta_v^{\max} \tag{17}$$

where θ denotes optimization variables in the upper-level formulation; v is the index of variables mentioned above, in this paper, $v = 1$ indicates the wind capacity and $v = 2$ indicates the PV capacity; $\theta_v^{\max}$ is the maximum planning installed capacity and $\theta_v^{\min}$ is the existing installed capacity.

In summary, the objective functions above and constraints constitute the model of optimal proportion of wind and PV capacity under low-carbon economy. A certain number of constraints can be chosen depending on the requirements of real power systems. The upper-level model is implemented in Matlab. The lower-level model is implemented in GAMS using CPLEX solvers. Matlab calls GAMS and retrieves data based on the methodology reported in [22].

3 Bilevel optimization algorithm

3.1 Overview of bilevel optimization algorithm

For practical power systems comprising several hundreds of generators, the TSPS is a MIP model required a large number of binary variables. The CPLEX solver in GAMS is applied to solve this problem to optimize the annual power system operation. However, if every different proportion of wind power and PV power capacity are calculated by the exhaustive method based on TSPS method, computation burden will be so large that the formulation cannot be applied into practical power systems.

Hence, the optimization process in provincial power systems can be divided into two sub-optimization problem named as upper-level optimization problem and lower-level optimization problem respectively. In the whole optimization process, reliability, efficiency and accuracy of optimization are considered comprehensively. The provincial system production simulation model based on time sequence which conforms to real power system operation is proposed in the lower-level model, in order to maximize benefits of energy conservation and emission reductions of the provincial grid. A simple model of optimal proportion of wind and PV capacity is developed in the upper-level problem, in which objective function is maximizing the benefit of energy conservation and emission reduction after solving the lower-level problem, some optimization method that is to solve the problem of no gradients could be used to update the particle optimization direction in upper-level according to this value.

Moreover, the OPS method was coined by Hooke and Jeeves in 1961 and it is a family of numerical optimization methods that do not require the gradient of the problem to be optimized. Its basic idea is as follows: the axial search and the pattern search are included in OPS method. New optimization point and direction which is conducive to minimize the function value are determined by the former search method after searching n^{th} axis directions. The latter search method makes the function value decrease faster in the direction of attachment between adjacent two optimization points. Hence OPS can be used on functions that are not continuous or differentiable.

Based on this, after the upper-level model is solved by the OPS method, the planning wind power capacity and planning PV power capacity are uploaded to lower-level model, then TSPS is applied to maximize benefit of energy conservation and emission reduction of the power grid. The result which is calculated by this bilevel optimization algorithm with high computational efficiency would be more realistic.

3.2 Flow of bilevel optimization algorithm

The flowchart of the proposed bilevel algorithm is shown in Fig. 1, in which S denotes optimum step, η denotes accelerated factor, α denotes step growth factor, β denotes step reduction factor, maxiter denotes the maximum iterations.

4 Case studies

An actual provincial power system is selected as the case study system. The methodology is utilized to optimize installed capacities of wind power and PV in a planning year.

4.1 Data

4.1.1 Time-sequence data collection

The normalized data of load curve, wind power curve and PV power curve during a typical day and the whole year in this province are obtained from Electricity Regulatory Authority of this province and are illustrated in Figs. 2 and 3 respectively.

As shown in Fig. 2, the maximum daily gap between peak load and valley load in the case system appears in the 328th day, with a value as high as 1756 MW. According to the information in Fig. 3, the theoretical utilization hours of wind power and PV power can be calculated as 1913 hours and 1550 hours, respectively. The utilization hour of wind power is larger than that of PV power because the output power of PV is always zero.

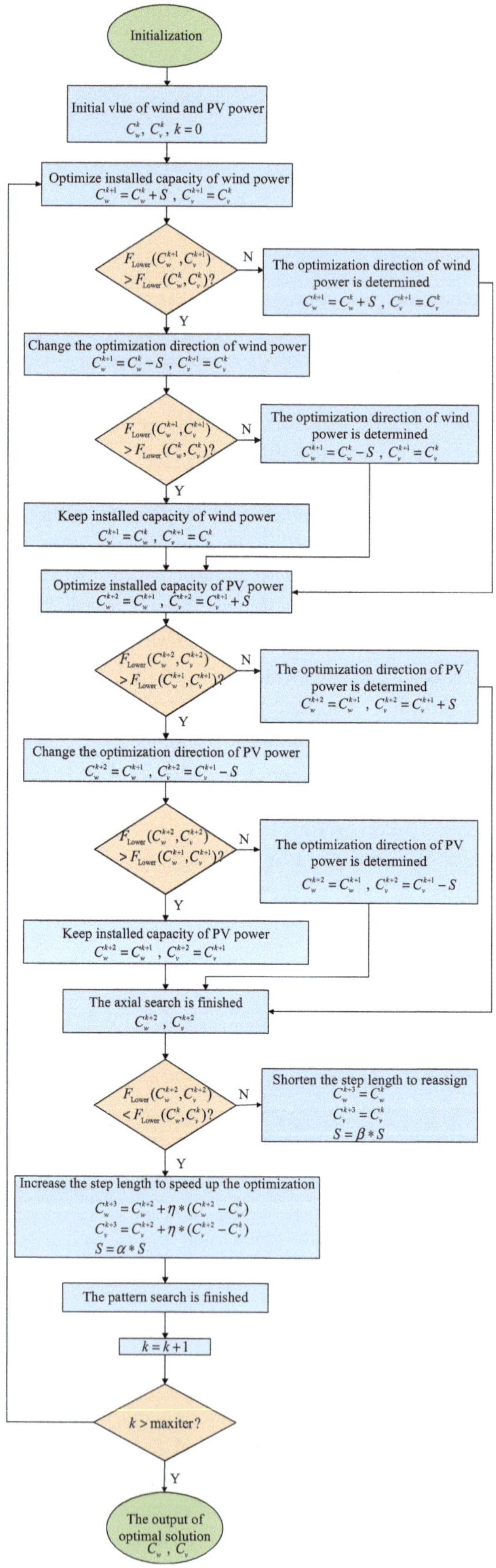

Fig. 1 Bilevel optimization algorithm

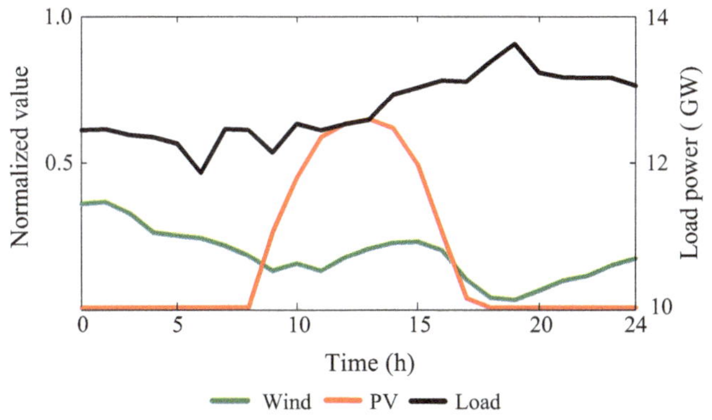

Fig. 2 The curve of load, wind power and PV power on a typical day

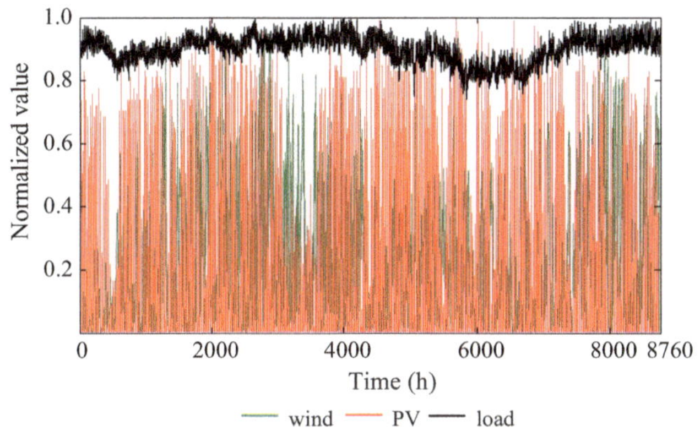

Fig. 3 The curve of load, wind power and PV power on an annual year

Table 1 Installed thermal capacity by technology

Technology	Unit capacity (MW)	Number of units	Minimum output (MW)
Condensate gas	200	2	120
	330	21	198
	660	8	396
Back pressure	60	1	30
Extraction turbo	200	11	120
	330	15	198

4.1.2 System data

Thermal generator units of the studied province can be divided into three types, i.e. condensate gas, back pressure and extraction turbo. The installed capacities and numbers of these three types of generators are presented in Table 1. The parameters of heating units and the coal consumption parameters of thermal units are presented in Table 2 and Table 3 respectively.

The installed capacity of maximum units (660 MW) is selected as the amount of spinning reserve. The heating

Table 2 The parameters of heating units

Technology	Unit capacity (MW)	Heat load (MW)	C_b	C_v
Back pressure	60	60	0.16	–
Extraction turbo	200	200	0.73	0.21
	330	330	0.89	0.02

period of this province starts on 25th October and ends on 25th April in the next year.

4.1.3 Planning target

At present, the installed capacities of wind power and PV power in this province are 2646.4 MW and 530.83 MW, which account for 11.67% and 2.34% of the total installed capacity respectively. Due to the limitations of geographical, government policy, economy and other factors, total installed capacity of wind power and PV power is expected to be less than 8000 MW in the planning year. The predicted maximum load in the planning year is assumed to be 14230 MW.

4.2 Illustrative examples

Apart from the methodology presented in this paper, two other algorithms are also utilized to optimize installed capacity of the case system. Three optimization results with different algorithms are illustrated in Fig. 4. PSO is an abbreviation of particle swarm optimization algorithm [23].

Figure 4 shows that if the typical day analysis method is utilized in the lower model, the amount of CO_2 emissions is increasing dramatically with 810000 tons per year because of the conservative property of the typical day analysis method. Neither the low-carbon economy of power system can be reflected by the typical day analysis method nor does the credibility of planning results can be guaranteed. Compared with PSO algorithm which is applied in the upper level model, OPS has a better computing accuracy by further reducing 20000 tons CO_2 emissions per year. In addition, from Fig. 4, it can be concluded that the higher proportion of PV power, the better energy conservation and emission reduction can be obtained. That is to say PV power is more excellent than wind power under low-carbon economy. Optimal capacities of wind power and PV power are 3900 MW and 2800 MW respectively and the corresponding proportion of wind and PV capacity is 1.39. However, the present proportion of wind and PV capacity is 4.98 and far larger than the optimal value. Therefore, PV stations should be strongly supported by the government.

The calculation times between different algorithms are presented in Table 4.

Table 3 The coal consumption parameters of thermal units

Technology	Unit capacity (MW)	*A* (ton)	*B* (ton)	*a* (ton/MW)	*b* (ton/MW)
Condensate gas	200	20	15	−20	400
	330	30	25	−35	380
	660	60	50	−45	370
Back pressure	60	6	4	−5	500
Extraction turbo	200	30	25	−45	370
	330	20	15	−20	400

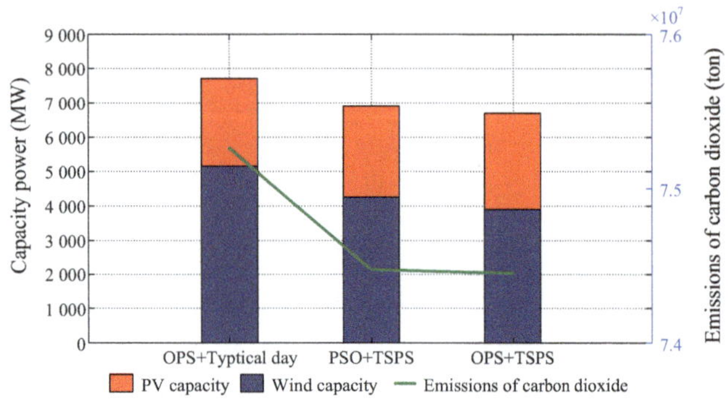

Fig. 4 The optimal solutions of different algorithms

Table 4 The comparisons of calculation times between different algorithms

Optimization algorithms	Optimization (step/MW)	Calculation times
The exhaustive algorithms	50	9216
The bilevel algorithms	50	326

In this paper, the average optimization time of TSPS in the lower-level model is about 15 mins, so the total computing time is reduced by 133350 mins by using the bilevel algorithms.

The further research under the case of optimal proportion of wind and PV is conducted (The optimal capacities of wind power and PV power are 3900 MW and 2800 MW) owing to TSPS is applied in the lower-level model. The average curtailment rate of wind and PV power and the average of start-up capacity of thermal power is presented in Fig. 5.

Figure 5 shows that renewable power balance ability off heating period is better than that of heating period because of its better flexibility and load regulation ability. The thermal unit maintenance schedule can be made according to this model, for instance, in the month with smaller start-up capacity of thermal power like September, more units could be arranged for maintenance. Planning and dispatching operation of thermal units can be made in different seasons according to this model.

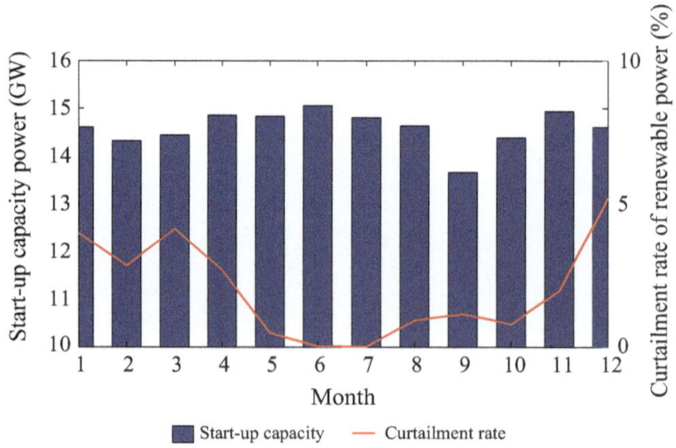

Fig. 5 The optimal solutions of different algorithms

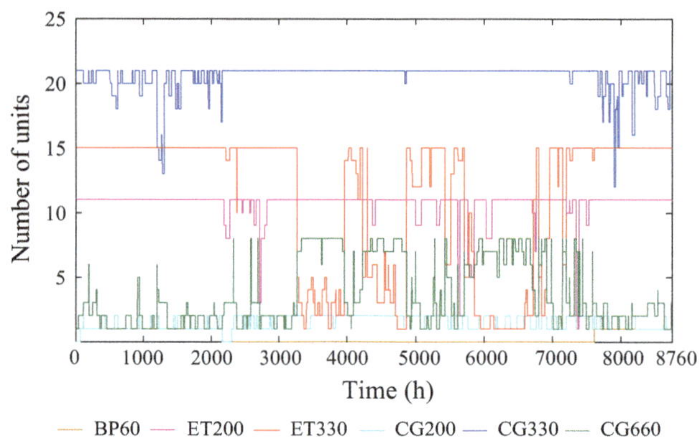

Fig. 6 The number of units in various thermal units

The operation numbers in various thermal units are presented in Fig. 6. The maintenance scheduling of certain kinds of units can be arranged precisely according to Fig. 6.

5 Conclusions

A novel optimization algorithm based on bilevel optimization under low-carbon economy is applied in this paper to optimize the proportion of wind and PV capacity in a provincial system. The TSPS model is applied in the lower-level model so that both the rationality and credibility of planning results could be improved. The OPS algorithm is applied in the upper-level model so that the efficiency and accuracy of optimization could be improved. Some guidance is given for the optimal proportion of wind and PV capacity power system dispatching as well as the relevant policies made by the government under low-carbon economy.

The future work will be focused on the optimal proportion of wind and PV capacity in the provincial power systems with consideration of the thermal power unit economy.

Acknowledgments This work is jointly supported by the research and application of evaluation of priority dispatching of wind/PV generation in multi-levels, State Grid Corporation of China (No. NY71-14-038), Jiangsu Provincial Graduate Education Innovation Project (No. KYLX_0431), the Fundamental Research Funds for the Central Universities (No. 2014B33314) and National Nature Science Foundation of China (No. 51407097). The authors appreciate the anonymous reviewer for the valuable comments, which is helpful to improve the article.

References

[1] Malla S (2009) CO_2 emissions from electricity generation in seven Asia-Pacific and North American countries: a decomposition analysis. Energ Policy 37(1):1–9
[2] Zhang N, Kang CQ, Liu JK et al (2013) Mid-short-term risk assessment of power systems considering impact of external environment. J Mod Power Syst Clean Energ 1(2):118–126
[3] Zhang Y, Yao F, Iu HHC et al (2013) Sequential quadratic programming particle swarm optimization for wind power system operations considering emissions. J Mod Power Syst Clean Energ 1(3):231–240
[4] Chinese Wind Energy Association (2013) China wind power capacity statistic 2013. Chinese Wind Energy Association, Beijing, China (in Chinese)
[5] The global new energy development report 2014. In: 8th China new energy international forum, Beijing, China, 7–8 Jun 2014
[6] Patel MR (2005) Wind and solar power systems: design, analysis and operation, 2nd edn. CRC, Boca Raton, FL, USA
[7] Sarkar S, Ajjarapu V (2011) MW resource assessment model for a hybrid energy conversion system with wind and solar resources. IEEE Trans Sustain Energ 2(4):383–391
[8] Safdarian A, Fotuhi-Firuzabad M, Aminifar F (2012) Compromising wind and solar energies from the power system adequacy viewpoint. IEEE Trans Power Syst 27(4):2368–2376
[9] Liu C, Cao Y, Huang YH et al (2014) An annual wind power planning method based on time sequential simulations. Automat Electr Power Syst 38(11):13–19 (in Chinese)
[10] Hong YY, Lian RC (2012) Optimal sizing of hybrid wind/PV/diesel generation in a stand-alone power system using Markov-based genetic algorithm. IEEE Trans Power Del 27(2):640–647
[11] Xu L, Ruan XB, Mao CX et al (2013) An improved optimal sizing method for wind-solar-battery hybrid power system. IEEE Trans Sustain Energy 4(3):774–785
[12] Das I, Bhattacharya K, Canizares C (2014) Optimal incentive design for targeted penetration of renewable energy sources. IEEE Trans Sustain Energ 5(4):1213–1225
[13] Geem ZW (2012) Size optimization for a hybrid photovoltaic-wind energy system. Int J Electr Power Energ Syst 42(1):448–451
[14] Markvart T (1996) Sizing of hybrid photovoltaic-wind energy system. Solar Energ 57(4):277–281
[15] Kellogg WD, Nehrir MH, Venkataramanan G et al (1998) Generation unit sizing and cost analysis for stand-alone wind, photovoltaic, and hybrid wind/PV systems. IEEE Trans Energ Conver 13(1):70–75
[16] Yang Q, Zhang JH, Liu ZF, et al (2008) A new methodology for optimizing the size of hybrid PV/wind system. In: Proceedings of the 2008 IEEE international conference on sustainable energy technologies (ICSET'08), Singapore, 24-27 Nov 2008, pp 922-927
[17] Xu P, Wang LZ (2014) An exact algorithm for the bilevel mixed integer linear programming problem under these simplifying assumptions. Comput Oper Res 41:309–318
[18] Wang QF, Guan YP, Wang JH (2012) A chance-constrained two-stage stochastic program for unit commitment with uncertain wind power output. IEEE Trans Power Syst 27(1):206–215
[19] GAMS Development Corporation, Washington, DC, USA (2005) GAMS–the solver manuals, version 23.3
[20] Hooke R, Jeeves TA (1961) "Direct search" solution of numerical and statistical problems. J ACM 8(2):212–229
[21] Zhou SL, Mao MQ, Su JH (2011) Short-term forecasting of wind power and non-parametric confidence interval estimation. Proceedings of CSEE 31(25):10–16 (in Chinese)
[22] Ferris MC (2005) MATLAB and GAMS: interfacing and optimization and visualization software. Computer Science Department, University of Wisconsin, Madison, WI, USA
[23] Kennedy J, Eberhart R (1995) Particle swarm optimization. In: Proceedings of the IEEE conference on neural networks, vol 4, Perth, Australia, 27 Nov-1 Dec 1995, pp 1942-1948

Yue YUAN received his B.E. and M.Sc. degrees in Electrical Engineering from Xi'an Jiaotong University, Xi'an, China, in 1987 and 1990, respectively. He then received the Ph.D. degree from Hiroshima University, Japan, in 2002. He joined the faculty of Xi'an Jiaotong University, China, in 1990. He has been with Hohai University, Nanjing, China, as a professor since 2003. Presently, he is also the Dean of the College of Energy and Electrical Engineering, Hohai University. His research interests include power system operations and optimization, renewable energy and distributed generation.

Yang CAO was born in the city of Yan Cheng of Jiangsu Province, China. He received the B.E. degree in electrical engineering in 2011 from the Hohai University, China. He is currently pursuing his Ph.D. degree in the College of Energy and Electrical Engineering at Hohai University, China. His research interests include smart grid and renewable power dispatch.

Xinsong ZHANG received the Ph.D. degree from Hohai University, Nanjing, China, in 2013. His research interests include power system control, renewable energy and energy storage systems.

Chun LIU is now with deputy director of the China Electrical Power Research Institute, Beijing, PRC. His research interests are in renewable power plan.

Yuehui HUANG is now with the China Electrical Power Research Institute, Beijing, PRC. Her research interests are in wind power dispatch.

Peng LI is now with the China Electrical Power Research Institute, Beijing, PRC. His research interests are in wind power integration and energy storage systems.

Kejun QIAN is now a Senior Engineer with Suzhou Power Supply Company, State Grid Corporation of China. His research interests include electric vehicles, smart grid and renewable energy systems.

24

A low-carbon oriented probabilistic approach for transmission expansion planning

Jing QIU, Zhaoyang DONG (✉), Junhua ZHAO,
Ke MENG, Fengji LUO, Kit Po WONG,
Chao LU

Abstract Following the deregulation of the power industry, transmission expansion planning (TEP) has become more complicated due to the presence of uncertainties and conflicting objectives in a market environment. Also, the growing concern on global warming highlights the importance of considering carbon pricing policies during TEP. In this paper, a probabilistic TEP approach is proposed with the integration of a chance constrained load curtailment index. The formulated dynamic programming problem is solved by a hybrid solution algorithm in an iterative process. The performance of our approach is demonstrated by case studies on a modified IEEE 14-bus system. Simulation results prove that our approach can provide network planners with comprehensive information regarding effects of uncertainties on TEP schemes, allowing them to adjust planning strategies based on their risk aversion levels or financial constraints.

Keywords Power system planning, Emission reduction, Dynamic programming, Risk management

J. QIU, J. ZHAO, K. MENG, F. LUO, Centre for Intelligent Electricity Networks (CIEN), University of Newcastle, Callaghan, NSW 2308, Australia
Z. DONG, School of Electrical and Information Engineering, University of Sydney, Sydney, NSW 2006, Australia
(✉) e-mail: zydong@ieee.org
K. P. WONG, School of Electrical, Electronics and Computer Engineering, University of Western Australia, Perth, WA 6009, Australia
C. LU, Department of Electrical Engineering, Tsinghua University, Beijing, China

1 Introduction

1.1 Transmission expansion planning

Transmission expansion planning (TEP) refers to comprehensive studies on determining the time, location, and type of adding new power transmission lines as well as the associated electrical components, in order to ensure the economic, secure, and reliable operations of a power system [1]. There are a variety of factors contributing to the necessity of TEP, including load growth, components' entry or decommissioning, technological improvement, policy incentive, etc. [2]

Following the advent of electricity markets in many countries, power industry has been transformed from a vertically integrated and regulated utility to an unbundled and liberalized structure. This deregulation and restructuring has resulted in fundamental changes to the power system TEP practices [3, 4]. For instance, the emergence of various self-interested market participants such as brokers and independent power producers has made planning difficult due to conflicting objectives and uncertainties in a market environment [5, 6].

Moreover, the power sector is one of the biggest emission sources and should take a key responsibility in promoting the coordinated development among economy, energy and environment by mitigating carbon emission [7, 8]. This concern on carbon emission mitigation is driving the need to explore power system planning practices under the mode of low-carbon economy, e.g. network planning to facilitate the integration of renewable energy and/or to encourage clean power outputs. Therefore, power system planning becomes more complicated under the impacts of different emission reduction policies.

1.2 Carbon pricing policies

Some key carbon pricing policies in the world are summarized as follows. In 2003 in the U.S., carbon emission allowances could be traded among American corporations under a voluntary scheme on Chicago Climate Exchange. In October 2003, the EU parliament approved a new emission trading scheme in order to meet its commitment made in the Kyoto Protocol. Following this, an EU emission trading scheme (ETS), which was the largest multinational, greenhouse gas emission trading scheme in the world, was enforced in February 2005. Under the EU ETS, a specific allowance for emission was allocated to each EU member and any excessive allowance could be sold to whom was in need of allowances. Some allowances were permitted to be transferred between countries through joint implementation (JI) or clean development mechanisms (CDM), but these transfers should be validated by the United Nations Framework Convention on Climate Change (UNFCCC). This type of EU ETS is also called cap-and-trade. In April 2007, an emission price of \$15 NZ per carbon equivalent was implemented. In September 2008, the New Zealand ETS was legislated, and it adopted all free allocation without caps. In July 2010, India introduced a nationwide carbon tax of about \$1.07 US on coal, as coal was a major fuel resource for power generation in India. In November 2011, China had a pilot test of carbon trading in seven provinces and a national trading is expected to start in 2016. In July 2012, the Australian government introduced a carbon price of \$23 AU per ton of emitted carbon equivalent, but this carbon price has been phased out.

1.3 Review of TEP

Conventional TEP models are formulated as minimizing investment costs of adding new lines. The widely used deterministic and static TEP model in a regulated power utility is defined as [6, 9, 10]:

Minimize

$$\sum_{i,j\in N} C_{ij}\eta_{ij} + \lambda \sum_{i\in N} r_i \tag{1}$$

subject to

$$\boldsymbol{S}^{\mathrm{T}}\boldsymbol{F} + \boldsymbol{g} + \boldsymbol{r} = \boldsymbol{L} \tag{2}$$

$$F_{ij} - B_{ij}\left(\eta_{ij}^0 + \eta_{ij}\right)\left(\theta_i - \theta_j\right) = 0 \tag{3}$$

$$\left|F_{ij}\right| \le \left(\eta_{ij}^0 + \eta_{ij}\right)\overline{F}_{ij} \tag{4}$$

$$0 \le \boldsymbol{g} \le \overline{\boldsymbol{g}} \tag{5}$$

$$0 \le \boldsymbol{r} \le \boldsymbol{L} \tag{6}$$

$$0 \le \eta_{ij} \le \overline{\eta}_{ij} \tag{7}$$

$$\eta_{ij} \text{ is integer, and } i,j \in N \tag{8}$$

where C_{ij} is the investment cost of a line added to corridor $i{-}j$; η_{ij}^0 and η_{ij} are the initial number of lines between $i{-}j$ and the number of new added lines; r_i is load curtailment at bus i due to insufficient transmission capacity, associated with a penalty factor λ; $\boldsymbol{S}$ is the branch incidence matrix; vectors $\boldsymbol{F}$, $\boldsymbol{g}$, $\boldsymbol{r}$, $\boldsymbol{L}$ denote active power flow, active power output of generators, load curtailment and predicted load, with elements of F_{ij}, g_i, r_i, L_i respectively; B_{ij} denotes the susceptance of a line between $i{-}j$; θ_i is the phase angle at bus i; $\overline{(\bullet)}$ denotes the upper bounds; N is the total bus number.

To obtain the value of r_i, usually an optimal power flow (OPF) model is employed to reschedule power generations and alleviate violations of network constraints. The objective of this OPF model is to minimize the total load curtailment [11]. Note this penalty factor for r_i helps the optimization process in (1) to find an economical planning solution without a loss of load.

However, the above model is more suitable for a vertically regulated power system, as they do not take into account the market interactions among various stakeholders, e.g. congestion costs caused by different marginal generation costs [12]. By contrast, the main objective of TEP in a new deregulated environment is to provide all stakeholders with nondiscriminatory access to cheap, secure and clean energy resources, subject to reliability and other criteria [13].

1.4 Contributions of this paper

The key contributions of the paper are:

1) A chanced constrained load curtailment index is proposed;
2) A risk based probabilistic TEP model is proposed with the consideration of planning uncertainties;
3) A novel hybrid solution algorithm in an iterative process is proposed to solve the formulated multistage programming problem.

2 Formulated probabilistic model

2.1 Define probabilistic load curtailment

In order to consider uncertainties in TEP, we propose a probabilistic formulation to reflect the level of load curtailment bounded by a threshold, $r_{\max}$. This load curtailment threshold is common in industrial practice. For instance, in Australia, expected energy not supplied (EENS) should be less than 0.002% of total energy consumption, which is also the network planning criterion used by the Australian Energy Market Operator (AEMO) [14]. In our probabilistic approach, the loss of load item in (1) is replaced by a percentage of having load curtailment over the threshold, as given in (9) and (10).

Note the meaning of λ_R is different from λ. In (1), λ is the cost of each unit of load loss, whereas λ_R is just a penalty factor and without special meaning, whose unit could be set as \$. With this penalty factor, any TEP schemes with or without excessive load curtailment can be included in the optimization process, helping search a bigger solution region and avoiding premature convergence to a local optima.

$$\sum_{i,j\in N} C_{ij}\eta_{ij} + \lambda_R R \tag{9}$$

$$R = \begin{cases} 0 & \sum_{i\in N_L} r_i \le r_{\max} \\ \dfrac{\sum_{i\in N_L} r_i - r_{\max}}{r_{\max}} & \sum_{i\in N_L} r_i > r_{\max} \end{cases} \tag{10}$$

When TEP takes into account a variety of uncertainties such as load and wind power output, a commonly used approach is using Monte Carlo (MC) simulations to sample uncertainty scenarios. After simulation stops, EENS is obtained as the average or mean of unsupplied power in the simulation after solving the optimal power flow problems. Planning schemes selected by this method are optimal in the statistical sense, leading to low-probability scenarios being discounted [6]. Moreover, in practice, TEP investment costs may be very high to guarantee that there is no excessive load curtailment at all times.

Alternatively, in this paper, we propose to find a planning scheme which can attain an acceptable probability of non-excessive load curtailment within system operational constraints. The second term in (9) is transferred into a probability version and this probability is required to be higher than a specified criterion α, as shown in (11). This probabilistic approach is rational particularly when network planners are subject to financial constraints or have different risk aversion levels. After that, a chanced constrained index ε is defined as shown in (12). This index becomes a new objective that should be included into the optimization process.

$$\Pr\left\{\sum_{i\in N_L} r_i \le r_{\max}\right\} \ge \alpha \tag{11}$$

$$\varepsilon = \begin{cases} 0 & \Pr\left\{\sum_{i\in N_L} r_i \le r_{\max}\right\} \ge \alpha \\ \alpha - \Pr\left\{\sum_{i\in N_L} r_i \le r_{\max}\right\} & \Pr\left\{\sum_{i\in N_L} r_i \le r_{\max}\right\} < \alpha \end{cases} \tag{12}$$

2.2 Carbon emission modelling

As summarized in Section 1.2, in general there are two types of carbon pricing policies: mode 1 is carbon tax; mode 2 is carbon trading [15], [16, 17]. In mode 1, only when the annual carbon emission exceeds free emission allowances, power generators will incur a cost for the excessive emission based on the carbon price. In mode 2, power generators can either sell their emission allowance surpluses or by the shortage in allowance from the emission trading market. Therefore, there will be some operating benefits for clean power generators that emit less than their emission allowance. On the other hand, power generators whose annual emission is greater than allowance will incur higher operating costs from buying extra emission allowance. Detailed mathematical formulations of the two emission pricing modes are given in (13) and (14). Note that in (14) $\Phi_{i,t}$ can be either positive or negative value, representing cost or benefit from emission trading.

$$\Phi_{i,t} = \begin{cases} 0 & \sum_{k=1}^{N_D} \delta_i D_{k,t} g_{i,k,t} \le A_{i,t} \\ \sum_{k=1}^{N_D} \delta_i D_{k,t} g_{i,k,t} - A_{i,t} & \sum_{k=1}^{N_D} \delta_i D_{k,t} g_{i,k,t} > A_{i,t} \end{cases} \tag{13}$$

$$\Phi_{i,t} = \sum_{k-1}^{N_D} \delta_i D_{k,t} g_{i,k,t} - A_{i,t} \tag{14}$$

where $\Phi_{i,t}$ and $A_{i,t}$ are the net annual emission and the annual free allowance for generator i in year t; δ_i is the emission coefficient of the generators; $g_{i,k,t}$ is the power output of generator i at load block k in year t; $D_{k,t}$ is the duration of load block k in year t; N_D is the total number of load blocks.

2.3 Uncertainties modelling

In this paper, uncertainties taken into account are wind power outputs, component working state of the power system, load growth and carbon price. Four types of probability density functions (PDFs) are used to model those uncertainties. Parameters of these PDFs can be obtained based on historical data. MC simulations are deployed to randomly generated scenarios composed by values from the four PDFs.

Wind speed V is generally modeled by a Weibull distribution [18, 19]. Wind power outputs can be derived by wind speed and a power curve of wind power generators. A piecewise function as given in (15) is widely used to model a power curve [20].

$$P_W = \begin{cases} 0 & 0 \leq V < V_{In} \\ P_W^{rate} \dfrac{V - V_{In}}{V_{rate} - V_{In}} & V_{In} \leq V < V_{rate} \\ P_W^{rate} & V_{rate} \leq V \leq V_{Out} \\ 0 & V_{Out} < V \end{cases} \quad (15)$$

where P_W is the power output of wind power generators, whose rated power output is P_W^{rate}; V_{In}, V_{rate}, V_{Out} are cut-in, rated and cut-out wind speeds, respectively.

The PDF of wind power output i.e. P_{pdf} is given in (16). Pr_W^{zero}, Pr_W^{rate} denote probabilities of zero and rated wind power output. $\delta(\bullet)$ is the Dirac delta function used to represent the discrete wind output PDF. $y(\bullet)$ can be expressed by either a fitted polynomial function or discrete samples [9].

$$P_{\text{pdf}}(P_W) = \begin{cases} \text{Pr}_W^{zero}\,\delta(P_W) & P_W = 0 \\ y(P_W) & 0 < P_W < P_W^{rate} \\ \text{Pr}_W^{rate}\,\delta(P_W - P_W^{rate}) & P_W = P_W^{rate} \\ 0 & \text{otherwise} \end{cases} \quad (16)$$

The availability of power system components is modelled by a Binomial PDF [21]. The probability Pr^{state} of a availability state is given in (17) and (18). Ω_{down}^{state}, Ω_{up}^{state} denote sets of components in unavailable and available states. r_{FOR} is the forced outage rate of components.

$$\text{Pr}^{\,state} = \prod_{i \in \Omega_{down}^{state}} (1 - \text{Pr}_i) \prod_{j \in \Omega_{up}^{state}} (\text{Pr}_j) \quad (17)$$

$$\text{Pr}_j = 1 - r_{FOR} \quad (18)$$

Uncertainties of load L are modelled by a Normal PDF with mean μ and standard deviation σ, as given in (19) [18].

$$P_{\text{pdf}}(l; \mu, \sigma) = \frac{1}{\sqrt{2\pi\sigma^2}} \exp\left[-\frac{(L - \mu)^2}{2\sigma^2}\right] \quad (19)$$

Uncertainties of carbon price ϕ are modelled by a Weibull PDF with scale parameter τ and shape parameter ρ, as given in (20) [15].

$$P_{\text{pdf}}(\varphi; \tau, \rho) = \frac{\rho}{\tau}\left(\frac{\varphi}{\tau}\right)^{\rho - 1} \exp\left[-\left(\frac{\varphi}{\tau}\right)^{\rho}\right] \quad (20)$$

2.4 Proposed probabilistic model

A multi-stage probabilistic TEP model is formulated as minimizing the total cost with the chance constrained index ε_t proposed in Section 2.1, as follows.

Minimize

$$\sum_{t=1}^{T} \sum_{i,j \in N} \frac{C_{ij}(\eta_{ij,t} - \eta_{ij,t-1})}{(1+\gamma)^{t-1}} + \sum_{t=1}^{T} \frac{\lambda_\varepsilon \varepsilon_t}{(1+\gamma)^{t-1}} + \sum_{t=1}^{T} \sum_{k=1}^{N_D} \sum_{i=1}^{N_G} \frac{D_{k,t} C_{Oi} g_{i,k,t}}{(1+\gamma)^{t-1}} + \sum_{t=1}^{T} \sum_{i=1}^{N_G} \frac{\varphi \Phi_{i,t}}{(1+\gamma)^{t-1}} \quad (21)$$

subject to

$$\sum_{i,j \in N} F_{ij,k,t} + \sum_{i \in N_G} g_{i,k,t} + \sum_{i \in N_L} r_{i,k,t} - \sum_{i \in N_L} L_{i,k,t} = 0 \quad (22)$$

$$F_{ij,k,t} - B_{ij}\left(\eta_{ij,t}^0 + \eta_{ij,t}\right)\left(\theta_{i,k,t} - \theta_{j,k,t}\right) = 0 \quad (23)$$

$$\left|F_{ij,k,t}\right| \leq \left(\eta_{ij,t-1} + \eta_{ij,t}\right)\overline{F}_{ij,k,t} \quad (24)$$

$$0 \leq g_{i,k,t} \leq \overline{g}_i \quad (25)$$

$$0 \leq r_{i,k,t} \leq L_{i,k,t} \quad (26)$$

$$0 \leq \eta_{ij,t} \leq \overline{\eta}_{ij} \quad (27)$$

$$\eta_{ij,t} \text{ is integer, and } i, j \subset N \quad (28)$$

Equation (21) is a model represented by net present value (NPV) with a discount rate γ. T is the total planning horizon. i, j are superscripts for buses. k, t are superscripts for load block and planning year respectively. The first term is the investment cost, and the second term is the chance constrained index at year t scaled by a penalty factor λ_ε. The third term is the operating cost of power generators, whose incremental cost of output is denoted by C_{Oi}. The total number of generators in the system is N_G. The fourth term is the cost of carbon emission, derived from carbon price and the net annual emission of power generators.

For completeness, we use a piecewise function to derive the quadratic losses in the DC power model, as given in (29)-(37). Note losses are not incorporated into the objective function. Instead, they are considered as the additional active power in (22) required from power generation, in order to satisfy the nodal balance due to the presence of losses.

The detailed linearization process is given as follow.

$$P_{ij,k,t}^{loss} \approx 2G_{\text{cond}ij}\left(1 - \cos\theta_{i-j,k,t}\right) \approx G_{\text{cond}ij}\theta_{i-j,k,t}^2 \quad (29)$$

$$\theta_{i-j,k,t}^2 \approx \sum_{h=1}^{H} \varpi(h)\Delta\theta_{i-j,k,t}(h) \quad (30)$$

$$\varpi(h) = (2h - 1)\frac{\theta_{i-j,t}^{\max}}{H} \quad (31)$$

$$\theta_{i-j,k,t} = \theta_{i-j,k,t}^{+} - \theta_{i-j,k,t}^{-} \quad (32)$$

$$\left|\theta_{i-j,k,t}\right| = \sum_{h=1}^{H} \Delta\theta_{i-j,k,t}(h) = \theta_{i-j,k,t}^{+} + \theta_{i-j,k,t}^{-} \quad (33)$$

$$0 \leq \theta_{i-j,k,t}^{+} \leq \chi\theta_{i-j,t}^{\max} \quad (34)$$

$$0 \leq \theta_{i-j,k,t}^{-} \leq (1 - \chi)\theta_{i-j,t}^{\max} \quad (35)$$

$$0 \leq \Delta\theta_{i-j,k,t}(h) \leq \frac{\theta_{i-j,t}^{\max}}{H} \quad (36)$$

$$0 \leq \Delta\theta_{i-j,k,t}(h) \leq \Delta\theta_{i-j,k,t}(h-1) \quad (37)$$

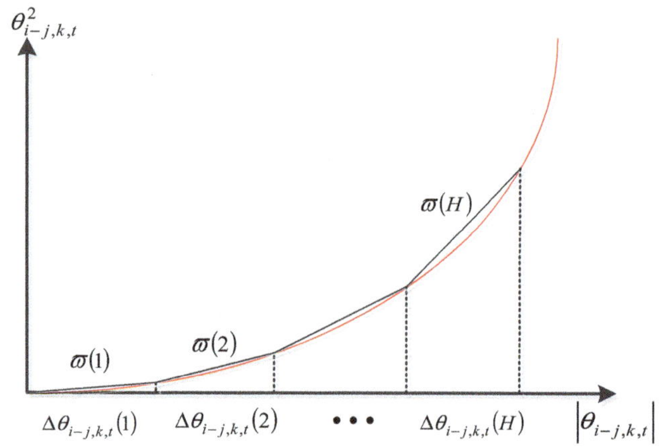

Fig. 1 Piecewise linear approximation of phase angle difference between $i-j$

Where $P^{loss}_{ij,k,t}$, $\theta_{i-j,k,t}$ are the active power losses and phase angle difference between $i-j$ at load block k in year t, repectively; $G_{\text{cond}ij}$ is the line conductance; the quadratic term $\theta^2_{i\text{-}j,k,t}$ is linearized by a piecewise function, as shown in Fig. 1; H is the total number of intersection, with a h-th intersection of $\Delta\theta_{i-j,k,t}(h)$, and the slope is $\varpi(h)$; two nonnegative slack variables $\theta^+_{i-j,k,t}$, $\theta^-_{i-j,k,t}$ are introduced to replace $\theta_{i-j,k,t}$, and they are also bounded by a binary variable χ to make this replacement effective, as shown in (32)-(35). Equations (36)-(37) can ensure the linear intersection on the left side is always filled up first.

3 Solution algorithms

Solution algorithms for TEP problems mainly fall into either mathematical programming classes or heuristic search classes. Mathematical programming methods have strict requirements on the model itself (e.g. the problem or the continuous relaxation of the problem should be convex) and can provide more clues on the quality of the final solution [22]. However, mathematical programming methods tend to be trapped by local optima in some cases. On the contrary, heuristic methods are suitable for stochastic global search, free from problem formulation difficulties and can escape from premature local optimal. The drawbacks of heuristic methods are: the quality of the solution cannot be guaranteed; and prohibitive computation efforts are required [22, 23, 24].

In this paper, the proposed TEP model is a dynamic optimization problem with a chance constrained reliability evaluation. To enhance the solution performance of the proposed model, a hybrid method based on decomposition is proposed as follows.

The overall stochastic programming problem can be divided into two subproblems: 1) the investment subproblem in the first and second terms of the defined objective; 2) the operating subproblem in the third and fourth terms of the defined objective. Note that the investment subproblem is targeted at a specific planning year, whereas the operating decisions are subject to the decisions from the investment subproblem and should be evaluated over multiple years thereafter [25].

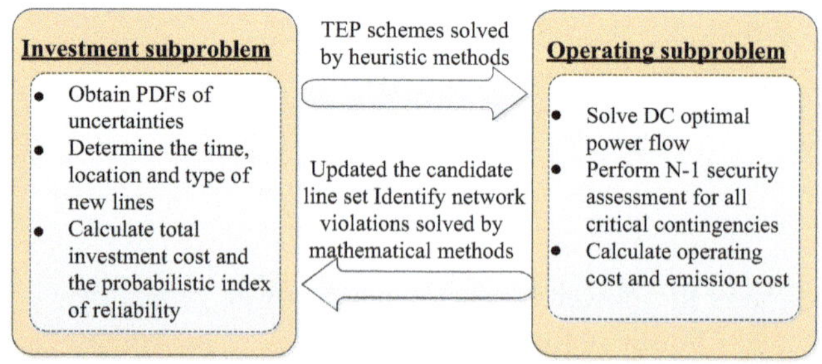

Fig. 2 An iterative solution algorithm for the decomposed two subproblems

Firstly, an evolutionary algorithm is deployed to help the optimization process in subproblem 1 to find a bigger set of expansion candidates. In this paper, differential evolution (DE) is used, as it is shown faster, simpler and more robust [6]. Secondly, the subproblem 2 can be soundly solved by a state-of-the-art nonlinear programming technique, such as the interior point (IP) method. IP is widely used in solving power system operation and dispatch problems, because the optimality of the solution found by IP is guaranteed [26]. Moreover, IP has higher efficiency in searching local optima [26], which means that IP can quickly locate local optima for the subproblem 2, adding them to new generation in the evolution process. As shown in Fig. 2, the two subproblems are solved in an iterative way to progressively converge to the final solution. The termination criterion can be defined by two ways: the maximum iteration number is reached; or the objective value does not improve for a few successive generations.

The main procedure of the proposed hybrid method is given as follows:

1) Determine the PDFs of wind speed, carbon price, load, and FOR based on historical data. Note that the PDF of wind power outputs is translated from the wind power curve in (15).
2) MC simulations are deployed to generate values of wind power outputs, load, component availability, and carbon price from the PDFs in step 1).
3) Initialize the population of the EA corresponding to the number of new added lines. Note that the quality of initial population has great impacts on the final solution. Therefore, in this process, uncertain features are neglected and some deterministic values are assigned. To be specific, carbon price is set to its

mean value, wind power output is set to the installed capacity scaled by its capacity factor, all components are set as available, and load is set as $(\mu + 3\sigma)$.

4) Using the initial values generated by step 3), denoted by $\eta^G = \{\eta_{ij,0}, \forall i, j \in N\}$, apply the IP to generate local optima, i.e. subproblem 2 is solved with this generation. In this process, IP can solve the DC optimal power flow (OPF) for each individual in the population. Note if network violations are identified, a linear programming is required to minimize the total load curtailment, i.e. re-dispatching of generation is not based on bids or marginal costs. Then calculate the operating and carbon emission costs and add the two local optima to η^{G+1}. Also, based on identify network violations, find the probability of $\Pr\left\{\sum_{i \in N_L} r_i \le r_{\max}\right\}$, and calculate the probabilistic index of reliability.
5) Start mutation and recombination process. Integers representing adding or reducing lines will be generated and recombined.
6) For each TEP scheme, the objective function in (21) is assigned as the fitness value. A newly generated child individual is compared with the parent, and replaces it if the fitness value of the child is smaller.
7) Terminate the algorithm if the stopping criterion is satisfied, otherwise go back to step 2).

4 Case studies

In this section, a series of numerical experiments are undertaken to demonstrate the performance of the proposed model on the modified IEEE 14-bus system. The 14-bus system initially has five power generators, and 20 transmission corridors.

In our paper, the five power generators in the base case are assumed to be fossil-fuel-fired, and the carbon emission coefficients for them are set as 1.2, 1, 0.8, 0.8 and 0.6 tCO_2/MWh respectively. Other power generation types such as hydro power or nuclear power could be included as future works. Moreover, for simplicity, power generators are assumed to receive 80% of their emission allowances for free based on their emissions in the base year, and free allowances will decrease linearly each year to 30% in the last planning horizon. Note that allocating emission allowances is a complex issue involving political motivation. The assumption made in this paper regarding emission allocation is in accordance with the EU ETS practice, which can be found in [27].

The total generation capacity of the base system is 720 MW, while the total load is 440 MW. Durations for load

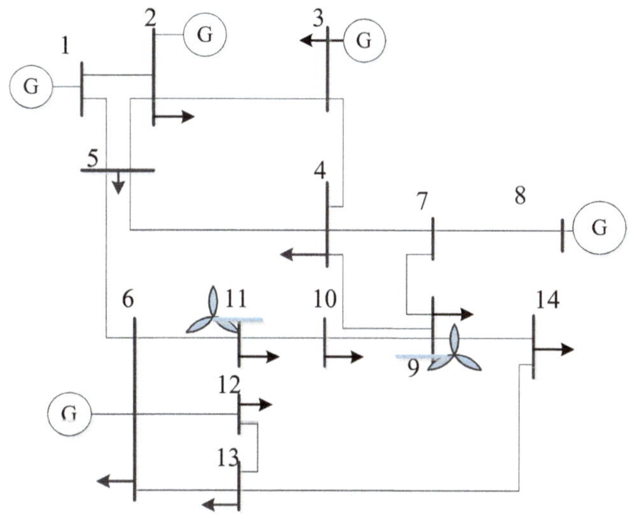

Fig. 3 Modified IEEE 14-bus system with two wind power generators

blocks are 20%, 50%, 29% and 1%. As illustrated in Fig. 3, two wind farms with capacities of 50 MW are assumed to be installed at buses 9 and 11. The two wind farms are assumed to be totally independent, and therefore the wind speeds for modelling their outputs are randomly generated from two different Weibull distribution functions. Parameters for modelling the wind power curve are $V_{In} = 4$, $V_{rate} = 10$, $V_{Out} = 22$m/s. To model the possibility of *N*-1 contingencies, transmission lines outages are modelled by a r_{FOR} of 1%. The mean value of carbon prices is set as $E[\varphi] = \$23/tCO_2$. The planning horizon is set as five years, and the annual load growth rate is 5%, with uncertainties of $\sigma/_{\mu} = 2\%$, i.e. $\mu = 0.05$, $\sigma = 0.001$.

The capacity of each candidate line is 100 MW, and up to four lines are allowed for each corridor. The investment cost is assumed to be 50 M$/100 km. We set the upper bound of load curtailment threshold as a percentage of the total annual energy demand, i.e. $r_{\max} = 0.1\%$. To make ε_t have a significant effect in the objective function, the penalty factor λ_ε is set big enough to ensure that planning schemes with considerable load curtailments will be eliminated in the optimization process during the heuristic search. The settings of α and λ_ε are 95% and $\$1 \times 10^9$.

In order to make a comparison and demonstrate the performance of the proposed probabilistic approach, three deterministic TEP studies with different carbon pricing policies are undertaken, i.e. deterministic TEP with no emission price, deterministic TEP in carbon mode 1 and mode 2 respectively with a carbon price of $\$23/tCO_2$. As illustrated in Table 1, investment costs for three deterministic models are all 378.25 M$, whereas the investment cost for Case 4 is the highest due to the volatility of emission market. For operating cost, Case 1 is the lowest as

Table 1 Comparisons of deterministic and probabilistic TEP with different carbon pricing policies

Case No	Total cost (M$)	Invest. cost (M$)	Operating cost (M$)	Emission cost (M$)
Case 1: deterministic TEP with no emission price	890.36	378.25	512.11	0
Case 2: deterministic TEP in mode 1 with $23/t$CO_2$	1119.22	378.25	615.62	125.35
Case 3: deterministic TEP in mode 2 with $23/t$CO_2$	1061.48	378.25	572.98	110.25
Case 4: probabilistic TEP in mode 2	1107.77	415.64	586.77	105.36

a result of no carbon pricing policies. Moreover, compared to Case 2, the operating cost in Case 3 is lower. This is because some power generators can earn benefits in the emission trading market if they have lower emission coefficient. Although the carbon price uncertainty increases the investment cost in Case 4, the total cost in Case 4 is lower compared to that in Case 2 (1,107.77 vs 1,119.22 M$), reflecting the operating benefits of carbon mode 2. It should be noted that, if stochastic features are removed, i.e. without uncertainties of wind power, load and carbon price, for zero load curtailment and selected line outages, the results of the probabilistic and deterministic TEP are identical, which validates the applicability of our approach. Furthermore, comparisons in Table 1 reveal that the planning schemes identified by deterministic TEP are greatly affected by carbon price uncertainties, leading to a high probability of load curtailment. This may imply that the deterministic TEP is insufficient, particularly when exposed to higher risks of uncertainties.

As mentioned in Section 2.1, the probability α is defined as the lower bound for each planning scheme the possibility of having load curtailment below the required threshold, and $\alpha \in [0, 1]$. This probability can be interpreted as a risk measure, which can be chosen by network planners. The higher value of α, the more important to choose a plan without excessive load curtailment, leading to a higher investment cost (more conservative). We set α as three different levels to run simulations of Case 4. The probabilities of load curtailment are given in Fig. 4. As seen, a plan chosen by a higher α is said to be risk averse, whereas a plan chosen by lower α is said to be risk preferring. Note a higher α also increases the probability of no load curtailment (see the intersections on the *Y*-axis in Fig. 4). In this way, our probabilistic approach can provide comprehensive information for network planners regarding the balance between minimizing investment costs and minimizing the risk of load curtailment in the face of uncertainties.

The detailed result of our probabilistic TEP in carbon mode 2 is given in left column in Table 2. As given in the right two columns in Table 2, carbon price uncertainties (i.e. high standard deviation) make a planning scheme expose to higher risk, requiring higher investment cost for

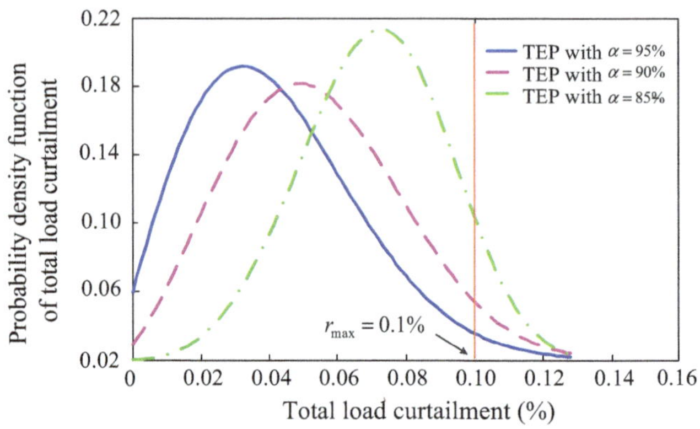

Fig. 4 Probability of load curtailment with different α values

meeting the α criterion (the initial α is set as 95%). As seen, if the standard deviation of carbon prices is set above $3/t$CO_2$, investment costs have to be increased as α becomes lower than 95%. Also, the same situation will happen when the mean of carbon prices are raised from $23/t$CO_2$ to $35/t$CO_2$, which implies that more use of intermittent renewable energy (e.g. wind power), will increase TEP uncertainties, thus requiring higher investment costs. These findings are also supported by simulation results in Fig. 5. According to Fig. 5, the total costs without carbon price or with fixed carbon price are always lower than total costs with carbon price uncertainties in both modes 1 and 2, regardless of the value of α.

To examine the effects of different upper bounds of the maximum load curtailment parameter r_{max}, we run simulations to obtain the total cost with different values of α. As shown in Fig. 6, the higher r_{max}, the lower total cost. This is more obvious with a conservative planning scheme, i.e. with a higher α. In addition, results of the total cost against total installed wind power capacity with different α values are given in Fig. 7. As seen, for a specific α, it is possible to obtain the optimal wind power capacity, i.e. the lowest total cost when integrating different wind power capacities into the system. For instance, when α is chosen as 95%, in order to obtain the lowest total cost, the optimal wind power capacity to be integrated into the power system should be about 60 MW. This finding is of significance for renewable energy integration analysis, as the rapid growth

Table 2 Final result against different carbon price characteristics

Lines added	Different carbon price characteristics ($/t$CO_2$)	α
$\eta_{1-2,t=2} = 1$; $\eta_{4-10,t=2} = 1$; $\eta_{7-9,t=2} = 1$; $\eta_{6-11,t=3} = 2$; $\eta_{10-11,t=3} = 1$; $\eta_{12-13,t=3} = 1$; $\eta_{14-9,t=3} = 1$; $\eta_{2-3,t=4} = 2$; $\eta_{3-4,t=4} = 3$; $\eta_{1-2,t=5} = 1$; $\eta_{1-5,t=5} = 1$; $\eta_{5-6,t=5} = 4$; $\eta_{6-11,t=5} = 2$; $\eta_{12-13,t=5} = 1$	Mean is 23; std. is 3	0.9575
	Mean is 23; std. is 5	0.9028
	Mean is 35; td. is 3	0.9329
	Mean is 35; std. is 5	0.8266

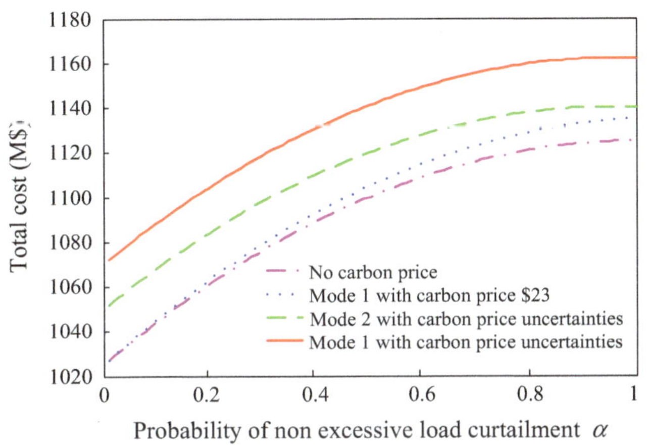

Fig. 5 Total cost against different values of α with different carbon uncertainties

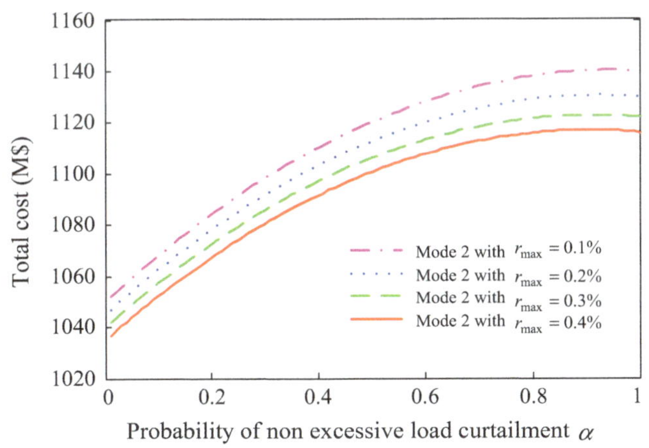

Fig. 6 Total cost against different values of α with different total load curtailment thresholds

of wind power will have significant impact on power system operations and planning [28]. Moreover, our approach can reveal the relationship between the overall system cost and the risk of load curtailment with the presence of wind power uncertainties. Therefore, our probabilistic TEP model serves a good indicative role in achieving low carbon economy, in terms of identifying a robust and cost-effective planning scheme.

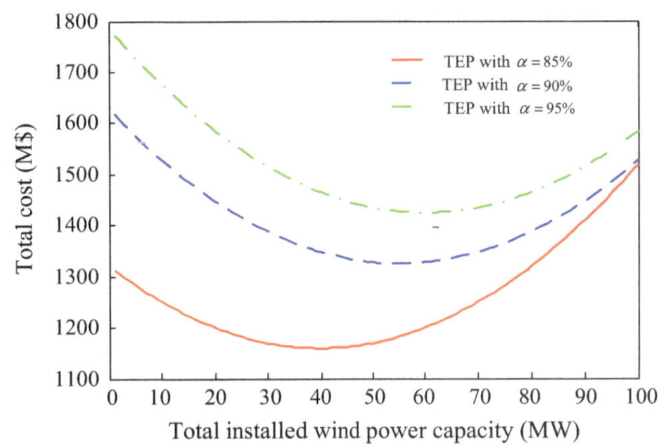

Fig. 7 Total cost against different total installed wind power capacity with different values of α

5 Conclusion

This paper has proposed a probabilistic TEP model with a chance constrained load curtailment index. Planning uncertainties such as wind power output, component availability, load, and carbon price are incorporated by a Monte Carlo simulation based approach. Our multi-stage planning objective is formulated as minimizing the total cost, including investment cost, operating cost, emission cost and a risk factor of load curtailment. For completeness, a piecewise approximation function is used to linearize the quadratic power losses. Meanwhile, a novel iterative solution algorithm combing heuristic search and mathematical programming is proposed to solve the formulated a dynamic optimization problem. The performance of the proposed approach is demonstrated by a modified IEEE 14-bus system. Simulation results have proved that our approach can give network planners an opportunity to trade-off between the overall cost and the probability of load curtailment in the presence of uncertainties. Our approach can also provide network planners with comprehensive information regarding effects of uncertainties on TEP schemes, allowing them to adjust planning strategies based on their risk aversion levels or financial constraints. Moreover, our approach can be used for renewable energy integration analysis in terms of long-term network planning. Therefore, our novel TEP approach is a risk-based, flexible decision tool, which is important for achieving low carbon economy through planning practices.

References

[1] Buygi MO, Balzer G, Shanechi HM, Shahidehpour M (2004) Market-based transmission expansion planning. IEEE Trans Power Syst 19(4):2060–2067

[2] Garces LP, Conejo LJ, Garcia-Bertrand R, Romero R (2009) A bilevel approach to transmission expansion planning within a market environment. IEEE Trans Power Syst 24(3):1513–1522

[3] Escobar AH, Gallego RA, Romero R (2004) Multistage and coordinated planning of the expansion of transmission systems. IEEE Trans Power Syst 19(2):735–744

[4] Zhang F, Hu Z, Song Y (2013) Mixed-integer linear model for transmission expansion planning with line losses and energy storage systems. IET Gen Trans & Dist 7(8):919–928

[5] Heejung P, Baldick R (2013) Transmission planning under uncertainties of wind and load: sequential approximation approach. IEEE Trans Power Syst 28(3):2395–2402

[6] Zhao JH, Dong ZY, Lindsay P, Wong KP (2009) Flexible transmission expansion planning with uncertainties in an electricity Market. IEEE Trans Power Syst 24(1):479–488

[7] Tan ZF, Ngan HW, Yang Wu, Zhang HJ, Song YH, Yu C (2013) Potential and policy issues for sustainable development of wind power in China. J Mod Power Syst Clean Energy 1(3):204–215

[8] Huang J, Xue F, Song XF (2013) Simulation analysis on policy interaction effects between emission trading and renewable energy susidy. J Mod Power Syst Clean Energy 1(2):195–201

[9] Yu H, Chung CY, Wong KP, Zhang JH (2009) A chance constrained transmission network expansion planning method with consideration of load and wind farm uncertainties. IEEE Trans Power Syst 24(3):1568–1576

[10] Xu Z, Dong ZY, Wong KP (2006) A hybrid planning method for transmission networs in a deregulated environment. IEEE Trans Power Syst 21(2):925–932

[11] Guo L, Qiu QW, Liu J, Zhou Y, Jiang LL (2014) Power transmission risk assessment considering component condition. J Mod Power Syst Clean Energy 2(1):50–58

[12] Roh JH, Shahidehpour M, Wu L (2009) Market-based generation and transmission planning with uncertainties. IEEE Trans Power Syst 24(3):1587–1598

[13] Maghouli P, Hosseini SH, Buygi MO, Shahidehpour M (2009) A multi-objective framework for transmission expansion planning in deregulated environments. IEEE Trans Power Syst 24(2): 1051–1061

[14] Australian Energy Market Operator (AEMO).

[15] Kazerooni AK, Mutale J (2010) Transmission network planning under security and environmental constraints. IEEE Trans Power Syst 25(2):1169–1178

[16] Chen QX, Kang CQ, Xia Q, Zhong J (2010) Power generation expansion planning model towards low-carbon economy and its application in China. IEEE Trans Power Syst 25(2):1117–1125

[17] Zhou X, James G, Liebman A, Dong ZY, Ziser C (2010) Partial carbon permits allocation of potential emission trading scheme in Australian electricity market. IEEE Trans Power Syst 25(1):543–553

[18] Ghofrani M, Arabali A, Etezadi-Amoli M, Fadali MS (2013) Energy storage application for performance enhancement of wind integration. IEEE Trans Power Syst 28(4):4803–4811

[19] Siano P, Mokryani G (2013) Probabilistic assessment of the impact of wind energy integration into distribution networks. IEEE Trans Power Syst 28(4):4209–4217

[20] Orfanos GA, Georgilakis P, Hatziargyriou ND (2013) Transmission expansion planning of systems with increasing wind power integration. IEEE Trans Power Syst 28(2):1355–1362

[21] Jaeseok C, Mount TD, Thomas RJ, Billinton R (2006) Probabilistic reliability criterion for planning transmission system expansions. IET Gen Trans & Dist 153(6):719–727

[22] Zhang H, Heydt GT, Vittal V, Quintero J (2013) An improved network model for transmission expansion planning considering reactive power and network losses. IEEE Trans. Power Syst. 28(3):3471–3479

[23] Tekiner H, Coit DW, Felder FA (2010) Multi-period multi-objective electricity generation expansion planning problem with Monte-Carlo simulation. Elec Power Syst Res 80(12): 1394–1405

[24] Zhang H, Vittal V, Heydt GY, Quintero J (2012) A mixed-integer linear programming approach for multi-stage security-constrained transmission expansion planning. IEEE Trans Power Syst 27(2):1125–1133

[25] Roh JH, Shahidehpour M, Fu Y (2007) Market-based coordination of transmission and generation capacity planning. IEEE Trans Power Syst 22(4):1406–1419

[26] Zhao JH, Wen FS, Dong ZY, Xue YS, Wong KP (2012) Optimal dispatch of electric vehicles and wind power using enhanced particle swarm optimization. IEEE Trans Indus Infor 8(4): 889–898

[27] European Commission on Climate Action.

[28] Xue YS, Cai B, James G, Dong ZY, Wen FS, Xue F (2014) Primary energy congestion of power systems. J Mod Power Syst Clean Energy 2(1):39–49

Jing QIU (S'12–M'14) is now a Research Associate at the Center for Intelligent Electricity Networks (CIEN), The University of Newcastle, Australia. He obtained his B.Eng. degree in control engineering from Shandong University, China, M.Sc. degree in environmental policy and management, majoring in carbon financing in the power sector, from The University of Manchester, U.K., and Ph.D. in electrical engineering from The University of Newcastle, Australia. His areas of interest include electricity market modeling, power system planning, renewable energy, carbon emission reduction and risk management.

Zhaoyang DONG (M'99–SM'06) obtained his Ph.D. degree from the University of Sydney, Australia in 1999. He is now Professor and Head of School of Electrical and Information Engineering, University of Sydney, Australia. He was previously Ausgrid Chair and Director of the Center for Intelligent Electricity Networks (CIEN), The University of Newcastle, Australia, and is now a conjoint professor there. He also held academic and industrial positions with the Hong Kong Polytechnic University and Transend Networks (now TasNetworks), Tasmania, Australia. His research interest includes Smart Grid, power system planning, power system security, load modeling, electricity market, and computational intelligence and its application in power engineering. Prof. Dong is an editor of the IEEE TRANSACTIONS ON SMART GRID, the IEEE TRANSACTIONS ON SUSTAINABLE ENERGY, IEEE POWER ENGINEERING LETTERS, and IET Renewable Power Generation.

Junhua ZHAO (S'04–M'07) obtained his B.E. and Ph.D. degrees from Xi'an Jiaotong University, China and The University of Queensland, Australia in 2003 and 2007 respectively. He then worked as a postdoctoral research fellow at the University of Queensland, Australia. He is now a Senior Lecturer in Center for Intelligent Electricity Networks, The University of Newcastle, Australia. His research interests include power system analysis and computation, smart grid, energy Internet, data mining and computational intelligence and their applications in power engineering.

Ke MENG (M'10) obtained Ph.D. from the University of Queensland, Australia in 2009. He is currently with the Centre for Intelligent Electricity Networks (CIEN), The University of Newcastle, Australia. His research interest includes pattern recognition, power system stability analysis, wind power, and energy storage.

Fengji LUO (M'13) obtained the B.S. and M.S. degrees in software engineering from Chongqing University, Chongqing, China, in 2006 and 2009, respectively. He received the Ph.D. degree in electrical engineering from the University of Newcastle, Australia, in 2013. Currently, he is the research associate of the Centre for Intelligent Electricity Networks, Australia. His research interests include computational intelligence applications, distributed computing, and power system operation & planning.

Kit Po WONG (M'87, SM'90, F'02), obtained M.Sc, Ph.D. and D. Eng degrees from the University of Manchester, Institute of Science and Technology, in 1972, 1974 and 2001 respectively. Since 1974, he has been with the School of Electrical, Electronic and Computer Engineering, The University of Western Australia, Perth, Australia, where he currently is a Winthrop Professor. From 2002 to 2011, he was also with Department of Electrical Engineering,The Hong Kong Polytechnic University, as Chair Professor and Head (2002–2007). He is a con-joint professor of The University of Newcastle, Australia. His current research interests include power system analysis, planning and operations, smart grids and renewable energy. Prof. Wong received three Sir John Madsen Medals (1981, 1982, and 1988) from the Institution of Engineers Australia, the 1999 Outstanding Engineer Award from IEEE Power Chapter Western Australia, and the 2000 IEEE Third Millennium Award. He was the General Chairman of IEEE/CSEE PowerCon2000. He is now Editor-in-Chief of IEEE POWER ENGINEERING LETTERS and was Editor-in Chief of IEE Proc. Generation, Transmission & Distribution. He is a Fellow of IEEE, IET, HKIE, and IEAust.

Chao LU obtained his B.E., Ph.D. from Tsinghua University, Beijing, China. He is now associate professor at Tsinghua University, China. His research interest includes power system stability, power system transients and power system planning.

25

Whole-system thinking, development control, key barriers and promotion mechanism for EV development

Chunlin GUO (✉), **Ching Chuen CHAN**

Abstract Electric vehicles (EVs) have received significant attention because of the potential energy savings and emission reductions they enable. However, current studies and demonstrations have focused mainly on specific technologies and equipment types, which cannot in themselves solve the predicaments facing EVs. This study points out that EVs will form a large, complex system that needs to be optimized over different aspects to compete with traditional vehicles. Therefore, whole-system thinking is needed to support the development control of EVs, with a broader scope than operational control, and the core issue is the interaction between EVs and power grid, including their coordinated development and operational management. For development control of EVs, a target system and a step-wise optimization method are presented, as well as the basic principles for designing the target system. There are two key barriers in EVs' development: the research and mass production of high-performance power batteries, and the formulation of a favorable mechanism to capture benefits and encourage development. To address the problems of EV development, a promotion method that combines franchising with moderate competition is proposed. The concepts and methods developed in this study can facilitate the research and development of EVs.

C. GUO, State Key Laboratory of Alternate Electrical Power System with Renewable Energy Sources, North China Electric Power University, Beijing, China
(✉) e-mail: gcl@ncepu.edu.cn
C. C. CHAN, Department of EEE, University of Hong Kong, Hong Kong, China

Keywords Electric vehicles, Charging infrastructure, Whole-system thinking, Key barrier, Development route

1 Introduction

The depletion of petroleum resources and worsening environmental pollution have encouraged alternatives to traditional vehicles based on combustion engines. As a consequence of technological improvements, the first mass-market electric vehicles (EVs) are now being introduced and greater market penetration can be expected. Now, EV research and policy are receiving substantial attention all over the world. Many countries have formulated their own policies and conducted different types of demonstration project to promote the development of EVs. Almost all international automotive giants such as GM, Ford, Daimler-Chrysler, and Toyota have invested huge resources in development of the next generation of EVs.

In China, a 'new energy' automotive industry focused on EVs is being fostered with government support as one of seven strategic emerging industries. The government is currently promoting a 'Ten Cities & Thousand Vehicles program' and issued an ambitious plan which recommends that the cumulative production of EVs should reach 500000 units by 2015. The local governments in Beijing, Shenzhen and Hangzhou, as well as enterprises such as the State Grid Corporation of China, are promoting business models such as battery rental and centralized management, and have conducted many demonstration projects.

However, even though significant resources have been invested to promote EVs, the development of the Chinese EV industry remains slow. By the end of 2013, only approximately 10000 new energy vehicles were delivered in 25 demonstration cities in China. The construction of charging infrastructure has also failed to progress as rapidly as hoped.

EV development in the US and Europe also remains unsatisfactory. The US Advanced Battery Consortium, composed by the 'Big Three' automotive manufacturers (GM, Ford, and Chrysler) and a few National Laboratories belonging to the US Department of Energy (DOE), established a set of performance targets of power batteries in the early 1990s. However, even the lowest mid-term goals are still challenging with today's technology. The DOE has reduced its research support for new energy vehicles several times.

Obviously, there are still a lot of problems that need to be dealt with before the large-scale deployment of EVs, and more research and development must be carried out to address them. Furthermore, as shown by a growing amount of research, the development of electric vehicles requires whole-system thinking to guide its top-level design. So, this paper discussed such issues of system level of EV development, including systematic thinking, development control, key barriers, etc. The concept and method gained here would be able to facilitate the research and development of EVs.

2 Necessity of whole-system thinking

2.1 The large complex system that EVs will form

2.1.1 Basic parts of the EV operating environment

Much literature has discussed the architecture and main issues related to EV system. The vehicle characteristics, the grid-connection or electric vehicle supply equipment, the building loads and generators, the local distribution system, and energy markets must operate well together in a complex and large EV system [1]. By virtue of being equipped with a battery, if a layer of communication exists between the system operator and the EV, the EVs are potentially capable of having their electricity requirements being served according to a pre-established schedule, and even of injecting energy back into the grid in cases where the electricity demand is high or intermittent generation output is low [2]. For example an operational framework for the implementation of a charging management model has been presented based on multi-agent system (MAS) technology [3].

All these authors point out that the relationships around EVs are very complex [4]. Figure 1 shows the overall structure of the operating environment of EV technology, which consists of three parts, including the EV, charging infrastructure and power grid.

The EV part of this system includes components such as the power battery, the converter, the driving motor, vehicle integration and compatibility. These are serially connected and form a relatively independent system. This part is

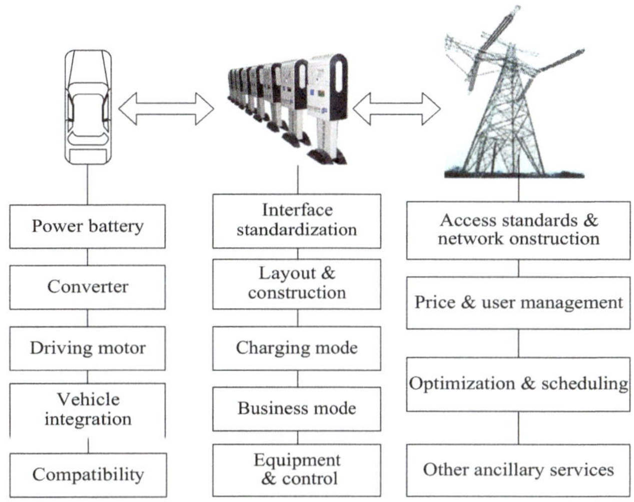

Fig. 1 Basic links in the EV operating system

primarily concerned with equipment and technology issues, where the power battery is both the weakest link and one of the bottlenecks for the entire system.

The charging infrastructure includes links components as interface standardization, layout and construction, charging model, business model, equipment types and the control system. This part is primarily related to business and operation.

There are two key issues.

1) The charging model, operating model and business model have not been clearly identified; thus, it is difficult to make decisions.
2) The infrastructure requires a high investment over a long period and is intertwined with many complex external factors, such as land utilization, resulting in complex relationships of benefits and restrictions.

A vertically integrated power grid includes components such as access standards and network construction, electricity pricing and user management, optimization and scheduling, and other ancillary services. Disaggregated power grids separate some of these responsibilities, such as pricing, but the essential functions are the same. This part is concerned with many complex problems of technology, control and operation. The core issue for EVs is to improve their convenience and maximize their system benefits, by coordinating EVs with the power grid.

2.1.2 Main couplings and performance requirements in the EV operating environment

As Fig. 2 shows, besides different physical devices, there are many entities in the EV system such as the generation companies (GENCOs), transmission system operators (TSO),

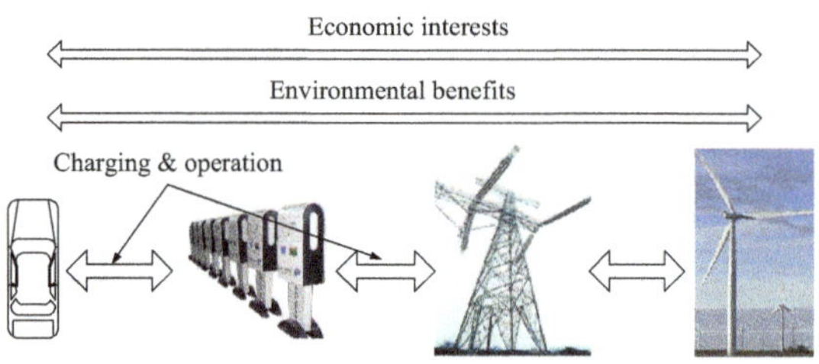

Fig. 2 Major coupling chains in EV system

distribution system operators (DSO), charging point managers (CPM), and EV users. And there are multiple relationships among different components and entities in the EV system, which are dominated by three coupling chains: charging and operating, environmental benefits and economic interests.

The charging and operating chain is composed of the hardware connections and operating interactions during the process of charging and power supply. This chain requires convenient charging and simple operations, so as to reduce the costs and time expended. The environmental benefits chain is based on energy conservation and emission reduction, which are both major driving forces for EV development. This chain requires the scheduling of new energy for global optimization. The economic interest chain connects the economic performance of the different sections. This chain must facilitate the formation of a sustainable business model for EVs and is the key issue that determines the viability of EV development.

The outcomes of these three coupling chains are convenience, environmental conservation and economy, which are the performance requirements of EV system. These couplings and performance requirements connect all parts of EV operation abovementioned to form a large, complex and interactive system.

2.2 Whole-system thinking for EV development and the challenges it brings

Because of the complicated structure and coupling relationships mentioned above, rather than an issue of technologies and devices, EV development is a problem of the construction and overall optimization of a large complex system, which needs whole-system thinking that will bring huge challenges to research and development of electric vehicles.

In earlier studies, EV and grid are often regarded independently, where EV simply consumes electricity from the power grid and pays as an ordinary consumer. For example, in comparisons of energy consumption, cost and environmental benefits of EVs and internal combustion engine vehicles (ICEVs), some articles conclude that pure electric cars and hybrid vehicles are suitable for short-term and medium-term development, and fuel cell vehicle are suitable for long-term development, while the driven distance greatly affects total CO_2 emissions and user cost [5–7].

However, as research deepens, people gradually realize that EV development involves optimization of the power grid and new energy, where EVs, charging infrastructure, the power grid and renewable power generation must be considered as a whole to achieve global optimization. First of all, in technical and operational aspects, smart charging, vehicle-to-grid (V2G), vehicle-to-home (V2H), battery replacement and many other technologies and operation schemes are proposed. For example, Ref. [4] puts forward an integrated method to study EV business models, and provides three morphological boxes of vehicle and battery, charging infrastructure, and system services, which have different characteristics and possible options and are helpful for the formulation and analysis of alternative applications and business models. Ref. [8] proposes a conceptual regulatory framework for EV charging, introduces EV charging managers and EV integrators, and analyzes the market functions and business relationships to support charging at home, at public charging on the street, and at specific charging station.

With respect to economics, power grid impacts, environmental benefits and other aspects, research more often takes a holistic viewpoint. Ref. [5] uses data from Western Australia to calculate the benefits of V2G in short-term market arbitrage, provision of spinning reserve, load following, and participating in demand-side management, and points out that currently EVs can be used for demand-side management but are not suitable for other services. Ref. [9] proposes a comprehensive method for assessing the impact of EV penetration on investment and network losses of two actual distribution networks, and pointed out that smart charging can reduce new investment by 60 %–70 % and decrease network losses by 40 %. Based on a stochastic model of Western European travel, Ref. [10] studied the influence of plug-in hybrid EVs (PHEVs) on Belgian electricity demand and showed that, although it is necessary to manage the influence through smart charging, time-of-use electricity price would lead to negative peak load. Other literature analyzed the costs and benefits of various methods of EV charging. A case study of fast charging in Germany shows that, a market-driven roll-out of fast charging infrastructure is unlikely to be profitable in Germany, and the EV adoption rate is the main risk factor to determine [11]. A stochastic optimization algorithm is presented in [12] with the aim of maximizing the use of renewable energy, using a Monte Carlo simulation of transportation patterns.

However, although the scientific community has begun to study EVs as a system, there are still many problems with these studies that need to be addressed. For example, a lot of literature takes either predetermined prices or an electricity market as the premise to analyze and optimize the user and grid benefits of EVs. However, what can finally be obtained this way are local economic and environmental benefits, which are different from global benefits and optimization from the point of view of the whole system described above. Studies focused on aspects of charging can also only reflect limited economic and environmental benefits.

In general, being a new form of large complex system, electric vehicle development is facing new historical conditions and challenges. First of all, electric vehicles are not simply a refit and transformation of fuel vehicles, but require revolutionary changes in the life-cycle of vehicles in our society, including production and operation. At the same time, sustainable development has become the theme of present times and requires a unified understanding of economic, environmental and social benefits of electric vehicles. Moreover, the EV production and operating environment involves many participants, with complex relations between their interests, and requires therefore a high level of judgment to be applied to achieve optimization of multiple values.

Therefore, not only in technology, but also in economic and environmental aspects, EV development requires whole-system thinking to guide its analysis and top-level design, so as to formulate a reasonable development route, business model, and policy system. This presents a huge challenge for the development and study of EVs.

2.3 Relationships between the EV system and the power grid

Among all the aspects of EV development, a harmonious relationship with the power grid is of key importance. Wide area controllers are necessary to manage grid-to-vehicle (G2V) or vehicle-to-grid (V2G) power transactions, that is, charging and discharging cycles of the plug-in EVs respectively [13]. In electricity markets, it is economically feasible to use electric vehicles to provide frequency modulation and spinning reserve service most likely through an aggregator to achieve sufficient scale for market participation. Frequency modulation services provide better financial returns than spinning reserve services [14]. Integration of EV charging and discharging facilities with RES gives a helpful synergy in micro grid systems [15].

To facilitate the relationship between EVs and the power grid, we should adhere to the following three principles.

1) The power grid should become a friendly access platform. This is one of the basic aims of power system development, and is accentuated by the current smart grid. The power grid should become a more open, safer and highly efficient energy exchange platform, making use of its huge terminal network and operating system to support EV development.
2) EVs should satisfy the norms and requirements of the power system, and become well-behaved loads. Compared to other loads, EVs have no special requirements for power quality. On the rare occasions when an EV is not available because of a power outage, the user can often choose alternate travel modes or can cancel his or her trip plans; generally, there would be no unavoidable great losses. In contrast, the enormous scale of the power system makes it difficult to estimate and manage the losses caused by breakdowns. Obviously, the safety and reliability of the power system should not be affected by EVs. Additionally, from the perspective of technical considerations, EVs can deal with harmonics and fluctuations that they cause by advanced converters and control technology.
3) Scheduling should be employed to ensure mutual benefit and sustainable development. EVs have excellent flexibility in their charging times; thus, EVs could be included in power system scheduling, lower the peak-valley difference and increase the consumption of new energy to provide additional benefits. As such, they would be useful resources under a competitive business model.

In short, the power system would provide the most essential support for EVs. A harmonious relationship with the power system has the potential to effectively accelerate EV development; otherwise, EV development will be impeded.

However, as some literature has pointed out, there are many preconceptions and barriers that inhibit a larger market penetration of EVs. The establishment of this mutually beneficial relationship is an arduous whole-system project, which involves multiple participants, in addition to complex technical and economic problems. Tremendous effort is required to reach consensus on the way forward, to form corresponding policies and regulations, and to put those policies and regulations into practical operation.

3 Realization of whole-system thinking by development control and the key barriers

3.1 Development control and the realization of whole-system thinking

In 1948, Wiener published Cybernetics: Or the Control and Communication in the Animal and the Machine and established control theory. This theory and the related methods have been widely used in almost all of the natural sciences and social sciences. From a control perspective, the realization of whole-system thinking requires the components, relationships between them and the overall performance of the EV system to achieve some goals, processes or trends, and is, in essence, a kind of development control. While, the formulation and implementation of a development path, business model and policy are essentially some important links and specifically realized means of development control. Therefore, we should take full advantage of system theory, cybernetics and their methods to study development control with whole-system thinking, and propose the basic ideas, implementation methods and key links of it. Then, the rules and methods obtained here will be instructive to the study of development path, business model, incentive policy for EVs.

Figure 3 shows the overall architecture of EV control. Two types of control are involved: the operational control of the physical system and the development control of the EV industry. The operational control (which lies within the dotted rectangle in the figure) comprises many controls of different layers and types, ranging from system to devices. Because numerous studies have explored this type of control, we will not provide further details here. The other type is the development control, and both types are supplementary to each other, constituting the complete control system for EVs.

Figure 3 shows an open loop/closed loop combined development control method using the target system as the core. First, the target system is established according to system requirements. Simultaneously, the system of basic principles, methods and standards is developed according to the technical and economic conditions, so that all of the scientific methods required are integrated and optimized. They are then used to formulate development planning and detailed implementation schemes, which achieve the control of the development process. Meanwhile, the closed loop control can adjust and refine the target system, according to the difference between the actual operational effect and the required effect.

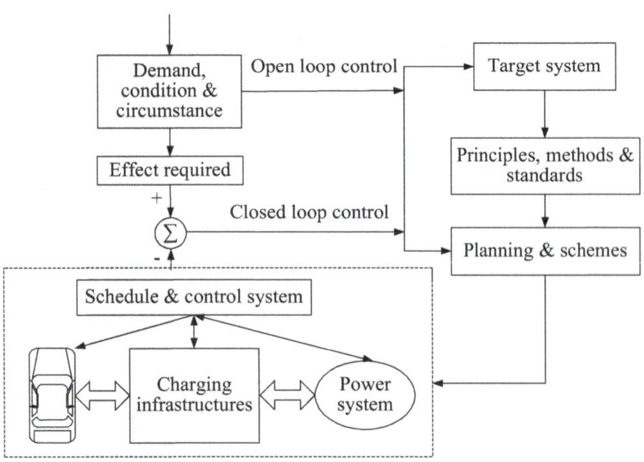

Fig. 3 Overall control architecture of EV development

It is viable to carry out the development control of EVs based on a target system. This control method is generally used in the development of large systems and in controlled economies it has been widely used in areas such as the national economy, industry construction and infrastructure construction. For example, this method is used by the power system to develop a mature control system. In China, there are more than 3000 regulations and standards for the electricity power industry, regulating all types of plans and schemes to realize the development objectives, including safety, stability and economics.

EV development is centered on industry construction and infrastructure construction; therefore, this control method can also be used in EV development. Further, the control method not only can draw on expertise from the aforementioned fields but also can assimilate EVs into the power system, while sharing mutual benefits from the start by integrating the objectives and principles of construction, so as to better use the power system to support and promote EV development.

It is also necessary to achieve effective development control of EVs by target system. First, this is determined by the complexity, uncertainty and partial uncontrollability of the large system. For small systems, it is usually very easy to define the control targets and they tend to be simple and clear; hence, the key challenge in control is the analysis of system characteristics. However, for complex, large systems, global control and optimization are difficult to realize using a limited number of targets and control variables; a target system that is more comprehensive and can efficiently control the key processes and characteristics becomes indispensable. Second, this is also needed for EVs to compete with fuel vehicles. Only if clear and superior targets of each part of the system are proposed and reached can EVs surpass fuel vehicles.

3.2 Requirements for the development control of EVs in competition with fuel vehicles

Generally, it must take a long time for a large, complex system to develop to maturity, and this process is often gradual, with development being a quasi-stable process and having the following characteristics.

1) The development is relatively slow; additionally, change is scarce, and the extent of change is limited during each stage.
2) The process can provide sufficient data and experience, which may facilitate the formulation of the basic models and methods for development analysis.

3) Because the investment is so massive, the construction period is long, and the losses engendered by adjustments for errors are significant and unacceptable. Its development control is primarily an open loop process, together with the closed loop process occasionally operating as a complement; significant alterations need to be avoided during the construction.

However, because of competition with traditional fuel vehicles, the external environment cannot allow EVs to develop in such a slow and gradual manner. Fuel vehicles have already reached maturity as a system, so EVs cannot develop widely without exceeding fuel vehicles in terms of convenience, economy and comprehensive benefits. Therefore, some special requirements are imposed on EV development.

1) EV technology must in a step-wise manner. Some sufficiently high objectives should be set at the beginning, so as to reach to economies of scale in construction as quickly as possible.
2) There are too many feasible development directions: thus, in the initial stage, the exploration and validation of the best direction and mode are more important strategic objectives than actual construction.
3) The development should enhance the usage of closed loop control: because of the absence of data and expertise, the analysis should be combined with closed loop verification to develop and refine various types of models, principles, methods and standards.

The analysis above leads to some basic principles for designing the target system for EV development.

1) The design must follow demand, and take the surmounting of fuel vehicles as the basic standard.
2) The development should be divided into different stages and be carried out step-by-step according to the goals.
3) The target system should cover all components and links shown in Fig. 1.
4) The targets relating to each link should be harmonized for coordinated development.
5) The charging infrastructure should be advanced suitably and should match with the environmental conditions.
6) The target system should be adjusted and refined according to observed changes during EV development.

3.3 Key barriers for EV development

There are more difficulties than envisaged need to be resolved to realize vigorous EV development and replace a notable segment of fuel vehicles. Many barriers to such an epochal transition into EVs are elucidated in recent literature, including the reluctance of existing petroleum firms and uncertain consumer attitudes. The introduction and penetration of EVs is confronted by several barriers such as immaturity of technology, fragmented infrastructure, incomplete standards and regulations, and skepticism of consumers towards an emerging technology [16]. If the subsidies provided by the government do not compensate the negative aspects of EVs' performance, they cannot be attractive and engender adequate incentive [17]. Lack of charging infrastructure and relatively high vehicle prices create challenges to EVs and affect government support policies [18].

To sum up, these difficulties are mainly covered by four aspects:

1) It is difficult to reduce the cost of EVs, because the environmental benefits are not urgent, direct, or easy to be quantified. Consequently, EVs cannot show obvious advantages; their development lacks sufficient impetus; and it is difficult to make decisions regarding EV development.
2) There are numerous viable charging models, operating models and business models, which cause difficulty in selecting a development path.
3) A sizeable and systematic demonstration project does not exist yet. Therefore, there are no fundamental data to sufficiently describe actual situations, so that it is not yet possible to verify EV studies and analyses.
4) The operating and industry chain is very long. There are many participants with complicated relationships, and the coordination among them is very difficult.

In general, there are two key barriers that are critical for the promotion of EV development. In terms of productivity or technology, the key barrier is the research and mass production of power batteries that can offer high energy density, power density and safety. Numerous researchers have conducted extensive studies on this topic.

In terms of production relationships and system construction, the key barrier is top-level design of development and benefit mechanisms, namely the formulation of a reasonable development path, benefit mechanisms, and regulating policies. Positive interactions must be promoted among the government, enterprises and the market. Appropriate coordination is also required among EVs, the charging infrastructure and the power grids, so as to form balanced economic interest chains, with reference to Fig. 2. The basic requirement is to reduce the uncertainty of the participants' profit and ensure a return on investment. Then, the following measures should be taken to promote the rational top-level design of the development and benefit mechanisms.

1) Conduct a thorough, comprehensive and realistic study of the development path for the entire, large EV system.

2) Carry out a global benefit analysis and top-level design to develop suitable mechanisms, improve economic impetus for development, coordinate economic relationships, and promote the formation of viable business models.
3) Carry out comprehensive demonstrations of a variety of models with explicit objectives, plans and sufficient scales, to verify the development direction, and refine the analysis models and methods.
4) Understand the methods and expertise of fuel vehicles and the power system to assist the development of EVs.
5) Enhance the discussion among different sectors to form the consensus for reciprocity, and then turn that consensus into basic regulations, standards and concrete actions.

4 Promoting EVs by macro motivation versus by micro intervention

4.1 Is macro motivation or micro intervention more suitable for current demands of EV?

As many papers discuss, both EV technology and charging infrastructure require huge investments and long periods of construction; while the return is inherently slow, the risk is very high, and substantial innovation is needed in technologies and business models. In the initial stage, the insufficiency of overall profit and the risk of investment will be most severe, so that they will be too difficult for individual participants to undertake. Therefore, it is clear that strong measures implemented using a reasonable promotion mechanism are for the development of EVs.

However, selecting the best promotion mechanism for EV development is a difficult problem for governments, although much literature has been written about it. The morphological method aims to support companies on a strategic level with a sound concept when developing a business model, to refine first ideas and approaches and list alternatives [4]. Two new electricity market agents, the EV charging manager and the EV aggregator, in charge of developing charging infrastructure and providing charging services are introduced [8]. Interaction between government policies and performance attributes of EVs also has a significant effect on the development of the EV market [17]. EV policies focused on early adopters and niche markets would create complementary system effects that will lead to increased EV market penetration and realization of intended societal benefits [18]. An aggregator participating in the reserve market could achieve a substantial reduction of its wholesale costs, even when a capacity price for available reserve is not considered [19].

To establish an effective support policy and promotion mechanism, the most first problem that should be solved is whether macro motivation or micro intervention should be taken. According to the analysis above, in EV development control, it is necessary to explore and verify a variety of directions and modes of development, realize balance of many different links, and make timely adjustment according to real-time changes. For these issues, because developing and changing policies is a long and difficult process, it is difficult to achieve good results using incentive methods based on micro intervention, while using macro regulations to motivate and guide enterprises as they develop liberally can help to reach reasonable benefit mechanisms.

However, in current circumstances, the power industry in many countries has completely or partially regulated utilities, and faces strong technical challenges. In order to realize the liberal development of EV enterprises in coordination with utilities, major changes in policy environment are needed. Moreover, due to the large investment and long cycle of returns in developing electric vehicles and charging infrastructure, promising methods to encourage enterprises to participate in the development of electric vehicles are franchising, build-operate-transfer (BOT) and related policy tools. A scheme is suggested below which combines franchising with moderate competition.

4.2 Promotion mechanism based on franchising

Franchising is a well-known commercial and policy tool. It can lower the uncertainty, ensure the interest of investors by a long-term mechanism, and provide sufficient impetus, so that it can strongly promote major construction projects and scientific and technological innovation. Therefore, franchising is widely practiced in fields such as telecommunications, energy and transportation. Meanwhile, patent rights also serve as a special franchise right in related to underlying technologies.

So, in the initial stage, franchise rights could be used to promote EV development, which would achieve the vigorous development of EVs. Then, in the mature stages, when the operating models and business models have been verified, more competition can be introduced to improve efficiency.

For example, in China, it might be declared that the initial stage of EV development will take 20 years. During this period, at most 6 EV manufacturers and 4 charging infrastructure operators will be able to obtain nationwide business licenses; further, all of these licenses will specify technologies developed in the demonstration projects. As a result, the enterprises that succeed in obtaining licences will be highly motivated, so as to powerfully promote EV development.

In the initial stage of EV development, franchising and demonstration should be combined together to carry out comprehensive demonstrations at scale, so as to overcome the problems of existing demonstration projects that are not comprehensive due to their limited scale. The following is the specific implementation mechanism proposed.

1) Choose several demonstration cities with populations of approximately 200000; at least 20000 EVs should be introduced to the populations in each city in the project period, which should be approximately 4 years.
2) Further, determine the charging model and operating model for each project before its commencement. Then, the charging infrastructure and the smart scheduling system should be constructed thoroughly, and the corresponding pricing policy must be established, to explore and form stable business models.
3) Take the first round of public bidding to determine the EV manufacturer, the charging infrastructure operator, as well as the technical and economic requirements. The manufacturer and operator must form a partnership for bidding, and they must jointly detail the operating methods, electricity pricing and other policies that they will implement. The bidding manufacturer must promise some basic items, such as the vehicle characteristics and the maximum price, the investment that will be made, in addition to appointing the minimal development team and investment, and providing security deposits for realization of these. In return, the government should guarantee the minimum sales of manufactured EVs that meet the quality requirements, and is responsible for policy making, supervision and assessment.
4) For the initial stage, it should be defined that the franchised operation will be the only producer of EVs and charging infrastructure, and all of the manufacturers and operators with successful demonstration projects and sufficient investments are eligible to bid for franchise rights.
5) In the first bidding above, the government obtains the right to permit several enterprises to use the technologies developed in the demonstration projects. Then, the use rights of these technologies can be sold to other proper enterprises via the second round of public bidding, and the income will be used for project construction. Finally, the government and investors may form an investment committee to supervise the project and provide the franchised enterprises with qualified personnel, technologies and other support.

The above mechanism has several advantages. 1) Franchising and guaranteed sales can lower risk and ensure the interests of the manufacturers and operators. 2) By taking full advantage of public bidding, the enterprises will be driven to improve technology and reduce costs, and the best balance can be found. 3) An interest community can be formed through the transfer of the technology use rights, which may solve numerous problems, including funding, supervision, evaluation, and technology sharing. Thus, a large-scale, systematic and comprehensive demonstration can be conducted and refined as shown as in Fig. 3, and EV development can be significantly promoted.

5 Conclusions and research suggestions

This paper investigated some important problems of EV development such as system architecture, whole-system thinking, development control, and promotion mechanism. The major conclusions are listed below.

1) EVs will lead to a complex system with multiple links and multiple coupling chains, which needs optimization in all aspects so as to realize the objective to surpass traditional vehicles. Therefore, whole-system thinking should be enforced in the development and control of EVs.
2) Among all the aspects of EV development, the core issue is the interaction between EVs and the power grid. The power grid should become a friendly access platform; EVs should become well-behaved loads; while coordination should be adopted to realize mutual benefit, so as to take full advantage of power grid to support and promote the development of EVs. However, the realization of these is an arduous system project that needs tremendous effort from each participant.
3) Development path, business model and policy all are important links and specific means of realizing development control. Therefore, we should take full advantage of system theory, cybernetics and their methods to guide study of development control with whole-systematic thinking.
4) A development control method with a target system as the core is presented. Because of the special requirement to compete with traditional vehicles, EVs must develop in a step-wise manner with strong closed-loop control. Moreover, the basic principles for designing the target system are given too.
5) There are two key barriers to the development of EVs. In terms of productivity and technology, it is the research and mass production of high-performance power batteries. In terms of production relationships and system construction, it is the top-level design of development and benefit mechanisms, namely the formulation of a reasonable development path, benefit mechanisms, and policies, where the basic requirement is to reduce the uncertainty and ensure a return on investment.
6) EV support policies and promotion mechanisms should be based on macroscopic incentives to guide

enterprises to develop liberally. Because of the regulation on power industry, appropriate changes in policy environment should been taken, so as to provide a suitable environment for the development of EVs.

7) Franchising was presented as the promotion mechanism most likely to achieve the vigorous development of EVs. Additionally, for the initial stage of development, a comprehensive demonstration mechanism was given that combines franchising with moderate competition.

Regarding the two key barriers above, there are two corresponding important research problems that need to be solved. The first one is battery technology and industry construction, where it is necessary to reduce the cost and enhance the performance of power batteries. New materials and structures are among the most important study directions. Only after the major breakthrough in these aspects, can the vigorous development of EV come, and can the issues of its application be solved finally. The manufacturing methods, energy management system, package technology and re-use or re-purposing of power batteries have important significance, too. Standardization system is also urgently required, so that disorderly development of incompatible products can be avoided.

The other one is the method to carry out reasonable system analysis and top-level design for EV development, so as to establish an excellent development and benefit mechanism. Although we have carried out a large number of studies and demonstrations on best development path and policy for EVs, the pace of industry development is not sufficient yet. This shows that there are still some blind areas in existing system analysis and top-level design of EVs, so more effective theories and methods are needed, and this is also a key research problem. Meanwhile, the combination of whole-system thinking and development control may have the potential to provide new ideas and tools for research in this area, and so is worthy of high attention.

Acknowledgments This work was partially supported National High Technology R&D Program of China (863 Program) (No. 2012AA050804).

References

[1] Garcia-Valle R, Peças Lopes JA (2013) Electric vehicle integration into modern power networks. Springer, New York, NY

[2] Ortega-Vazquez MA, Bouffard F, Silva V (2013) Electric vehicle aggregator/system operator coordination for charging scheduling and services procurement. IEEE Trans Power Syst 28(2):1806–1815

[3] McArthur SDJ, Davidson EM, Catterson VM et al (2007) Multi-agent systems for power engineering applications–Part I: concepts, approaches, and technical challenges. IEEE Trans Power Syst 22(4):1743–1752

[4] Kley F, Lerch C, Dallinger D (2011) New business models for electric cars—a holistic approach. Energ Policy 39(6):3392–3403

[5] Mullan J, Harries D, Bräunl T et al (2012) The technical, economic and commercial viability of the vehicle-to-grid concept. Energ Policy 48:394–406

[6] Faria R, Moura P, Delgado J et al (2012) A sustainability assessment of electric vehicles as a personal mobility system. Energ Conver Manag 61:19–30

[7] Silva C, Ross M, Farias T (2009) Evaluation of energy consumption, emissions and cost of plug-in hybrid vehicles. Energ Conver Manag 50(7):1635–1643

[8] San Román TG, Momber I, Rivier M et al (2011) Regulatory framework and business models for charging plug-in electric vehicles: infrastructure, agents, and commercial relationships. Energ Policy 39(10):6360–6375

[9] Pieltain Fernández L, San Román TG, Cossent R et al (2011) Assessment of the impact of plug-in electric vehicles on distribution networks. IEEE Trans Power Syst 26(1):206–213

[10] Geth F, Willekens K, Clement K, et al. (2010) Impact-analysis of the charging of plug-in hybrid vehicles on the production park in Belgium. In: Proceedings of the 15th IEEE mediterranean electrotechnical conference (MELECON'10), Valletta, Malta, 25–28 Apr 2010, pp 425–430

[11] Schroeder A, Traber T (2012) The economics of fast charging infrastructure for electric vehicles. Energ Policy 43:136–144

[12] Pantoš M (2011) Stochastic optimal charging of electric-drive vehicles with renewable energy. Energy 36(11):6567–6576

[13] Mitra P, Venayagamoorthy GK (2010) Wide area control for improving stability of a power system with plug-in electric vehicles. IET Gener Transm Distrib 4(10):1151–1163

[14] Rong LL, Wen FS, Xue YS (2013) Economic analysis of ancillary service provision by plug-in electric vehicles. Autom Electr Power Syst 37(14):43–49 (in Chinese)

[15] Xiao XN, Chen Z, Liu N (2013) Integrated mode and key issues of renewable energy sources and electric vehicles' charging and discharging facilities in microgrid. Trans China Electrotech Soc 28(2):1–14

[16] Steinhilber S, Wells P, Thankappan S (2012) Socio-technical inertia: understanding the barriers to electric vehicles. Energ Policy 60:531–539

[17] Zhang X, Wang K, Hao Y et al (2013) The impact of government policy on preference for NEVs: the evidence from China. Energ Policy 61:382–393

[18] Green EH, Skerlos SJ, Winebrake JJ (2014) Increasing electric vehicle policy efficiency and effectiveness by reducing mainstream market bias. Energ Policy 65:562–566

[19] Bessa RJ, Matos MA (2013) Optimization models for EV aggregator participation in a manual reserve market. IEEE Trans Power Syst 28(3):3085–3095

Chunlin GUO was born in Hunan Province in China, on February 4, 1975. He received the B.Sc. degree and Ph.D. degree in electrical engineering from Tsinghua University, Beijing, China, in 1998 and 2003, respectively. Now, he is an associate professor of the School of Electrical and Electronic Engineering, North China Electric Power University, Beijing, China. His research interests include electric vehicle charging and its interaction with grid, analysis and control of FACTS and power quality control.

Ching Chuen CHAN has over 10 years of industrial experience and over 35 years of academic experience. He is currently the honorary professor and the former head of department of Electrical and Electronic Engineering, the University of Hong Kong, Hong Kong, China. He is the founding president of the International Academy for Advanced Study, China, the cofounder and the rotating president of the World Electric Vehicle Association, and the president of the Electric Vehicles Association of the Asia Pacific. Dr. Chan is a fellow of IEEE/IEE, and an academician of the Royal Academy of Engineering, U.K., the Chinese Academy of Engineering, the Ukraine Academy of Engineering Sciences, the Institution of Electrical Engineers (IEE), and the Hong Kong Institution of Engineers. He has delivered lectures on electric vehicles worldwide. He was named Father of Asian Electric Vehicles by the magazine Global View in 2002. He was also named BPitamaha (Grandfather) of Electric Vehicle Technology in India at the IEEE Conference on Electric and Hybrid Vehicles in 2006. He received the IEE International Lecture Medal in 2000.

Uncertainty-averse TRANSCO planning for accommodating renewable energy in CO_2 reduction environment

Chunyu ZHANG, Yi DING (✉), Qi WANG, Yusheng XUE, Jacob ØSTERGAARD

Abstract The concern of the environment and energy sustainability requests a crucial target of CO_2 abatement and results in a relatively high penetration of renewable energy generation in the transmission system. For maintaining system reliability and security, the transmission company (TRANSCO) has to make strategic planning to handle the uncertainty challenges from the intermittent renewable energy resources. In this paper, a stochastic multi-period multi-objective transmission planning (MPMOTP) model is proposed to reduce correlated uncertainties from renewable energy generation, conventional generation, demand-side variations, market price volatility, and transmission configuration. Three objectives, i.e. social CO_2 reduction benefit, energy purchase and network expansion cost and power delivery profit, are optimized simultaneously by a developed two-phase multi-objective particle swarm optimization (MOPSO) method. The feasibility and effectiveness of the proposed uncertainty-averse MPMOTP model have been verified by the IEEE 24-bus test system.

Keywords CO_2 reduction, Renewable energy, Uncertainty, Multi-objective planning, TRANSCO, MOPSO

C. ZHANG, Y. DING, Q. WANG, J. ØSTERGAARD,
Center for Electric Power and Energy, Technical University of Denmark, 2800 Kgs. Lyngby, Denmark
(✉) e-mail: yding@elektro.dtu.dk
Y. XUE, State Grid Electric Power Research Institute (SGEPRI), Nanjing 210003, China

1 Introduction

The theme of energy sustainable development and conservations is widely recognized around the world, while the electric power industry is regarded as the major CO_2 emission sector with the traditional fossil-dependent production. In the deregulated environment, CO_2 reduction has already become the most significant concern involved in the decision making process within the multi-layer architecture of electricity generation, trading, transmission and distribution, even in the retail aspect, respectively dominated by generation companies (GENCOs), market operator (MO), transmission companies (TRANSCOs), distribution companies (DISCOs) and retailers.

From the last decade, numerous literatures and projects have already been carried out to demonstrate the feasible solutions to reduce CO_2 emission, which can be categorized into three schemes, i.e. technological CO_2 abatement, market-oriented based CO_2 trading and alternative energy production. As firstly discussed in [1] and further investigated in [2], on perspective of GENCOs and power system operation, CO_2 capture and storage (CCS) is addressed as a most promising technology for CO_2 reduction. On basis of the low-carbon economy analysis in [3], for the conscious of the economic factors in day-ahead energy market and cap-and-trade carbon emission market, a flexible operation model is proposed in [4] to trade off the maximum profit with adaptive carbon emission for a generating unit combining with proper coordination of generation schedule, CCS schedule and market bidding strategies. Alternative energy production could be highly efficient energy resources or more environmental energy conversions, in which renewable energy is leading this role for CO_2 mitigation from the current to the future, a huge amount of wind and photovoltaic (PV) energy will widespread in the transmission system to replace the conventional generation, as reviewed in [5].

However, the intermittent characteristics of the diverse renewable energies will derive huge uncertainties to the transmission system operation. Therefore, a strategic uncertainty-averse TRANSCO planning is necessary to ensure the adequacy and reliability of the power system operation for accommodating renewable energy in CO_2 reduction environment.

Various approaches for handling uncertainties in transmission planning process have been summarized in [6]. In [7], the market uncertainties are identified as possible future scenarios solved by the mixed integer linear programming (MILP) on a flexible transmission planning model, and assessed with reliability and security criterions referring to the indices of expected energy not supplied (EENS), expected cost of interruptions (ECOST), and interrupted energy assessment rate (IEAR) [8]. Subsequently, a stochastic MILP approach complemented with risk aversion was proposed in [9]. Furthermore, to assess the robustness of expansion plans, a Brazil test grid is presented [10] to illustrate the stochastic approach is more suitable than the traditional deterministic method. Additionally, a stochastic multi-objective optimization framework is proposed in [11] to take security constrains into concern for transmission planning. In the other perspectives, a congestion surplus [12] is identified in a multi-objective transmission planning (MOTP) model for dealing with the risk of the network congestions. Recently, for integrating the large-scale wind power, a probabilistic MOTP model equipped with risk-control strategies is developed by [13] to avoid transmission overloading. Extensive uncertainty studies are allocated in [14] and [15] to conduct the distributed energy resources (DERs) impacts on distribution systems.

Till now, the uncertainties faced by the TRANSCO are normally considered as independent factors, however, immersed in the future smart grid surroundings, as a profit chasing commercial player, an extra pressure for the TRANSCO is to fully understand the high correlations among these uncertainty diversities hidden in the transmission planning process to flexibly balance the intermittent recourses and stochastic consumptions.

In this paper, the wind power and PV energy are selected to represent the renewable energy caused uncertainties. The effort of demand response (DR) is concerned in the demand side to intensify the consumption behavior. To assess the circumvent uncertainties in terms of the output of wind and PV generating units, demand response related load fluctuation, conventional generation units, market price volatility, and transmission path deployment, the correlation coefficient matrix is introduced to handle the dependency of the uncertainties in the paper. A stochastic MPMOTP model is proposed for this uncertainty-averse TRANSCO planning, incorporated with the following objectives: 1) maximize the social benefit via CO_2 reduction, 2) minimize the TRANSCO cost of energy purchase and network expansion, 3) maximize the profit of power delivery. A two-phase MOPSO schema is employed to be the solver. The application of the proposed MPMOTP model is demonstrated on the IEEE 24-bus testing system to show its feasibility.

2 Modeling of uncertainties

Aiming to fulfill the critical target of CO_2 reduction, in energy purchase and transmission processes, the TRANSCO has the natural attribute to take reactions to hedge the uncertainties associated with the government policy and the huge amount of renewable energy integration. The proper planning strategy can help the TRANSCO to maintain the power systems operated in an economic efficient and secure condition, constrained by various new uncertainty boundaries. In this paper, various uncertainties are taken into concerns, i.e. renewable energy generation (wind and PV), demand-side variations, conventional units' production, market factor, and transmission path configuration. Precisely, the correlation characteristics of these uncertainties are taken into account and formulated as follows.

2.1 Correlation of uncertainties

The indication of stochastic variables for wind, PV, facilities configuration and demand identification can be seen as uncertain factors according to the forecast deviations, the measurement errors, the unpredictable system contingencies or the electricity market price volatilities. Therefore, the probabilistic analysis is the proper approach to handle the variations of these uncertainties. However, the uncertainty variables could be dependent to each other, e.g. the weather conditions can impact wind/PV generation and household consumption, simultaneously. Hence, in this work, the correlation coefficient matrix is used to illustrate the considerable dependency of the uncertainty variables. To define the degree of the dependence among variables, each correlation coefficient should be assigned in $[-1, 1]$, where -1 and 1 indicate the perfect positive and negative relationship between the related variables.

As described in [16], the correlation coefficient $\varphi_{X,Y}$ between two uncertainty variables X and Y can be expressed as

$$\varphi_{X,Y} = \frac{E[(X-\mu_X)(Y-\mu_Y)]}{\sigma_X \sigma_Y} \tag{1}$$

where E represents the expected value operator; (μ_X, μ_Y) and (σ_X, σ_Y) are the expected value and standard deviations of X and Y.

Once the probability density function (PDF) of each uncertainty variable is addressed, the correlation coefficient matrix $\boldsymbol{C}$ can be obtained,

$$\boldsymbol{C} = \begin{bmatrix} \varphi_{11} & \varphi_{12} & \dots & \varphi_{1n} \\ \varphi_{21} & \varphi_{22} & \dots & \varphi_{2n} \\ \vdots & \vdots & & \vdots \\ \varphi_{n1} & \varphi_{n2} & \cdots & \varphi_{nn} \end{bmatrix} \tag{2}$$

correspondingly, the refined uncertainty variable can be re-denoted as $\boldsymbol{Z}$ with the expression

$$\boldsymbol{Z} = \boldsymbol{H} + \boldsymbol{C}\widetilde{\boldsymbol{X}} \tag{3}$$

where $\widetilde{\boldsymbol{X}}$ is the forecasted values of uncertainty variables, such as wind power or PV output; $\boldsymbol{H}$ represents the initial value of these variables, which can be achieved from the historical data.

2.2 Wind energy uncertainty

The wind energy is converted from wind speed and behaved with the fluctuating and stochastic characteristics. As discussed in [17], the wind speed is normally followed a Weibull distribution, the PDF function is adopted to the wind variations, following

$$f(v) = \frac{k}{c}\left(\frac{v}{c}\right)^{k-1} \mathrm{e}^{-\left(\frac{v}{c}\right)^{k}} \tag{4}$$

where v, k and c are respectively represent the wind speed, shape factor and scale factor, $k > 0$, $v > 0$ and $c > 1$.

Then, the wind power output can be formulated as

$$P_w = \begin{cases} 0 & 0 \le v < V_{ci} \\ \overline{P_w}(A + Bv + Cv^2) & V_{ci} \le v < V_r \\ \overline{P_w} & V_r \le v < V_{co} \\ 0 & v \ge V_{co} \end{cases} \tag{5}$$

where the parameters of A, B and C can be defined in [17]; V_{ci}, V_r and V_{co} are the cut-in wind speed, rated wind speed and cut-out wind speed, respectively; $\overline{P_w}$ is the rated power of a wind unit.

2.3 PV energy uncertainty

The production of PV unit is dominated by the illumination intensity, a number of investigations have shown the PDF of solar radiation is following the Beta distribution [18],

$$f(R) = \frac{\Gamma(\alpha+\beta)}{\Gamma(\alpha)\Gamma(\beta)}\left(\frac{R}{R^{\max}}\right)^{\alpha-1}\left(\frac{R}{R^{\max}}\right)^{\beta-1} \tag{6}$$

where Γ is the Gamma function; α and β are the shape parameters; R is the illumination intensity with the maximum value of $R^{\max}$.

The relationship between the illumination intensity and the output of a PV unit can be described as [18]

$$P_s = \begin{cases} \overline{P_s}\left(\frac{R}{R_r}\right) & 0 \le R \le R_r \\ \overline{P_s} & R > R_r \end{cases} \tag{7}$$

where R is the illumination intensity with the rated value R_r and $\overline{P_s}$ is the rated output power of the solar unit.

2.4 Conventional energy uncertainty

For the conventional generation unit, considering the capacity expansion or withdrawal, the unit output can be defined as the chance constrained probability distribution,

$$\Pr\left(P_{gi} - \overline{P_{gi}} \le 0\right) \ge \gamma_i \tag{8}$$

where P_{gi} is the possible value of conventional generation at i^{th} bus with the maximal capacity limitation $\overline{P_{gi}}$ and quantified by a specified probability γ_i.

2.5 Market price uncertainty

Since the electricity price could vary in multiple planning periods, the forecasted locational marginal pricing (LMP) of ith bus can be adopted to meet the predicted demand, and denoted as λ_i. The LMP is assumed to follow the discrete probability distribution,

$$\Pr(C_{\text{LMP}} = \lambda_i) = \varepsilon\zeta_i \tag{9}$$

where C_{LMP} is the value of LMP; ζ_i is the occurrence probability of λ_i and ε is a random variable represents the possible volatility of LMP, while ε is within the boundary of [0.9, 1.1], $\sum\zeta_i = 1$, $i = (1, 2, \ldots, N_{\text{LMP}})$.

2.6 Demand-side uncertainty

The fast growth of smart grid and intelligent control technology will offer a good opportunity to apply DR and flexible consumption in the demand side. Therefore, the uncertainty can be decomposed to be two portions, i.e. DR and load variation.

In this work, the active DR is assumed to be with the price-driven scheme. A Gaussian distribution is applied to describe the effect of the elastic demand, which is bounded by the maximum and minimum credible values [19],

$$f(\varsigma) = \begin{cases} 0 & \varsigma < \varsigma^{\min} \\ \frac{1}{\sqrt{2\pi}\sigma_d}\mathrm{e}^{-\left[(\varsigma-\mu_d)/\sqrt{2}\sigma_d\right]^2} & \varsigma^{\min} \le \varsigma \le \varsigma^{\max} \\ 0 & \varsigma > \varsigma^{\max} \end{cases} \tag{10}$$

where σ_d and μ_d represent the standard deviation and mean value of the demand elasticity ς, which responds to λ_i, and the volume of DR can be formulated as

$$P_{ei} = \overline{P_{ei}} \frac{\varsigma\left(\lambda_i - \overline{\lambda_i}\right)}{\overline{\lambda_i}} \tag{11}$$

where $\overline{P_{ei}}$ is the settled initial value of the elastic demand and $\overline{\lambda_i}$ is the fixed price on basis of the historical archive.

In addition, according to the predicted demand $\overline{P_{li}}$, the load variation $\widetilde{P_{li}}$ at i^{th} bus is imposed to follow the normal distribution $N\left(\mu_i, \sigma_i^2\right)$ for exhibiting the uncertainty of the natural load growth. Here, μ_i is the expectation of the forecast load and σ_i is the standard deviation.

Eventually, the assembled demand yielded in this paper can be given by

$$P_{di} = \overline{P_{li}} + \widetilde{P_{li}} - P_{ei} \tag{12}$$

2.7 Transmission line uncertainty

Typically, the availability of the existing and candidate transmission lines can utilize the (0–1) distribution to represent the line uncertainty, where 0 indicates the line is in failure (or maintenance) status, while 1 shows the line is in the normal operating state.

3 Uncertainty-averse TRANSCO planning

3.1 Uncertainty characterization

For properly addressing the uncertainties mentioned above in the TRANSCO planning progress, the scenario-based stochastic programming approach is employed here to handle the issued uncertain conditions. A scenario is a sequence of time-based transmission system state, consisted of renewable energy, conventional generation, active demand, electricity price and transmission network. In this paper, the Monte Carlo simulation (MCs) method is applied to generate the set of numerous transmission system scenarios.

The repetitive process of the MCs method is built on the random sampling and statistical analysis. Generally, identifying the PDF of each uncertainty variable (as explained in Section 2) is the initial part, then further step is to attain some random samples via the random number generator (RNG). Consequently, the output values of these variables can be calculated in a deterministic model. For shortening the time consuming computation, a well-known scenario reduction technique [20] is introduced to eliminate the non-essential scenarios.

3.2 TRANSCO MPMOTP model

The TRANSCO considered in this model can acquire electricity from differing types of energy resources, including wind units, PV units and fossil-oriented conventional generation units. Three objectives are emphasized in this proposed MPMOTP model. The maximum benefit of the social CO_2 reduction is pursued as the first objective, shown as O_{cr} in (13). In order to minimize the operation costs, the TRANSCO has to make a decision on composing the energy volumes purchased from various resources, as well as the network investment for suiting the liability of power transmission. This goal is treated as the second objective and denoted as O_{pp} in (14). The third objective is to maximize the profit from the demand-side consumption, which is consisted of the positive electricity selling revenue and the negative load shedding penalty, formulated as O_{rd} in (15).

minimize

$$O_{cr} = \rho^{CO_2} \sum_{\omega} \sum_{t} \left[\sum_{g} P_{gt\omega} \xi_g - \left(\sum_{w} P_{wt\omega} + \sum_{s} P_{st\omega} \right) \xi_{se}^{Av} \right] \eta_\omega \tag{13}$$

$$O_{pp} = \sum_{\omega} \sum_{t} \left[\sum_{w} \lambda_{wt\omega}^{Wind} P_{wt\omega} + \sum_{s} \lambda_{st\omega}^{PV} P_{st\omega} + \sum_{g} \lambda_{gt\omega}^{CG} P_{gt\omega} + \sum_{\ell} \lambda_{\ell t\omega}^{Line} P_{\ell t\omega} \right] \eta_\omega \tag{14}$$

$$O_{rd} = \sum_{\omega} \sum_{t} \sum_{i} \left(\lambda_{it\omega}^{LS} P_{dit\omega}^{LS} - \lambda_{it\omega} P_{dit\omega} \right) \eta_\omega \tag{15}$$

s.t.

$$\sum_{w \in i} P_{wt\omega} + \sum_{s \in i} P_{st\omega} + \sum_{g \in i} P_{gt\omega} - P_{dit\omega} = \sum_{ij, j \neq i} P_{ijt\omega}, \quad \forall i, t, \omega \tag{16}$$

$$P_{ijt\omega} = \frac{\theta_{it\omega} - \theta_{jt\omega}}{x_{ij}}, \quad \forall i, j, t, \omega \tag{17}$$

$$-P_{ijt\omega}^{\min} \leq P_{ijt\omega} \leq P_{ijt\omega}^{\max}, \quad \forall i, j, t, \omega \tag{18}$$

$$P_{wt\omega}^{Wind,\min} \leq P_{wt\omega} \leq P_{wt\omega}^{Wind,\max}, \quad \forall w, t, \omega \tag{19}$$

$$P_{st\omega}^{PV,\min} \leq P_{st\omega} \leq P_{st\omega}^{PV,\max}, \quad \forall s, t, \omega \tag{20}$$

$$P_{gt\omega}^{CG,\min} \leq P_{gt\omega} \leq P_{gt\omega}^{CG,\max}, \quad \forall g, t, \omega \tag{21}$$

$$0 \leq P_{dit\omega} \leq P_{dit\omega}^{\max}, \quad \forall i, t, \omega \tag{22}$$

$$0 \leq P_{dit\omega}^{LS} \leq P_{dit\omega}, \quad \forall i, t, \omega \tag{23}$$

$$-\pi \leq \theta_{it\omega} \leq \pi, \quad \forall i, t, \omega \tag{24}$$

where t is the indices of time periods within the set T; ω is the index of possible scenarios set Ω generated by MCs; η_ω is the occurrence probability of the scenario ω; i, j, w, s, g are the index or indices of the mapping of the bus set B with the wind unit set W, the PV unit set S and the conventional power unit set G, i.e. $i, j \in B$, $w \in W$, $s \in S$, $g \in G$, where $\{W, S, G\} \subseteq B$; $P_{wt\omega}$, $P_{st\omega}$, and $P_{gt\omega}$ are the purchased power from wind units, PV units and conventional generation units corresponding to the individual marginal price of $\lambda_{wt\omega}^{Wind}$, $\lambda_{st\omega}^{PV}$ and $\lambda_{gt\omega}^{CG}$; ξ_g is the carbon intensity of each type of the fossil-based generation unit, while ξ_{se}^{Av} is the average value of the whole society involved carbon emission intensity; ρ^{CO_2} is the carbon emission tariff; $P_{\ell t\omega}$ is the nominated expansion capacity of the transmission lines with the marginal price $\lambda_{\ell t\omega}^{Line}$ for each possible scenario, and the term $\sum_l \lambda_{\ell t\omega}^{Line} P_{\ell t\omega}$ represents the TRANSCO investment cost, correspondingly; $P_{dit\omega}$ is the load at bus i, while $P_{dit\omega}^{LS}$ is the potential load shedding with the penalty price $\lambda_{it\omega}^{LS}$; $P_{ijt\omega}$ is the power flow through the line i-j; $\theta_{it\omega}$ is the voltage phase angle at i^{th} bus.

In this MPMOTP model, the DC optimal power flow (OPF) is used for the specific intention on various sources of uncertainty in the transmission level, equations (16) guarantee the power balance at each bus. Equations (17) enforce the power flowing through the line i-j, and further impose the capacity limits in (18). Constraints (19), (20) and (21) limit the production of the wind power, PV power and conventional generation unit within the particular maximum and minimum values, respectively. Likewise, the constraints (22) ensure the demand of each bus is bounded in the individual upper limit. Constraints (23) imply the capacity of the possible load shedding at i^{th} bus is limited to the actual demand $P_{dit\omega}$. Constraints (24) set the voltage angle bounds for each bus.

4 Methodology

In this section, the well-developed two-phase MOPSO algorithm is introduced to properly handle the proposed MPMOTP model, since it is a non-convex nonlinear mixed integer problem associated with the uncertainties' penetration.

4.1 PSO algorithm

In general, the particle swarm optimization (PSO) algorithm [21] is a population-based self-adaptive method sorted as one of the heuristic methods. Incorporating with the components of particle and swarm, PSO encourages the local and global exploration of the problem space to obtain better convergence, in which a particle denotes the potential optimal solution and a swarm contains a set of particles. Each particle moves towards a multiple dimensions space to seek a possible solution experienced by the decisions of itself and its neighbors. In the searching space, the searching route of a particle can be recognized as the velocity (m) and position (n). The updating rule of PSO will steer the particle swarm to gather in a more promising area with better objective value.

4.2 Two-phase MOPSO schema

The primary aim of MOPSO is to find an optimal tradeoff between several competing objectives for which usually no single optimal solution exists that minimizes all objective function values simultaneously.

To illustrate the MOPSO algorithm served for the proposed TRANSCO MPMOTP model, in each scenario ω of a specified time period t, the complex cumulative model can be solved as an independent MOTP problem. The pseudo-codes of the MOPSO calculation procedures are shown in Fig. 1, in which archive set A is the vital feature for storing a better Pareto front approaching the optimal solution and hanging out the particles with the best global positions.

Further developed in [22], a two-phase MOPSO schema is proposed in the optimization process which can notably balance the convergence speed and solution diversity. In the first phase, the Sigma method [21] is dedicated to accelerate the convergence and obtain an approximated Pareto front, then an ideal optimal particle method [22] is contributed to facilitate the diversity of the solution in the second phase. The compiling keynotes of the two-phase MOPSO method are specifically indicated in Fig. 2.

4.3 MPMOTP planning implementation

In order to indicate the application of the two-phase MOPSO-based programming for solving the proposed MPMOTP model, the illustrative flow chart is shown in Fig. 3. Due to the MOPSO algorithm is essentially a non-constrained heuristic method, for evading the constraint violations, the possible calculation risks coupling with the constraints (16)–(24) are handled by a traditional scheme [22]. In addition, a mutation operator is performed for keeping the efficiency of the Pareto front. Considering a bunch of solutions obtained from the multi-objective problem, but none of them occupy a priority to the others, therefore, a fuzzy decision making approach [23] is taken to select the final solution.

5 Case study

The proposed uncertainty-averse MPMOTP model is applied to the IEEE 24-bus test system for a long term planning of continuous 15 years with the interval of 5 years, i.e. three periods. The initial network data can be found in [24]. 41 candidate lines are occupied for this

P1: Initialize the swarm ψ (size $\hbar$) in searching space, and the PSO coefficients ϖ, c_1 and c_2 [18], maximum iterations Υ, and iteration τ=0. Set the position and velocity of each particle randomly in ψ;

P2: Decode the particles to obtain the *networks* and calculate the objective functions (13a)-(13c);

P3: Find the initial non-dominated solutions and store them in archive A;

P4: For pth particle in τ=1,2,…,$\hbar$

P4.1: Select the global best position p^{τ}_{gp} from the archive A refer to the solution's diversity;

P4.2: Update position and velocity of each particle according to [21]

$$m^{\tau+1}_{pq} = \varpi \cdot m^{\tau}_{pq} + c_1 \cdot r_1 \cdot (p^{\tau}_{pq} - n^{\tau}_{pq}) + c_2 \cdot r_2 \cdot (p^{\tau}_{gp} - n^{\tau}_{pq})$$
$$n^{\tau+1}_{pq} = n^{\tau}_{pq} + m^{\tau}_{pq}$$

P4.3: Decode the particles to get the network structure and calculate the objective functions (13a)-(13c);

P4.4: Retain the particles within the search space in case they go beyond their boundaries. *If the particle position moves beyond its boundaries, the value of its corresponding boundary will be taken*;

P4.5: Update the local best position p^{τ}_{pq}, *if the current location is dominated by its local best location, then the previous location is maintained, otherwise, the current location is set as the personal best location. If the particles are mutually non-dominated, one particle is selected randomly*;

P4.6: Update the archive A to store the non-dominated solutions from a combined population of the swarm and the archive.

P4.7: Iteration τ+1.

P5: Archive A contains optimal solutions for this MOTP problem.

Fig. 1 Pseudo-codes of the MOPSO calculation procedures for an independent MOTP

K1: Apply the Sigma method to facilitate the convergence and obtain a rough Pareto front;

K2: Update a set of potential global best positions in each iteration τ using the available non-dominated solutions in archive A;

K3: Gain the corresponding particle p^{τ}_{pq}. If f^{*}_{z} is the previous optimal value of the zth objective;

K4: Promote the diversity of the solution with ideal optimal particle method. For a multi-objective optimization problem, the ideal optimal particle $p^{\tau,*}_{pq}$ is decided by

$$\min \phi(f(p^{\tau,*}_{pq})) = \sqrt{\sum_{z=1}^{Z} (f_z(p^{\tau,*}_{pq}) - f^{*}_{z})^2}$$

K5: Output the global best position $p^{\tau,*}_{gq}$ (the optimal value for each objective) according to the ideal optimal particle $p^{\tau,*}_{pq}$.

Fig. 2 Keynotes of the two-phase MOPSO method

system, and the new added lines remain the same parameters as described in [24]. The active power limit of each transmission line is set to 170 MW. For each bus, the expected natural growth rate of the connected generator or load is assumed to be 25% per 5-year. The parameters of

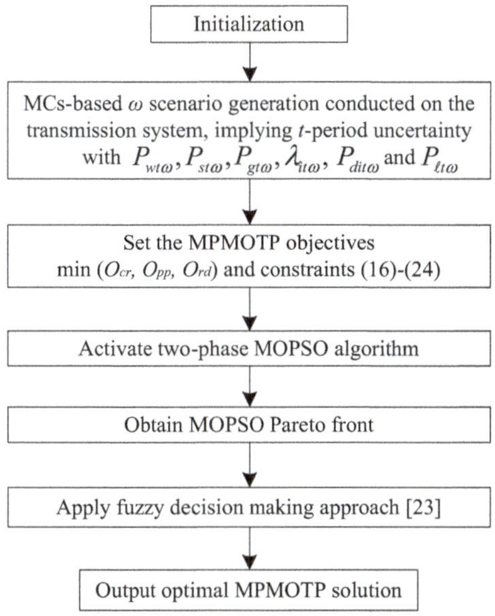

Fig. 3 Flowchart of the proposed MPMOTP model

Table 1 The parameters of the two-phase MOPSO

MOPSO	Parameters
Swarm size	$\hbar = 160$
Coefficient	$\varpi = 0.9$, $c_1 = 2.2$, $c_2 = 3.0$, $r_1 = 0.6$, $r_2 = 0.8$
Iteration	$\Upsilon = 100$

the two-phase MOPSO are indicated in Table 1, the physical parameters of wind and PV unit are summarized in Table 2, and the economic assumptions of carbon emission tariff and energy purchase price are described in Table 3.

Referring to [16], the clustered inter-dependency of the uncertainties $P_{wt\omega}$, $P_{st\omega}$, $P_{gt\omega}$, $\lambda_{it\omega}$, $P_{dit\omega}$ and $P_{\ell t\omega}$ is reflected in the coefficient matrix $\boldsymbol{C}$, and imposed as the same formula for simplicity at each bus,

$$\begin{matrix} P_{wt\omega} \\ P_{st\omega} \\ P_{gt\omega} \\ \lambda_{it\omega} \\ P_{dit\omega} \\ P_{\ell t\omega} \end{matrix} \begin{bmatrix} 1 & 0.021 & 0.011 & 0.278 & -0.292 & 0.048 \\ 0.021 & 1 & 0.019 & 0.386 & -0.352 & 0.027 \\ 0.011 & 0.019 & 1 & 0.576 & -0.125 & 0.133 \\ 0.278 & 0.386 & 0.576 & 1 & -0.863 & 0.001 \\ -0.292 & -0.352 & -0.125 & -0.863 & 1 & 0.896 \\ 0.048 & 0.027 & 0.133 & 0.001 & 0.896 & 1 \end{bmatrix}$$

To illustrate the effectiveness and adaptability of this proposed TRANSCO planning model, differing conditions should be compiled into the test system, therefore, three case studies are concluded in this section. Case 1 is assumed that no renewable energy unit is explored in this system, only conventional units are devoted into expansion process. In Case 2, for simulating the impacts of the wind and PV units, a 240 MW wind farm and a 160 MW PV

Table 2 The physical parameters of wind and PV unit

Type	Parameters
Wind unit	$V_{ci} = 4$ m/s, $V_r = 15$ m/s, $V_{co} = 22$ m/s
	$k = 2$, $c = 5.5$
PV unit	$R_r = 1$ kW/m^2
	$\alpha = 1.8$, $\beta = 4.5$

Table 3 The economic parameters of energy and CO_2 emission

Factors	Parameters
CO_2 emission	$\xi_g = 0.85$ Ton/MW, $\xi_{se}^{Av} = 0.38$ Ton/MW, $\rho^{CO_2} = 20$ \$/Ton
Load shedding	$\lambda_{it\omega}^{LS} = 900$ \$/MW
Energy purchase	$\lambda_{wt\omega}^{Wind} = 150$ \$/MW, $\lambda_{st\omega}^{PV} = 200$ \$/MW, $\lambda_{gt\omega}^{CG} = 350$ \$/MW, $\lambda_{lt\omega}^{Line} = 450$ \$/MW

plant are considered to connect with bus 3 and 7 and increased by 30% per period, respectively. Emphasis on performing the critical penetration of wind and PV energy, in Case 3, two-times capacity is assumed to inject into the same buses with 60% growth rate for each period. The simulation results are conducted in Table 4, in which C_{inv} represents the investment cost of network expansion.

The planning schemes of the Case 1 concentrate on covering the uncertainty of the conventional generation growth to meet the demand variety, the significant investment of network persists in every planning period. The high fossil reliance inspires a high CO_2 charge versus the social expect of reducing the CO_2 emission, also causes an incremented price burden to the consumer (e.g. 2672.37 k\$ in the 3rd period), while the net income of the TRANSCO is increasing slowly from 60.15 k\$ (1st period) to 185.26 k\$ (3rd period).

Regarding the same situation for Case 2 and 3, on preliminary planning stages, the striking point is that a higher expenditure on energy purchase and branch update is visible to enhance the network more tightly to mitigate the plenty uncertainties mentioned in Section 2. However, according to the remarkable amount of wind and PV energy integration, the goal of social CO_2 elimination is achieved. Highlighted in the 3rd period of Case 3, the embedding capacity of renewable energy has touched upon 2048.00 MW, approximately dominating half of the energy supply (4334.49 MW). Comparing with Case 1, the decrement of CO_2 reduction relieves a notable social benefit (15.53 k\$), it can be also observed that, starting from the 2nd period, the TRANSCO cost of energy stocking and network reinforcement is fairly decreased in Case 2 and 3. Further observation is that, not only as an uncertainty bearer, but as a beneficiary, accommodating huge quantity of renewable energy can facilitate the TRANSCO to stabilize the investment expectations and hedge the business risks, e.g. in the 3rd period of Case 3, the net profit is growing dramatically to 737.69 k\$.

Moreover, results from the deterministic investment minimization transmission planning (IMTP) model without considering uncertainties are presented in Table 5. The single objective is to minimize C_{inv}, while the load, wind and PV output, conventional generation, as well as the LMPs are assumed to be fixed according to the predicted value. Comparing with the MPMOTP schemes (shown in Table 4), for each case, the TRANSCO investment increases significantly period by period, new lines are

Table 4 The MPMOTP planning schemes for various cases

Planning		1st 5-year	2nd 5-year	3rd 5-year
Case 1	Schemes	1–5(1), 14–16(1)	6–10(1), 12–23(1), 15–21(1)	6–7(1), 7–8(1), 6–10(1), 15–24(1)
	O_{cr}(k\$)	56.28	77.13	89.61
	O_{pp}(k\$)	1325.27	1864.53	2487.11
	O_{rd}(k\$)	1385.42	1977.45	2672.37
	C_{inv}(k\$)	208.52	595.47	1072.01
Case 2	Schemes	1–5(1), 7–8(1), 14–16(1)	3–11(1), 6–7(1), 10–12(1)	3–8(1), 5–10(1), 6–10(1)
	O_{cr}(k\$)	51.58	71.34	83.51
	O_{pp}(k\$)	1372.98	1527.58	1895.42
	O_{rd}(k\$)	1391.56	1803.17	2366.72
	C_{inv}(k\$)	539.64	613.74	658.14
Case 3	Schemes	1–5(1), 3–11(1), 7–8(1), 14–16(1)	3–8(1), 6–7(1), 16–17(1)	4–5(1), 6–10(1), 17–18(1)
	O_{cr}(k\$)	49.87	68.34	74.08
	O_{pp}(k\$)	1401.95	1378.56	1458.16
	O_{rd}(k\$)	1417.82	1751.38	2195.85
	C_{inv}(k\$)	945.73	635.27	647.52

Table 5 The IMTP planning schemes for each case

Planning		1st 5-year	2nd 5-year	3rd 5-year
Case 1	Schemes	1–5(1), 14–16(1)	7–8(1), 12–23(1), 15–21(1)	6–7(1), 7–8(1), 16–17(1), 15–24(1)
	C_{inv}(k$)	185.08	559.20	927.11
Case 2	Schemes	1–5(1), 14–16(1)	7–8(1), 10–12(1), 14–16(1)	3–8(1), 6–2(1), 7–8(1), 6–10(1)
	O_{rd}(k$)	279.56	611.23	971.14
Case 3	Schemes	1–5(1), 14–16(1), 3–8(1)	6–7(1), 6–10(1), 14–16(1)	3–8(1), 6–10(1), 16–17(1), 17–18(1)
	O_{rd}(k$)	452.91	655.33	1095.06

added to cope with overload lines, but can not clearly perform the effects of the variability in production of wind and PV units.

Particularly in Case 1, ignoring the wind and PV penetration, the periodical investment is much lesser than MPMOTP plan, respectively declined 23.44, 36.27, and 144.90 k$ for each 5-year. That means the extra amounts are requested as the uncertainty-averse expenses on conventional generation, LMP volatility, demand response, and line availability.

For Case 2 and 3, the TRANSCO investment could not exhibit a specific trend as for three periods. However, concluded from the results of both MPMOTP model and IMTP model, the further observation shows that, the total 15-year investment tends to be an equivalent amount. In Case 2, the MPMOTP 15-year investment is 1811.52 k$, while that is 1861.93 k$ in IMTP. Accordingly, the value is 228.52 and 2203.30 k$, respectively in Case 3. That means, for a long term perspective, if a precise total amount control of investment is allocated, the proposed MPMOTP model can not only handle the heterogeneous uncertainties, but also ensure the robustness of phased investment in the strategic TRANSCO planning process.

6 Conclusion

Incorporating the ambition of the CO_2 reduction, an uncertainty-averse MPMOTP planning model is proposed to handle the multiple uncertainties from renewable energy, conventional generation, market price, load deviation and network deployment. In this paper, the virtue of uncertainty codependency is evaluated by the correlation coefficient matrix and contributed to optimize three TRANSCO concerned objectives. Associated with an introduced two-phase MOPSO solving algorithm, the proposed model is implemented and applied on the IEEE 24-bus test system. The results show that, considering a variety of uncertain conditions, the released planning schemes can be feasibly and effectively put forward to issue the transmission network with high stable and reliable intention of the TRANSCO.

Acknowledgments The authors gratefully acknowledge the financial supports of Danish national project iPower and the great contributions of Danish Energy Association and DONG Energy involved in this task.

References

[1] Chen Q, Kang C, Xia Q et al (2010) Modeling flexible operation mechanism of CO_2 capture power plant and its effects on power-system operation. IEEE Trans Energy Convers 25(3):853–861
[2] Ji Z, Chen Q, Xia Q et al (2013) Low-carbon power system dispatch incorporating carbon capture power plants. IEEE Trans Power Syst 28(4):4615–4623
[3] Chen Q, Kang C, Xia Q et al (2012) Optimal flexible operation of a CO_2 capture power plant in a combined energy and carbon emission market. IEEE Trans Power Syst 27(3):1602–1609
[4] Chen Q, Kang C, Xia Q et al (2010) Power generation expansion planning model towards low-carbon economy and its application in China. IEEE Trans Power Syst 25(2):1117–1125
[5] Zhang C, Ding Y, Nordentoft NC et al (2014) FLECH: a Danish market solution for DSO congestion management through DER flexibility services. J Mod Power Syst Clean Energy 2(2):126–133
[6] Buygi MO, Shanechi HM, Balzer G et al (2006) Network planning in unbundled power systems. IEEE Trans Power Syst 21(3):1379–1387
[7] Zhao JH, Dong ZY, Lindsay P et al (2009) Flexible transmission expansion planning with uncertainties in an electricity market. IEEE Trans Power Syst 24(1):479–488
[8] Goel L, Gupta R, Ercan MF (2001) Including higher order outages and well-being considerations in sub-transmission system reliability worth assessment. Electr Power Syst Res 60 (1):39–47
[9] Delgado D, Claro J (2013) Transmission network expansion planning under demand uncertainty and risk aversion. Electr Power Energy Syst 44(1):696–702
[10] Cedeño EB, Arora S (2011) Performance comparison of transmission network expansion planning under deterministic and uncertain conditions. Electr Power Energy Syst 33(7):1288–1295

[11] Akbari T, Rahimi-Kian A, Bina MT (2012) Security-constrained transmission expansion planning: a stochastic multi-objective approach. Electr Power Energy Syst 43(1):444–453

[12] Wang Y, Cheng ZH, Wang C et al (2008) Pareto optimality-based multi-objective transmission planning considering transmission congestion. Electr Power Syst Res 78(9):1619–1626

[13] Zheng J, Wen F, Ledwich G et al (2014) Risk control in transmission system expansion planning with wind generators. Int Trans Electr Energy Syst 24(2):227–245

[14] Shaaban MF, El-Saadany EF (2014) Accommodating high penetrations of PEVs and renewable DG considering uncertainties in distribution systems. IEEE Trans Power Syst 29 (1):259–270

[15] Samper ME, Vargas A (2013) Investment decisions in distribution networks under uncertainty with distributed generation-part I: model formulation. IEEE Trans Power Syst 28(3):2331–2340

[16] Plerou V, Gopikrishnan P, Rosenow B et al (2002) Random matrix approach to cross correlations in financial data. Phys Rev E 65:0066126

[17] Soroudi A, Caire R, Hadjsaid N et al (2011) Probabilistic dynamic multi-objective model for renewable and non-renewable distributed generation planning. IET Gener Transm Distrib 5(11):1173–1182

[18] Liu Z, Wen F, Ledwich G (2011) Optimal siting and sizing of distributed generators in distribution systems considering uncertainties. IEEE Trans Power Deliv 26(4):2541–2551

[19] Doostizadeh M, Ghasemi H (2012) A day-ahead electricity pricing model based on smart metering and demand-side management. Energy 46(1):221–230

[20] Leite da Silva AM, Arienti VL (1990) Probabilistic load flow by a multilinear simulation algorithm. IEEE Proc 137(4):276–282

[21] Reyes-sierra M, Coello CAC (2006) Multi-objective particle swarm optimizers: a survey of the state-of-the-art. Int J Comput Intell Res 3:287–308

[22] Hu G, Mao Z, He D (2011) Multi-objective optimization for leaching process using improved two-stage guide PSO algorithm. J Cent South Univ Technol 18(4):1200–1210

[23] Zangeneh A, Jadid S, Rahimi-Kian A (2011) A fuzzy environmental-technical-economic model for distributed generation planning. Energy 36(5):3437–3445

[24] Fang R, Hill DJ (2003) A new strategy for transmission expansion in competitive electricity markets. IEEE Trans Power Syst 18(1):374–380

Chunyu ZHANG received the B.Eng. and M.Eng. degrees from North China Electric Power University, Beijing, China, in 2004 and 2006, respectively, both in electrical engineering. Since then he joined National Power Planning Center of China as a senior engineer till 2012. He is currently pursuing the Ph.D. degree at the Center for Electric Power and Energy, Technical University of Denmark (DTU), Denmark. His research interests include power systems planning and power market innovation.

Yi DING received the B.Eng. degree from Shanghai Jiaotong University, China, and the Ph.D. degree from Nanyang Technological University, Singapore, both in electrical engineering. He is an Associate Professor in the Department of Electrical Engineering, Technical University of Denmark (DTU), Denmark. His research interests include power systems reliability/performance analysis, and smart grid performance analysis.

Qi WANG received the M.Sc. degree in electrical control theory and engineering from North China Electric Power University, Beijing, China, in 2008. Since then she joined East China Electric Power Design Institute as a senior engineer till 2013. She is currently pursuing the Ph.D. degree at the Center for Electric Power and Energy, Technical University of Denmark (DTU), Denmark. Her research interests are power systems economics, stochastic programming, robust optimization and complementarity modelling.

Yusheng XUE received his Ph.D. degree in Electrical Engineering from the University of Liege (Belgium) in 1987. He became a Member of Chinese Academy of Engineering in 1995. He is now the Honorary President of State Grid Electric Power Research Institute (SGEPRI), State Grid Corporation of China. His research interests include nonlinear stability, control and power system automation.

Jacob ØSTERGAARD received the M.Sc. degree in electrical engineering from the Technical University of Denmark (DTU), Lyngby, Denmark, in 1995. He was with Research Institute of Danish Electric Utilities for 10 years. Since 2005, he has been Professor and Head of Centre for Electric Technology, DTU. His research interests cover smart grids with focus on system integration of renewable energy and distributed energy resources, control architecture for future power system, and flexible demand.

Cascading failure analysis of power flow on wind power based on complex network theory

Yushu SUN, Xisheng TANG (✉)

Abstract Cascading failure is a potential threat in power systems with the scale development of wind power, especially for the large-scale grid-connected and long distance transmission wind power base in China. This introduces a complex network theory (CNT) for cascading failure analysis considering wind farm integration. A cascading failure power flow analysis model for complex power networks is established with improved network topology principles and methods. The network load and boundary conditions are determined to reflect the operational states of power systems. Three typical network evaluation indicators are used to evaluate the topology characteristics of power network before and after malfunction including connectivity level, global effective performance and percentage of load loss (PLL). The impacts of node removal, grid current tolerance capability, wind power instantaneous penetrations, and wind farm coupling points on the power grid are analyzed based on the IEEE 30 bus system. Through the simulation analysis, the occurrence mechanism and main influence factors of cascading failure are determined. Finally, corresponding defense strategies are proposed to reduce the hazards of cascading failure in power systems.

Keywords Complex network theory (CNT), Cascading failure, Wind power, Power flow analysis model, Defense strategies

Y. SUN, X. TANG, Institute of Electrical Engineering, Chinese Academy of Sciences, Beijing, China
(✉) e-mail: tang@mail.iee.ac.cn
Y. SUN
e-mail: yushusun@mail.iee.ac.cn

1 Introduction

It is proved that almost all the blackouts are caused by cascading failures, which are triggered initially by single or multiple disturbances in certain circumstances, such as long time overload or stability issues in bulk power system [1]. For example, the "8·14" blackout in North America in 2003 was induced by five 345 kV AEP-DPL ground fault lines in MISO. Then, the transmission line cascading trips took place to make the failure propagation, and eventually power outage occurred [2, 3].

With the rapid development of wind power around the world, its penetration in the power grid grows higher and higher. Different from the distributed utilization of wind power in North America and Europe, large-scale grid-connected and long distance transmission of wind power, also known as wind power base (WPB), is inevitable in China due to the uneven distribution of wind resources and load centers. The intermittent and stochastic of output power of wind farms, as well their easy tripping under abnormal conditions, will increase the occurrence probability of cascading failures in power systems.

Recently, several typical methods on cascading failure have been researched [4–7], including pattern search, macroscopic evaluation, complex system methods and complex network theory. Heuristic search is often used by the pattern search method. However, it is hard to use practically due to the quite low search precision and efficiency. The research object of macroscopic evaluation method is based on sample data, so the evaluation is not accurate enough and lack of universality. The model for complex systems just reveals the mechanism of cascading failures qualitatively, which are hard to describe the details of an actual power system. From the perspective of network, CNT can be used to analyze the impacts of power grid topology structure on cascading failure, which provides a new perspective and method. CNT has been widely used in power system cascading failure analysis [8–11].

Complex network is a complicated system, which contains a high number of individuals and the interaction among them. A typical network can be expressed as many nodes and edges, where every node represents individual and every edge stands for the mutual relation of two individuals. For power system, power plants, substations (or buses) and consumers can be considered to be the nodes, and the transmission lines are used as the edges. Then, the basic topology of complex power system will be acquired. CNT is used to study the structure characteristics of power system and dynamic propagation behaviors of failure. Its purpose is to explain the cause and development law of power blackout, and then effective measures and suggestions can be proposed to prevent the occurrence of power accident. The main feature of CNT is that the analysis model is easy to be built, and can reflect internal relations and operating characteristics of the system. In addition, it has good application flexibility and adaptability for power system analysis.

In general, the cascading failure analysis based on CNT can be classified into the following steps, including load definition, boundary condition determination, failure mode selection, network failure evaluation and corresponding defense strategy. As to the definition of load, the node load is defined primarily by the total number of the shortest paths that pass through the node, i.e., node betweenness [12–17]. Wang and Rong defined the relevant functions of the node degree as the node load [18–21]. But the above load definitions lose sight of the electrical specification of the grid [22] and power transmission doesn't follow the Kirchhoff's Law. In addition, Bompard and Xu et al. used the extended betweenness [23–25] or electrical betweenness [26–29] based on power transfer distribution factor (PTDF) as the load in the power grid. However, the extended or electrical betweenness identifies the criticality of lines from a structural point of view without considering operational states of power system. As to boundary conditions, the capacity $C_i = (1 + \alpha)L_i$ of node i is supposed proportional to its initial load L_i [12–17], where $\alpha(>0)$ is the tolerance parameter. However, the upper limit of the capacity is considered while the lower limit is often ignored in simulation. As to failure mode selection, removing special nodes or edges is often adopted in traditional grid failure analysis [30, 31]. But the type of failure becomes diversification due to wind power integration. As to network failure evaluation, many of the existing evaluation indicators [28, 32, 33] are just suitable for general complex networks which can not accurately reflect electrical characteristics before and after malfunction in power system. As to defense strategy,many measures are provided to suppress cascading failure [34–36] in previous researches, but the ones which are suitable for wind power integration should be proposed.

In this paper, a power system cascading failure power flow analysis model based on CNT is established to study the influence of WPB on power system cascading failure. In the model, power flow is used as the load to exactly describe the distribution of voltage and power in power grid. The upper and lower operating limits of power system are both considered under condition of wind power integration. Three typical evaluation indicators including connectivity level, global effective performance and percentage of load loss (PLL) are presented to evaluate network performance before and after malfunction. Accordingly, some effective defense strategies are proposed to reduce the impacts of wind power on the stability and the security of power system.

The studies indicate that influences of wind power integration on power system cascading failure are analyzed, and the harm extent is quantified compared with traditional fault. In the different simulation scenarios, the dominant factors of cascading failure caused by wind power are determined, and then the corresponding defensive strategies are put forward. It provides a reference for the prevention and control of grid-connected wind power system cascading failure, and promotes the rapid development of large-scale grid-connected wind power.

The remains of this paper are organized as follows. Section 2 introduces the topology principles and basic characteristics of CNT considering wind power integration. The cascading failure analysis model is presented in Section 3. In Section 4, the influences of node removal, grid current tolerance capability, different wind power instantaneous penetrations and wind farms coupling points on power grid are explored through a case study. Conclusions are given in Section 5.

2 Complex network theory with wind power integration

2.1 Topology principles and methods

CNT method can reflect the network topology, operating characteristics and other elements of power grid from the static and dynamic perspective. In order to analyze the characteristics of the grid accurately, a grid-connected wind power system directed-weighted network model based on undirected-unweighted and undirected-weighted network model [37–39] is established.

1) Only the high-voltage transmission grids are considered, while the internal main wirings of power plants, substations and distribution network are ignored.

2) The power plants, buses and customers are considered as generation nodes, transmission nodes and load nodes, respectively. They have their own characteristic parameters and voltage level. Wind power is added to the traditional grid, which is equivalent to PQ node to distinct from conventional power generation node.
3) The weight of power generation node is defined as the rated output of the generator. The maximum load demand is considered as the weight of load node. And the weight of the transmission node is equivalent to its capacity.
4) The network edges are denoted as the branches of the transmission lines and transformers, the weight as the impedance of the line and the direction as that of power flow.
5) The transmission lines connected the same two nodes are merged, and the branch-path of the parallel capacitors are neglected. In order to make the network structure diagram simplification, self-loop and excessive edges are eliminated in the grid.

Through the above mapping process, the power grid can be modeled as a weighted, directed and sparse connected graph with N nodes and K edges.

2.2 Basic characteristics

1) Node degree and its distribution

The degree of a node is the total number of the edges attached to it, usually represented by k. Degree distribution is the ratio of nodes in the network with degree k. The degree of nodes can measure importance of nodes in the grid, while degree distribution reflects the uniformity of the network.

2) Clustering coefficient

Assuming that node i has degree k_i, and there are t_i edges among its nearest nodes. For the nearest nodes, $k_i(k_i - 1)/2$ edges can be acquired under the condition of a complete graph. The clustering coefficient of node i can be defined as

$$\gamma_i = \frac{t_i}{k_i(k_i - 1)/2} \tag{1}$$

Clustering coefficient indicates the clustering and tightness of nodes in the grid.

3) Electrical distance

Electrical distance is the electric potential difference caused by a unit of current transmitted from node i to node j, namely, is equivalent impedance between two nodes.

$$d_{ij} = \left|Z_{ii} + Z_{jj} - 2Z_{ij}\right| \tag{2}$$

where Z_{ii} and Z_{jj} are the self-impedances of node i and j, respectively; Z_{ij} is the mutual impedance between node i and j. The size of the electrical distance reflects electrical connection between different nodes in the power grid.

4) Electrical betweenness

According to the power flow transmission characteristics, the electrical betweenness of edge (m, n) is defined as:

$$B_e(m,n) = \sum_{i\in G, j\in L} w_{ij}|I_{mn}(i,j)| \tag{3}$$

where G and L are the sets of generation nodes and load nodes, respectively; w_{ij} are the weights of power transmitted from node i to j; $I_{mn}(i, j)$ is the generated currents in the edges (m, n), when a unit of current is transmitted from the generation node i to the load node j. In addition, $I_{mn}(i, j)$ are also known as the power transfer distribution factors (PTDFs) to reflect the sensitivity of the power flowing in each line for a power injection/withdrawal at a couple of nodes.

The electrical betweenness of node k can be calculated as:

$$B_n(k) = \begin{cases} \frac{1}{2}\left(\sum\limits_{l\in F(k)} B_e(k,l) + \sum\limits_{j\in L} w_{kj}\right) & k \in G \\ \frac{1}{2}\left(\sum\limits_{l\in F(k)} B_e(k,l) + \sum\limits_{j\in G} w_{kj}\right) & k \in L \\ \frac{1}{2}\sum\limits_{l\in F(k)} B_e(k,l) & k \notin G,\ k \notin L \end{cases} \tag{4}$$

where $F(k)$ is the set of nodes that are connected to node k.

Electrical betweenness can reflect the influence of the capacity and distribution of generators and load, even quantify the contribution of the nodes and edges to the power flow transmission.

3 Power system cascading failure power flow analysis based on CNT

According to the previous analysis, this paper adopts power flow as the load to reflect the distribution of voltage, active power and reactive power at each node and line in the grid network. Over-voltage, under-voltage and over-current are considered as boundary conditions of power grid. Once the network failure occurs, the power flow of the grid is calculated again. Time after time, the process is repeated until there is no failure. In addition, connectivity level, global effective performance and percentage of load loss are used to evaluate network performance before and after malfunction. Finally, some defense strategies are proposed.

3.1 Power flow calculation

The voltage and current in power grid can be obtained through the following power flow calculation equation.

$$S = \Gamma(U) \tag{5}$$

where S is the given value of the injection power of a certain node; $\Gamma(U)$ is the function of the corresponding impedance of S and node voltage U.

3.2 Voltage and current setting values

The voltage and current setting values can reflect the boundary values including the upper and lower limits of voltage and current in power system. The detailed description is as:

$$U_c = (1+\alpha)U_0 \tag{6}$$

$$I_c = (1+\alpha)I_0 \tag{7}$$

where U_0 and I_0 are the rated voltage and current; U_c and I_c are the setting values of voltage and current which mean the operating limits of voltage and current of power system; $\alpha(>-1)$ is the tolerance parameter of the network, which represents the capacity of the network to resist interference.

3.3 Voltage and current constraints

The power grid has a good operation under the boundary conditions of voltage and current. If power system over-voltage, under-voltage or over-current occurs, the corresponding nodes or lines will be removed.

$$U_{c,\min} \leq U \leq U_{c,\max} \tag{8}$$

$$I \leq I_{c,\max} \tag{9}$$

where U and I are the voltage and the current in power system; $I_{c,\max}$ is the upper operating limit of the current while $U_{c,\min}$ and $U_{c,\max}$ are the lower and upper operating limits of the voltage.

3.4 Network evaluation indicators

In order to measure the damage of cascading failure in the gird, and evaluate robustness of the power network, three improved indicators are proposed to analyze power system cascading failure.

1) Connectivity level

Connectivity level is defined as the ratio of the number of nodes in the largest supply area before and after malfunction. It reflects the connectivity and splitting degree of power system when the failure occurs. The smaller connectivity level indicates the worse robustness of the network.

$$g = \frac{N'_{G,L}}{N_{G,L}} \tag{10}$$

where $N_{G,L}$ and $N'_{G,L}$ are the numbers of nodes in the largest connected region which can maintain normal power supply before and after malfunction.

2) Global effective performance

Global effective performance represents the average tightness between generation nodes and load nodes, and reflects the efficiency and loss of energy in power transmission.

$$E = \frac{1}{N_G N_L} \sum_{i \in G, j \in L} \frac{1}{d_{ij}} \tag{11}$$

where subscript G and L are the sets of generation nodes and load nodes; N_G and N_L are the total numbers of generation nodes and load nodes; d_{ij} is the shortest electrical distance between generation node i and load node j.

3) Percentage of load loss

Percentage of load loss reveals transmission capacity of the whole power grid, which is used to measure the breadth of failure propagation. The more load loss, the greater the damage to the power system.

$$p_{\mathrm{PLL}} = \sum_{j \in G_1} L_j \Big/ \sum_{k \in G_0} L_k \tag{12}$$

where p_{PLL} is the value of percentage of load loss (PLL); subscript G_1 and G_0 are the sets of failed and all transmission nodes; L_k and L_j are the loads of node k and failed node j.

3.5 Calculation process

Figure 1 shows the flow chart of the process.

1) Calculate the initial power flow in the grid, and the boundary conditions of current and voltage of lines and nodes are expressed as the definition of node capacity $C_i = (1+\alpha)L_i$.
2) Remove the nodes in the network or wind power integration.
3) Recalculate the power flow in the grid, and take out the nodes or edges whose voltage or current are out-of-limit. Repeat it until there is no failure happened in the network.
4) Evaluate the impacts of network failures on power system depending on evaluation indicators.
5) Propose defense strategies for network failures.

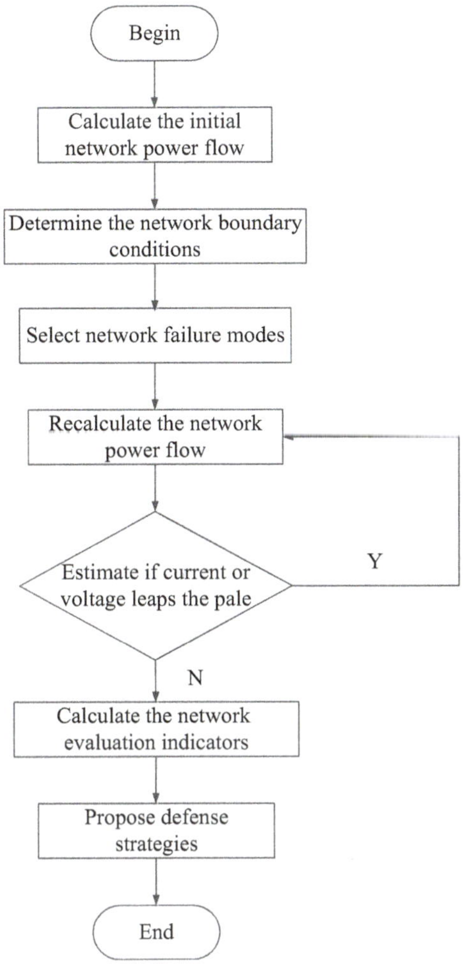

Fig. 1 Flow chart of power grid failure analysis process based on CNT

4 Case study

In this paper, an IEEE 30 bus reference power system is utilized to analyze the influence of node removal and wind power on power system. According to the CNT model with wind power integration from Section 2, the IEEE 30 bus system can be described as a weighted, directed and sparse connected graph with 30 nodes and 41 edges, as shown in Fig. 2.

4.1 Impacts of node removal on power system

Under the conditions of the fixed voltage and the current boundary, Fig. 3 shows the results by removing all the nodes in order in the network. The average value of connectivity level is 0.6922. The maximum one of 0.9667 indicates no failure happened after removing a certain node. The minimum one of 0 means the total collapse of the entire network triggered by cascading failure after an important node removal. The average, maximum and

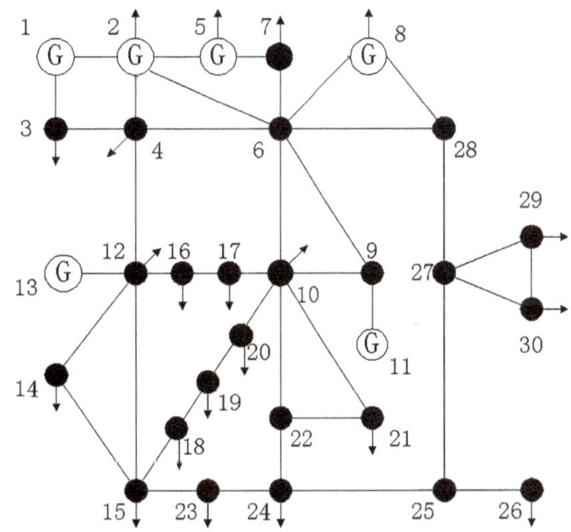

Fig. 2 IEEE 30 bus system topology structure

minimum values of the global effective performance are 1.2345, 1.87 and 0. When the global effective performance cuts to the minimum, it represents that removing a particular node leads to the interruption of the whole power transmission network. The average value of percentage of load loss is 0.3157. The maximum value of 1 denotes that all the network loads are lost due to a certain node removal while there is no load loss for the minimum value of 0. That is, the power grid can maintain the normal operation after taking out a non-significant node.

As can be seen, different node removal will lead to different impacts on the power grid. For instance, the system will not be malfunction when nodes 7 and 17 are removed, while the removal of nodes 1 and 6 will lead to a chain reaction of failures, and even the paralysis of the entire network. Among all the nodes in power system, node 1 has the largest generation capacity, while node 6 has the maximum node degree and electrical betweenness that play an important role in power transmission. To distinguish, the nodes 1 and 6 are considered to be critical nodes [36, 40]. In summary, critical node is an important factor which makes cascading failure occur. To strengthen the protection of critical nodes can reduce the occurrence probability of cascading failures including the advancement of the technics and compatibilizing modification.

4.2 Impact of wind power on the power system

1) Comparative analysis of the impact of wind power and node removal

The generation node 13 with smaller generation capacity in IEEE 30 bus system is replaced by a wind farm with the power curve shown in Fig. 4. The comparison results of the impacts of grid-connected wind and node removal in power system are shown in Fig. 5.

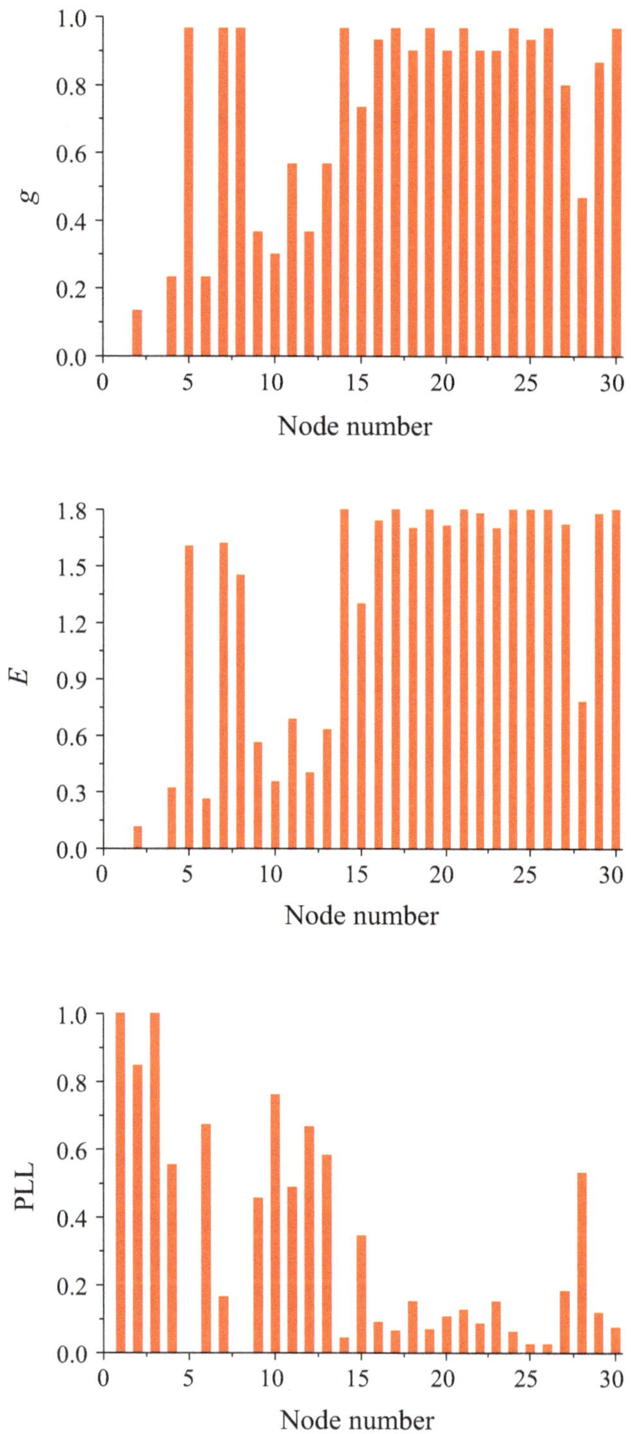

Fig. 3 Results of node number removing all the nodes in order in the network

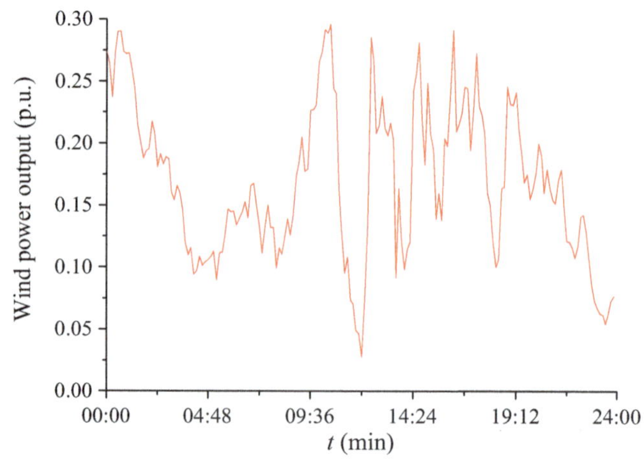

Fig. 4 The power curve of a typical day of a wind farm

Figure 5 shows that system evaluation indicators change with the change of the wind power output. Drastic fluctuations of wind power make rapid changes of connectivity level, global effective performance and percentage loss of load. In other words, the more severe power fluctuations, the greater the harm of wind power to the system, which can be seen from the time period of 10:00:00 to 11:00:00 and so on. In addition, through the comparisons of wind power integration and node removal, connectivity level and global effective performance of the former ones are 80% greater than that of the latter. Meanwhile, the percentage of load loss of the former is 80% less than that of the latter. As a result, the impacts of wind power with certain capacity in power network are less than that of the average of node removal on the same constraints.

2) Impact of grid current tolerance capability

In order to analyze the relationship of boundary conditions of power grid operation and cascading failure, the impact of grid current tolerance capability in the power system is studied.

Supposing that the generation node 13 is replaced by certain capacity of wind farm (Fig. 4), the effects of wind power on the system can be observed from Fig. 6 by adjusting the grid current tolerance capability. It can be seen that the nodes and lines in the grid withstand higher level current, and the impact of wind power weakens with the increase of the current setting coefficient. In other words, with the current setting coefficient increased, the numbers of the failed nodes in the network decrease, and the level of power transmission among nodes is enhanced which can be illustrated by the increase of connectivity level and global effective performance. When the current setting coefficient is over 1.8, both connectivity level and global effective performance reach their maximum values. That is, there is no failure occurred in the power system. The same conclusions can be obtained for the percentage of load loss. The network load loss decreases with the increase of the grid current tolerance capability. And there is no load dropped out of the power system when the current tolerance capability increases to 1.8. Thus the

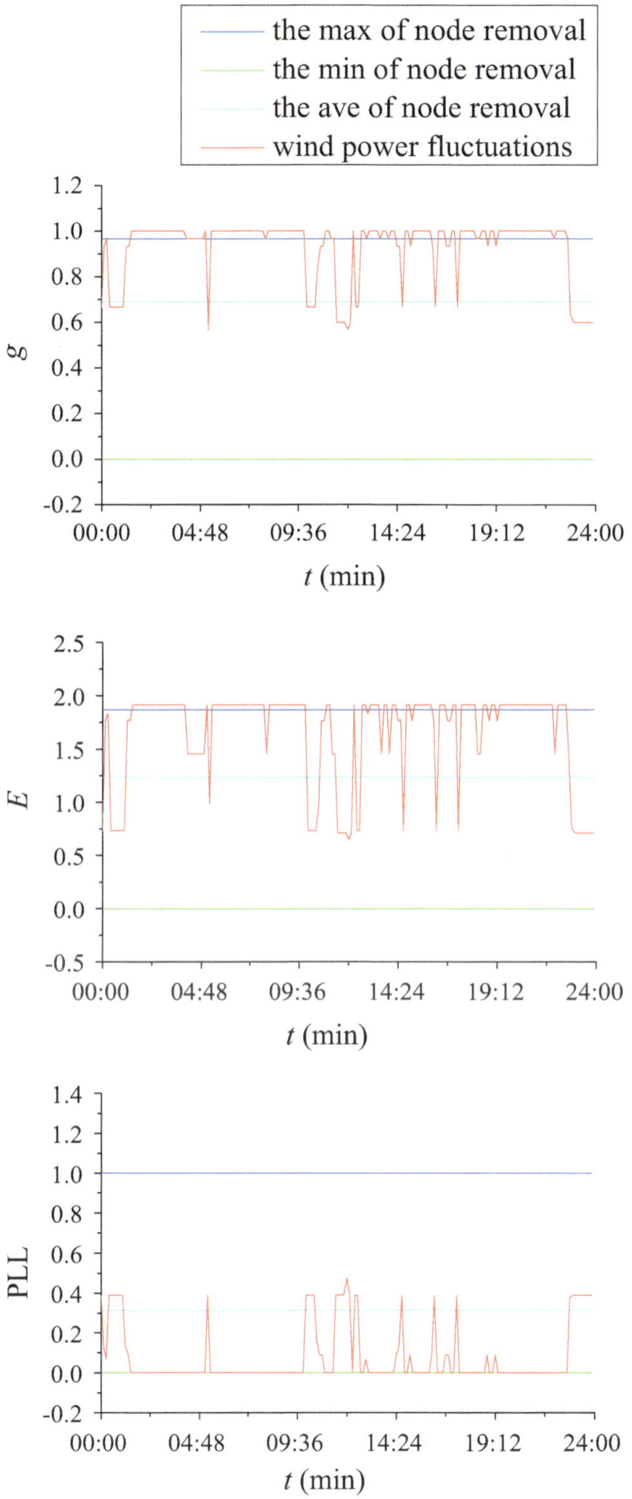

Fig. 5 Contrastive analysis of the impact of wind power and node removal on power system

cascading failure can be reduced and even avoided by the strengthened of grid current tolerance capability through the series current limiting reactor, the upgrading of equipment and lines and so on.

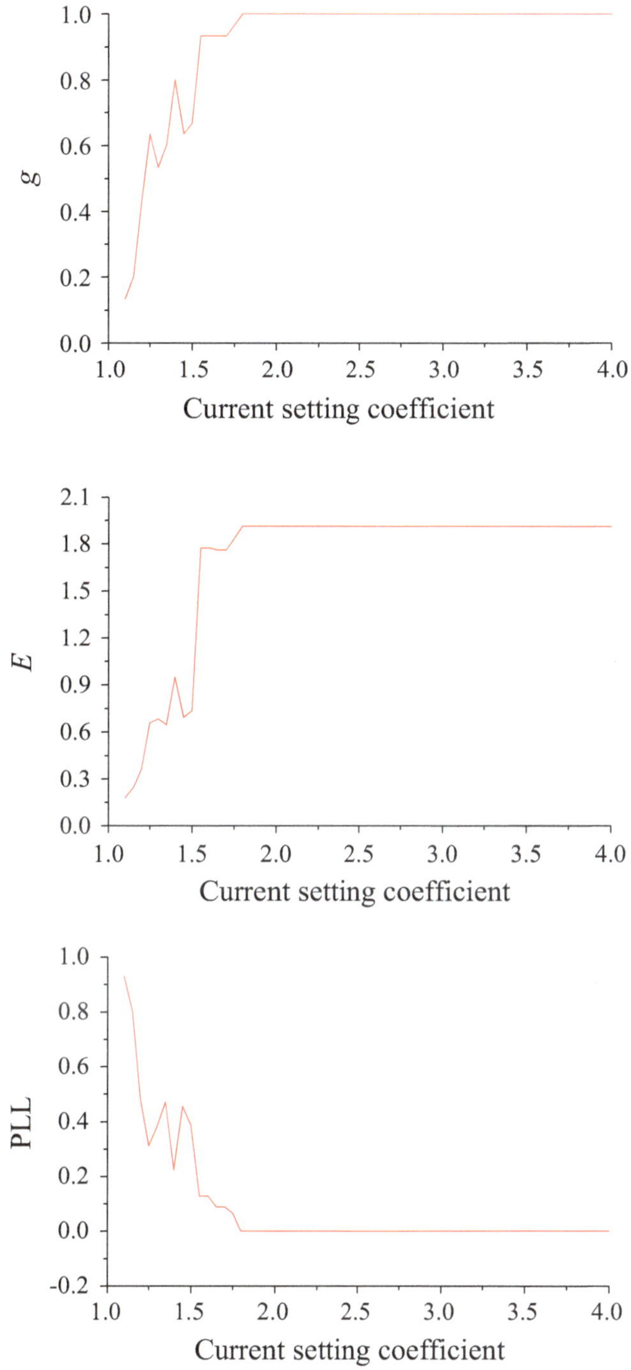

Fig. 6 Impacts of grid current tolerance capability change on power system

3) Impact of wind power instantaneous penetration

The instantaneous penetration of wind power can be described as the ratio of the present wind power production to the present system load [41].

Displaced the traditional generator with wind farm (Fig. 4) in node 13, Fig. 7 illustrates the simulation results of connectivity level, global effective performance and

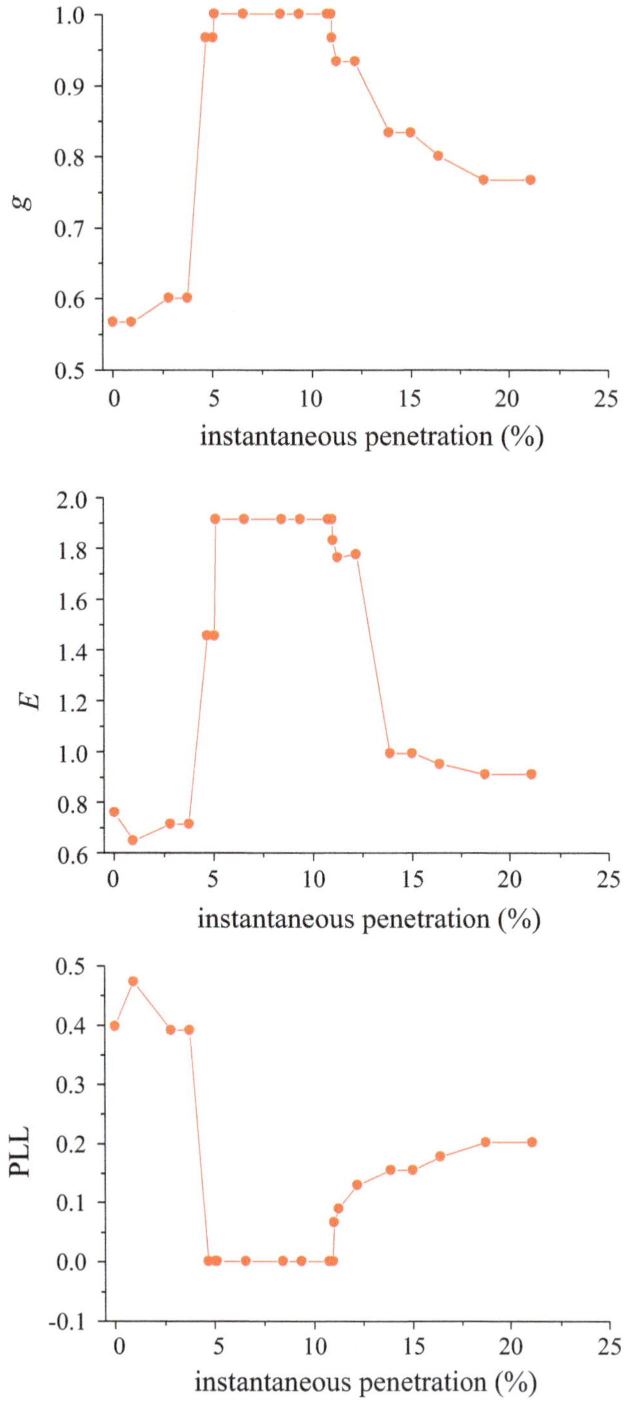

Fig. 7 Impact of wind power instantaneous penetration on power system

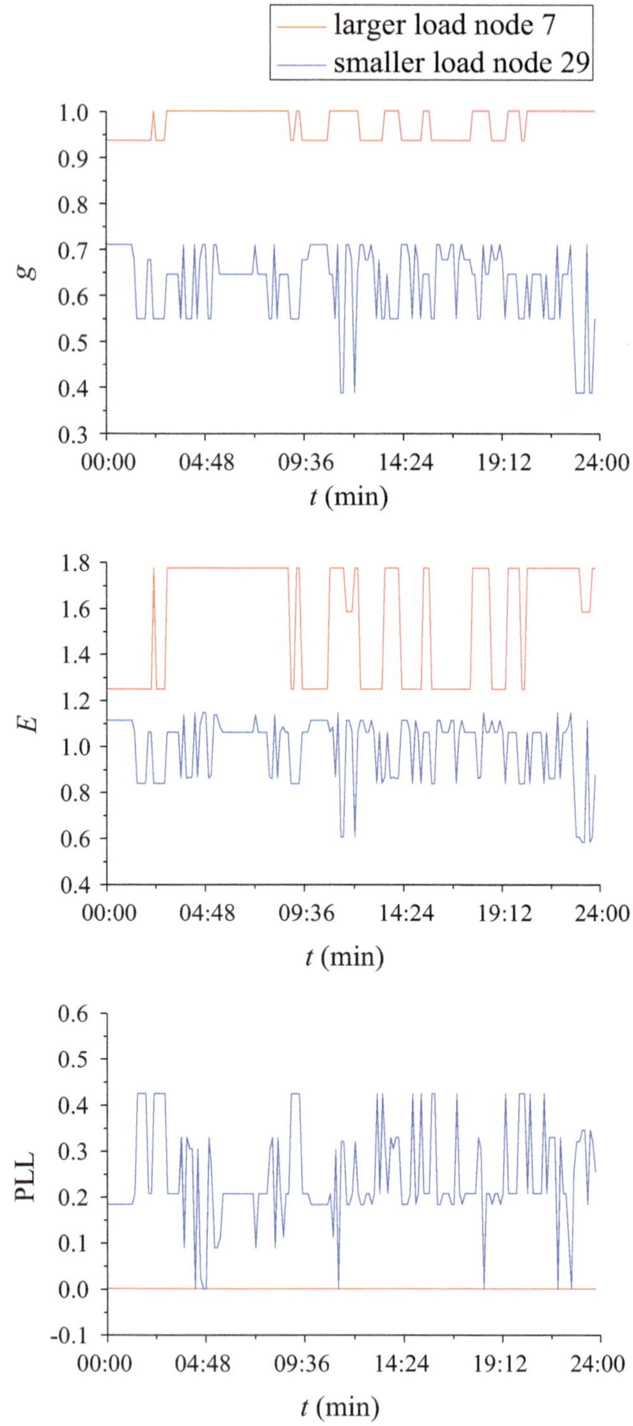

Fig. 8 Impacts of different load points of wind power integration on power system

percentage of load loss respectively with wind power instantaneous penetration of 0 %, 0.93 %, 2.79 %, 3.72 %, 4.66 %, 5.03 %, 5.08 %, 6.53 %, 8.40 %, 9.34 %, 10.74 %, 10.93 %, 10.98 %, 11.21 %, 12.15 %, 13.86 %, 14.97 %, 16.37 %, 18.72 %, and 21.06 %.

When the wind power instantaneous penetration is lower, connectivity level and global effective performance are relatively low while the percentage of load loss is high, That is to say, the low output power of a certain wind farm will deteriorate the security operation of power system. With the output power of the wind farm increasing, connectivity level and global effective performance increase and the percentage of load loss decreases. When the

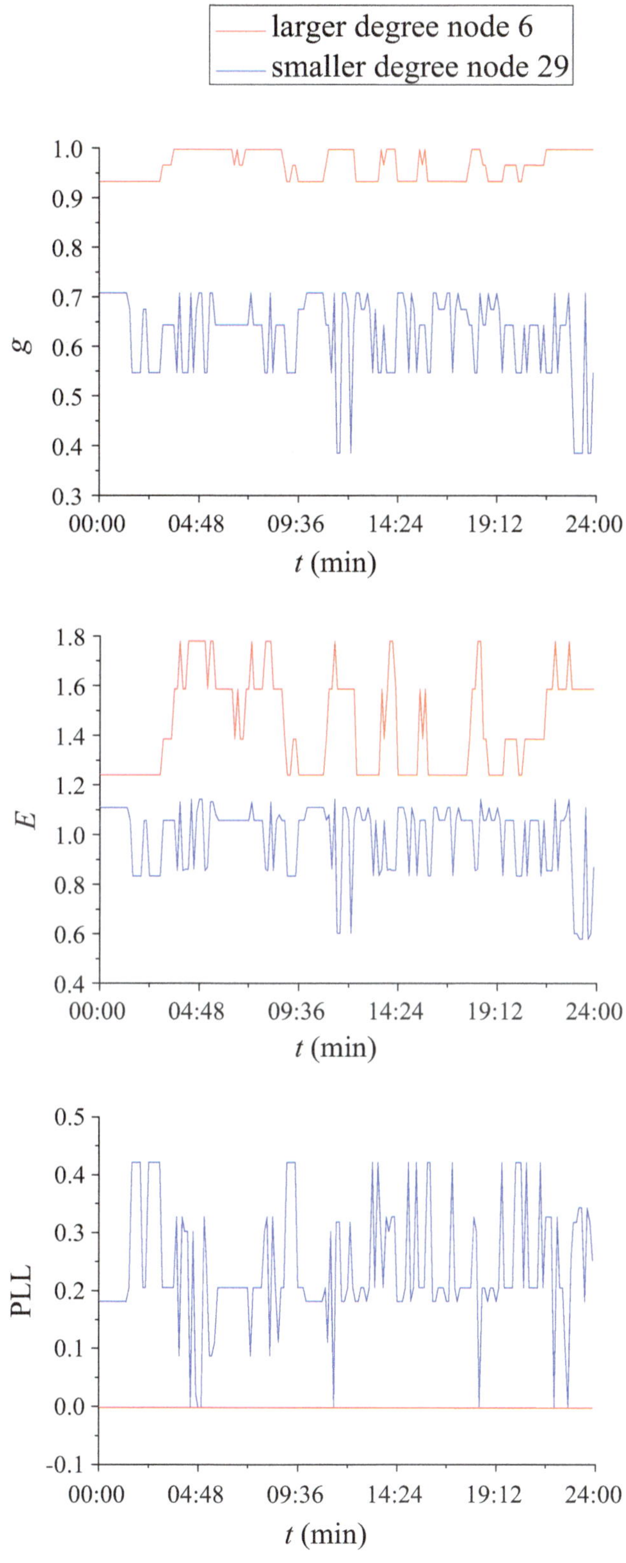

Fig. 9 Impacts of different degree points of wind power integration on power system

instantaneous penetration of wind power locates in the range of 5.08 %–10.93 %, both connectivity level and global effective performance reach the maximum values while percentage of load loss reduced to the minimum of zero. Power system in this situation is also robust enough with little chance of cascading failure.

However, when wind power instantaneous penetration grows still higher to 21.06 %, connectivity level and global effective performance of the network decrease, while the percentage of load loss increases. This reflects that the power system becomes more vulnerable to cascading failure with the increase of wind power instantaneous penetration.

It can be seen from the above simulations that different wind power instantaneous penetrations have different impacts on power system performance. Therefore, proper wind power install capacity plan is important for the reliable operation of power network. What's more, wind farm active power control, such as the utilization of energy storage, can mitigate nodes or line failures and then the possibility of cascading failure.

4) Impacts of wind power coupling locations

To analyze the impacts of different points of wind power coupling with the power grid, a typical wind farm in north China (Fig. 4) is added as the 31st power node in the IEEE 30 bus system, where the locations of the wind farm is set at the nodes with rather larger load or degree, respectively.

It can be seen from Fig. 8 and Fig. 9 that the larger degree and load nodes of the points of wind farm integration, the better of connectivity level and global effective performance, as well no load loss in the power system. Therefore, the selection of greater load and degree nodes for wind farm coupling point has less influence on power system performance.

5 Conclusion

With the development of complex network theory, it provides a new perspective to study and analyze cascading failures which will be deteriorated with the increasing of wind power instantaneous penetration. In this paper, improved CNT network topology principles and methods considering wind farms integration are presented. A cascading failure power flow analysis model is established with three typical evaluation indicators including connectivity level, global effective performance and percentage of load loss. The IEEE 30 bus system is taken as an example to investigate comparatively the influence of node attacks and wind power integration in the power system, as well grid current tolerance capability, wind power instantaneous penetrations and wind farm grid coupling points in the power grid. Some key conclusions can be obtained as follows:

1) The failures of the critical nodes will lead to a series of chain reaction, and strengthening the protection of these nodes can reduce the occurrence probability of cascading failure due to wind power integration.
2) The normal operation of power system is subject to grid current tolerance capability, and cascading failure can be reduced and even avoided by the strengthened of grid current tolerance capability.
3) The instantaneous penetration of wind power affects system cascading failure performance, while appropriate wind power install capacity and active power control can reduce or avoid cascading failure probability.
4) The coupling point of wind power with power system is sensitive to cascading failure, while the selecting of larger degree and heavy load nodes for wind power is conducive to system stable operation.

Acknowledgments This work was financially supported by a grant from the National Basic Research Program of China (973 Program) (No. 2012CB215204), and the Key Project of the CAS Knowledge Innovation Program "Research and demonstration of the coordinated control system based on multi-complementary energy storage" (No. KGCX2-EW-330).

References

[1] Bak P, Tang C, Wiesenfeld K (1987) Self-organized criticality: an explanation of 1/*f* noise. Phys Rev Lett 59(4):381–384
[2] U.S.-Canada power system outage task force (2004) Final report on the August 14, 2003 blackout in the United States and Canada: causes and recommendations
[3] Gan DQ, Hu JY, Han ZX (2004) Thoughts for some international blackout in 2003. Autom Electr Power Syst 28(3):1–4, 9 (in Chinese)
[4] Deng HQ, Ai X, Zhao L (2006) Grid cascading failure analysis review. Modern electric power 23(6):10–20 (in Chinese)
[5] Sun K, Han ZX, Cao YJ (2005) Comments on complex grid cascading failure model. Power Syst Technol 29(13):1–9 (in Chinese)
[6] Lu ZX (2005) Reliability research of the grid complexity and blackouts. Autom Electr Power Syst 29(12):93–97 (in Chinese)
[7] Zhan Y, Cheng HZ, Xiong HG (2005) Review of the grid cascading failure. Electr Power Autom Equip 25(9):93–98 (in Chinese)
[8] Jonnavithula S, Billinton R (1997) Topological analysis in bulk power system reliability evaluation. IEEE Trans Power Syst 12(1):456–463
[9] Solé RV, Rosas-Casals M, Corominas-Murtra B et al (2008) Robustness of the European power grids under intentional attack. Phys Rev E 77(2):026102/1–026102/7
[10] Holmgren A (2006) Using graph models to analyze the vulnerability of electric power networks. Risk Anal 26(4):955–969
[11] Rosato V, Bologna S, Tiriticco F (2007) Topological properties of high-voltage electrical transmission networks. Electr Power Syst Res 77(2):99–105
[12] Holme F, Kim BJ (2002) Vertex overload breakdown in evolving networks. Phys Rev E 65(6):066109/1–066109/8
[13] Kim BJ, Yoon CN, Han SK (2002) Attack vulnerability of complex networks. Phys Rev E 65(5):056109/1–056109/14
[14] Lai YC, Motter AE, Nisllikawa T (2004) Attacks and cascades in complex networks. Lect Notes Phys 650:299–310
[15] Motter AE, Lai YC (2002) Cascade-based attacks on complex networks. Phys Rev E 66(6):065102/1–065102/4
[16] Latora V, Marchiori M (2001) Efficient behavior of small-world networks. Phys Rev Lett 87(19):198701/1–198701/4
[17] Crucitti P, Latora V, Marchiori M (2004) Model for cascading failures in complex networks. Phys Rev E 69(4):045104/1–045104/4
[18] Wang JW, Rong LL, Zhang L (2009) A model for cascading failures in complex networks with a tunable parameter. Mod Phys Lett B 23(40):1323–1332
[19] Wang JW, Rong LL (2009) Cascade-based attack vulnerability on the US power grid. Saf Sci 47(10):1332–1336
[20] Wang WX, Chen GR (2008) Universal robustness characteristic of weighted networks against cascading failure. Phys Rev E 77(2):026101/1–026101/5
[21] Wang JW, Rong LL (2009) Vulnerability of effective attack on edges in scale-free networks due to cascading failures. Int J Mod Phys C 20(8):1291–1298
[22] Cotilla-Sanchez E, Phines PDH, Barrows C et al (2012) Comparing the topological and electricalstructure of the North American electric power infrastructure. IEEE Syst J 6(4):616–626
[23] Arianos S, Bompard E, Carbone A et al (2009) Power grid vulnerability: a complex network approach. Chaos 19:013119/1–013119/6
[24] Bompard E, Wu D, Xue F (2011) Structural vulnerability of power systems: a topological approach. Electr Power Syst Res 81(7):1334–1340
[25] Bompard E, Masera M, Napoli R et al (2009) Assessment of structural vulnerability for power grids by network performance based on complex networks. In: Critical information infrastructure security, LNCS 5508: Proceedings of the 3rd international workshop on critical information infrastructures security (CRITIS'08). Rome, Italy, 13–15 Oct 2008, pp 144–154
[26] Xu L, Wang XL, Wang XF (2010) The application of betweenness in the critical line identification of power system. Proc CSEE 30(1):33–39 (in Chinese)
[27] Xu L, Wang XL, Wang XF (2010) Propagation mechanisms and effective defense of power system cascading failure based on betweenness. Proc CSEE 30(13):61–67 (in Chinese)
[28] Wang K, Zhang BH, Zhang Z et al (2011) An electrical betweenness approach for vulnerability assessment of power grids considering the capacity of generators and load. Phys A 390(23/24):4692–4701
[29] Bompard E, Pons E, Wu D (2012) Extended topological metrics for the analysis of power grid vulnerability. IEEE Syst J 6(3):481–487
[30] Mohsenian-Rad AH, Leon-Garcia A (2011) Distributed internet-based load altering attacks against smart power grids. IEEE Trans Smart Grid 2(4):667–674
[31] Salmeron J, Wood K, Baldick R (2004) Analysis of electric grid security under terrorist threat. IEEE Trans Power Syst 19(2):905–912
[32] Albert R, Jeong H, Barabási AL (2000) Error and attack tolerance of complex networks. Nature 406:378–382
[33] Yu K, Rong LL, Wang JW (2009) A new attack on scale-free networks based on cascading failures. Mod Phys Lett B 23(20/21):2497–2505

[34] Wu ZX, Peng G, Wang WX et al (2008) Cascading failure spreading on weighted heterogeneous networks. J Stat Mech 5:P05013/1–P05013/14
[35] Li P, Wang BH, Sun H et al (2008) A limited resource model of fault-tolerant capability against cascading failure of complex network. Eur Phys J B 62(1):101–104
[36] Su HL, Li Y (2012) Exploring the vulnerability of the components from the complex network characteristics of power system. Autom Electr Power Syst 36(23):12–17 (in Chinese)
[37] Sun K (2005) Complex networks theory: a new method of research in power grid. In: Proceedings of the 2005 IEEE/PES transmission and distribution conference and exhibition: Asia and Pacific, Dalian, China, 14–18 Aug 2005, 6 pp
[38] Cao YJ, Chen XG, Sun K (2006) Identification of vulnerable lines in power grid based on complex network theory. Electr Power Autom Equip 26(12):1–5, 31 (in Chinese)
[39] Saedy M, Kelley B (2012) Foundations of ultra-low power scale free sensor networks for cluster to cluster communications. IEEE Sens J 12(9):1890–2881
[40] Chen XG, Sun K, Cao YJ et al (2007) Identification of vulnerable lines in power grid based on complex network theory. In: Proceedings of the 2007 IEEE Power Engineering Society general meeting, Tampa, FL, USA, 24–28 Jun 2007, 6 pp
[41] Tarnowski GC (2011) Coordinated frequency control of wind turbines in power systems with high wind power penetration. PhD Thesis, Technical University of Denmark

Yushu SUN works as a research assistant at the Institute of Electrical Engineering of Chinese Academy of Sciences. His research interests include renewable energy generation, energy management platform and hybrid energy storage coordinated control in power system.

Xisheng TANG is currently a professor at the Institute of Electrical Engineering of Chinese Academy of Sciences. His research interests are in power system stability and control, distributed generation and energy storage technologies, application of power electronics in power system.

The impact of key parameters on the cycle efficiency of multi-stage RCAES system

Bin LIU, Laijun CHEN, Shengwei MEI (✉), Feng LIU, Junjie WANG, Sixian WANG

Abstract Due to the uncertainty and anti-peaking nature, large scale integration of renewable energy imposes great challenges to the operation and dispatch of power systems. Compressed air energy storage (CAES) system provides new ideas to solve this problem as its characteristics of fast regulating, flexible location and long-service life. Especially, regenerative compressed air energy storage (RCAES) system is widely concerned as its capability of heat recovery in the compression process. The cycle efficiency is a key indicator of RCAES system which can be significantly impacted by the key parameters of the systems including compression ratio, exhaust air pressure of throttle (EAPT) and the maximum working pressure (MWP) of compressed air storage vessel (CASV). However, current research mostly focuses on the thermodynamic process and few studies have focused on the impact of key parameters on RCAES system. Based on the efficiency evaluation method which was formulated through the electrical-mechanical-thermal dynamic process and measurable parameters, the impact of key parameters on the cycle efficiency of RCAES system is analyzed in this paper and a practical RCAES design scheme is adopted for case study.

Keywords RCAES, Efficiency evaluation, Parameters analysis, Energy conversion, Energy storage

B. LIU, L. CHEN, S. MEI, F. LIU, State Key Lab of Power Systems, Department of Electrical Engineering, Tsinghua University, Beijing 100084, China
(✉) e-mail: meishengwei@tsinghua.edu.cn
J. WANG, S. WANG, Key Laboratory of Cryogenics, Technical Institute of Physics and Chemistry, Chinese Academy of Sciences, Beijing 100190, China

1 Introduction

Integration of large-scale renewable energy has imposed great challenges on power system operation due to its uncertainty and anti-peaking nature. The load regulating capacity was seriously deficient in Gansu power grid with 5.409 GW integration of wind power by the end of 2011 [1]. In order to effectively cope with the uncertainty and anti-peaking nature of renewable energy, viable option includes the innovative schedule strategy to optimize system reserves, Energy Storage System (ESS) and so on [2–7].

Chemical battery, pumped hydro, flying wheel, superconducting and compressed air are main patterns of energy storage currently. Chemical battery is the most mature one but with high cost and environmental risk. Pumped hydro energy storage is widely applied in power systems. However, its location depends on the geographical environment. Large-scale applications of flying wheel or superconducting energy storage are very limited in power system practically [8]. Compressed air energy storage (CAES) system is widely concerned in recent years due to its characteristics of high-capacity, long-service life and flexible location of which Huntorf in Germany and McIntosh in USA are representative systems globally [9].

Most existing CAES systems are with combustion process which could improve the cycle efficiency with compressed air. However, such CAES system relies greatly on natural gas supply [10]. To overcome the above disadvantages, regenerative compressed air energy storage (RCAES) system is concerned in energy storage area both in science and engineering due to its capability of reusing compression heat in expansion process. The RCAES system can conspicuously improve the system efficiency without any greenhouse gases emission. The expected efficiency of the typical advanced adiabatic compressed air energy storage system can be regarded as a kind of RCAES system under construction by RWE in Germany is 70 % [11–13].

The technology project "Research of Key Technology and Engineering Application of Compressed Air Storage" (simply as RCAES project) in China is started in 2012. One of the key technologies to be concerned in the project is how to reuse the heat released in the compression process with high efficiency and explore a viable solution to the prevalent waste of renewable energy in power systems.

The cycle efficiency is a key indicator of RCAES system and can be significantly impacted by the key parameters of the system. Different kind of efficiency definitions are compared in [14] and the thermodynamic effect of advanced adiabatic compressed air energy storage system is analyzed in [15]. However, neither the efficiency evaluation method nor the impact of the parameters is concerned by any of the aforementioned references.

Based on the electric-mechanic-thermal dynamic process and measurable parameters of RCAES system, we proposed an efficiency evaluation method of RCAES system in [16] previously. One advantage of the proposed method is that it can be extended to study the impact of key parameters on the cycle efficiency of RCAES system, which is to be fulfilled in this paper. The key parameters of a RCAES system in this paper includes compression ratio, exhaust air pressure of throttle (EAPT) and the maximum working pressure (MWP) of compressed air storage vessel (CASV).

The remaining of this paper is organized as follows. Firstly, the efficiency evaluation method proposed in [16] will be briefly reviewed and the analytical method will also be formulated theoretically. Then the typical design scheme of RCAES project is studied. Finally, the conclusions are drawn.

2 Efficiency evaluation of RCAES system

2.1 Main assumptions

A simplified structure of RCAES system is showed in Fig. 1 including four main parts: compression, thermal energy storage (TES), CASV and expansion. The main assumptions in the system are as follow:

1) Ideal gas during the operation;
2) The air mass flow rate in compression and expansion process is known and constant in operation;
3) The isothermal model is adopted for CASV in which the temperature is the same as ambient circumstances;
4) The temperature and pressure of compressed air after throttling become constant.

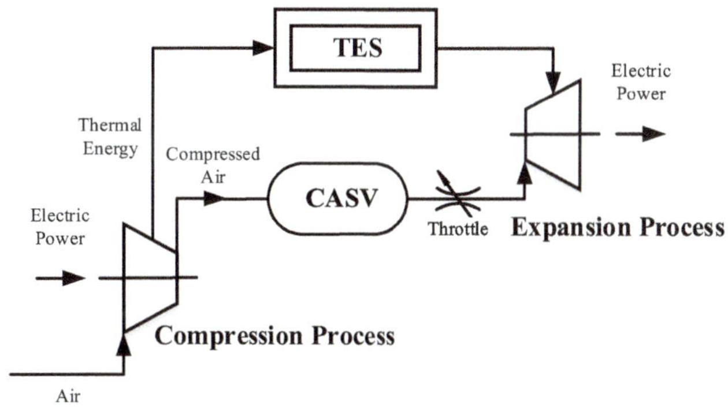

Fig. 1 Design structure of RCAES system

To facilitate the energy conversion process, without loss of generality, we further assume that:

1) The compression process starts when the pressure of CASV reaching its minimum value p_{str}^{min} and ends when the pressure reaching its maximum value p_{str}^{max}. Besides, EAPT denoted by p_{eap} is equal to p_{str}^{min}.

2) The compressor consists of N stages of which stage 1 to stage $N-1$ are steady stages and the last stage, i.e. stage N is an unsteady stage. The exhaust air pressure of stage $N-1$ denoted by $p_{cou,N-1}$, is lower than the MWP of CASV, i.e. $p_{str}^{min} < p_{cou,N-1} < p_{str}^{max}$, while the exhaust air pressure of the last stage denoted by $p_{cou,N}$ is larger than the MWP of CASV, i.e. $p_{cou,N} > p_{str}^{max}$.

Hence, all the stages of compressor will work simultaneously and the previous $N-1$ stages are steady while the last stage is unsteady with time-variant pressure with the same value as CASV. Once the above assumptions are not satisfied for a practical RCAES system, the compression process can be formulated in a multi-period way with the same methodology. Furthermore, TES can be controlled to keep the inlet air temperature of all compression stages the same value except the 1st one in practical engineering and the corresponding temperature is denoted by T_{cin} in this paper.

2.2 Efficiency formulation of the whole RCAES system

Traditional and general efficiency evaluation method for CAES system is indicated by the ratio of the output electric energy and input electric energy. However, this method is merely applicable to systems already in operation and can hardly be extended to analyze or optimize the key parameters of the system. To overcome the deficiency, we proposed an efficiency evaluation method based on the energy conversion of compression and expansion process in [16].

2.2.1 Energy conversion in compression process

In compression process, the RCAES system is charged with electric power of which one part is stored in high-pressure air as molecular potential energy and the other part is stored in high-temperature water as thermal energy. However, the close relationship between electric power and compression work can be illustrated with the compression efficiency η_{com}. The compression efficiency is defined as the ratio of compression work and electric power charged in the compression process and is regarded as a known parameter when the system efficiency is analyzed in this paper.

As the isothermal model is adopted for CASV, the operation time (denoted as t_{com} with the unit of *s*) of the compression process can be illustrated as a function of p_{str} based on the state equation of ideal gas under normal circumstances which is expressed as follows:

$$t_{com} = \frac{V_{str}\left(p_{str} - p_{str}^{min}\right)}{q_{com}R_g T_{str}} \tag{1}$$

where $p_{str}, V_{str}, T_{str}$ are pressure, volume and temperature of CASV with units of $\mathrm{Pa}, \mathrm{m}^3, \mathrm{K}$ separately; q_{com} is the exhaust air rate of compressor with unit of kg/s; $R_g = 287.1\,\mathrm{J/(kg \cdot K)}$ is a gas constant of air under normal circumstances [17].

When the p_{str} reaches the MWP of CASV, the value of t_{com} equals the operation time of compression process.

In the compression process, the compression work of steady stages and unsteady stage should be calculated separately as the compression power of unsteady stage changes dynamicly in accordance with the pressure of CASV.

The total compression power (denoted by P_{std} with the unit of kW) of steady stages $1 \sim N-1$ remain unchanged and can be expressed as follows [18]:

$$P_{std} = \frac{q_{com}R_g\gamma}{10^3 \times (\gamma - 1)}\left(\lambda^{\frac{\gamma-1}{\gamma}} - 1\right)[T_{c1} + (N-2)T_{cin}] \tag{2}$$

where λ is the ratio of each stage of compressor satisfying $\lambda > 1$; T_{c1} is the inlet temperature of the 1st stage and γ is the polytropicexponent of each stage.

The total compression work (denoted by W_{std} with the unit of kWh) of steady stages can then be illustrated as follows after the multiplication of (1) and (2):

$$\begin{aligned} W_{std}(p_{str}) &= P_{std}t_{com}/3600 \\ &= \frac{V_{str}\gamma\left(p_{str} - p_{str}^{min}\right)\left(\lambda^{\frac{\gamma-1}{\gamma}} - 1\right)}{3.6 \times 10^5 T_{str}(\gamma - 1)}[T_{c1} + (N-2)T_{cin}] \end{aligned} \tag{3}$$

Similarly, the last stage of compressor is unsteady of which the exhaust air pressure is the same with CASV. The compression power (denoted by $P_{utd}(p_{str})$ with the unit of kW) can be expressed as [18]:

$$P_{utd}(p_{str}) = \frac{q_{com}R_g T_{cin}\gamma}{10^3(\gamma - 1)}\left[\left(\frac{p_{str}}{p_{cin,N}}\right)^{\frac{\gamma-1}{\gamma}} - 1\right] \tag{4}$$

where $p_{cin,N}$ is the inlet pressure of stage *N*.

Then, the compression work (denoted by $W_{utd}(p_{str})$ with the unit of kWh) can be obtained after integrating the compression power on time basis.

$$\begin{aligned} W_{utd}(p_{str}) &= \int \frac{P_{utd}}{3600}dt = \int \frac{V_{str}P_{utd}/3600}{q_{com}T_{str}R_g}dp \\ &= \alpha_1\left[\left(\frac{p_{str}}{p_{cin,N}}\right)^{\frac{2\gamma-1}{\gamma}} - \left(\frac{p_{str}^{min}}{p_{cin,N}}\right)^{\frac{2\gamma-1}{\gamma}}\right] - \alpha_2\left(p_{str} - p_{str}^{min}\right) \end{aligned} \tag{5}$$

where $\alpha_1 = \frac{T_{cin}V_{str}\gamma^2\lambda^{N-1}p_{cin,1}}{3.6\times10^{-5}T_{str}(2\gamma-1)(\gamma-1)}$, $\alpha_2 = \frac{T_{cin}V_{str}\gamma}{3.6\times10^{-5}T_{str}(\gamma-1)}$.

In summary, the compression power and work can be expressed as a function of p_{stor}:

$$\begin{aligned} P_{com}(p_{str}) &= P_{std} + P_{utd}(p_{str}) \\ W_{com}(p_{str}) &= W_{std}(p_{str}) + W_{utd}(p_{str}) \end{aligned} \tag{6}$$

In (6), P_{std} denotes the total compression power of steady stages which is independent of p_{str} and $P_{utd}(p_{str})$ denotes the compression power of unsteady stage changing along with p_{str}. The total compression power $P_{com}(p_{str})$ in the compression process is a summation of compression powers of steady and unsteady stages and thus a function of p_{str}. The total compression work $W_{com}(p_{str})$ can be explained similarly.

Then, the electric energy (denoted by E_{com}^{max} with the unit of kWh) charged during the compression process can be expressed as:

$$E_{com}^{max} = \frac{W_{com}\left(p_{str}^{max}\right)}{\eta_{com}} \tag{7}$$

2.2.2 Energy conversion in expansion process

In the expansion process, the high-pressure air changes into constant-temperature and isopiestic air after throttling, and then drives the turbine to generate electric power after exchanging heat with TES system. Similar to the compression process, the relationship of expansion work and electric energy can be illustrated by the expansion efficiency η_{tur} which is defined as the ration of discharged electric power and the expansion work in this paper. The expansion efficiency is also regarded as a known parameter when analyzing the system efficiency in this paper.

Due to the throttling effect, all stages are steady in expansion process. Supposing the tempreture of inlet air and polytropicexponent are indicated by T_{tin} and μ

separately, the total expansion power (denoted by P_{tur} with the unit of kW) can be expressed as follows [18]:

$$P_{tur} = -\frac{Mq_{tur}R_gT_{tin}\gamma}{10^3(\mu-1)}(\phi^{\frac{\mu-1}{\mu}}-1) \tag{8}$$

where M is the total number of stage in the turbine; ϕ is the expansion ratio of each stage satisfying $\phi<1$; q_{tur} is the exhaust air rate of turbine with the unit of kg/s.

The expansion time (denoted as t_{tur} with the unit of s) can be obtained in the same way as described in 2.2.1 showed as follows:

$$t_{tur} = \frac{(p_{str}^{max} - p_{str})V_{str}}{q_{tur}R_gT_{su}} \tag{9}$$

Hence, the expansion work (denoted by W_{tur} with the unit of kWh) can be expressed as a function of p_{str} and showed as follows:

$$W_{tur}(p_{str}) = P_{tur}t_{tur}/3600$$

$$= \frac{MT_{tin}\mu V_{str}(p_{str}^{max} - p_{str})}{3.6\times10^5T_{str}(\mu-1)}(1-\phi^{\frac{\mu-1}{\mu}}) \tag{10}$$

The electric energy (denoted by E_{tur}^{max} with the unit of kWh) discharged by the RCAES system can then be expressed as the multiplication of expansion work and expansion efficiency:

$$E_{tur}^{max} = W_{tur}(p_{str})\eta_{tur} \tag{11}$$

Finally, the efficiency of RCAES system can be calculated as (12) and the details can be found in Appendix A.

$$\eta_{RCAES} = \frac{E_{tur}^{max}}{E_{com}^{max}} = \frac{W_{tur}(p_{str}^{min})}{W_{com}(p_{str}^{max})}\eta_{com}\eta_{tur} \tag{12}$$

According to the above formulas, the efficiency of the RCAES system is a nonlinear function of λ, p_{eap} and p_{str}^{max}.

2.3 The impact of key parameters on the efficiency of RCAES systems

2.3.1 The impact of compression ratio λ

When the compression ratio increases, more electric energy will be charged into the RCAES system and more thermal energy will be released by the compressor and absorbed by the TES system in compression process. Consequently, more thermal energy will be reused by the turbine and more electric energy will be discharged in the expansion process.

The relationship between the inlet air temperature of compressor T_{cin} and the exhaust air temperature in steady stage, denoted by T_{cou}, can be showed as follows:

$$T_{cou} = T_{cin}\lambda^{\frac{\gamma-1}{\gamma}} \tag{13}$$

To establish the relationship of T_{cou} and the inlet air temperature of the turbine T_{tin}, a reasonable and applicable technique in engineering is assuming the gap being a constant, indicated by ΔT, which represents the efficacy of the TES. The smaller the value of ΔT is, the better the efficacy of the TES is. Therefore, the relationship of T_{cou} and T_{tin} can be expressed as follows:

$$T_{tin} = T_{cou}\lambda^{\frac{\gamma-1}{\gamma}} + \Delta T \tag{14}$$

By substituting (14) into (12), we can get the mathematical expression of the efficiency of RCAES system (see Appendix).

Additionally, to ensure all of the compressor working properly, the exhaust pressure of the stage $N-1$ should be no more than the MWP of CASV showed as follows:

$$p_{cin,1}\lambda^{N-1} < p_{str}^{max}$$

Otherwise the last stage, i.e. stage N will be useless. Similarly, stage N of the compressor should be larger than the MWP of CASV showed as follows implying that the maximum pressure of CASV can only be reached when all stages work simultaneously.

$$p_{cin,1}\lambda^{N} > p_{str}^{max}$$

Therefore, the compression ratio λ should satisfy

$$\sqrt[N]{\frac{p_{str}^{max}}{p_{cin,1}}} < \lambda < \sqrt[N-1]{\frac{p_{str}^{max}}{p_{cin,1}}} \tag{15}$$

2.3.2 The impact of EAPT p_{eap}

On one hand, the variety of EAPT will affect the operation time of both compression and expansion process according to (1) where $p_{str}^{min} = p_{eap}$ under given assumptions. On the other hand, as the exhaust air pressure of the turbine's last stage $p_{tou,M}$ and total stage number of the turbine M are both fixed, the relationship of the expansion ratio ϕ can be expressed as

$$\phi = \sqrt[M]{\frac{p_{tou,M}}{p_{eap}}} \tag{16}$$

In the above formulation, the reason that $p_{tou,M}$ can be treated as a constant is that the exhaust air pressure of the last stage of turbine and the air pressure of ambient circumstances are approximately equal in the practical engineering applications.

By substituting (16) into (12), we can get the expression of the efficiency of RCAES system over p_{eap} (see Appendix).

2.3.3 *The impact of the MWP of CASV* p_{str}^{max}

Similar to the impact of EAPT, the MWP of CASV will affect the efficiency of RCAES system indirectly. As the system efficiency is a nonlinear function of p_{str}^{max}, the specific impact will be analyzed in case study.

Additionally, p_{str}^{max} also should satisfy the following constraint to guarantee the efficacy of every stage of the compressor, i.e., the MWP of CASV can be reached only when all stages work simultaneously.

$$p_{cin,1}\lambda^{N-1} < p_{str}^{max} < p_{cin,1}\lambda^{N} \tag{17}$$

3 Case study

3.1 System introduction

The typical design scheme of the RCAES project showed in Fig. 2. According to Fig. 2, the compressor consists of 5 stages and the turbine consists of 3 stages. The basic parameters are showed in Table 1 under the comprehensive consideration of the operation performance and construction cost of the RCAES system.

3.2 Efficiency of RCAES system with basic parameters

According to (1), the compression process can be divided into two periods. The first one takes 3.65 h and the second one takes 1.36 h. The operation time of the expansion process is 1.14 h according to (9). Fig. 3 and Fig. 4 show the compression power/energy process changing with p_{str} in the compression process.

Table 1 Parameters of practical RCAES system

Parameter	Value	Parameter	Value
T_{c1}	25 °C	$p_{str}^{min}(p_{eap})$	2.5 MPa
$p_{cin,1}$	0.1 MPa	p_{str}^{max}	10 MPa
η_{com}	75 %	λ	3
η_{tur}	75 %	ϕ	1/3
T_{str}	20 °C	q_{com}	0.46 kg/s
ΔT	40 °C	q_{tur}	2.41 kg/s
γ	1.4	μ	1.4
$p_{tou,M}$	0.11 MPa	V_{str}	100 m^3
T_{cin}	86.35 °C		

As showed in Fig. 3 and Fig. 4, the working pressure of CASV p_{str} increases over compression power P_{com} and compression work W_{com} nonlinearly. In the compression process, the maximum electric power and electric energy charged into the RCAES system is 283.90 kW and 1407.55 kW separately. The electric power discharged by the RCAES system remains 498.17 kWh and the maximum electric energy discharged is 567.92 kW. According to (12), the efficiency of RCAES system is 40.35 % in the given case.

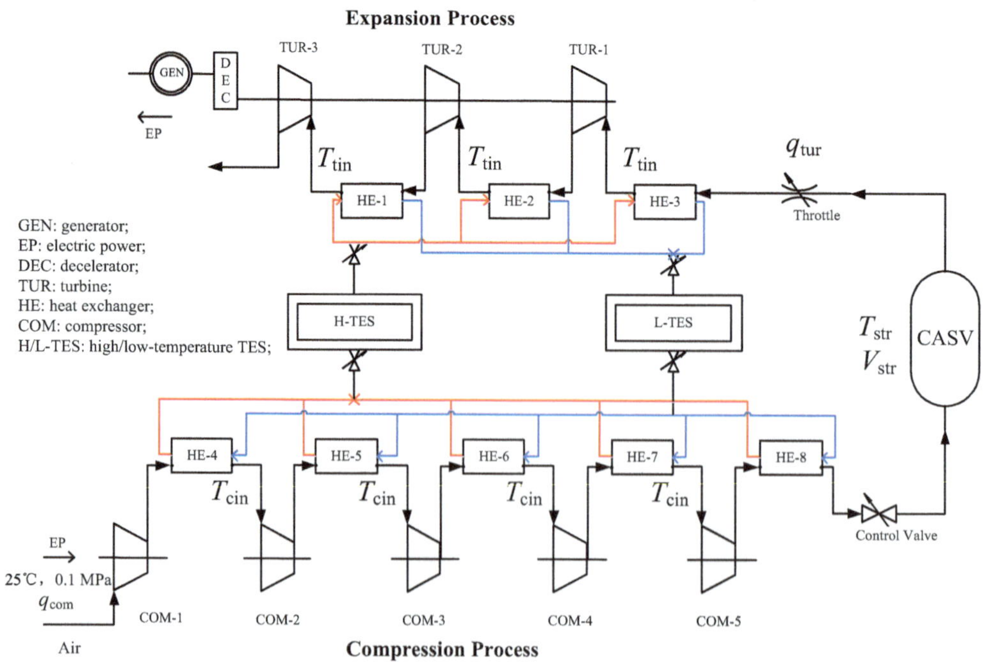

Fig. 2 Design scheme of RCAES project

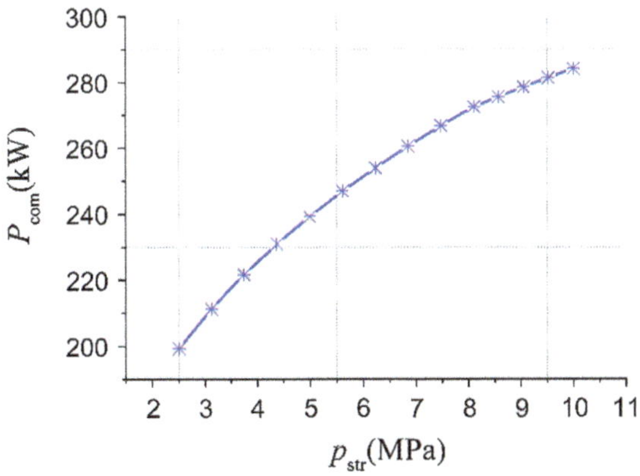

Fig. 3 Compression power P_{com} with different p_{str} in compression process

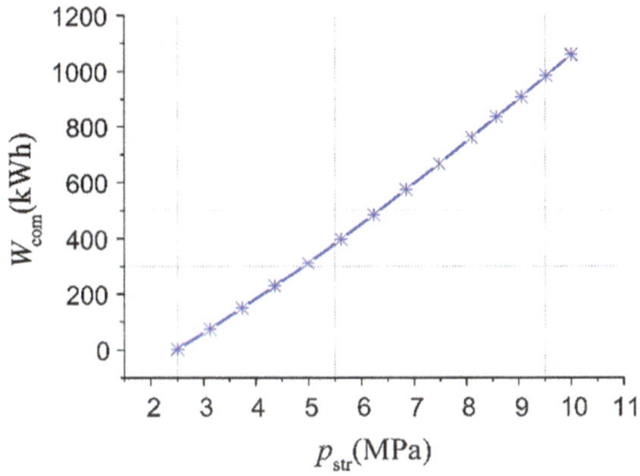

Fig. 4 Compression work W_{com} with different p_{str} in compression process

3.3 The impact of compression ratio

In the formulation of this part, all parameters excluding the compression ratio λ of the RCAES system are given as basic parameters given in Table 1 and λ must satisfy the following condition according to (15).

$$2.5119 \approx \sqrt[N]{\frac{p_{\text{str}}^{\max}}{p_{\text{cin},1}}} < \lambda < \sqrt[N-1]{\frac{p_{\text{str}}^{\max}}{p_{\text{cin},1}}} \approx 3.1623$$

Thus, λ will be fixed as the following given values successively:

$\lambda = 2.6, 2.7, 2.8, 2.9, 3.0, 3.1$

The electric energy charged in compression process and discharged in expansion process with different λ are given in Fig. 5 and Fig. 6 and both of them are monotonically increasing with λ. The efficiency of RCAES system, i.e., the ratio of $E_{\text{tur}}^{\max}$ and $E_{\text{com}}^{\max}$ is given in Fig. 7.

As showed in Fig. 7, the RCAES system's efficiency increases approximately linearly over λ. This is partly because of the narrow feasible region of λ under the given design scheme, although the nonlinearity is indicated by the formulas in Appendix.

As the compression ratio λ increases, more thermal energy is released, resultantly, raising the temperature of H-TES. Accordingly, the rising inlet air temperature of the turbine implies that more electric energy will be discharged in the expansion process. The simulation result shows that

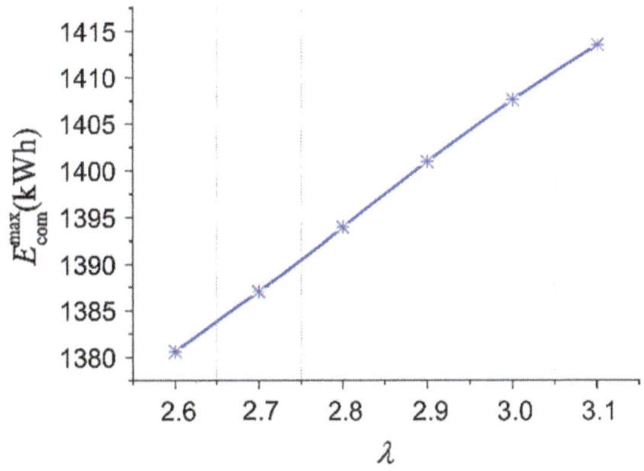

Fig. 5 Electric energy exchanged with different λ in compression processes

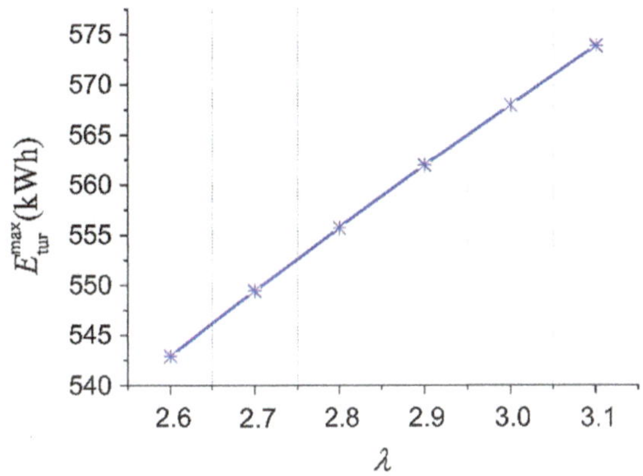

Fig. 6 Electric energy exchanged with different λ in expansion processes

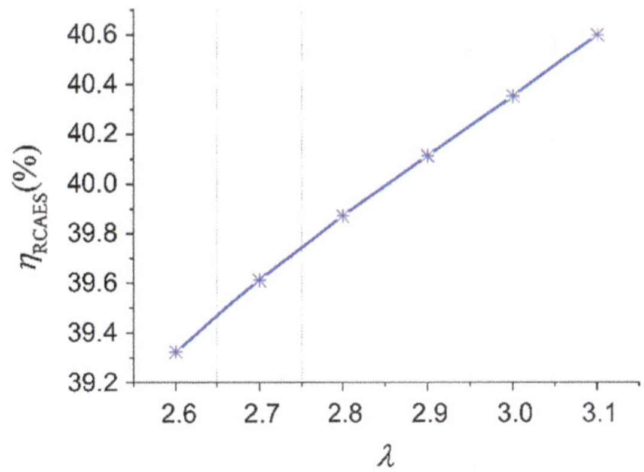

Fig. 7 Efficiency of RCAES system with different λ

the incremental electric energy discharged in expansion process is larger than the incremental electric energy charged in compression process as λ increases, i.e., the efficiency of the RCAES system is increased.

3.4 The impact of EAPT

Although p_{eap} can be arbitrary value in the range of 0.1 MPa ~ 10 Mpa theoretically, p_{eap} will be fixed as the following given values successively in our formulation:

$$p_{eap} = 2.5\,\text{Mpa}, 3\,\text{Mpa}, 3.5\,\text{Mpa}, \cdots, 6.5\,\text{Mpa}, 7\,\text{Mpa}$$

The simulation result is showed in Fig. 8 and Fig. 9, in which E_{com}^{max} and E_{tur}^{max} decrease over p_{eap} linearly and nonlinearly separately. As the change rate of E_{tur}^{max} is smaller than E_{com}^{max}, the efficiency of the system increases nonlinearly and there is less energy loss with bigger value of p_{eap}, which is consistent with the simulation result in Fig. 10.

The energy charged into RCAES system is stored in TES as high-temperature water and in CASV as high-pressure air. Given the volume of the CASV system, more energy will be stored with higher pressure. However, the high-pressure air becomes isothermal and isopiestic after throttling resulting in the decrement of compressed air's capability of doing work. Furthermore, more energy loss will be caused with larger difference of the pressure.

3.5 The impact of the MWP of CASV

According to (17), p_{str}^{max} must satisfy the following condition:

$$8.1\,\text{MPa} = p_{cin,1}\lambda^{N-1} < p_{str}^{max} < p_{cin,1}\lambda^{N} = 24.3\,\text{MPa}$$

Thus, p_{str}^{max} will be fixed as the following given values successively:

$$p_{str}^{max} = 10\,\text{MPa}, 12\,\text{MPa}, \cdots, 22\,\text{MPa}, 24\,\text{MPa}$$

Fig. 11 and Fig. 12 show that the electric energy charged in compression process and discharged in expansion process increases approximately linearly over p_{str}^{max}. This results in nonlinear decrease of the efficiency of the system which is showed in Fig. 13.

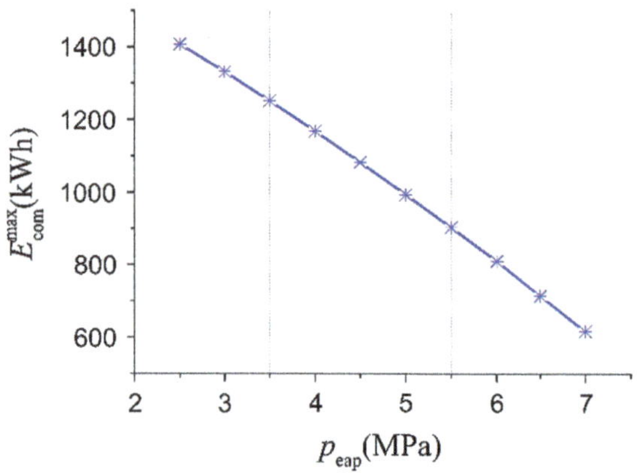

Fig. 8 Electric energy exchanged with different p_{eap} in the compression process

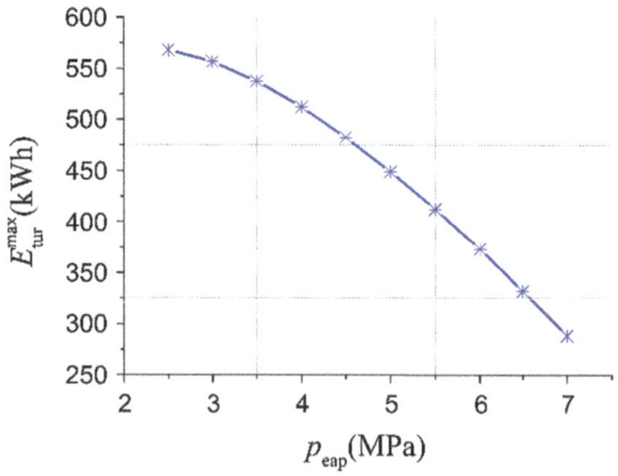

Fig. 9 Electric energy exchanged with different p_{eap} in the expansion process

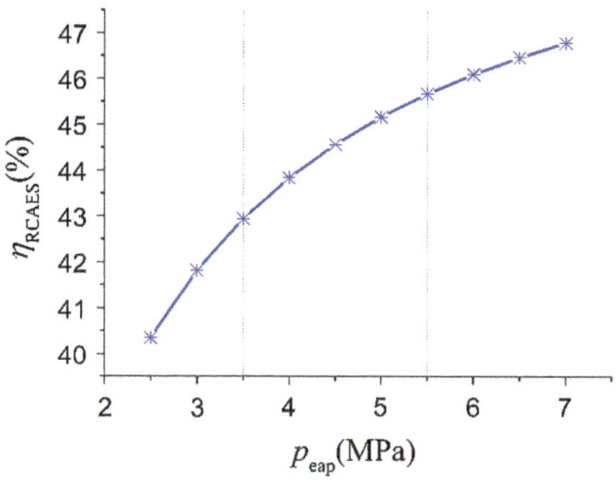

Fig. 10 Efficiency of RCAES system with different p_{eap}

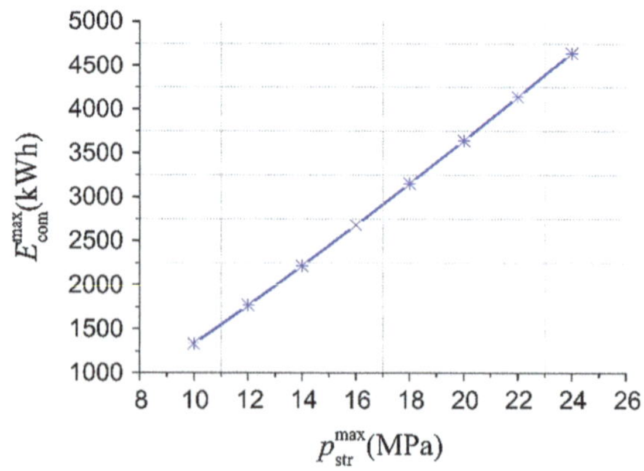

Fig. 11 Electric energy exchanged with different p_{str}^{max} in the compression process

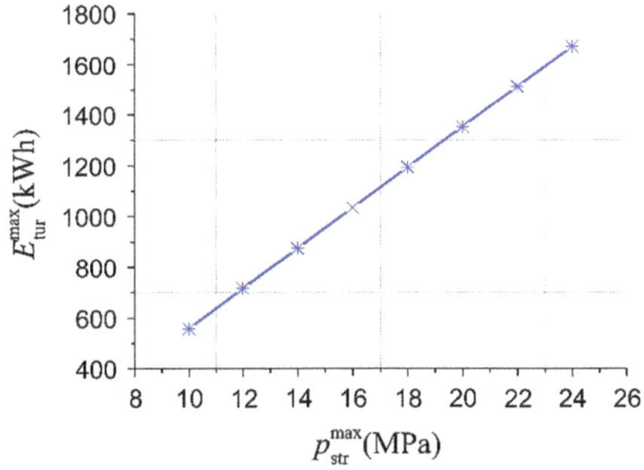

Fig. 12 Electric energy exchanged with different p_{str}^{max} in the expansion process

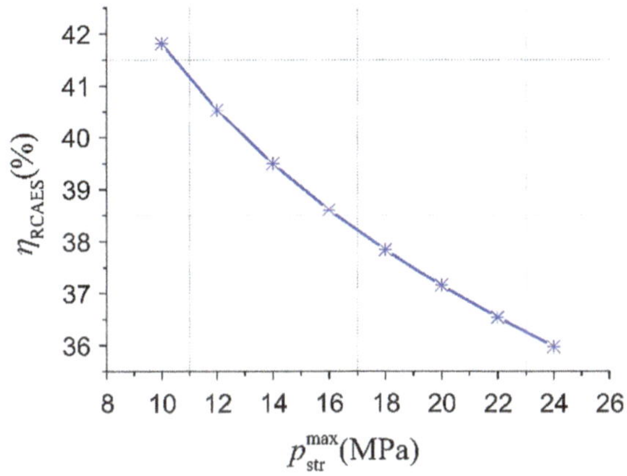

Fig. 13 Efficiency of RCAES system with different p_{str}^{max}

If p_{eap} is fixed, the compressed air's capability of doing work will decrease after throttling which results in lower efficiency of the RCAES system. The simulation result also indicates that the loss is larger when MWP of the CASV is higher.

According to all of the above simulation results, higher compression ratio, higher EAPT and lower MWP of CASV in certain range are helpful to improve the cycle efficiency of the RCAES system.

4 Conclusions

The RCAES system is widely concerned in recent years. Based on the efficiency evaluation method, the impacts of key parameters including compression ratio, EAPT and the MWP of CASV on the cycle efficiency of RCAES system are studied in this paper. The simulation results reveal the variation of the system efficiency over these parameters. This paper provides useful information to design the RCAES system. In addition, other key indicators of the RCAES system, e.g. maximum electric power, operation hour, maximum capacity should also be considered in practical engineering applications. The efficiency evaluation model proposed in this paper can also be used to study these influencing factors.

Acknowledgments This work is supported by National Science Foundation of China (51321005), Technology Projects of State Grid Corporation of China (SGCC) (KJ-2012-627).

Appendix

According to part 2.2, the efficiency of RCAES system can be obtained by subsisting (3), (5) and (10) into (12):

$$\eta_{RCAES} = \frac{c_1 z_1}{c_2 z_2 + c_3 z_3 - c_4 z_1}$$

where

$$c_1 = \frac{MT_{tin}\mu}{(\mu-1)}\left(1-\phi^{\frac{\mu-1}{\mu}}\right);\ c_2 = \frac{\gamma[T_{c1}+(N-2)T_{cin}]}{(\gamma-1)}$$

$$c_3 = \frac{T_{cin}\gamma^2 p_{cin,1}}{(2\gamma-1)(\gamma-1)};\ c_4 = \frac{T_{cin}\gamma}{(\gamma-1)}$$

$$z_1 = p_{str}^{max} - p_{str}^{min};\ z_2 = \left(\lambda^{\frac{\gamma-1}{\gamma}}-1\right)\left(p_{str}^{max}-p_{str}^{min}\right)$$

$$z_3 = \lambda^{N-1}\left[\left(\frac{p_{str}^{max}}{\lambda^{N-1}p_{cin,1}}\right)^{\frac{2\gamma-1}{\gamma}} - \left(\frac{p_{str}^{min}}{\lambda^{N-2}p_{cin,1}}\right)^{\frac{2\gamma-1}{\gamma}}\right]$$

According to part 2.3.1, the efficiency of the RCAES system over λ can be obtained by replacing T_{tin} in c_1 with (14):

$$\eta_{RCAES} = \frac{c_1' z_1}{c_2 z_2 + c_3 z_3 - c_4 z_1}$$

where $c_1' = \frac{M(T_{cin}\lambda^{\frac{\gamma-1}{\gamma}}+\Delta T)\mu}{(\mu-1)}\left(1-\phi^{\frac{\mu-1}{\mu}}\right)$, $c_2 \sim c_4$ and $z_1 \sim z_3$ remain unchanged.

According to part 2.3.2, replace ϕ in c_1 with (16), the efficiency of the RCAES system over p_{eap} can be obtained as follows:

$$\eta_{RCAES} = \frac{c_1'' z_1}{c_2 z_2 + c_3 z_3 - c_4 z_1}$$

where, $c_1'' = \frac{M(T_{cin}\lambda^{\frac{\gamma-1}{\gamma}}+\Delta T)\mu}{(\mu-1)}\left(1-\left(\sqrt[M]{\frac{p_{tou,M}}{p_{eap}}}\right)^{\frac{\mu-1}{\mu}}\right)$, $c_2 \sim c_4$ and $z_1 \sim z_3$ remain unchanged.

References

[1] Wang NB (2012) Transmission and consumption of large-scale wind power. China Electric Power Press, Beijing (in Chinese)

[2] Wang QF, Guan YP, Wang JH (2012) A chance-constrained two-stage stochastic program for unit commitment with uncertain wind power output. IEEE Trans Power Syst 27(1):206–215

[3] Bertsimas D, Litvinov E, Sun XA et al (2013) Adaptive robust optimization for the security constrained unit commitment problem. IEEE Trans Power Syst 28(1):52–63

[4] Jiang XD, Xie YG (2011) Impact of energy storage system on the unit commitment problem with volatile wind power. Autom Electr Power Syst 35(5):19–24 (in Chinese)

[5] Chakraborty S, Senjyu T, Yona A et al (2009) Generation scheduling of thermal units integrated with wind-battery system using a fuzzy modified differential evolution approach. In: Proceedings of the 15th international conference on intelligent system applications to power systems (ISAP'09), Curitiba, 8–12 Nov 2009, 6 pp

[6] Liu X, Xu Z, Wong KP (2013) Recent advancement on technical requirements for grid integration of wind power. J Mod Power Syst Clean Energy 1(3):216–222

[7] Xu G, Xu L, Yao L (2013) Wind turbines output power smoothing using embedded energy storage systems. J Mod Power Syst Clean Energy 1(1):49–57

[8] Zhang WL, Qiu M, Lai XK (2008) Application of energy storage technologies in power grids. Power Syst Technol 32(7):1–9 (in Chinese)

[9] Zhang XJ (2011) Investigation on compressed air storage energy storage system. PhD Thesis, Graduate University of the Chinese Academy of Sciences, Institute of Thermophysics, Beijing (in Chinese)

[10] Inage S (2009) Prospects for large-scale energy storage in decarbonised power grids. International Energy Agency, Paris

[11] Kim YM, Lee JH, Kim SJ et al (2012) Potential and evolution of compressed air energy storage—energy and exergy analysis. Entropy 14(8):1501–1521

[12] Bullough C, Gatzen C, Jakiel C, et al (2004) Advanced adiabatic compressed air energy storage for the integration of wind energy. In: Proceedings of the European wind energy conference (EWEC'04), London, 22–25 Nov 2004, 8 pp

[13] Ekman CK, Jensen SH (2010) Prospects for large scale electricity storage in Denmark. Energy Convers Manag 51(6):1140–1147

[14] Liu WY, Yang YP, Song ZP (2005) Optimization and simulation of different CAES systems. J Eng Thermophys 26(S1):25–28 (in Chinese)

[15] Kim YM, Favrat D (2010) Energy and exergy analysis of a micro-compressed air energy storage and air cycle heating and cooling system. Energy 35(1):213–220

[16] Liu B, Chen LJ, Mei SW et al (2014) An efficiency evaluation method of multi-stage RCAES system. Adv Technol Electr Eng Energy 33(8):1–6 (in Chinese)

[17] Li JP (1991) Dynamics of pneumatic transmission system. South University of Technology Press, Guangzhou (in Chinese)

[18] Shen WD, Tong JG (2007) Engineering thermodynamics, 4th edn. Higher Education Press, Beijing (in Chinese)

Bin LIU received the B.S. degree and M.S. degrees both in electrical engineering from Wuhan University, Wuhan, China and China Electric Power Research Institute, Beijing, China, in 2009 and 2012 respectively. He's now pursuing his Ph.D. degree in the Department of Electrical Engineering, Tsinghua University, Beijing, China. His main research interest includes power system optimization with renewable energy, modeling and analysis of compressed air energy storage system.

Laijun CHEN received the B.Eng. degree and Ph.D. degrees both in electrical engineering from Tsinghua University, Beijing, China, in 2006 and 2011 respectively. He is now a research assistant in the same department. His research work concerns parallel computation of electromagnetic transient in power system, modeling and analysis of compressed air energy storage system.

Shengwei MEI (SM'05) received the B.S. degree in mathematics from Xinjiang University, Urumqi, China, the M.S. degree in operations research from Tsinghua University, Beijing, China, and the Ph.D. degree in automatic control from the Chinese Academy of Sciences, Beijing, China in 1984, 1989, and 1996, respectively. He is currently a Professor of Department of Electrical Engineering at Tsinghua University. His research interests include power system analysis and control.

Feng LIU (M'12) received the B.S. and Ph.D. degrees in electrical engineering from the Department of Electrical Engineering at Tsinghua University, Beijing, China, in 1999 and 2004 respectively. He is currently an Associate Professor of Department of Electrical Engineering at Tsinghua University. His research concerns power system stability analysis and control.

Junjie WANG received the B.S. degree in school of mechanical engineering from Xi'an Jiaotong University, Xi'an, China, in 1983. He's now a professor in the Key Laboratory of Cryogenics, Chinese Academy of Sciences, Beijing, China. His main research interest includes energy storage system and cryogenics engineering.

Sixian WANG received the B.S. degree in school of Energy and Power Engineering from Huazhong University of Science and Technology, Wuhan, China, in 2011. He's now pursuing his Ph.D. degree in the Key Laboratory of Cryogenics, Chinese Academy of Sciences, Beijing, China. His main research interest focuses on energy storage system including compressed air energy system and liquid air energy system.

Permissions

All chapters in this book were first published in JMPSCE, by Springer; hereby published with permission under the Creative Commons Attribution License or equivalent. Every chapter published in this book has been scrutinized by our experts. Their significance has been extensively debated. The topics covered herein carry significant findings which will fuel the growth of the discipline. They may even be implemented as practical applications or may be referred to as a beginning point for another development.

The contributors of this book come from diverse backgrounds, making this book a truly international effort. This book will bring forth new frontiers with its revolutionizing research information and detailed analysis of the nascent developments around the world.

We would like to thank all the contributing authors for lending their expertise to make the book truly unique. They have played a crucial role in the development of this book. Without their invaluable contributions this book wouldn't have been possible. They have made vital efforts to compile up to date information on the varied aspects of this subject to make this book a valuable addition to the collection of many professionals and students.

This book was conceptualized with the vision of imparting up-to-date information and advanced data in this field. To ensure the same, a matchless editorial board was set up. Every individual on the board went through rigorous rounds of assessment to prove their worth. After which they invested a large part of their time researching and compiling the most relevant data for our readers.

The editorial board has been involved in producing this book since its inception. They have spent rigorous hours researching and exploring the diverse topics which have resulted in the successful publishing of this book. They have passed on their knowledge of decades through this book. To expedite this challenging task, the publisher supported the team at every step. A small team of assistant editors was also appointed to further simplify the editing procedure and attain best results for the readers.

Apart from the editorial board, the designing team has also invested a significant amount of their time in understanding the subject and creating the most relevant covers. They scrutinized every image to scout for the most suitable representation of the subject and create an appropriate cover for the book.

The publishing team has been an ardent support to the editorial, designing and production team. Their endless efforts to recruit the best for this project, has resulted in the accomplishment of this book. They are a veteran in the field of academics and their pool of knowledge is as vast as their experience in printing. Their expertise and guidance has proved useful at every step. Their uncompromising quality standards have made this book an exceptional effort. Their encouragement from time to time has been an inspiration for everyone.

The publisher and the editorial board hope that this book will prove to be a valuable piece of knowledge for researchers, students, practitioners and scholars across the globe.

List of Contributors

Ping HE
South China University of Technology, Guangzhou 510640, China
Zhengzhou University of Light Industry, Zhengzhou 450002, China

Fushuan WEN and Gerard LEDWICH
Queensland University of Technology, Gardens Point, Brisbane, Australia

Yusheng XUE
State Grid Electric Power Research Institute, Nanjing 210003, China

Xiaonan HAN, Xin AI and Yingyun SUN
State Key Laboratory for Alternate Electrical Power System with Renewable Energy Sources, North China Electric Power University, Changping District, Beijing 102206, China

Dailin YAO, San Shing CHOI and King Jet TSENG
Center for Smart Energy Systems, School of Electrical and Electronic Engineering, Nanyang Technological University, Singapore 639798, Singapore

Tek Tjing LIE
Auckland University of Technology, Auckland 1102, New Zealand

Xiaoge LIU and Zhao XU
The Hong Kong Polytechnic University, Hung Hom, Hong Kong

Kit Po WONG
The University of Western Australia, Crawley, WA 6009, Australia

Yi DING
School of Information and Control, Nanjing University of Information Science & Technology, Nanjing, China
Department of Electrical Engineering, Technical University of Denmark, Lyngby, Denmark

Lin CHENG
Department of Electrical Engineering, Tsinghua University, Beijing, China

Yonghong ZHANG
School of Information and Control, Nanjing University of Information Science & Technology, Nanjing, China

Yusheng XUE
State Grid Electric Power Research Institute, Nanjing, China

Zulu ESAU and Dilan JAYAWEERA
Department of Electrical and Computer Engineering, Curtin University, Perth, WA, Australia

Shuo YANG, Weisheng WANG, Chun LIU and Yuehui HUANG
Renewable Energy Department, China Electric Power Research Institute, Beijing 100192, China

Xiaoqiang GUO, Huaibao WANG, Zhigang LU and Baocheng WANG
Yanshan University, Qinhuangdao 066004, China

Xianliang TENG, Zonghe GAO, Yingyuan ZHANG, Hua HUANG and Lili LI
NARI Group Corporation, Nanjing 211102, China

Tingting LIANG
Zhangjiakou Wind and Solar Power Energy Demonstration Station Co. Ltd, Hebei, China

Y. V. PAVAN KUMAR
Department of Electrical Engineering, Indian Institute of Technology Hyderabad (IITH), Hyderabad, India

Ravikumar BHIMASINGU
Department of Electrical Engineering, Indian Institute of Technology Hyderabad (IITH), Hyderabad, India

Mingtao YAO, Ning ZHANG, Wei DUAN and Jian ZHANG
School of Electrical Engineering, Beijing Jiaotong University, Beijing 100044, China

Zhaoguang HU
State Grid Energy Research Institute, Beijing 102209, China

Mingshen WANG, Yunfei MU, Hongjie JIA and Xiaodan YU
Key Laboratory of Smart Grid of Ministry of Education, Tianjin University, Tianjin 300072, China

Jianzhong WU and Janaka EKANAYAKE
School of Engineering, Institute of Energy, Cardiff University, Wales CF10 3BN, UK

Xiuping YAO
State Grid Xinjiang Electric Power Company, Urumqi 830002, China

Di WU, Milad JAVADI and John Ning JIANG
Electricity and Energy Reliability and Risk Laboratory, School of Electrical and Computer Engineering, University of Oklahoma, Norman, OK, USA

Gangui YAN, Xudong ZHENG, Gang MU, Junhui LI, Yang CUI and Yue WANG
School of Electrical Engineering, Northeast Dianli University, Jilin 132012, China

Hao HUANG
Customized Energy Solutions, Philadelphia, PA 19102, USA

Fangxing LI
University of Tennessee, Knoxville, TN 37996, USA

Dilan JAYAWEERA and Syed ISLAM
Curtin University, Perth, WA 6102, Australia

Jiaming LI, Jinyu WEN and Xingning HAN
State Key Laboratory of Advanced Electromagnetic Engineering and Technology, School of Electrical and Electronic Engineering, Huazhong University of Science and Technology, Wuhan 430074, China

Shaoyun GE, Li XU and Hong LIU
Key Laboratory of Smart Grid, Tianjin University, Tianjin 300072, China

Jian FANG
Electric Power Research Institute, Guangzhou Power Supply Company, Guangzhou 510000, China

Yang ZHANG, Fang YAO, Herbert Ho-Ching IU, Tyrone FERNANDO, Kit Po WONG
The University of Western Australia, Perth, WA, Australia

Yuanzheng LI and Mengshi LI
Department of Electric Power Engineering, South China University of Technology (SCUT), Guangzhou 510641, Guangdong, China

Qinghua WU
Department of Electric Power Engineering, South China University of Technology (SCUT), Guangzhou 510641, Guangdong, China
Department of Electrical Engineering and Electronics, University of Liverpool, Liverpool L69 3GJ, UK

Xiaoming DONG, Chongqing KANG and Ning ZHANG
The State Key Lab of Power Systems, Department of Electrical Engineering, Tsinghua University, Beijing 100084, China

Huaguang YAN and Junxia MENG
The China Electric Power Research Institute, Beijing 100192, China

Xinsheng NIU and Xin TIAN
The State Grid Shandong Power Economic Research Institute, Jinan 250001, China

Yue YUAN and Yang CAO
College of Energy and Electrical Engineering, Hohai University, Nanjing 211100, China

Xinsong ZHANG
College of Electrical Engineering, Nantong University, Nantong 226019, China

Chun LIU, Yuehui HUANG and Peng LI
China Electrical Power Research Institute, Beijing 100192, China

Kejun QIAN
Suzhou Power Supply Company, State Grid Corporation of China, Suzhou 215004, China

Jing QIU
Centre for Intelligent Electricity Networks (CIEN), University of Newcastle, Callaghan, NSW 2308, Australia

Zhaoyang DONG
School of Electrical and Information Engineering, University of Sydney, Sydney, NSW 2006, Australia

Junhua ZHAO, Ke MENG and Fengji LUO
Centre for Intelligent Electricity Networks (CIEN), University of Newcastle, Callaghan, NSW 2308, Australia

Kit Po WONG
School of Electrical, Electronics and Computer Engineering, University of Western Australia, Perth, WA 6009, Australia

Chao LU
Department of Electrical Engineering, Tsinghua University, Beijing, China

Chunlin GUO
State Key Laboratory of Alternate Electrical Power System with Renewable Energy Sources, North China Electric Power University, Beijing, China

Ching Chuen CHAN
Department of EEE, University of Hong Kong, Hong Kong, China

Chunyu ZHANG, Yi DING, Qi WANG and Jacob ØSTERGAARD
Center for Electric Power and Energy, Technical University of Denmark, 2800 Kgs. Lyngby, Denmark

Yusheng XUE
State Grid Electric Power Research Institute (SGEPRI), Nanjing 210003, China

Yushu SUN and Xisheng TANG
Institute of Electrical Engineering, Chinese Academy of Sciences, Beijing, China

Bin LIU, Laijun CHEN, Shengwei MEI and Feng LIU
State Key Lab of Power Systems, Department of Electrical Engineering, Tsinghua University, Beijing 100084, China

Junjie WANG and Sixian WANG
Key Laboratory of Cryogenics, Technical Institute of Physics and Chemistry, Chinese Academy of Sciences, Beijing 100190, China